Environmental Indicators and Shellfish Safety

Environmental Indicators and Shellfish Safety

Edited by

Cameron R. Hackney
and
Merle D. Pierson

SPRINGER-SCIENCE+BUSINESS MEDIA, B.V

Originally published by Chapman & Hall in 1994
Softcover reprint of the hardcover 1st edition 1994

Library of Congress Cataloging in Publication Data

Environmental indicators and shellfish safety / edited by Cameron R. Hackney and Merle D. Pierson,

p. cm.

"An AVI book."

Includes bibliographical references (p.) and index.

ISBN 978-1-4613-5843-5 ISBN 978-1-4615-2035-1 (eBook)

DOI 10.1007/978-1-4615-2035-1

1. Shellfish as carriers of disease. 2. Shellfish—Microbiology.

3. Shellfish fisheries—Sanitation. I. Hackney, Cameron Ray, 1951-

II. Pierson, Merle D.

RA602.S6E6 1993

614.4' 3—dc20 92-30335

CIP

British Library Cataloguing in Publication Data available

Preface

The abundant shellfish populations of the rich brackish waters of the coastal areas have been an important source of food from prehistoric times. Productive shellfish-harvesting waters have been decreasing at an alarming rate in recent times. Currently, of the fifteen million acres of national estuarine waters available for shellfish harvesting, only 64% are classified as approved and the amount of acreage available is decreasing annually. The remaining areas are classified as prohibited, conditionally approved or restricted. Despite the growing problems, the shellfish industry is of vital importance to the economy of many coastal areas.

The safety of molluscan shellfish has been brought into question. The executive summary of the National Academy of Sciences, Institute of Medicine's Report on Seafood Safety, states that "available data from CDC and the Northeast Technical Services Unit (NETSU) of the Food and Drug Administration for 1978–87, as well as literature reports, suggest that the greatest number of seafood-associated illnesses are from raw molluscan shellfish harvested in waters contaminated with raw or poorly treated human sewage. The majority of these illnesses have unknown etiologies suggestive of Norwalk or Norwalk-like agents that cause human viral gastroenteritis". In addition, there have been numerous reports on safety problems of raw molluscan shellfish. The safety of raw molluscan shellfish is determined by using environmental indicators of fecal pollution. With respect to shellfish, the indicator concept evolved early this century. Originally, the levels of indicators were related to contamination with *Salmonella typhi*, the causative agent of typhoid fever. In time, the safety concerns shifted from enteric bacterial pathogens, such as *S. typhi*, to enteric viruses; however, the indicator system that evolved is still being used to assess the safety of the harvesting waters and the molluscan shellfish. Concern over the adequacy of the indicator system has been noted in several reports. The current environmental indicator system cannot differentiate between fecal pollution from humans and animals. Yet, most of the illnesses associated with molluscan shellfish are most likely caused by enteric viruses, which are strictly of human origin. In addition, it cannot be used to assess safety with respect to naturally occurring marine pathogens, such as the members of the genus *Vibrio*. Many individuals in academia, government and the industry have questioned the current environmental indicator system.

In recognition of the importance of health and economic issues, seed money for a "shellfish water standards research" was appropriated in fiscal year 1989 by the 100th congress. The main objective for the use of this seed money was to obtain background or framework information necessary to begin field work and new methodology research effectively for a full four-year national indicator study. Among the top priorities was to develop a comprehensive literature review of environmental indicators and shellfish safety. It was felt that it is important to bring together the available information on environmental indicators and shellfish safety so that decisions on the efficacy of current indicators could be made. The editors' goal was to have nationally recognized experts write each chapter, and this book is the result of this work. The book begins with a history of shellfish safety. This chapter was written by the former head of the U.S. FDA's Shellfish Sanitation branch and contains much information here, available before only as FDA files. The second chapter is on microbial and chemical indicators of pollution. This is perhaps the most comprehensive chapter ever written on the subject. The author of this chapter, Howard Kator, has been the chairman of the ISSC microbiology committee for several years and has had considerable research experience on the relationship between environmental indicators and pathogens. The next four chapters examine shellfish safety with respect to enteric viruses and bacterial pathogens and their relationship to environmental indicators. These chapters give the characteristics of the organisms, the illness, their survival in the environment, the methodology to recover the organisms and their relationship to environmental indicators. The next three chapters concentrate on the methodology, including rapid methods and the emerging methodology. The phenomenon of cell injury is reviewed, along with viable but nonculturable cells. Next, changes in indicator and pathogen populations during handling or storage of shellstock and during processing of shellfish are reviewed. The last four chapters review sanitary surveys of growing waters, depuration and relaying of molluscan shellfish, using indicator information for managing risks, epidemiologic studies, and statistical sampling of growing waters and shellfish meats. The chapter on sanitary surveys will be especially useful to state and federal regulators. It provides an easy-to-follow guide for conducting sanitary surveys. The editors anticipate this book to be useful for anyone dealing with microbial aspects of environmental indicators and safety. It will be extremely useful to federal and state regulatory agencies and scientists in the environmental area. In addition, it could be used as a textbook on environmental quality.

Contents

1

Historical Overview

J. David Clem

U.S. Food and Drug Administration,
1624 Craig Lane, McLean, VA 22101

EARLY SHELLFISH CONTROL EFFORTS

Written records do not indicate clearly when man first became aware that edible shellfish could transmit communicable disease organisms. Descriptions of the Roman empire noted that emperors enjoyed feasting on oysters that had been transported over long distances and kept in cold storage for such occasions. Illnesses were ascribed to these feasts but no common source was recognized.

In the old testament (Leviticus, Chapter 11) the laws were listed for the types of food that may be eaten. Regarding clean and unclean food sources, it was written that "Of the various creatures that live in the water, you may eat the following: whatever in the seas or in river waters has both fins and scales you may eat. But of the various creatures that crawl or swim in the water, whether in the sea or in the rivers, all those that lack either fins or scales are loathsome for you, and you shall treat them as loathsome. Their flesh you shall not eat, and their dead bodies you shall loathe. Every water creature that lacks fins or scales is loathsome for you." The origin of this admonition is lost in time but is thought to be associated with the insanitary conditions where shellfish were commonly harvested.

In 1603, oysters were incriminated when King Henry IV of France became ill with intestinal disorders (Reille 1907). A French physician, Pasquier, wrote a book in 1816 entitled *The Oyster from the Medical Point of View.* He observed that workmen had laid down 60,000 oysters in a moat around an old citadel. The moat received sewage from a French garrison. Persons eating these oysters after suitable incubation became ill with typhoid fever (Fisher 1927). Keep in mind that these observations occurred before Pasteur advanced the germ theory of disease transmission.

Not until the late 19th century did scientific inquiry firmly establish the link between consumption of shellfish and enteric illness. Dr. Chantemesse presented before the Academie of Medecine in Paris, France, a report on an oyster-related outbreak of enteric disease that affected 14 persons. The illnesses occurred in a small town of Saint Andre-de-Sangoins, where there had been no previously reported enteric fever. Other members of the households where the illnesses occurred did not partake of the oysters and did not suffer. The common source of the outbreak was traced to an imported barrel of oysters (Bulstrode 1896).

Dr. Chantemesse recommended, as a sanitary control measure, placing oysters for several weeks in clean seawater to free them of impurities before consumption. It is of interest to note that in France today the preferred method of control is to hold shellfish in "claires" (man-made tidally submersible ponds) and "degorgeorirs" (wet storage ponds or pools). Typically, the oysters spend several months in claires and two days in degorgeoirs (Furfari 1981). The classification of growing areas in France is determined primarily by the levels of indicator organisms in shellfish.

Sir Charles Cameron of Dublin, Ireland, made the observation that enteric fever was sometimes propagated by means of oysters taken from the estuaries of rivers into which sewage of a large town was deposited. In 1879, when he stated his theory before the British Association along with bacteriological data to support the theory, the Association thought it was a joke. For the next 17 years, other medical authorities from various localities around England reported sporadic illnesses attributed to eating oysters taken from sewage-polluted waters (Newsholme 1896).

Dr. Newsholme of Brighton, England, described antidotal evidence of the consumption of sewage-contaminated shellfish being related to 56 of his typhoid fever patients. He had seen these patients personally between 1894 and 1896 and offered exhaustive epidemiological evidence to support his observations that the only common factor among these patients was the consumption of sewage-contaminated oysters and mussels, 36 and 20 cases, respectively.

Dr. Newsholme recommended that controls be placed on the places where shellfish were being harvested and stored. He suggested that legislation be enacted patterned after the existing milk legislation, that had been passed to prevent typhoid fever from contaminated milk supplies. His findings were met with skepticism and officials were reluctant to restrict the harvest of shellfish for fear of hurting the trade.

Bulstrode and his associates conducted an extensive study of the shellfish industry in England (Bulstrode 1896). They found that it was a common practice to lay oysters near the mouths of rivers for fattening, where the crops could be conveniently managed and harvested. The sites were often located near urban populations and exposed to raw sewage.

In their study, they made sanitary surveys of the growing areas, fattening sites, and along shorelines adjacent to these shellfish activities. Detailed maps were prepared that indicated shellfish-growing areas known to be free of pollution. These findings were widely disseminated and publicized.

A call for action to protect the public health came in the form of an editorial in the *British Medical Journal* in October, 1896. It was concluded, given the unbridled oyster-related cases of typhoid fever, that "the only efficient means of protecting the public is for the Local Government Board to close those oyster beds which are open to contamination until the objectionable conditions have been removed, just as the local authorities have at present the power of closing a milk shop which is regarded as a source of danger . . . and we see no reason why the central sanitary authority should not have the power to close an oyster bed which may broadcast disease over the whole kingdom" (Editorial 1896).

Still more outbreaks were reported associated with shellfish consumption. One particularly striking outbreak involved mostly children who had been on an outing to Exmouth, England. There they had eaten raw cockles. After returning to their home of Exeter, 58 children developed enteric fever. Upon further investigation, it was discovered that a total of 85 persons had become ill after eating cockles from the same source between July 1 and September 11, 1899. The cockles were harvested from around a sewage outfall that was known to be downstream of homes where there were several active enteric cases (Editorial 1899).

Because of the adverse publicity given to these outbreaks in England, the oyster trade was languishing. Parliament had under consideration legislation to govern the harvesting and layings of oysters but jurisdictional disputes stalled the bill. The British Government did advise that members of the public who are foolish enough to eat shellfish without a knowledge of their source will suffer from enteric fever.

"We hope that in the next session, Parliament will provide a really efficient remedy for a great and serious evil. And we particularly urge that this remedy would not be restricted to oysters but extended to other shellfish" (Editorial 1899).

By the turn of the century, health officials in Europe had widely accepted the theory first advanced by Sir Cameron of Dublin in 1879. That is, the typhoid bacillus may survive its transit through a sewer to its outfall, and through the water of a river or bay to an oyster bed where it may be retained for some time in the stomach of an oyster, and that this oyster, if eaten raw, may induce typhoid fever in a susceptible subject (Allen 1899).

Over the period 1904–1909 the city of Birmingham, England, thoroughly investigated the reported typhoid cases. Primary cases for this period numbered 855, of which 124 were attributed to eating mussels. The peak of the mussel cases and the quantity of mussel shipments to the Birmingham market coincided.

Bacteriological examinations were conducted of mussels received in the markets of Birmingham. During a one-year period, there were 65 samples of mussels collected that were representative of 22 sources. The principal investigator designed a classification system based on the quantitative bacteriological methods that he had developed. The samples were examined for the total number of microorganisms, the number of spores of *Bacillus enteritidis sporogenes*, the number of glucose fermenters, and the number of Streptococci.

From the results of the bacteriological analysis they found a general correlation between the quantity of microorganisms and the relative purity of harvest sites associated with the 22 mussel sources. A guide was established that suggested a mussel with greater than ten spores of *B. enteritidis* should be viewed with grave suspicion. *Streptococci* organisms found in a large portion of the 65 samples were considered to be a good indication of grossly polluted mussels. No other density standards were given, but the data suggested some general ranges for guidance (Buchan 1910).

At this time, apparently, no British national laws had been enacted because the officials from Birmingham criticized strongly the current food control legislation as being inadequate to protect public health. They concluded that traditional inspection and examination methods were impossible to ensure safe mussels and other shellfish. The officials in Birmingham reviewed carefully the possible remedies that would improve the sanitary quality of shellfish.

After considerable consultations they outlined some potential remedies: (1) control the sources of pollution that contaminate shellfish waters; (2) seize and destroy shellfish that are found to be from polluted waters; (3) completely remove sources of pollution that pollute shellfish waters; (4) require shellfish to be subject to a cleansing process; (5) require all shellfish to be subject to a heat sterilization process; and (6) restrict the source of harvesting to only waters of known sanitary quality. From these range of remedies the last one was considered to be the most practical method that would ensure the stability of the trade.

U.S. EARLY EXPERIENCES

During the late 19th century, shellfish were being consumed with impunity and the trade enjoyed bountiful harvests and a growing demand. Oysters had found inland markets through transportation on improved U.S. railroad networks. There developed a substantial trade in oyster shellstock transported in barrels to distant markets from the east and Gulf coast areas. The per capita consumption of oysters reached its peak during this period.

Until about the mid-1890s, there was little concern that molluscan shellfish may transmit communicable diseases. The common understanding was that disease germs were transmitted from person to person via an airborne route. The United States medical community noted with interest the experiences occurring in England, where practicing physicians began to note among their patients a commonality between eating shellfish and enteric diseases. The British were careful to discount the usual causes of enteric fever such as previous contact with an infected person, insanitary household conditions, and contaminated water or milk (Broadbent 1895). Apparently, U.S. health officials did not pay heed to the warnings posted in Great Britain or did not believe that sanitary conditions warranted taking corrective measures.

The first well-documented outbreak of shellfish-related illnesses in the United States occurred at Wesleyan University, Middletown, Connecticut, in 1894. No cases of typhoid fever had been reported at the university, nor in the surrounding community, prior to this outbreak. Between October 20 and November 9, 26 cases of illnesses were reported. Twenty-three cases were diagnosed as typhoid fever, 13 were seriously ill, and 4 resulted in death. Only male students were affected.

Because of the large number of illnesses relative to the student population, an extensive investigation was conducted by health officials and, in particular, by Professor Conn at the university. All probable sources of infection were sought. It was discovered that well water used by many people could not account for the illnesses, nor the city water supply served the whole community, including the university. The cases were scattered randomly among campus and off-campus housing units. The investigators found that one common factor was that the male students belonged to or were associated with three fraternities. At the beginning of the school year, these three fraternities joined forces to sponsor an initiation banquet.

Plumbing in the fraternities was inspected and found to be of approved construction and operation. The menus for the joint banquets were reviewed. There were four common food items, that included celery, fruit, ham, and oysters. It was discovered that the first three food items were also distributed to other persons and eaten without ill-effects. The investigation was narrowed to the common denominator, oysters. The oysters were served raw on the half shell. There were 100 male students that attended the banquets and ate raw oysters.

A local dealer had supplied the oysters and, in turn, had obtained his supply directly from an oyster supplier. The period under suspicion was two weeks prior to the onset of illness of the first case, October 20. Typhoid fever has a mean period of onset of 14 days and may range between 7 and 21 days. The identity of the shipment was determined and traced back to the point of origin.

The oyster supplier first harvested the oysters from Long Island Sound, a large body of water between Long Island, New York, and Connecticut. After harvest, the oysters in question were deposited in a creek for a day or two to freshen and fatten. A sanitary survey of the relay site disclosed that within 300 ft of the oyster relay site was a private sewer outlet. The sewer served an individual home where it was learned that two cases of typhoid fever were convalescing. According to medical reports, the two cases suffered from high fevers and requested to see a physician on October 11. One of them died subsequently.

Further inquiries through correspondence with students that were guests at Wesleyan fraternities uncovered further cases on other college campuses. At Amherst and Yale, there were cases consistent with the circumstances under investigation for the Wesleyan banquets (Conn 1894).

Professor Conn, who conducted the Wesleyan outbreak, commented: "If one had planned beforehand a series of experiments designed to prove the possibility of oysters as distributing typhoid, it would have hardly been possible to have desired a more satisfactory series of conditions than those which have been obtained in this outbreak."

Typhoid fever is a serious communicable disease. Approximately 10% of the cases in reported outbreaks have resulted in death. Modern-day antibiotics will cure patients of the infection, but, of course, they were not available in the early 20th century. The disease causes a generalized infection and, left untreated, is characterized by a continued fever lasting four weeks. The causative organism is *Salmonella typhi*, but earlier references used the names *Eberthella typhosa*, *Bacillus tyhposus*, *B. typhoid abdominalis*, or *B. typhi*. The organism appears in the blood early, usually about one week after exposure. *S. typhosa* may be found in both the feces and urine of an infected person. At least 50% of the cases continue to discharge typhoid bacilli in their feces for three weeks after the onset of the disease. Some may become chronic carriers, while a small number of healthy persons may harbor the organism and become what are known as healthy carriers. Man is the only reservoir for the causative microorganism (Maxcy 1956; Benenson 1985).

In the light of the convincing Wesleyan epidemiological investigation conducted by Professor Conn, questions were raised about the viability of *B. typhi* in seawater and oysters. Follow-up bacteriological testing was done by Foote to determine the survival of *B. typhi* under the conditions noted in the outbreak. His studies concluded that *B. typhi abdominalis* will live in unsterilized salt or brackish water for eight days at temperatures of 40°F to below freezing, while in warmer water the organism diminishes rapidly after the first week and cannot be detected after three weeks.

In oysters his experiments did not answer the question of multiplication, but did show a progressive decrease of *B. typhi abdominalis*, which could be found even after thirty days. Experimental results indicated that *B. typhi abdominalis* lives longer in the juice and stomach of the oyster than in the water where oysters grow (Foote 1895).

With continued reporting of shellfish-borne infections at the turn of the century from abroad and widely separated reports in the United States, officials became sufficiently aroused to survey the sanitary conditions of the shellfish industry. Phelps (1911) studied the relaying of polluted oysters to clean waters in Narragansett Bay, Rhode Island. Rhode Island was one of the few states to have a shellfish sanitation control program. At the time of the study, the Rhode Island Shell Fish Commission had established a guide of greater than five *B. coli* per cubic centimeter in oysters for certifying beds, and greater than ten was a sufficient ground to condemn beds.

Initially, Phelps collected oyster samples from two polluted areas: Bullocks Point and Conimicut Point. Batches of these oysters were placed in the Kickimuit River (presumably a clean body of water), where the oysters were resampled four days later. The Bullocks Point oysters initial level of 22.8 was reduced to less than one and the Conimicut Point oysters from 71.2 to 2.8 of *B. coli* per cubic centimeter.

The study conducted by Phelps suggested that relaying of shellfish from polluted waters to clean waters may be feasible for the shellfish industry and protect public health. Earlier, Phelps (1908) had published a method for the enumeration of *B. coli* in oysters. The concept of a score was advanced that was derived simply by taking the reciprocal of the highest positive dilution and summing the total. This method of enumeration, also known as the Phelps index, was arbitrary in the case of skips or inversions and gave low results compared to the MPN method used years later.

There were no generally accepted standard methods for the bacteriological examination of shellfish meats, liquor or seawater, which made comparison of bacteriological data conducted in different laboratories difficult, if not impossible. The American Public Health Association became interested in this situation and invited the few existing state shellfish control agencies and the federal government to form a committee to explore this problem. Apparently, the states of Rhode Island, New York, New Jersey and Maryland as well as the Public Health Service and the Bureau of Chemistry (now the Food and Drug Administration) used methods developed independently for their bacteriological studies of shellfish-growing areas. In 1909, APHA appointed a committee to develop a satisfactory "standard" method for the examination of oysters and other shellfish (Hunter 1928).

This APHA committee made a preliminary report and provided a system for scoring the bacteriological results of fermentation tubes serially diluted for the presence of *B. coli.* The method preliminarily adopted was based on examining five oysters using only the liquor drained from the shell. The results were expressed using the Phelps index but renamed as the American Public Health Association Method of Scoring Oysters for *B. coli.* In simple terms, the score was determined by summing the following values.

1. If present in 1.0 cc but not in 0.1 cc, the value was 1.
2. If present in 0.1 cc but not in 0.01 cc, the value was 10.
3. If present in 0.01 cc but not in 0.001 cc, the value was 100 etc.

The committee was not prepared to give any sanitary significance to the score but hoped to be able to within a year, when further studies would be completed (Whipple 1911).

In an effort to gather sufficient representative data, the Bureau of Chemistry conducted an extensive study covering the Atlantic and Gulf coast areas (Stiles 1911). During the period 1908–1910, about 1000 samples of oysters, quahogs, soft clams, and seawater mussels were collected and examined bacteriologically. Most of the emphasis was on oyster-growing areas.

The study concluded that "oysters intended for human consumption should be grown in water free from pollution, and the greatest care should be exercised to keep such a product in a clean and wholesome condition until consumed". This represented an official position of the federal government since Stiles was at that time a Bureau of Chemistry senior official.

In 1906, the landmark Pure Food and Drugs Act was passed. One of the first tasks for the Bureau of Chemistry under this law was the investigation of the oyster industry. Few guidelines had been developed at this juncture to assist inspectors or sanitary engineers in conducting sanitary surveys of shellfish-growing areas or in inspection of handling plants.

The enumeration of bacteria using the scoring (Phelps index) system was troublesome to bacteriologists for it gave inaccurate readings. McCrady provided the first scientific approach to expressing the number of bacteria when fermentation methods were used. He explained in exquisite detail how the numerical value of the bacterial density is derived when the results are given in terms of the number of positive tubes in each serial dilution (McCrady 1915).

Certain assumptions must be made in order to apply probability theory to the results. The first is that bacteria are randomly distributed throughout the sample. Secondly, the bacteria exist independently and, thirdly, one organism will produce a positive test. Calculation of the bacterial density can now be performed using the relationship between the number of the bacteria and the probability of a percentage positive test. Given that bacteria grow in a geometric progression, the best statistic to use in describing this phenomenon is the Poisson distribution.

An example of the formula used to derive the estimated density of the bacteria in a sample from the number of positive fermentation tubes is as follows:

$$1 - \left(\frac{V-1}{V}\right)^x = \frac{p}{p+q}$$

where

V = sample volume in milliliters
p = probability of the number of positive tubes
q = probability of the number of negative tubes
x = numerical density of bacteria in 1 ml

In the simplest of cases, consider that a single dilution of fermentation tubes is used in 1 ml volumes. Assume that the sample volume (V) is 100 ml and that the number of tubes positive (p) in the single dilution represents 50% (half) of the tubes. Therefore, $p = 0.5$ and $q = 0.5$. Intuitively, if one-half of the 1 ml tubes are positive by straight arithmetic, one would suppose the estimated number of bacteria in the sample to be 50. However, this ignores the geometric growth characteristics of bacteria. Placing the assumed values in the above formula and solving for x will result in a value of 69, correctly estimating the bacteria density most likely to be present. The term used for this statistical method later became known as the most probable number of bacteria, abbreviated as the MPN value per volume of sample. The MPN value is usually expressed per 100 ml for water or 100 g for shellfish meats. The McCrady MPN method was slow to be adopted by control officials and was not officially recognized until 1946 in the Public Health Service guidelines for the sanitary control of shellfish (U.S. Public Health Service 1946).

Reports of shellfish-related typhoid fever continued to occur in the United States. Noteworthy was an outbreak that occurred during November and December 1915, involving 50 cases (Brooks 1916). There was an alarming increase in typhoid fever among several cities in New York state adjacent to a common railroad line. There was no common water supply and the cases were distributed over a number of milk routes so that neither milk nor water could be the factor responsible for the outbreak. The only common food source to the cases was oysters. There were six wholesale oyster dealers in Binghamton who supplied oysters for this section of the state. Nearly all the cases had eaten oysters which were supplied by two of the wholesalers.

The two companies had obtained their oysters from two suppliers in Maryland. They, in turn, had received their oyster supply from over 22 shucking houses in Maryland. A follow-up investigation of Maryland officials concluded that it was impossible to locate with certainty the source of the infected oysters responsible for the outbreak (Brooks 1916). Such a situation leads public health authorities to incriminate broad areas involved, even though the majority may not have been the actual cause of the outbreak. The inability to trace suspect lots of shellfish through the supply line back to the growing area has been a weakness in the NSSP control procedures.

Studies on several aspects of shellfish sanitary control concerning relaying, hibernation and cleansing of oysters were under way. Still, the major health concern was with oysters and not with other commercial species of molluscan shellfish. As public health officials instituted better sanitation control, improved water supplies, educated the public and, especially, food handlers about personal hygiene, the prevalence of typhoid fever in the United States began to fall. There were no reported outbreaks of typhoid fever related to shellfish between 1916 and 1924 (Hunter 1928).

Some states had developed well-defined shellfish control programs by this time. The state of New Jersey had been surveying its coastal waters with a 42 ft gasoline-powered boat since 1912. The boat had its own laboratory for the bacteriological and chemical analysis of oysters and water. The crew consisted of a captain and bacteriologist. Their findings supported the fact that certain waters were so badly polluted that oysters or clams would be a menace to public health. The waters were condemned for shellfish harvesting (Bacon 1922).

The state of New Jersey granted permission for the removal of shellfish during the cold winter months, when oysters were believed to be in a partial state of hibernation and feeding was much curtailed. Also, relaying was permitted (transplant) from condemned areas to other waters whose purity was approved by the N.J. Health Department. As states, in general, were beginning to close or restrict many productive growing areas and clean shellfish areas were being depleted, the oyster industry began to have trouble finding clean dependable harvest areas to meet the demand.

GENESIS OF THE NATIONAL SHELLFISH SANITATION PROGRAM

On about December 5, 1924, the Chicago city health department noted an excessive prevalence of typhoid fever. The cases were among some of Chicago's more prominent and wealthy citizens. Considerable publicity was given to the rise in typhoid fever. On a close examination of 67 municipal health department statistics, it was noted that the cities of Washington, D.C.,

Chicago, and New York had excessive prevalence of typhoid fever, and in ten other cities there was a markedly excessive prevalence.

Investigations conducted by the municipalities found that a large majority of the cases in each of these cities was in persons who within the three or four weeks preceding the onset of their illness had eaten raw oysters. There was no other common source identified likely to be operative in the spread of typhoid infection. There was quick and decisive action taken by the involved health officers to issue warnings not to eat raw oysters. The sale of oysters in the United States plummeted. The sales of other seafood were also affected.

The Public Health Service at the request of local health departments sent epidemiologists to assist in the investigation of the widespread outbreaks. A carefully constructed case history questionnaire was used to interview as many cases as possible in the three prime cities affected and in the ten cities where possible. Forty-one case histories were gathered in Washington, D.C., and it was found that 23 had eaten raw shellfish within the time period consistent with the incubation of typhoid fever. Similarly, 69 of the 80 cases interviewed in Chicago had only one common factor—eating raw shellfish. Two hundred and sixty-three case histories were made in greater New York city and 176 were found to have eaten raw shellfish.

The investigators questioned how unique eating raw oysters was among the general population. From limited data they estimated that less than 20% of the total population of the city ate raw oysters, and that less than 10% ate raw shell oysters within 30 days of November 1924. Therefore, it was concluded that the number of cases associated with the consumption of raw shell oysters was highly significant and believed to be responsible for many of the cases arising in November and December of 1924.

From the food histories it was discovered that the common source of oysters was most often traced to a production company operating in the vicinity of West Sayville, New York. Other shipments were closely related to this same location but could not definitely be pinpointed. The plant in West Sayville stored its oyster supply in a protected enbayment, where floats were secured alongside the packing plant (Lumsden 1925).

This multifoci outbreak of typhoid fever involved several hundred cases, with highly significant epidemiological evidence that singled out polluted raw oysters from one supplier in New York as the cause. This incident brought down the sales of oysters dramatically and rallied local, state, and federal officials to take some corrective action. The oyster industry appealed to the federal government to bring about some form of government program that would provide public health assurances for the interstate shipment of oysters.

Because of the requests for action, and widespread alarm and warnings issued by state authorities in New York, Washington, D.C., and Chicago not to eat raw oysters, the Surgeon General of the United States Public Health

Service called a conference on February 19, 1925. The conference was attended by federal, state and municipal authorities and representatives of the shellfish industry (Frost 1925). This conference formulated the basic principles of the National Shellfish Sanitation Program.

The conference adopted and recommended to the Surgeon General the following resolutions as a basis for permanent oyster control:

1. "The beds on which shellfish are grown must be determined, inspected, and controlled by some official State agency and the United States Public Health Service."
2. "The plants in which shellfish are shucked or otherwise prepared or packed by the shipper must be inspected and controlled by some official State agency and the United States Public Health Service."
3. "The freedom from typhoid bacilli of the workers who handle shellfish must be determined by some official governmental agency."
4. "Failing to secure sufficiently high standards to protect the people as regards beds, floating and plumping practices, and methods of shucking, or other methods of preparation, a satisfactory practical method of pasteurization or other heat treatment, or chemical, or biological method which produces satisfactory results, must be installed and operated under proper governmental supervision."
5. "There must be such governmental supervision and such a trade organization as will make plain the source of shellfish and will prevent shellfish from one source being substituted for those from another source. This will be chiefly a problem of the individual State."
6. "The methods of shipping must be supervised, inspected, controlled, and approved by the proper official Federal and interstate agency."
7. "The methods of storing, displaying for sale, and dispensing must be determined, inspected, and controlled by the proper State or city agency."
8. "The product must conform to an established bacterial standard and must meet Federal, State, and local laws and regulations relative to salinity, water content, and food proportion, and must conform to the pure food laws standards."

In addition, the conference urged that the Surgeon General appoint a committee to review procedures, provide advice and take prompt action in conformity with the resolutions above. From the transcripts of the conference it appeared that many involved people were not satisfied with the current Standard Methods adopted by the American Public Health Association. The conference asked that the committee provide some oversight on the proposed standard methods before they would be made a part of the oyster control program.

The Surgeon General did adopt the recommendations of the conference and appointed an 18-member committee with W. H. Frost, Surgeon, U.S. Public Health Service, as chairman. After deliberations with experts, affected industry members, local, state and federal officials, the committee submitted its first report on September 28, 1925. The report noted that the few standards recommended have received more than a two-thirds majority of the committee; however, they advised that there was a need for collection of additional data upon which to propose specific standards.

The committee noted that the scope of their deliberations was to comprise shellfish of commercial importance. These would include oysters, hard clams (quahogs or littleneck clams), soft clams, razor clams, saltwater mussels, and scallops. But the committee stated that the sanitary significance of these species differs because of their differences in biology, in quantities consumed, and in customs of production, distribution and food preparation. Consequently, the committee developed standards and administrative procedures specific to the oyster industry.

The committee was reluctant to state specific criteria for the classification of oyster harvest waters. They advised, in brief, that sanitary surveys should be made of the harvest areas and at such frequencies as to discover any changes that may affect the sanitary conditions. The committee report stated: "The survey should also include bacteriological examinations sufficient to aid and check the judgments formed from other data, and the findings in each survey should be made a matter of permanent record."

At this time, there was a general understanding that estuarine dynamics caused currents and dispersion that provided natural safeguards of dilution and time (for natural purification). However, because representative data were lacking, it was not possible for the committee to give specific guidelines to establish areas for harvesting safe oysters. The committee deferred to the judgments of local officials to make the important decisions in determining boundaries for approved harvest areas.

The committee noted the need to be vigilant about close discharges of human sewage to harvest areas. It was noted that "In sparsely populated areas, even where there is no direct discharge of sewage from the shore, careful attention should be paid to the likelihood of contamination in coves and inlets from sources which would be insignificant in relation to a large and well-mixed body of water."

The issue of a bacteriological standard for shellfish-growing waters proved to be too difficult for the committee to address in the short period available to them. They noted that application of the existing U.S. Public Health Service drinking water standard for interstate commerce to shellfish waters would be too stringent. The committee did make this observation: "It is considered that the waters should ordinarily not show the presence of *B. coli* in 1 cc amounts, tests for *B. coli* being made in 10 cc, 1 cc, and 0.1 cc amounts, according to the Standard Methods of the American Public Health Association." (Frost 1925). Note that this density would approximate an MPN value of less than 23 per 100 ml.

Other control practices addressed by the committee concerned the hibernation of oysters below 50°F, relaying of oysters from polluted waters to clean waters with a holding period of 15 days, and the processing of polluted oysters in chlorinated seawater. Again, the committee noted that insufficient data were available to draw definite conclusions. They encouraged the concerned state and federal agencies to undertake such studies that would permit additional data upon which reliable standards could be established.

Strict measures were proposed for the practice of floating oysters and wet storage facilities. This one common industry practice had been responsible for many of the outbreaks reported to health officials in the United States and in Europe. The administrative procedure adopted was to require the issuance of permits by local control officials to wet storage facilities following the strict guidelines laid down by the committee.

The committee struggled to establish a bacteriological standard for market oysters. They noted the rather wide use of *B. coli* as an index of fecal contamination, but found that it related only roughly to the presence of disease-producing germs. They, therefore, gave the following advice:

"In shellfish freshly removed from their beds the numbers of *B. coli* present are, in a general way, proportionate (though not in any fixed ratio) to the numbers present in the waters overlying the beds. Moreover, it appears from the evidence available that bacteria of the *B. coli* group do not tend to multiply in shell stock after removal from the water so long as the shellfish are kept alive and in a thriving condition. Consequently, the bacteriological examination of fresh shellfish, in the shell, affords a fair index of the probable cleanliness or contamination of the waters from which the shellfish were taken and of the probable safety of such shellfish for food."

The committee members were cautious in their recommendations regarding criteria for market shellfish. They acknowledged that bacterial examinations of shucked oysters collected in the marketplace are subject to such a variety of interpretations that they are unreliable as a means of judging the presence or absence of dangerous contamination.

A bacteriological standard for fresh shucked oysters, that is, as packed in the processing plant or before, was recommended. It was the same standard that was in accordance with Standard Methods of the American Public Health Association, which would permit oysters to have a *B. coli* score of not more than 50. This equates to an approximate MPN of 2400 per 100 g.

Tonney (1926) examined 856 oyster shipments as they were received in the Chicago markets and found that 71% would pass a *B. coli* score of 140. He recommended on the basis of their data that a tentative *B. coli* score of not more than 140 for shucked oysters and a *B. coli* score of not more than 50 for shell oysters (shellstock) and hard clams be officially set for the Illinois market. He also suggested that three successive results in noncompliance be required before regulatory action be taken against a shipper.

The interest in conducting research on the fate of *B. typhosus* and *B. coli* in oysters gained momentum. Even though a number of investigators had studied the question, there were still conflicting opinions about the viability of *B. typhosus* in oysters and even whether the oyster can harbor the organism at all (Kinyoun 1925). There was a wide divergence of opinion. Kinyoun found that *B. typhosus* could be recovered 15 days after oysters had been fed the organism. Other investigators disputed such findings. The quest for a reliable, efficient bacteriological index of fecal contamination was undertaken by a number of agencies.

Two years after the initial conference, it was observed that the states had been slow to put into place the recommendations of the Surgeon General (Fisher 1927). The oyster industry, anxious to improve sales, was cautioned not to take advantage of the certification scheme and lead the public to think that state and federal agencies certify the quality of all oysters offered to the public. Some of the industry had been using misleading advertising, which made it look like that certification meant inspection employed by the meat inspection program. "Certification" was meant to convey only that mechanisms were in place to control selected phases of production and only indirectly related to oyster quality.

The Public Health Service continued to sponsor meetings and encouraged state agencies to adopt and enforce the provisions of the National Shellfish Sanitation Program (NSSP). When the guidelines of the NSSP were first published, it was entitled "Minimum Requirements for Endorsement of State Shellfish Control Measures and Certifications for Shippers in Interstate Commerce." A progress report on the status of state programs was made by the Public Health Service in 1930. It was reported that 12 states had good shellfish control programs, while seven were not entirely satisfactory. However, conditions were not deemed to be sufficient to warrant a withdrawal of approval by the Public Health Service (USPHS 1930*a*).

Technical reports evaluated by APHA and the standing Committee on Shellfish Sanitation resulted in some revisions to the initial 1925 report on controls. A bacteriological definition for a "grossly polluted" shellfish-growing area was adopted. It stated that where there is a demonstration of organisms of the *coli-aerogenes* group in water in numbers exceeding 10 per cubic centimeter, that is, giving a score in excess of 50, it should be considered as evidence of gross pollution (USPHS 1930*b*).

In the same report it was noted that many states had adopted a tentative bacteriological standard, which specified that not more than 50% of the water samples in a given area should show the presence of organisms of the *coli-aerogenes* group in 1 cc portions. This equates to an MPN of approximately 70 per 100 ml.

Guidelines for the control of oysters in "hibernation" were made more restrictive. It was recommended that the states certify oyster harvesting only for the period that the temperature of overlying waters continuously and consistently remained at or below 41°F. Further, the bacteriological quality of oysters taken from such areas must be consistently low. "Low" was defined as a score of not over 5 as a general rule and rarely over 50 *B. coli.* A score of 5 would be approximately in the range 140–250 MPN per 100 g of shellfish meat. The adoption of the MPN statistic for the enumeration of bacteria when reading serially diluted fermentation tubes was recommended. The MPN method was acknowledged to be more scientific, but control officials were reluctant to change because they would be unable to compare past score results with the MPN statistic.

In the late 1920s and early 1930s, commercial cleansing plants were operated. A plant was constructed in Newburyport, Massachusetts, for soft shell clams. Up to 1000 bushels of oysters were processed at a commercial plant located at West Sayville, New York. These facilities used chlorine to disinfect the water supply so that it would retain a 1 ppm chlorine residual. The shellfish were submerged in large tanks of chlorinated seawater for 24–48 h.

The state of Virginia built a demonstration cleansing plant to test for its possible application to the Chesapeake Bay oyster commercial industry (Messer 1937). At the time this plant was built, Messer stated that there were about a dozen in operation around the world.

Early work by Wells (1916) showed that experiments purified polluted Chesapeake oysters placed in water which had been treated with calcium hypochlorite. He found that remarkable purification results were obtained within 24 h based on the elimination of *B. coli.* The interest in developing cleansing plants for the purification of shellfish was prompted by the investigations of Dodgson in England, a number of investigators in France, and Wells (1929) in the United States.

Carmelia (1921) conducted large-scale experiments on Raritan Bay oysters. The polluted oysters were placed in scows located in Great Kills, New York, and then filled with seawater at high tide. The seawater was chlorinated by adding calcium hypochlorite solution directly to the process water and mechanically mixed for 10–15 min. Free chlorine was held to a level that would only be a trace. They found among 13 runs reductions in the *B. coli* content of oysters of 40–99% with 24 h of treatment. Carmelia stated that such a process should be applied to all market shellfish, regardless of origin, to ensure an additional safety factor to public health.

It was believed necessary to experiment first with purification processes at specific locations before the operation could be certified by local control officials. There was a reluctance to accept the results of other cleansing plants as being directly applicable to other state shellfish operations. Therefore, cleansing plants did not become widely adopted and found only minimal acceptance with the shellfish industry. In large part, economic incentives and public health pressures were lacking for the industry to become sufficiently interested to invest in constructing cleansing plants.

Cleansing plants went into disfavor as it was discovered that chlorination of seawater above 1 ppm was interfering with the feeding activity of the oyster. By the late 1930s, interest in further development of purification facilities, once thought as important a public health control measure as pasteurization of milk, had faded out (Stimson 1938).

Perry (1928) made an extensive bacteriological study of the oyster industry in Maryland between 1925 and 1927. His study involved the collection of samples at various stages of production from the harvest areas to the market. Analyses were conducted on 1132 oyster samples and 2615 water samples. One significant finding was that stratification of water may cause excellent bacterial quality at the surface, while grossly polluted water may exist at the bottom. He, thus, suggested routine bottom samples be taken to judge pollution conditions.

The APHA committee on standard methods remained active and reported on the evaluation of new studies (Gorham 1933). They reaffirmed the reliance on the *coli-aerogenes* group as the indicator of choice. Again there was the advice that the methods recommended for bacteriological analyses were applicable to only oysters. The recommended standard method explicitly used only the shell liquor in the analysis.

TRENDS IN SHELLFISH SANITATION AND STANDARDS

After 13 years of concentrated studies and the development of state control programs, the U.S. Public Health Service noted the progress that had been made in shellfish sanitation control practices (Old 1938). Nearly all the important shellfish-producing areas along the Atlantic and Gulf coasts had been subject to sanitary surveys by state and federal agencies. Shellfish harvest areas were generally classified as "approved," "grossly polluted restricted," and "moderately polluted restricted" areas.

The consensus view on bacteriological criteria applied to these classified areas was summarized by Old. For approved areas the bacteriological quality of the overlying water quality was accepted if the *coli-aerogenes* group was found in 50% or less of the 10 cc portions, or a score of approximately 0.3. For the grossly polluted areas the general standard was the presence of the *coli-aerogenes* group in 50% or more of the 0.1 cc portions, or a score of 32.

TABLE 1-1. Comparison of Coliform Organism Scores with the Most Probable Number of Coliform Organisms Per 100 ml

Score	MPN	Score	MPN
1	20	41	1300
2	45	50	2400
3	78	140	3500
4	130	230	5400
5	230	320	9200
14	330	410	16,000
23	490	500	24,000
32	790	1400	35,000

For the moderately polluted classification less than 50% of the 1 cc portions, or a score of 3.2, was often used as the standard. It was also recognized that when less than 50% of the 1 cc portions were positive, an area may be considered approved. (Note that the last value equates to an MPN value of approximately less than 70 MPN per 100 ml.)

For the convenience of control officials that had been using the score method to report results, the 1946 manual provided a useful table (Table 1-1).

There was dissatisfaction among control officials over the present use of the APHA bacteriological standard methods of analysis of shellfish-growing waters and the associated standards used by the various state control agencies. At issue was the use of the broad-spectrum index of pollution, the *coli-aerogenes* group and the use of the score method to report the density of bacteria. Most officials desired a better index of fecal contamination. Adopting the use of the "most probable number" for reporting bacteriological results, as originally proposed by McCrady in 1915, was favored. Also, there was disagreement on the value of examining only oyster shell liquor as opposed to examining combinations of liquor and body meat.

The APHA committee recognized the shortcomings of the Standard Methods that had been in use since the 1920s. Perry, referee of the APHA Committee on Bacteriological Methods, had made extensive studies and questioned knowledgeable investigators in the United States and certain foreign countries about the standards presently in use. The research results were requested for review by the committee members. In the interim the committee members concluded the following:

1. "New procedures need to be developed that include at least such edible mollusks as oysters, clams, and mussels."

2. "*Escherichia coli* rather than the colon group should be the index of pollution for both shellfish and shellfish waters."
3. "A new procedure should include methods for the examination of shellfish waters as well as shellfish."
4. "The whole oyster rather than just the shell liquor should be examined."
5. "*Escherichia coli* results should be expressed as most probable numbers rather than as a score."
6. "Certain recommendations should be made in regard to amount of pollution which should ordinarily be tolerated."

The work of the APHA committee carried considerable weight, but there was a lack of data and a lack of consensus to carry forward the recommendations. Perry had made bacteriological studies on the oyster industry in Maryland for over ten years (Perry 1939). His studies supported the use of the modified Eijkman medium for selective growth at 46°F for the enumeration of *Escherichia coli.* He was a strong advocate for changing the index from the *coli-aerogenes* group to *E. coli.* Unfortunately, just five other state agencies had been able to conduct only limited studies on the proposed methods. Methods development and research into bacterial indicators of fecal contamination proceeded at a slow pace.

The New York State Conservation Department requested assistance from the U.S. Public Health Service in studying the sanitary quality of clam-harvesting areas of Great Kills to Princess Bay sectors of Raritan Bay, a large estuary between New York's Staten Island and New Jersey. There resulted in an in-depth bacteriological survey of prime clam-harvesting areas of Raritan Bay that had been closed since 1925. A comprehensive cooperative study was planned for the collection and bacteriological examination of growing waters and clams between April and June 1941. Three cooperating state laboratories participated in the study (Kehr 1941).

This study was a large effort to evaluate the existing sanitary conditions associated with bacteriological results using the current Standard Methods recommended by the American Public Health Association. Given the known sewage pollution inputs that affect the sanitary quality of Raritan Bay waters, the authors present a rationale for computing the likely coliform density. The comparison between the theoretical and the actual results during the period of the study was in fairly good agreement.

From literature reviews of waterborne outbreaks of illnesses, Kehr and others computed an average ratio of pathogen to coliform density in the incriminated drinking water supplies. From these data they extrapolated the number of potential cases of typhoid fever that may result from the ingestion of raw clams harvested from the present waters of Raritan Bay. There were a number of assumptions made that were derived from the study results, literature search, current knowledge, and collective judgment of the experts and officials involved in the study.

Some of the factors that permitted these extrapolations were as follows. The amount of clam production in pounds of edible meat was estimated. Based on the bacteriological results of the study, an average coliform ratio of 1.3 for clam meat to the overlying water quality was calculated. After a review of the water supply typhoid fever outbreaks, it was summarized that an average attack rate was 1.5% among the exposed population. The authors made the assumption that one bacterium of *E. typhosa* from aged and diluted sewage would cause illness. It was estimated from the clams produced, presumably, eaten raw, that consumers would ingest 400 million coliforms. At that time, the morbidity of typhoid fever in New York was 0.05 cases per 1000 population per year.

From the above statements the estimated number of typhoid fever cases that may result from eating raw Raritan Bay clams was calculated to be 36 per year. Apparently, Kehr and others felt this number of cases to be unacceptable. The actual bacteriological quality of the overlying water in the "open" areas classified by the New York Conservation Department was an average of 80.9 MPN coliform per 100 ml. The standard used by this department for approved areas was stated to be 50% of the average ratio of 1 ml portions of water which can be positive for coliforms.

After due consideration of the actual data and the theoretical calculations of typhoid fever infection potential from eating raw Raritan Bay clams, the authors took a pragmatic approach. They concluded that "The most feasible method of reducing the danger of infection from the ingestion of raw hard clams is through the adoption of an arbitrary standard which would reduce to a satisfactory degree the estimated total coliform content of the annual production of Raritan Bay hard clams." . . . "Therefore, 70 coliforms per 100 ml, on the basis of averaged results from waters overlying the shellfish beds, is recommended as the limiting allowable standard for the taking of hard clams to be eaten raw."

Kehr and his colleagues pointed out that the limiting value of 70 was a compromise between the potential dangers of intestinal infections and the comparative rarity of actual infections due to shellfish from approved sources under the controls that have been exercised. Even though this 1941 study was limited, it provided one of the few thorough and rigorous public health evaluations of the NSSP bacteriological standard for approved growing areas.

The APHA committee (Gibbard 1943) reported on a recommended revision to the standard procedures for the bacteriological examination of shellfish that had been in use with minor modifications since 1922. The committee members clearly provided analytical procedures that, in part, required the examination of the entire oyster, adopted the MPN procedure after the tables published by Hoskin (1934), emphasized the examination of both water and shellfish, and recommended confirmation for *Escherichia coli*, since such examinations may become the basis of legal action.

The Public Health Service followed these recommendations and incorporated them in the 1946 issue of the *Manual of Recommended Practice for Sanitary Control of the Shellfish Industry* (USPHS 1950). The published approved-area bacteriological standard was a median coliform content not in excess of 70 MPN per 100 ml of water samples collected from a portion of the area most probably exposed to fecal contamination. The number of samples was not specified, but language indicated a "large" number should be collected on the line bordering the moderately polluted area. The manual advised further that sampling stations should be located no further apart than 1000 ft on the line between approved and moderately polluted areas.

The bacteriological standard of moderately polluted restricted areas was stated to be a median coliform MPN of between 70 and 700 per 100 ml. The manual advised further that the associated sanitary survey should show that the contamination responsible for the coliforms was probably of human origin. Furthermore, the manual would allow the harvesting of oysters from moderately polluted areas when the water temperature remained below 41°F and the coliform content of the oysters did not exceed 20 MPN per 100 ml. This practice was permitted in recognition that oysters hibernate under certain cold water conditions and do not feed.

The manual stipulated that the bacteriological quality of oysters should follow these general guidelines. "When an MPN value of 230, or more, of coliform organisms per 100 ml of sample persists in oyster shell stock sampled at the growing areas or in shell stock or shucked oysters at the point of shucking, it should be interpreted as an indication of unfavorable conditions or practices surrounding the production and handling of the product and should necessitate investigation and improved control measures on the part of the supervisory agency, providing that in occasional samples an MPN value of 2400 coliform organisms per 100 ml may be tolerated in some samples." The manual advised: "If this occurs in more than two consecutive samples, corrective measures shall be enforced by the regulatory authority."

For shellfish other than oysters, the manual advised that an MPN value of 2400, or more, coliform organisms per 100 ml, occurring in successive samples of shellfish, should be questioned by the regulatory agency.

Shellfish control officials and the shellfish industry have long sought the elusive goal of an enforceable bacteriological standard for market shellfish. The quest for such a standard has been frustrated by the limitations of the methodologies, high natural variability in bacteriological quality, differences among shellfish species, geographic and seasonal differences, a variety of industrial processing practices and other uncontrolled factors. Nonetheless, considerable resources have been expended to develop meaningful and feasible bacteriological standards for market shellfish. However, the *primary* public health control measure has been and continues to be the proper classification of approved harvesting waters using the bacteriological standard of "70" as the adjunct to the sanitary survey information (Hunt 1977).

A serious challenge came to the NSSP in 1961, when there was an outbreak of infectious hepatitis traced to the consumption of raw oysters. During January–March 1961, 80 cases of infectious hepatitis occurred in Mississippi and Alabama which were linked to the consumption of sewage-contaminated oysters harvested from the mouth of the Pascagoula River. This was the first incidence where the virus of infectious hepatitis had been implicated in the transmission of disease via shellfish consumption in the United States (Mason 1962). Following this outbreak, another one occurred in the New York–New Jersey area of Raritan Bay involving raw clams (CDC 1961).

Infectious hepatitis, now referred to as viral hepatitis A, is caused by a 27 nm picornavirus with characteristics of an enterovirus. Man and chimpanzees are known to be its reservoirs. Diagnosis is usually established by demonstration of immunoglobulin (IgM) antibodies against hepatitis A virus. The incubation period is 15–50 days, with an average of 29 days. The onset of illness is usually abrupt with fever, abdominal discomfort, followed within a few days by jaundice. Convalescence is usually prolonged, but rarely does infectious hepatitis result in fatalities.

The immediate question raised by shellfish control officials when the Pascagoula outbreak came to light was whether the NSSP bacteriological standards were adequate to prevent the transmission of enteric viruses. Investigations of the Mississippi and New Jersey outbreaks disclosed that the waters where the shellfish had been harvested were grossly polluted with sewage and were seriously out of compliance with the NSSP growing-area standards. But these episodes did give pause to the adequacy of the standards and begged for more substantiation and research.

Larkin (1982) provides a summary on the state of knowledge with respect to the potential enterovirus hazards in raw shellfish. The conclusion is about the same as the pioneering work of Kehr in 1941: Apply the NSSP standards for approved growing areas properly and the consumer of raw shellfish from these waters will be reasonably safe from enteric bacterial or enteric virus infection.

Jensen (1962) describes the rather large shellfish sanitation research effort that was initiated by the Public Health Service after the Pascagoula incident. Three research centers were established and expanded in Rhode Island, Alabama, and Washington. Investigations were conducted on the depuration (purification) process, enterovirus survival in the seawater environment and other important topics of public health concern. The results of this research were evaluated and presented to the state shellfish control officials and the public, with a view to update the NSSP standards and guidelines.

The current NSSP standards, after some arduous times in workshops and meetings with state control officials, have evolved into a revised *NSSP Manual of Operations*. Through the aegis of the Interstate Shellfish Sanitation Conference (formed in 1982 at a meeting held in Annapolis, Maryland), the approved-area standards have been revised with more specific criteria (FDA 1989). There are now two recognized sampling schemes that may be applied: (1) sampling under adverse pollution conditions, and (2) systematic random sampling data analysis strategy. The first one requires at least 15 samples, while the second one requires 30 samples.

The bacteriological standards for approved growing areas provides a choice of either total coliform or fecal coliform criteria: "The total coliform median or geometric mean MPN of the water does not exceed 70 per 100 ml and the estimated 90th percentile does not exceed an MPN of 230 per 100 ml for a 5-tube decimal dilution test (or an MPN of 330 per 100 ml for a 3-tube decimal dilution test)."

"The fecal coliform median or geometric mean MPN of the water does not exceed 14 per 100 ml and the estimated 90th percentile does not exceed an MPN of 43 per 100 ml for a 5-tube decimal dilution test (or an MPN of 49 per 100 ml for a 3-tube decimal dilution test)." The fecal coliform group is defined as the bacteria of the coliform group which will produce gas from lactose in a suitable multiple-tube procedure liquid medium (EC or A-1) within 24 ± 2 h at $44.5 \pm 0.2°C$ in a water bath.

A restricted-area classification has been established for moderately polluted areas that may be used for the source of shellfish for relaying or depuration (purification) processes. The bacteriological standards for restricted depuration areas requires one of the following:

1. "The total coliform median or geometric mean MPN of the water does not exceed 700 per 100 ml and not more than 10% of the samples exceed an MPN of 2300 per 100 ml for a 5-tube decimal dilution test (or an MPN of 3300 per 100 ml for a 3-tube decimal dilution test)."
2. "The fecal coliform median or geometric mean MPN of the water does not exceed 88 per 100 ml and not more than 10% of the samples exceed an MPN of 260 per 100 ml for a 5-tube decimal dilution test (or an MPN of 300 per 100 ml for a 3-tube decimal dilution test)." (A thorough review leading to the establishment of this standard was prepared by the Shellfish Sanitation Branch, Food and Drug Administration, Washington, D.C., January 6, 1986, Interpretation number S.S. 38.)

A word of caution is raised in the use of MPN values, as they are estimates. Woodward (1957) provides a useful set of tables describing the confidence range of MPN values. An MPN is not an exact number to enumerate the density of bacteria in fermentation tubes. For example, for the five-tube multiple dilution test, the 95% confidence limits cover approximately 24–324% of the MPN. As applied to an MPN value of 70, the 95% confidence range would be about 17–230 MPN. This widespread range led Woodward to remark that the "MPN estimates are not precise and many of them are inherently improbable."

STATE SHELLFISH CONTROL AGENCIES

In 1925 the Committee on the Sanitary Control of Shellfish reported to the United States Surgeon General specific and lengthy measures to be taken to ensure that safe shellfish are produced and shipped from each state. It may be of interest to note that the initial report encompassed controls for oysters, hard clams, soft clams, razor clams, saltwater mussels, and scallops. In the first set of controls the committee addressed controls only for oysters, with the caveat that controls to other varieties should be subject to certain modifications.

The responsibilities to be adopted by the State Shellfish Control Agency were broad, complex, and fairly specific. There was no stipulation as to the type of state agency but only that the agency that took responsibility had the statutory authority to issue the required ordinances and codes. In some states the Public Health Department took the responsibility, while in others conservation agencies received the responsibilities. A few states divided the responsibilities among two or more agencies. It took several years for all coastal states to put into effect regulatory controls that fully satisfied the recommendations of the 1925 committee report.

All coastal states have either actual or potential shellfish harvest areas and, therefore, were expected to have a satisfactory program for sanitary control. The states, in order to comply with the intent of the cooperative program, must have laws and rules to permit them to designate waters where safe shellfish may be harvested. Early control standards were subjective, and stipulated simply that shellfish come from beds which have been carefully examined and found free from any justifiable suspicion of dangerous contamination with disease-producing microorganisms.

State regulatory agencies must have laboratories that have the capability to conduct bacteriological examinations of shellfish-growing waters and, in some instances, shellfish meats. These examinations must conform to the standard methods recommended by the NSSP. To ensure comparative, consistent results, laboratory evaluations are conducted by the Food and Drug Administration and, on occasion, split samples are distributed to the cooperating official state laboratories for analysis.

Not only do producing coastal states play a role in the NSSP, but consuming states, sometimes referred to as inland states, have a part. The FDA issues a list of state-certified interstate shellfish shippers for the information of all food control officials. This list is to be used to check that only "certified" shellfish from approved sources are reaching the marketplace. Inland receiving markets have a responsibility to use the list by checking each package label for the certification number unique to each state-certified shellfish shipper.

Other responsibilities for state shellfish control agencies are contained in the more than 150 page *NSSP Manual of Operations* (FDA 1989).

U.S. SHELLFISH INDUSTRY ROLE

Involvement of the U.S. shellfish industry on interstate commerce matters began in 1925, when the committee recommended the resolutions that became the basis of permanent oyster control. It was then that representatives from the shellfish industry participated in the formulation of sanitary controls for interstate shipment of shellfish. The shellfish industry was represented by members from the largest U.S. trade association, Oyster Growers and Dealers' Association of North America.

In a report subsequently adopted by the Surgeon General of the United States, recommendations concerning the sanitary control of the shellfish industry were detailed (Frost 1925). Among eight comprehensive controls, one pointed out a critical industry responsibility. It was stated that "there must be such governmental supervision and such trade organization as will make plain the source of shellfish and will prevent shellfish from one source being substituted for those from another source. This will be chiefly a problem of the individual state."

The role of shellfish trade organizations in the formulation of state and federal controls has proven to be of considerable assistance in arriving at reasonable and practical regulations. Where trade organizations exist, their involvement has been sought. However, the trade is divided along regional interests, and often segmented as to the type of business, such as harvesting, processing, and/or distribution. There are a number of customary industry practices, market demands and distribution considerations that have made it difficult for some trade organizations to reach a consensus among its members.

Of particular concern has been the need to trace harvested shellfish from their original harvest site. Should there be any problem reported concerning such shellfish, it is in the interest of the industry to reduce the areas affected to its lowest common denominator. This issue has been a difficult one for the industry and state regulatory agencies to develop workable measures that will apply on an interstate basis. Reliable tagging of shellstock, shellfish in the shell, with information as to the date, location and amount harvested, has been a continuing problem to enforce.

The shellfish industry has accepted certain responsibilities within the context of the cooperative program for interstate sanitary control of shellfish. Historically, it has agreed to obtain shellfish from state-designated harvest areas, process shellfish only in plants that meet sanitary standards, label shellfish containers with unique certification numbers issued by state food control agencies and make available such records as necessary to assure the control agents a proper origin and distribution of shellfish.

Adherence to these industry obligations go to the core of the program established by the U.S. Surgeon General. Many of the basic tenants continue today within the framework of the Interstate Shellfish Sanitation Conference (ISSC). The ISSC will be explained later in this chapter.

IMPORTED SHELLFISH

Imported fresh and fresh-frozen shellfish may enter the United States without being certified under the terms of the NSSP; however, many state food control authorities would rule such sources to be unapproved under state law and prohibit distribution or sale. Additionally, imported shellfish, whether certified or noncertified, must meet the regulatory requirements of the Federal Food, Drug, and Cosmetic Act as to the safety, wholesomeness and proper labeling.

Countries that desire to import fresh or fresh-frozen shellfish to the United States are advised by the Food and Drug Administration to develop a program equivalent to the NSSP for these products. There is no federal requirement that the countries do so, but by following the NSSP guidelines the involved countries may then be recognized as having suitable controls to certified imported fresh and fresh-frozen shellfish.

In order to recognize officially foreign imported shellfish under the provisions of the NSSP, the source country must negotiate a bilateral agreement (known as a Memorandum of Understanding—MOU) with the FDA. Once a country has an effective MOU, certified shellfish may be shipped and the foreign firms will be listed on the FDA's published list of Interstate Certified Shellfish Shippers.

The first country to enter into an agreement was Canada in 1948. Prior to that, some Canadian shippers were listed by the PHS even though there was no established administrative mechanism to ensure adherence to the NSSP standards (Bidwell 1947). The Food and Drug Administration conducted some bacteriological and chemical sampling of imported shellfish, but this was universally agreed as an inadequate control measure to protect public health.

The FDA now has MOUs with Australia, Chile, England, Iceland, Japan, the Republic of Korea, Mexico, New Zealand, and Canada for the sanitary control of fresh and fresh-frozen shellfish under the terms of the NSSP. Consequently, the NSSP requirements that apply to the production of shellfish in the United States for interstate shipment apply equally to those shellfish shipped under the terms of the MOUs. The ISSC has established specific requirements for interstate shipment of shellfish products and will appear in the 1990 revision of the *NSSP Manual of Operations, Part II* (Dressel 1990). Matters such as the physical condition of shipments, container labeling, product temperatures, and bacteriological quality are addressed.

FEDERAL ROLE

Since the inception of the NSSP, the U.S. Public Health Service has had the primary responsibility of administering and coordinating federal activities for the sanitary control of fresh and frozen shellfish which are shipped in interstate commerce. Other agencies that have had a role include the Department of Commerce, the Bureau of Commercial Fisheries (now the National Marine Fisheries Service), the Environmental Protection Agency, and the Food and Drug Administration, before the merger of 1968.

The Public Health Service has had the ongoing responsibility to evaluate the effectiveness of state shellfish sanitary control programs, to publish a list of state-certified interstate shellfish shippers, to conduct research into health hazards associated with the consumption of shellfish, to develop improved methods, and to provide technical assistance and training for state and foreign shellfish control personnel. In addition, the Public Health Service maintains relationships with foreign countries that ship fresh and frozen shellfish to the United States and evaluates these programs on a periodic basis.

Workshops of federal, state shellfish regulatory officials and representatives from the shellfish industry have been held to review administrative and technical procedures of the NSSP. These workshops were held in 1954, 1956, 1958, 1961, 1964, 1968, 1971, 1974, and 1977. To arrive at an agreement on important issues, the workshop procedure was to gain a consensus from the participants. Once an issue had consensus, the matter was then reflected in the next revision to the *NSSP Manual of Operations*. However, as the workshops grew in size, and the issues became more contentious, this method of deciding important revisions to the NSSP Manual proved to be unsatisfactory.

REORGANIZATION ACT OF 1968

To create greater efficiencies in the then "Department of Health, Education and Welfare," it was decided to consolidate all related food control programs within the Public Health Service and transfer them to the Food and Drug Administration. Also, there was concern that the growing responsibilities of the then independent "Food and Drug Administration" should come under closer supervision within the Department. With the approval of Congress under the Federal Reorganization Act of 1968, the administration of the shellfish sanitation program, among others, was transferred from other parts of the PHS to the Food and Drug Administration (FDA), and at the same time the FDA was reorganized under the Public Health Service.

A smooth transition was made since FDA had been a part of the NSSP, beginning with the 1925 program. However, there was concern among state officials and industry members that the program would be made more regulatory in nature under FDA's control. It was apparent that the cooperative voluntary spirit of the program had suffered in the conduct of the workshops and federal–state relations suffered over the difficulties state control programs were experiencing in attempting to meet the guidelines of the NSSP. As a continuation of the PHS responsibilities, FDA maintained the evaluation of state shellfish control programs. During the early 1970s, FDA made evaluation reports on state control programs that indicated several states had unsatisfactory ratings.

States with unsatisfactory ratings by the procedures of the NSSP would lose the endorsement of the federal government, now resting with the FDA. In the entire history of the NSSP to this time, no state had actually lost its federal endorsement. Letters were issued to the noncomplying states, requesting that they either correct their deficiencies or lose endorsement.

Loss of endorsement meant that FDA would no longer list those state-certified interstate shellfish shippers, effectively cutting off their interstate commerce. This was a strong sanction and one that the states knew would create an extreme hardship on their industry. Still, some states did not make the corrections, because either they were unwilling or they were unable to do so. The matter reached a crisis point. The question reached higher administrative levels within FDA and the Department of Health, Education, and Welfare.

It was advised by general counsel (legal office of the department) that the procedures of the NSSP were insufficient for the FDA to take action to withdraw endorsement from a state. Withdrawal of endorsement would have considerable economic consequences and place FDA in a regulatory situation that might not be legally supported. In the future, if FDA were to carry out the sanction of withdrawal of endorsement, the agency would need to promulgate federal regulations. The FDA, without alternatives, turned to the task and published the proposed National Shellfish Safety Program regulations in 1975 (FDA 1975).

In response to the proposal, there were many written and oral comments to the effect that the regulations would be too costly for the benefits to be received, that the present NSSP program was adequate, and that FDA had overreached its authority. State and federal control officials then set about to look for alternative mechanisms to maintain the principles of the NSSP and lessen the perceived economic impact the proposed federal regulations may have had upon states and the affected industry.

A model food control program, in effect since 1950, for the interstate shipment of milk was examined. With a few modifications, FDA officials and some key state officials met to formulate what is now known as the Interstate Shellfish Shippers Conference (ISSC). This model was found acceptable to the federal agencies, state shellfish control agencies and the shellfish industry. The Food and Drug Administration signed an agreement (Memorandum of Understanding) with the Chairman of the ISSC on March 14, 1983. On February 26, 1985, FDA announced in the Federal Register the decision to withdraw the proposed NSSP regulations and recognize the formation of the ISSC as a suitable alternative to maintain the principles of the *NSSP Manual of Operations* (FDA 1985). Thus, the ISSC would provide the forum for deliberating changes in the manual and for determining whatever sanctions may be necessary to obtain compliance with the voluntary program.

SUMMARY

The British experience in the development of shellfish controls at the turn of the century was repeated later in America. As history shows, it has been difficult to convince the state officials and industry members of the need to apply uniform national standards to local shellfish production areas. It is expected that the National Shellfish Sanitation Program (NSSP) will be challenged in the future. The safety of eating raw shellfish is relative to the ability of the control officials to enforce reasonable standards and to the adherence of industry members to the principles and standards of the NSSP. For more than 60 years the NSSP has appeared to provide the proper balance between protecting public health and maximizing the availability of a valuable natural resource—shellfish.

By a slow, unsure process of consensus among the federal officials, state officials and members of the shellfish industry, the NSSP administrative and microbiological standards have been challenged and revised in the light of new knowledge. Even though federal regulations were believed necessary in 1975, the development of a stronger, unified voluntary program administered through the creation of the Interstate Shellfish Sanitation Conference (ISSC) in 1982 has since been shown to be a successful public health federal–state cooperative arrangement.

2

Microbial and Chemical Indicators

Howard Kator and Martha Rhodes

Department of Biological Science, School of Marine Science and Virginia Institute of Marine Science, College of William and Mary, Gloucester Point, VA 23062

MICROBIOLOGICAL INDICATORS

Microbiological Indicators of Fecal Pollution for Use in Shellfish-Growing Waters—General Comments

Historically, recognition of feces as a source of waterborne disease transmitted by the fecal–oral route occurred in Europe in the mid to late 1800s. In the early 1900s it was generally recognized that microorganisms were more sensitive indicators of fecal contamination than chemical measures and that direct detection of pathogens to assess water quality was fundamentally limited because of difficulties associated with detection of multiple pathogens and their unpredictable occurrence. Use of surrogate microorganisms was considered as a reasonable means to detect the presence of fecal contamination in receiving waters. This review is generally restricted to indicators found exclusively in feces, sewage or septage that have applicability to shellfish waters. Microorganisms found in sewage or septage which can grow saprophytically in natural waters are not included. Examples of such microorganisms are *Aeromonas hydrophila*, *Klebsiella pneumoniae*, and *Pseudomonas aeruginosa*.

Although bacteria have been and are now the approved fecal indicator organism, use of fungi, and specifically the yeasts, should be briefly mentioned as they have seen limited advocacy as candidate indicators of fecal contamination. Simard (1971) discussed the use of "pink yeasts" as a possible indicator of fecal pollution, presenting various observations associating densities of yeasts with sewage pollution. On the basis of *in vitro* experiments which demonstrated the superior survival of *Candida albicans* compared to selected bacterial pathogens exposed to seawater, Jamieson, Madri, and Claus (1976) suggested this organism might be a useful indicator of fecal pollution in estuarine waters. Buck (1977), in a review of *C. albicans* as a candidate indicator of health hazards, cited deficiencies in information concerning its survival characteristics in natural receiving waters and the lack of rapid and selective enumeration methods. A "Standard Test Method" is now available based on membrane filtration and a selective medium designated mCA (ASTM 1987). Cabelli (1979) eliminated *C. albicans* from serious consideration as a health effects indicator because its detection in feces is inconsistent and at low densities. Hood (1983) failed to observe a relationship between levels of the yeast *Rhodotorula rubra* in fresh Gulf-of-Mexico oysters and clams with the classification status of the harvest waters, but observed that populations increased during storage, suggesting possible utility as an indicator of product quality.

Sources of Indicators to Shellfish-Growing Waters

Feces

Since the early work of Escherich, who identified *Bacillus coli* (now *Escherichia coli*) as a dominant bacterium in feces, considerable effort has been expended to identify the dominant bacteria associated with the gastrointestinal tracts of humans and warm-blooded animals. Microorganisms now used as indicators of fecal contamination are necessarily (but not exclusively) inhabitants of the alimentary tracts of humans and warm-blooded animals and are not numerically dominant components of the microbiota (Savage 1977). The bacterial composition of mammalian feces, which can vary with host species, age, diet, and geographic location (Savage 1977; Feachem et al. 1983), is dominated by obligate anaerobic bacteria belonging to major taxa that include *Bacteroides*, *Bifidobacterium*, *Clostridium*, and *Eubacterium*. Commensal facultative anaerobes include the lactobacilli, *Enterobacteriaceae*, and various cocci belonging to *Enterococcus* and *Streptococcus*. Concentrations of these fecal organisms range from 10^5–10^{11} organisms/g in the 150 g/day feces produced by humans in industrialized societies (Feachem et al. 1983). It is estimated that up to 40% of feces is composed of microbial cells (Savage 1977). Considering that obligate anaerobes are numerically dominant in human feces (Holdeman, Good, and Moore 1976; Moore and Holdeman 1974), use of aerobic recovery methods and facultative anaerobes as fecal indicators may appear somewhat paradoxical. There is little doubt that this reflects the latter's ease of recovery from aquatic environments compared to the more rigorous procedures necessary for recovery of obligate anaerobes and the presumption of poor survival in oxygen-containing environments. Human and animal feces also may harbor bacteriophages that utilize bacteria as hosts. Bacteriophages that use *E. coli* as a host, e.g., somatic coliphages and F-specific coliphages, and *Bacteroides fragilis* phages will be discussed in subsequent sections.

Domestic and feral animals are important potential sources of fecal contamination in urban and rural areas (Feachem et al. 1983; Geldreich 1972). The microbial composition of domestic animal feces has been examined to catalog dominant genera (Barnes 1986), to compare ratios of common indicators with those in humans (Geldreich and Kenner 1969), and to identify indicator microorganisms unique to animals (e.g., Cooper and Ramadan 1955; Geldreich and Kenner 1969; Wheater, Mara, and Oragni 1979). Similar data for feral warm-blooded animals living adjacent to or in shellfish-growing waters (e.g., harbor seals, Calambokidis, McLaughlin, and Steiger 1989; sea lion, Oppenheimer and Kelly 1952) are rare. Information relating the effects of animal presence on indicator and pathogen levels in adjacent estuarine receiving waters in the absence of gross pollution is needed.

Wastewaters

Anthropogenic sources of fecal organisms to shellfish-growing areas include discharges of treated municipal sewage, discharges of partially treated or raw sewage that occur owing to mechanical failure, combined sewer systems, failing septic systems, and vessel discharges. Bypassing, the release of untreated sewage during periods of high rainfall, is not uncommon. Levels of bacterial indicators in effluents from a properly functioning sewage treatment plant using disinfection will be considerably reduced compared to those in the influent (Miescier and Cabelli 1982). However, disinfection is not a stoichiometric process and its effectiveness varies with waste composition, influent volume, susceptibility of target organisms, contact time and flow. Other sources include agricultural and stormwater runoff, which can transport fecal wastes from humans, and domestic and feral animals to shellfish-growing areas. Resuspension and transport of contaminated sediments may also be a source to growing waters. Discharges from commercial and recreational vessels into marine waters, tributaries, and growing areas have been cited as sources of fecal contamination (Faust 1982).

Nonpoint Source Runoff

Much of the literature concerning indicators in shellfish-growing waters seems to have developed around the concept that fecal pollution is primarily derived from large point sources of sewage or river discharges. These rivers may integrate multiple point sources and typify conditions encountered in some highly urbanized northeastern coastal cities. Of all the municipal sewage facilities discharging into marine and estuarine waters, 42% are located in the northeast and these account for 60% of the total volume discharged (Shigenaka and Price 1988). Diffuse or nonpoint pollution is now recognized as a major source of fecal contamination to all types of receiving waters (Geldreich et al. 1968; Gilliland and Baxter-Potter 1987; Faust 1976). Indeed, in the mid-Atlantic and Southeast regions, significant proportions of the total shellfish-growing acreage are closed to direct harvesting presumably because of runoff from rural, agricultural, and wildlife sources (Leonard, Broutman, and Harkness 1989). Improperly functioning septic systems are implicated as important sources of fecal contamination. It has been estimated that 25–30% of the households in the United States use on-site septic systems for treatment of wastewaters and that at least 50% of these are operating marginally (Canter and Knox 1985). Specific issues of concern are the degree of treatment afforded by such systems, which is minimal and does not inactivate viruses, and the potential migration of effluent to estuarine waters because of inadequate drainfields, poor soil characteristics, or leaching through subsurface groundwater that can be affected by tidal fluctuations. Resolution of these issues awaits the application of rigorous and imaginative experimental approaches that facilitate tracing this contamination.

The public health significance of coliform levels in growing areas contaminated by diffuse sources remains unclear. The paradigm of a diluted sewage effluent does not apply in this scenario, because sources and ratios of indicators to pathogens are expected to vary considerably more than in large sources of domestic sewage. Fecal coliform densities that exceed the growing-area standard are not uncommon in Chesapeake Bay and other estuaries lacking identifiable point sources of human fecal pollution. Although direct evidence is lacking, it is believed conditions in estuaries can promote indicator survival (Erkenbrecker 1981). Contributing factors include inorganic and organic nutrient loading, high suspended solids, elevated temperatures, the presence of fine-grained organic-rich sediment and poor tidal flushing. Indeed, the need to identify and validate indicator systems to assess the sanitary quality of growing areas impacted by nonpoint sources may be the greatest challenge to sanitarians and shellfish microbiologists since the adoption of the coliform growing-area standard. An evaluation of indicators to index health risk in nonpoint-source-impacted areas must also consider the effects of microbial food webs and the estuarine environment on the removal and enumeration of allochthonous bacteria, the possible extra-enteral origin of indicators, and contamination derived from multiple sources including wild and domestic animals. A paucity of data to evaluate these concerns has engendered criticism of the current indicator and its validity as a public health standard in these environments. However, it appears reasonable in these environments to determine if indicator data are consistent with the results of shoreline surveys.

Environmental Factors Affecting the Fate of Allochthonous Indicators of Fecal Contamination

Effects of various environmental parameters on indicator survival have been examined by many investigators, but it is often difficult to unequivocally assess their "true" significance. Concerns arise with studies based on the use of *in vitro* experiments, laboratory-adapted strains, artificial or filtered menstrua, and exposure devices incapable of preventing external contamination or containment of test strains (Roper and Marshall 1979; Anderson, Rhodes, and Kator 1983). Given the probability that the microbiota can be a major factor affecting bacterial numbers, it seems unwise to generalize based on experimental designs that exclude this component. This is not to discredit the value of *in vitro* studies, which are useful to establish basic principles, but ecological hypotheses should be tested under conditions that duplicate open or natural systems as closely as possible. Another concern focuses on methods used for preparation of test cells. A majority of survival studies involved procedures for preparation of test cells known to result in sublethal stress or be physiologically debilitating. Culture age, growth conditions, and laboratory manipulations are important factors contributing to sublethal injury. Use of cells pregrown in rich media and harvested during the exponential phase of growth, cold-shocking, and harvesting by repeated centrifugation in unfavorable solutions are known to compromise physiological indices, increase sublethal stress, reduce adenylate levels, reduce enzyme activities, produce changes in membrane integrity leading to leakage of cell constituents, and render cells sensitive to free radicals or heavy metals (Postgate 1967; Strange 1976; Anderson, Rhodes, and Kator 1979; Granai and Sjogren 1981; Rhodes, Anderson, and Kator 1983). Failure to recognize these factors as sources of experimental bias has led to incorrect inferences of causality. Similar reservations apply to using laboratory-adapted strains because these cells may exhibit survival characteristics that are "atypical" compared to "fresh" fecal isolates (Anderson, Rhodes, and Kator 1979).

In view of the importance accorded to test cell preparation, a relevant hypothesis that should be tested concerns the effect of indicator origin (exposure prehistory) on its fate: Are there differences in the survival of indicator cells prepared under laboratory conditions and cells derived from "natural" sources such as sewage treatment plant (STP) effluent, septic tank effluent, agricultural or stormwater runoff? The STP or septic system provides a nutrient-rich environment, at times maintained at a higher temperature than the receiving waters, and exposes cells to varying degrees of toxic materials or disinfectant-mediated injury. Could it be that indicator bacteria from diffuse natural sources are physiologically adapted to adverse nutrient concentrations, intermittent lack of water, unfavorable temperatures, etc.?

Such adaptation could alter indicator persistence and recoverability compared to laboratory-grown cells.

Finally, it is difficult to see how investigations that treat the effects of environmental factors as if they were independent operators can lead to developing basic principles of indicator cell survival. Treatments that isolate single variables will only provide information concerning the potential role of that variable. In the environment survival will be affected by the interaction of that factor with perhaps more dominant processes. Laboratory studies must be augmented by admittedly more complex and variable *in situ* exposure studies that integrate physicochemical and biological factors. Conversely, *in situ* experiments must be carefully interpreted because of the potential for undetected processes that may affect test cell densities. Examples include breaching of exposure devices by autochthonous organisms owing to design or structural damage (Roper and Marshall 1979; Anderson, Rhodes, and Kator 1983) and physical penetration through membrane "pores" by bacterivorous nannoflagellates (Cynar, Estep, and Sieburth 1985) or other procaryotes (Li and Dickie 1985). Chambers may exhibit highly variable "bottle" effects because of nutrients contributed by fouling communities on chamber surfaces, blooms caused by exclusion of autochthonous organisms or predators, and attenuation of incident light. Finally, estuaries are unique dynamic systems, characterized by temporal and spatial heterogeneity. Generalizations based on experiments performed in Puget Sound may not be applicable to Chesapeake Bay or Gulf-of-Mexico growing waters.

The following sections summarize important aspects of the effects of physical, chemical, and biological factors on indicator survival, primarily under estuarine and marine conditions.

Physical and Chemical Factors

Temperature

At first glance, the literature dealing with the role of temperature on the *in situ* survival of *E. coli* in saline waters appears equivocal. Investigators have reported both positive (Anderson, Rhodes, and Kator 1983; Rhodes and Kator 1988) and negative (Faust, Aotaky, and Hargadon 1975; Vasconcelos and Swartz 1976; Lessard and Sieburth 1983) correlations of temperature and survival. This confusion can be resolved if it is understood that temperature has both indirect and direct effects on indicator fate. It can have a direct effect on bacterial activity, so that under appropriate conditions multiplication may occur at warmer seasonal temperatures. *E. coli* cells prepared under conditions to minimize stress initially exhibit multiplication (large negative values of the mortality rate coefficient, k) in membrane-filtered water (0.2 μm) at elevated temperatures in Chesapeake Bay (Rhodes and Kator 1988).

Maximum negative values of k correspond to increases in viable counts of approximately 1.0–1.5 log units. Although it may be thought that low environmental temperatures ($< 10°C$) would reduce coliform metabolic activity based on Q values, thereby favoring persistence, low environmental temperatures do not appear to have this effect as claimed by some authors. The indirect effects of low environmental temperature will be addressed in the context of sublethal stress but exposure of bacterial cells to low temperatures is known to compromise cell envelope integrity and physiological indices (Strange 1976). Indeed, positive values of k were obtained at temperatures below 10°C in filtered estuarine water (Rhodes and Kator 1988).

An indirect and highly significant relationship between temperature and indicator recovery and enumeration is the effect of temperature on the development of sublethal stress. Sublethal stress reflects the inability of a microorganism to be cultured in a medium because of prior injury, impairment or damage that resulted from exposure to unfavorable environmental conditions. Sublethally stressed cells are particularly sensitive to recovery methodologies that use selective temperatures, lack resuscitative protocols or use inhibitory substances to enhance media specificity and selectivity (Hackney, Ray, and Speck 1979). The development of sublethal stress in *E. coli* as a function of both temperature and salinity was explored by Anderson, Rhodes, and Kator (1979) and Rhodes, Anderson, and Kator (1983). Sublethal stress and mortality are inversely related to temperature. At temperatures below 10°C, transiently acute or progressive development of sublethal stress is detectable using a variety of techniques. These include differential counts on selective vs. nonselective recovery media and an assay technique (Anderson, Rhodes, and Kator 1979) based on the observation that sublethally stressed cells require a longer period of time to produce an electrochemically induced potential difference compared to nonstressed cells [electrochemical detection time (EDT)]. In practical terms, sublethally stressed cells may not be detected using enumeration methods that incorporate selective procedures. The effect of exposure to estuarine water on enumeration efficiency using experimental and approved recovery methods (Rhodes, Anderson, and Kator 1983) causes progressive stress over time even with favorable environmental temperatures. Selective methods such as the direct M-FC procedure, an approved method (APHA 1989), exhibit poor enumeration efficiency. The rosolic acid in this medium significantly reduces the recoverability of chlorine-injured fecal coliforms from sewage (Presswood and Strong 1978). Although resuscitation procedures in nonselective media have been employed to diminish the impact of selective enumeration, such measures may not be completely effective.

Sublethal stress is an important phenomenon that must be considered if conventional selective enumeration procedures are used for the recovery of indicator cells. Results from survival studies with *E. coli* where viable counts were not "corrected" for the effects of sublethal stress tend to overestimate mortality (underestimate cell densities), with the error being most significant at temperatures below 10°C. As a result, the responses of cells to physical parameters or treatments where sublethal injury was unrecognized must be interpreted with caution because the observed mortality may have been incorrectly attributed to another variable and not the enumeration process. These concerns reinforce the notion that sublethal stress is an important factor affecting indicator choice when viable recovery methods are used and observed cell densities are to have regulatory significance.

Temperature-induced sublethal stress also renders cells sensitive to other stressors. Mackey and Derrick (1986) presented evidence that cold-shock sensitized *E. coli* to very low concentrations of hydrogen peroxide and, as a consequence, reduced the recovery of viable cells on an organic-rich medium. Postgate (1967) cautioned against chilling of samples and cultures as a practice that promoted the death of stressed cells. Jackson (1974) observed loss of viability of *Staphylococcus aureus* and increased sensitivity to a selective medium at 5°C and noted that *E. coli* became more sensitive to violet red bile agar when exposed at this temperature. The implications of these observations apparently remain unheeded because Standard Methods (APHA 1992) suggests icing of microbiological samples if they cannot be processed within 1 h. A systematic examination of sample storage parameters to optimize recovery under a variety of seasonal temperature regimes should be part of the indicator evaluation process.

Indirect effects of temperature on coliform fate arise from the influence of seasonal temperature on the densities, composition and activities of the indigenous microbiota (Verstraete and Voets 1976; Anderson, Rhodes, and Kator 1983; Rhodes and Kator 1988). The role of the microbiota on indicator persistence and survival has been until recently a topic of considerable speculation. Thus, although the activities of antagonistic substances, parasitic and lytic microorganisms, and protozoans were recognized as potentially important factors controlling indicator abundance (Mitchell and Yankofsky 1969; Mitchell 1972; Mitchell and Chamberlin 1975; Drake and Tsuchiya 1976), it is only recently that microbial ecologists have identified bacterivory as important carbon and energy pathways (Wright and Coffin 1984). The importance of bacteria in coastal waters as food for heterotrophic microflagellates has been demonstrated (Fenchel 1982; Anderson and Fenchel 1985). Using diffusion chambers, Awong, Bitton, and Chaudhry (1990) observed that densities of genetically engineered strains of *E. coli* were significantly reduced in nonsterile lake water compared to cells in filter-sterilized water. Gonzalez, Sherr, and Sherr (1990) observed bacterivory of *E. coli* and *E. faecalis* by natural assemblages of estuarine flagellates and ciliates incubated in Whirl-pak bags. Both predator groups exhibited grazing preference for large bacterial cells but ciliates ingested *E. faecalis* at a higher rate than *E. coli*. Studies with diffusion chambers, perhaps somewhat artifactual in terms of community development and enhanced encounters between predator and prey, have revealed the potential influence of the microbiota on *E. coli* survival in estuarine water (Rhodes and Kator 1988). Generally, times of maximum decline in nonfiltered water coincided with maximum densities of heterotrophic flagellates or other predators (Rhodes and Kator 1988). Although predation and other microbially mediated removal processes, and the "intrinsic" growth responses of enteric bacteria were direct functions of temperature, the net combined effect of increased temperature was indicator removal. It is evident that failure to recognize both direct and indirect effects of temperature has led to the erroneous conclusion that elevated temperature, *per se*, does not favor coliform survival.

Salinity

Efforts to determine the effect of salinity on indicator survival constitutes a small and somewhat inconclusive literature (Carlucci and Pramer 1960; Faust, Aotaky, and Hargadon 1975; Orlob 1956; Vasconcelos and Swartz 1976). Results reported, which are generally for *E. coli*, range from no effect of salinity on *E. coli* viability to reduced survival with increasing salinity. This literature remains problematic for many of the same reasons noted for temperature studies. The combination of all-or-none viable counting methods with factors known to be stressors, e.g., selective recovery methods, artificial seawater with possible trace toxicants, harsh inocula preparation methods, contributed to the apparent disappearance of test cells. These problems were avoided with techniques that measured graded responses and used cell preparation techniques that minimized injury (Anderson, Rhodes, and Kator 1979). Graded responses included EDT in selective and nonselective media, β-galactosidase specific activity, and growth rate. Increased salinity was accompanied by decreased recovery of viable cells, an increase in sublethal stress expressed both in terms of larger EDT values and as the difference in cell recovery on selective versus nonselective media, and as a marked reduction in β-galactosidase specific activity (Anderson, Rhodes, and Kator 1979). Cells exposed to 30 psu seawater for 2–9 days exhibited an apparent mortality about 9 times greater when recovered in EC medium than the same cells recovered in TSB. The idea that exposure of *E. coli* to saline water yields cells with altered physiological properties, thereby requiring modified enumeration methods, has been suggested. Dawe and Penrose (1978) observed that salinity-induced debility in coliforms is reduced by incorporation of seawater into the recovery medium. Gauthier, Munro, and Mohadjer (1987) increased recovery of *E. coli* cells adapted to seawater by including sodium chloride in an enumeration medium, although the degree of response was strain- and time-dependent. Munro, Gauthier, and Laumond (1987) observed that *E. coli* starved in seawater manifested a variety of physiological and structural responses including loss of β-galactosidase activity, increased activity of other enzymes, and altered sensitivities to antibiotics, phages, and heavy metals. Increased enzyme activity toward 4-methylumbelliferyl heptanoate observed during starvation of *E. coli* in seawater was presumably a response to low nutrient conditions (Fiksdal et al. 1989). *E. coli* cell envelope composition and functional characteristics are different in cells grown in an estuarine-water-based medium compared to those grown in a standard medium (Chai 1983). The effect of high osmolarity in *E. coli* is to repress the synthesis of OmpF porin protein, a protective mechanism to reduce cell permeability to naturally occurring detergents, i.e., bile salts (Nikaido and Vaara 1987). This adaptative process retains membrane permeability, albeit reduced, to nutrients with molecular weights of 100–200. An indirect consequence of reduced membrane permeability under hyperosmolar conditions could be to alter significantly or reduce the uptake of molecules used in direct cell viability assays such as tetrazolium salts. Gauthier et al. (1991) discussed the significance of cell preparation methods on *E. coli* survival in seawater. Procedures causing loss of intracellular K^+ and glutamate affect the ability of cells to regulate osmotic pressure and reduce their resistance and survival in seawater. Restoration of intracellular K^+ and glutamate restores regulation of cellular osmotic pressure and enhances survival in seawater. In summary, the studies mentioned support an emerging hypothesis that the survival of *E. coli* in seawater is an active process involving physiological adaptation to an adverse environment. Recovery of adapted cells may be optimized using a resuscitative environment that allows cells to readapt to conditions of lower osmolarity.

Strange (1976) in his monograph on stress noted that osmotic shock causes loss of cell viability, reduces active transport of solutes, and releases a variety of metabolites and enzymes. Roth, Leckie, and Dietzler (1988) demonstrated that a large proportion of *E. coli* cells exposed to hyperosmotic conditions are not recoverable on a nonselective medium. However, if the cells are osmotically upshocked in the presence of betaine (*N,N,N*-trimethylglycine, a naturally occurring and ubiquitous nitrogen-containing compound produced by microorganisms, animals, and plants), the cells remain culturable. Betaine accelerates uptake of ATP and reduces intracellular ATP that accumulates in osmotically upshocked cells and facilitates protein synthesis. Conjugative transfer of a plasmid between donor and recipient *E. coli* cells was also enhanced by glycine betaine in autoclaved sediments (Breittmayer and Gauthier 1990). Roth, Leckie, and Dietzler (1988) developed a resuscitation method using a medium incorporating betaine, chloramphenicol (to prevent changes in cell density during resuscitation), ammonium, and glucose as nitrogen and carbon sources. At this writing we are unaware if this method has been evaluated with estuarine samples.

Adsorption and Sedimentation

The effect of estuarine particulates on the persistence of autochthonous bacteria continues to be an interesting area of research. Rubentschik, Roisin, and Bieljansky (1936) suggested that adsorption of *E. coli* to particulates with subsequent deposition in sediments is an important aspect of self-purification in salt lakes. Weiss (1951) investigated *E. coli* adsorption on river and estuarine silts and concluded that adsorption enhances the bacterial sedimentation rate and is a function of particle type and size. Although flocculation of silts increases in seawater, the adsorptive capacity of silts toward bacteria decreases with increased salinity and the bacteria desorb. Mine, Curran, and Wilson (1986) concluded that removal of fecal coliforms from estuarine water by deposition was directly related to the concentration of naturally occurring suspended solids. A similar relationship was not observed in seawater and attributed to the differences in the depositional behavior of suspended solids in the two systems. Roper and Marshall (1974, 1979) observed that *E. coli* adsorb to sediments at high electrolyte concentrations and desorb below a critical concentration. They hypothesized that in low electrolyte (freshwater) systems electrostatic forces allow bacteria to exist as stable colloidal dispersions. As salinity is increased, flocculation and sedimentation of bacteria and particles will occur. This straightforward model can be complicated by organics present on the microorganism's "surface." Thus, adsorbed bacteria can adhere very strongly to sediment particles owing to production of organic exopolymers (Marshall 1985). Viruses also adsorb to estuarine sediments. Labelle and Gerba (1979) observed $> 99\%$ adsorption of various enteric viruses to sediments at salinities of 1–35 psu. Alterations in salinity and pH produce small but variable effects on adsorption and desorption of most virus types examined. Because enteric viruses did not readily desorb, they concluded that viral transport would be dependent on particle resuspension and transport.

Based on the above observations, it is not surprising that densities of fecal indicator bacteria (Erkenbrecker 1981; Shiaris et al. 1987) and enteric viruses (LaBelle et al. 1980; Rao et al. 1984) are elevated in estuarine sediments compared to overlying waters. Sediments from an urban shellfish-growing area in the Chesapeake Bay contained fecal coliforms at densities two orders of magnitude larger than in the water column (Erkenbrecker 1981). Variable but generally smaller values of the ratio of fecal coliform densities in sediment to water were observed in a small subestuary subject to nonpoint pollution (Kator and Rhodes 1989). LaBelle et al. (1980) found no correlation between the densities of bacterial indicators and viruses in estuarine waters. There was a positive correlation between the numbers of viruses and fecal coliforms in sediments.

In vitro and *in situ* experiments (Gerba and McLeod 1976; Roper and Marshall 1974, 1979; Perez-Rosas and Hazen 1988) have shown the positive effect of marine or estuarine sediments on *E. coli* persistence. Similarly, adsorption to estuarine sediments enhances the survival of enteroviruses (Smith, Gerba, and Melnick 1978; Toranzo, Berja, and Hetrick 1982) and bacteriophage T7 (Bitton and Mitchell 1974). Protective effects are attributed primarily to accumulation of nutrients on particle surfaces and reduced predation (Roper and Marshall 1978*b*) and antibiosis. As previously mentioned, Roth, Leckie, and Dietzler (1988) reported the organic compound betaine minimizes osmotic stress and restores the culturability of *E. coli* exposed to seawater. Le Rudulier and Bouillard (1983) have shown that betaine concentrations as low as 1 mM eliminate the osmotic stress in *K. pneumoniae*, *S. typhimurium* and *E. coli* over a range of high NaCl concentrations (0.65–1.0 M). There was no evidence that betaine is utilized as a growth substrate in *K. pneumoniae*. Betaine is a ubiquitous and a relatively abundant compound in benthic organisms and is an important substrate in fermentation pathways coupled to methanogenesis in marine sediments (King 1984, 1988; Heijthuijsen and Hansen 1989). The apparent enrichment and persistence of *E. coli* (or other enteric bacteria) in sediments may be augmented by the osmotolerance afforded by betaine or other osmolytes (Munro et al. 1989; Gauthier and Le Rudulier 1990; Ghoul, Bernard, and Cormier 1990). The effects of anaerobic sediments, which are sources of inorganic nutrients and small-molecular-weight fermentation products, on indicator persistence is obviously a research issue requiring further study.

The accumulation and enhanced survival of sewage microorganisms in sediments have been employed as a justification for questioning the validity of using a water quality criterion as the basis for a classification of shellfish-growing waters (Shiaris et al. 1987). Sediments provide a more integrated recent history of fecal pollution than overlying waters, which reflect transient pollution events. Relationships between indicators in sediment and river water were examined in a study by Matson, Hornor, and Buck (1978). Sediment deposition, resuspension, and transport were identified as processes that significantly affect densities of indicators (and enteric pathogens) in overlying waters (Matson, Hornor, and Buck 1978; LaBelle et al. 1980; Erkenbrecker 1981). Thus, the absence of microorganisms from the water column may not reflect a diminished health hazard. Metcalf, Slanetz, and Bartley (1973) noted that fecal coliform concentrations and the likelihood of isolating salmonellae in the water column were functions of tidal stage, suggesting an association between tidal currents and the resuspension and transport of particulate material. Particle-associated pathogen transport is emphasized by greater viral accumulation in shellfish exposed to resuspended sediment as compared to those placed in a system containing undisturbed sediment (Landry et al. 1983).

Light

Direct lethal effects of light on enteric bacteria in seawater have been demonstrated under *in situ* (Cornax et al. 1990; Gameson and Saxon 1967; Gameson and Gould 1975; Bellair, Parr-Smith, and Wallis 1977; Fujioka et al. 1981) and *in vitro* (Gameson and Gould 1975; Fujioka et al. 1981; Kapuscinski and Mitchell 1981; McCambridge and McMeekin 1981; Fujioka and Siwak 1987) experimental treatments. Tartera, Bosch, and Jofre (1988) compared the sensitivities of bacterial and viral indicators to UV light under *in vitro* conditions. Coliphage f2 and a *Bacteroides fragilis* bacteriophage are more resistant than *E. coli* and *E. faecalis.* Male-specific coliphages are significantly more resistant to UV radiation than indicator bacteria (Havelaar 1986, 1987). Barcina et al. (1989) reported that artificial visible light causes "progressive dormancy" in *E. coli* in flasks of river water based on comparisons between viable and nonculturable cells counts and metabolic measurements. Glucose uptake was inhibited by light and the authors suggested that light-mediated injury could possibly affect recovery using culturable methods.

The overall importance of these studies to the mortality of indicator organisms in shellfish-growing waters remains somewhat equivocal. This is because estuaries where shellfish are produced, for example in the eastern United States and Gulf of Mexico, are highly turbid and can be dominated by complex microbial food webs. The importance of these factors resides in the attenuating capacity of suspended and dissolved material toward lethal wavelengths of light and the light-stimulating effect on components of the microbiota that can lead to enhanced bacterial mortality. Attenuation of UV-B (280–320 nm) light may occur in the uppermost layer of the water column, although the degree of attenuation in coastal waters is influenced by local properties that affect light absorption such as dissolved organic matter, chlorophyll concentration and particulate load (Calkins 1982). Interactive effects of light and the autochthonous microbiota have been demonstrated in fresh and estuarine waters under *in vitro* conditions (McCambridge and McMeekin 1981; Barcina et al. 1986). A series of *in situ* experiments were performed in the Chesapeake Bay using diffusion chambers specially modified to maximize light penetration and employing light and dark, filtered and nonfiltered treatments (Rhodes and Kator 1990). Compared to cells suspended at 1.0 cm below the water surface, mortality from sunlight was essentially insignificant at 25.0 cm except during periods (fall, winter) of minimal light attenuation. Cells at 1.0 cm also exhibited significant sublethal stress. During the warm seasons, significantly greater mortality occurred in the presence of light and the microbiota than with either alone. Enhanced mortality of the combined treatment can be attributed to stimulation of predation, by sunlight, light-dependent release of antagonistic substances, or formation of photochemically induced toxicants. Therefore, in turbid shellfish-growing waters sunlight-induced injury and mortality result from direct and indirect effects, whose relative influence will be a function of local conditions.

Organic Compounds and Nutrients

In a review of factors affecting the survival of enteric microorganisms in marine and estuarine environments, Mitchell and Chamberlin (1975, 241) concluded a section on "Nutrient deficiencies" as follows: "These results coupled with laboratory findings tend to suggest a significant role for nutrients in determining survival of enteric bacteria in seawater." Until recently, the literature did not refute this conclusion, but did little to amplify it or provide data supporting this hypothesis. Fortunately, this area has been the focus of a number of research groups such as Lopez-Torres, Hazen, and Toranzos (1987), who describe a positive statistical association between nutrient concentrations and persistence of *E. coli*. *E. coli* starved in seawater manifest a variety of physiological and structural responses including changes in enzyme activity, and altered sensitivities to antibiotics, phages, and heavy metals (Munro, Gauthier and Laumond 1987). Fiksdal et al. (1989) concluded that the increased hydrolase activity toward 4-methylumbelliferyl heptanoate by *E. coli* starved in seawater is an adaptation to nutrient limitation. Studies are still needed to evaluate the direct effects of inorganic nutrients and organic carbon on enteric survival, especially in nutrient-rich shellfish-growing areas that are not well flushed, where high productivity associated with detrital food webs, allochthonous inputs, and organically enriched fine sediments may provide nutrient conditions favoring indicator persistence.

Toxic Compounds

Although the lethality of disinfectants toward indicator bacteria is well described, the effects of toxicants present in shellfish-growing waters and sediment on these microorganisms is a rather impoverished area of research. This is somewhat surprising, considering that many estuarine areas are polluted by high levels of toxic heavy metals, xenobiotics and other materials. Runoff from agricultural and urban areas contains a variety of pesticides of varied toxicity and persistence characteristics. Elevated concentrations of zinc have been found in coastal sediments impacted by sewage discharges

(Bruland et al. 1974). Effluents from kraft pulp mill plants may elicit a broad range of toxic effects (Sodergren 1989). Oil shale process waters are lethal to coliforms and *E. faecalis* under *in vitro* exposure conditions (Adams and Farrier 1982). Jones and Cobet (1975) reported the toxicity of naturally occurring heavy metals in Caribbean seawater to enteric bacteria. Widespread use of organotin compounds as antifouling paints has been recognized as contributing to high tributyltin concentrations in estuarine waters and sediments frequently exposed to vessels. Pettibone and Cooney (1986) concluded that organotin compounds are not acutely toxic to *E. coli* and *E. faecalis* isolates at naturally occurring concentrations but act as stressors. Chai (1983) noted that *E. coli* grown in a medium containing estuarine water manifested changes (adaptations) in cell envelope composition, changes that afforded decreased sensitivity to bacteriophage infection and colicins but also rendered the cells more sensitive to heavy metals and detergents. In a series of experiments to evaluate the effects of 10 mM Zn^{2+} on bacteria and selected coliphages, the toxicity of zinc to *E. coli* was increased in the presence of 0.1 M or higher sodium chloride (Babich and Stotzky 1978). *In vitro* expósure of *E. coli* to estuarine water containing various toxic chemicals alters cell envelope protein composition, detectability of plasmids, and affects carbohydrate and amino acid metabolism (Palmer et al. 1984). Creosote contamination of estuarine sediments is detrimental to microbial communities, as reflected in reduced secondary production and biomass (Koepfler and Kator 1986). Although definitive information is lacking, sediments may be sources of chemical and biological stressors toward indicator bacteria, though it is likely the degree of effect will be site-specific.

Biological Factors

Biological factors that affect indicator fate have been discussed or mentioned in other sections. A complete survey of this literature is beyond the scope of this review. Biological interactions between indicators (bacteria or viruses) and components of the microbiota are complex and definitive *in situ* rate measurements describing interactive processes are lacking. Processes effecting removal through predation, parasitism or lytic activity (e.g., Berk et al. 1976; Roper and Marshall 1977, 1978*c*; Enzinger and Cooper 1976; Anderson, Rhodes, and Kator 1983; Rhodes and Kator 1988) or antibiosis (Sieburth and Pratt 1962; Moebus 1972; Aubert, Pesando, and Gauthier 1975; Toranzo, Barga, and Hetrick 1982; Girones, Jofre, and Bosch 1989), and competition for nutrients (Jannasch 1966) have been described. A need exists to determine the relative importance of biological interactions compared to other processes that affect indicator survival or removal. Because biological processes in estuaries are highly variable in temporal and spatial dimensions, studies to assess these effects must be performed with sufficient replication and seasonal coverage to ensure the collection of representative data. Technical impediments to *in situ* experimentation that require resolution are uncontrolled contamination of exposure chambers by autochthonous microorganisms, inadequate sample volumes, and the need to unequivocally differentiate test cells from the autochthonous cells. Garcia-Lara et al. (1991) proposed use of [^{3}H]thymidine-labeled cells to evaluate the overall contribution of grazing and lytic processes on indicator mortality. Analytical approaches combining rapid direct viable counting methods with those permitting unequivocal identification of test organisms (e.g., fluorescent antibody, gene probes, etc., Roszak and Colwell 1987*a*) may also provide solutions.

Issues That Affect Indicator Usefulness

The coliform group of indicator bacteria has traditionally been the basis for the microbiological growing-area standard for shellfish waters since the early 1990s. Since that time, researchers have identified critical deficiencies in its use as an indicator of fecal contamination or potential health risk in aquatic systems (e.g., Berg 1978; Cabelli 1978*b*; Dutka 1973; Shear and Gottlieb 1980). Responses to some of these criticisms are evident in a measured progression to adopt the most fecal-specific coliform organisms as approved indicators. Thus, the fecal coliform has supplanted the total coliform group, and *Escherichia coli* may replace the fecal coliform. Recent studies challenge the fundamental assumption that *E. coli* is the most efficient predictor of enteric disease risk in marine waters (Cabelli et al. 1983). Other workers have identified processes whose effects on indicator densities seriously question the validity of the coliform group as the basis for the growing-water standard. Differences in resistance to disinfection and persistence characteristics of bacterial indicators compared to the most prevalent viral agents causing shellfish-associated disease continues to focus attention on the adequacy of bacterial indicators to protect public health.

As a result, interest in other indicators of fecal pollution has intensified, although information concerning the survival characteristics and recovery methodologies of alternate indicators is for the most part restricted to freshwater environments. Thus, the following statement is made in Section 9010 A (General Discussion) of Standard Methods (APHA 1992) with regard to pollution in ". . . tidal estuaries and other bodies of saline water," that "In the following sections, applications of specific techniques to saline water are not discussed because the methods used for fresh waters also can be used satisfactorily with saline waters." Contrary to Standard Methods (APHA 1992), there is no a priori basis to conclude that enumeration methods for freshwater will be effective in shellfish-growing waters. In fact, much of the literature suggests that this assumption is unwarranted.

The Indicator Concept

Use of microbiological water standards to minimize the transmission of enteric disease through the consumption of raw molluscan shellfish has been a rather successful public health strategy, eliminating major outbreaks of gastroenteritis caused by salmonellae. The indicator concept, first elaborated for drinking water, where contamination was assessed on the basis of an operational response using a multiple-tube enumeration procedure, was later applied to the waters of Raritan Bay in the early 1900s following incidents of typhoid associated with the consumption of raw clams (Kehr et al. 1941). A target density of total coliforms was derived based on the dilution of a large point source of domestic sewage with a sufficient volume of water to yield a theoretical final ratio of indicator to bacterial pathogen. Dilution was anticipated to lower the pathogen density to a value yielding an unknown but significantly reduced potential public health risk. Use of a water-based, rather than shellfish-based, standard was supported by coliform data showing that bivalves (i.e., hard clams) grown in water at or below that standard indicator level would not bioconcentrate pathogens to densities exceeding a presumed minimum infective dosage. Adoption of a surrogate (indicator) for the pathogen *Salmonella typhi* was expedient for reasons that included the absence of an accurate and selective method for the recovery of this organism and the inability to predict its occurrence in sewage effluent. Moreover, Kehr et al. (1941) recognized the standard as an arbitrary index that does not index a predetermined level of risk, requiring verification through epidemiological investigation. "It is believed, therefore, that the most favorable method of reducing the danger of infection from the ingestion of raw hard clams is through the adoption of an arbitrary standard that would reduce to a satisfactory degree the estimated coliform content of the annual production of Raritan Bay hard clams" (Kehr et al. 1941, 94).

A number of basic assumptions were implicit to the indicator concept. It was assumed that the index or standard was applicable only to a diluted waste effluent, that there existed a constant density ratio of indicator to the pathogen of concern (i.e., *S. typhi*) in the effluent, and that the ratio was conserved in the environment, at least within the immediate vicinity of a discharge (Kehr et al. 1941). However, the authors recognized that if pollution sources departed from these conditions, reliance on the bacteriological standard to reflect health risk was unwarranted. Analysis of contributing sources, the interdiction of judgement, and skepticism concerning the ability of a numerical standard value to offer unequivocal health protection were aspects of their thinking.

Many authors (Bonde 1977; Cabelli 1978*a*; Berg 1978; Wheater et al. 1980) have considered the characteristics that an ideal indicator should possess. Most have agreed that no single indicator is "ideal" or can be universally applied. Indicators possess unique characteristics whose applicability depends on the question being asked and the specific environment involved. Thus, an indicator that is not very resistant to chlorination (a poor indicator of disinfection) would be inappropriate for sewage effluents but could be perfectly adequate in a nonpoint-source-polluted estuary. Very persistent indicators such as *Clostridium perfringens* spores or coprostanol may not be appropriate in a nonpoint-source area but can be useful as indices of sewage plume dispersion or transport of particulate-bound material.

One paradigm of an indicator in shellfish waters would be to provide a relationship to index health risk. However, epidemiological studies to establish predictive relationships are lacking and the standards now used can reflect the presence of fecal contamination and disease-causing "potential". As noted already, this usage is based on an assumed quantitative relationship between coliform bacteria and an enteric pathogen in diluted sewage from a large population. However, in nonpoint-impacted estuarine and marine receiving waters, the constancy of this association may not hold because of variation inherent in small multiple sources, which are by nature intermittent, stormwater runoff, the interaction of both indicators and pathogens in the environment, and fecal pollution from nonhuman sources.

Indicators may also serve to provide information concerning source or "age" of pollution in nonpoint-impacted watersheds. An indicator can be used as an investigatory tool to provide information relevant to fecal source, i.e., location within a watershed, human versus animal sources or animal type, and the age of pollution or its "freshness." Indicators inappropriate as indices of health risk may be useful for this purpose.

Persistence

Sanitary engineers and microbiologists have expended considerable effort to identify environmental factors and processes that affect densities of enteric bacteria and their recovery. Much of the early literature on coliform survival suggested that "dieoff" (or "decay") was the only functional response of coliforms exposed to marine or estuarine environments. This was primarily attributed to the "bactericidal" property of seawater (Ketchum, Ayers, and Vaccaro 1952). Immediate decreases in concentrations of coliform organisms discharged from a sewage treatment plant can result from physical dilution through turbulent transport and mixing, processes that presumably occur over time periods of minutes or hours. At some distance from the discharge where the effluent becomes dynamically passive, and in addition to dilution, light, temperature, bacterivory, antagonism, inhibition, sedimentation, and autecological responses become important factors affecting indicator fate.

Use of the term "dieoff" to describe changes in indicator densities over time is inappropriate. The apparent reduction of recoverable counts from estuarine waters is not only the result of "true" cell death, i.e., cells that become nonviable, but is also a function of

1. enumeration methodology,
2. physiological adaptation to an adverse environment, and
3. complex interactions of physical, biological, and chemical processes.

The net effect of these processes will be a function of their degree of interaction, relative dominance and the unique characteristics of each environment.

Initially, dilution was considered to be the major physical factor affecting the densities of indicator bacteria, i.e., indicator bacteria were treated as "quasi" conservative elements. We now recognize that the approved coliform indicator for shellfish-growing areas does not behave as a conservative element, and can exhibit changes in density that are functions of (1) physiological responses to a variety of physical and chemical properties of the environment that may be expressed as loss of viability, change in culturability, persistence or aftergrowth, and (2) complex interactions between the indicator and components of the microbiota. In estuaries, where entrainment and flushing characteristics increase contact times between indicator organisms and the environment, these are important processes affecting indicator densities.

Nonfecal Origin

Other deficiencies of coliform indicators include new evidence for the extra-enteral origins of some species that comprise the fecal coliform group (Lopez-Torres, Hazen, and Toranzos 1987), the capacity of organisms defined as fecal coliforms to persist or even grow in aquatic environments under favorable temperature and nutrient regimes (Santo Domingo, Fuentes, and Hazen 1989; Hazen 1988; Knittel et al. 1977; Rhodes and Kator 1988; Verstraete and Voets 1976), the inability of the indicator to differentiate vertebrate source, the effect of environmental exposure on recoverability (Rhodes, Anderson, and Kator 1983; Xu et al. 1982), and the recognition that allochthonous bacterial indicators are prey to microbial grazers (Anderson, Rhodes, and Kator 1983; McCambridge and McMeekin 1979, 1980; Rhodes and Kator 1988).

If these criticisms promote uncertainty about the validity of the fecal coliform indicator, they must also diminish confidence in its use as a numeric standard applicable to all shellfish-growing waters. Arguably, other bacterial indicators may be found that offer advantages in terms of source specificity, determination of source age, or recovery methodology. However, considering the important influence of the environment on bacterial indicator fate and recovery, and the inherent regional and temporal variations that characterize dynamic estuarine environments, it is difficult to find credible the proposition that a standard based on a bacterial indicator can be "universally applied."

Recoverability

Another important concept concerning the fate and culturability of enteric organisms following environmental exposure was described by Xu et al. (1982). These workers noted a differential between numbers of viable cells recovered from seawater, using a direct measure of viability and the viable count obtained using nonselective traditional culture methods. It was demonstrated that a significant proportion of cells starved in seawater rapidly enter a "nonrecoverable" stage, i.e., do not grow in standard media, but remain viable, as shown by a direct viable count assay (Kogure, Simidu, and Taga 1979). The phenomenon of nonculturability has been observed by other workers (Lopez-Torres, Hazen, and Taranzos 1987; Munro et al. 1987; Roth, Leckie, and Dietzler 1988; Martinez, Garcia-Lara, and Vives-Rego 1989; Desmonts et al. 1990; Garcia-Lara et al. 1991). Culturable cell counts drop significantly within 2–4 days exposure. At these times, culturable counts are 20% or less of total direct viable counts. Although significant numbers of culturable cells may remain (depending upon the initial cell density), these responses suggest that the "true" numbers of indicator organisms could be significantly underestimated because of the low culturable densities found in approved shellfish waters. Garcia-Lara et al. (1991) employed the thymidine-labeling method of Servais, Billen, and Vives-Rego (1985) to demonstrate that actual mortality rates of *E. coli*, *S. typhimurium* and *S. faecium* (*E. faecium*) in natural seawater were an order of magnitude lower than mortality rates measured by culturable counts. This method involves measuring loss of radioactivity from the trichloroacetic-acid-insoluble fraction (containing the DNA from intact cells) after addition of [^{3}H]thymidine-labeled cells to seawater. *E. faecium* exhibited the lowest rate of disappearance as a culturable count. Desmonts et al. (1990) combined the direct viable count (DVC) procedure of Kogure, Simidu, and Taga (1979) with an indirect fluorescent-antibody (IFA) for *Salmonella* spp. *Salmonella* spp. were detected in all samples of raw domestic and chlorinated wastewaters by IFA-DVC at comparatively high densities, even when culturable cells were not found. Of the IFA detected cells in chlorinated wastewater, 5–31.5% were viable by DVC. The public health significance of these studies remains to be evaluated, but the relatively high counts of salmonellae observed in disinfected effluents should promote an examination of the disinfection process. Grimes and Colwell (1986) demonstrated that cells of an enterotoxigenic strain of *E. coli* became nonculturable on a BHI-synthetic seawater medium after 13 h exposure to ocean water and remained virulent, based on retention of plasmids coding for virulence genes and subsequent rabbit ileal loop assays. A detailed discussion of the viable but nonculturable phenomenon can be found in Grimes et al. (1986) and Roszak and Colwell (1987*a*). Recognition of this physiological adaptation to starvation has done much to stimulate rethinking the fate of allochthonous bacteria in marine and estuarine environments as well as the efficacy of traditional enumeration methods. Entry of a significant proportion of an enteric indicator population during exposure to seawater into a nonrecoverable stage would seriously question the validity of bacterial indicators.

However, numerous reports also demonstrate coliforms (generally *E. coli*) can persist for extended periods. Awong et al. (1990) found both wild-type and genetically engineered strains of *E. coli* survived for periods of 10–30 days, albeit with gradual reductions in culturable densities, in filtered and nonfiltered lake water. Flint (1987) reported that *E. coli* survived in sterile river water for 260 days in the absence of predators and that direct and cultural counts were similar over the first 40 days of exposure. Fiskdal et al. (1989) observed both rapid and slow declines in culturable counts of *E. coli* in filtered seawater. Extended culturability (ca. 5 months) of *E. coli* in seawater was reported by Munro et al. (1989). Munro et al. (1989) suggested that genetic differences in osmoregulatory capabilities, acting in response to the effects of test cell pretreatment and exposure conditions, are responsible for differences in survival and culturability.

We are uncertain about the causes of variation concerning entry into a nonculturable state. Some evidence suggests that these may be related to the properties of test strains, inocula preparation methods (e.g., Gauthier et al. 1991) or to water quality properties such as the concentration or type of organic matter. Because of the importance of this phenomenon in terms of indicator selection, its consequences on culturable enumeration methods, and recent reports that document its occurrence, a rigorous and standardized evaluation of factors that control entry of cells into the nonculturable state is desirable.

INDICATORS, THEIR DETECTION AND ENUMERATION

Traditional Indicators for Water

Coliforms

Historically, coliforms have evolved to become approved indicators of sanitary water quality for drinking, surface and estuarine receiving waters.

During this period, methods for their recovery and enumeration have been under continual scrutiny, yielding refinements in selectivity, specificity, and efficiency characteristics. Over the last several decades, the effects of environmental exposure on coliform recovery have been recognized and repair procedures evaluated. The 18th edition of Standard Methods (APHA 1992) recognizes injury as a factor affecting the recovery of coliform indicator bacteria, suggesting a variety of steps to enhance recovery.

Although the total coliform group can be found in most current compilations of the approved microbiological methods, the overwhelming thrust of literature dealing with indicators, relationships between indicators and pathogens and the epidemiological verification of indicators is rather clear in authenticating the extrafecal origins and growth potentials of members of this group, the superior specificity of the fecal coliform as an indicator of feces, and its lack of correlation with pathogens and health effects. Based on these facts, one must conclude that although the group remains an accepted indicator and standard for drinking water, its continued application to shellfish-growing waters is not justified.

Although, as previously noted, the validity of total coliforms as an indicator of fecal contamination in natural shellfish waters is equivocal, use of the total coliform group to classify shellfish-growing waters is still recognized by the National Shellfish Sanitation Program (USFDA 1990) and, therefore, methods for its recovery are included in this review. Approved methods (APHA 1992) for enumeration of the coliform group of bacteria from shellfish waters include multiple-tube fermentation and membrane filtration techniques. Each method has its unique characteristics and advantages. The multiple-tube fermentation method (MPN) is comparatively simple to perform but mechanically repetitious, and using three or five tubes per dilution (most commonly used) provides a relatively imprecise estimate of the density of coliforms in a given sample. However, the MPN method does allow for processing samples that contain relatively high amounts of suspended particulates, chlorine, or other toxic chemicals compared to membrane filtration.

The total coliform group consists of all aerobic and facultative anaerobic, gram-negative, non-spore-forming rod-shaped bacteria that ferment lactose with gas and acid formation within 48 h at 35°C (APHA 1992). Approved methods call for incubation first in lauryl sulfate tryptose broth, although lactose broth may be used, followed by transfer of tubes producing gas to brilliant-green lactose bile broth, etc. Approved MPN-based methods do suffer from interferences caused by the presence of microorganisms that inhibit gas production or are antagonistic to coliforms (Evans et al. 1981). Coliform masking can be intensified with sublethally stressed cells and was demonstrated with seawater samples (Olson 1978). Some coliforms may be anaerogenic and, therefore, tubes with growth should be confirmed through additional and time-consuming testing (Olson 1978; Evans et al. 1981).

Another source of interference are organisms such as *Aeromonas* spp., which produce gas in standard media or can overgrow coliforms. These effects can lead to an underestimation of coliform densities or failure to detect their presence.

Total coliforms may also be enumerated using a membrane filter approach (Geldreich 1981). The latest edition of Standard Methods (APHA 1992) offers an approved total coliform membrane method (using either M-Endo or LES Endo agars) and also presents a method for the recovery of injured total coliform bacteria as well as an abbreviated discussion of stress. Avila et al. (1989) recommended M-Endo as the most suitable medium for the recovery of coliforms from seawater by membrane filtration. The membrane method offers better reproducibility, sensitivity, and precision than the MPN approach (Clark et al. 1951). However, the obvious advantage of sample concentration leads to problems associated with high turbidity, injury owing to adsorption of chlorine or other toxics by the filter, and concentration of noncoliform bacteria. Interference caused by the physical presence of the background microorganisms or even the production of coliform-specific bacteriocin-like substances has been considered (Means and Olson 1981; Burlingham et al. 1984). The volume filtered in saline waters may be restricted because of high particulate concentrations, which interfere with colony formation and counting. Geldreich (1981) and Hartman, Reinbold, and Saraswat (1966) consider basic problems associated with membrane-filtration-based methods.

Fecal Coliform

The fecal coliform group, the most frequently used indicator of fecal contamination to classify shellfish-growing waters, replaced the total coliform group as a more specific indicator of fecal pollution in natural waters. However, it also possesses characteristics that have engendered criticism of the total coliform group. The fecal coliform group is operationally defined (APHA 1985) and includes genera that are not restricted to fecal habitats and may grow in saline receiving waters. The standard procedure for enumerating fecal coliforms does not differentially resolve component genera. By extension, if the numerical value of the total coliform standard was not based on a quantitative assessment of health risk, then the same must be said of the fecal coliform standard, the latter derived by correlative analysis of paired total and fecal coliform data from growing areas (Hunt and Springer 1974). The discrete value chosen also reflects a methodological bias dictated by the constraints of the MPN distribution table. Compared to the enterococci, the fecal coliform indicator (and *E. coli*) was considered an inferior indicator of sewage pollution because of large reductions in its density that occur during sewage treatment, its greater sensitivity to chlorination, and its higher rate of "dieoff" (Miescier and Cabelli 1982). *E. coli* has been shown capable of persistence for periods of extended duration, or even aftergrowth in estuarine waters. It is noteworthy that shellfish harvested from approved growing areas in the Gulf of Mexico during periods of seasonally warm temperatures may be rejected because fecal coliform levels in the meats exceeded the market guideline owing to multiplication of fecal coliforms after harvesting (FDA 1984). Recent reports note how its entry into a nonculturable but viable physiological state, its physiological adaptation to seawater, and the effects of sublethal stress can lead to significant underestimation of its numbers in seawater. Sublethal stress can have a significant effect on bacterial recovery at environmental temperatures below 10°C, in waters of high salinity, in the presence of toxic chemicals, or due to light-induced damages.

An alternative to the APHA MPN method (APHA 1985) is the approved 24 h modified A-1 MPN method (Hunt and Springer 1978; APHA 1985). For better precision a membrane-filter direct counting method based on the use of mFC agar is approved for use (APHA 1989) but is subject to the same limitations discussed for the total coliform membrane filter method. Use of rosolic acid may be inhibitory to injured coliforms and may be omitted (Geldreich 1981). Pagel et al. (1982) found that the recovery of fecal coliforms on mTEC agar, a medium designed for selective recovery of *E. coli*, is superior in overall performance characteristics to mFC. An alternate filter-based approach is the hydrophobic grid membrane filter (e.g., Sharpe 1981) coupled with an appropriate selective medium such as that described by Entis and Boleszczuk (1990). These filters accommodate an increased range of counts, i.e., three orders of magnitude compared to ordinary membrane filters, and provide considerably better precision than the MPN method. Waxed grid lines, used to divide the membrane into hundreds of independent compartments or cells, confine colony growth, prevent spreading and inhibition by background bacteria.

Unapproved or Emergent Methods for Enumeration of Coliforms and Fecal Coliforms

The methods discussed in this section have been developed to enumerate the target microorganisms within a significantly shorter time interval than required for approved APHA MPN methods (APHA 1985, 1989). The methods can be arbitrarily classified into approaches that are modifications of the existing enumeration procedures but use different media or procedures, those that combine different media with a variety of techniques to detect activity or densities of viable cells, and those that are based on direct detection of cellular nucleic acids. Additional variations may be obtained by combining these three categories. A recent review by Boardman, McBrayer, and Kohlhepp (1989) discusses new methods for detection of coliform bacteria and other indicators found in marine waters.

Kampelmacher et al. (1976) compared the existing APHA MPN enrichment method for detection of total coliforms with three enrichment media used in Europe and also compared the direct recovery of *E. coli* from the same media. One medium, formate glutamate, was found to result in higher recoveries and significantly fewer false positives after 48 h than LST broth.

A variety of rapid enumeration methods for coliforms and fecal coliforms are referred to in Standard Methods as Special Techniques (APHA 1992). A table of methods with appropriate references is provided (APHA 1992) and include radiometric, enzyme assay, electrochemical, impedance, gas chromatographic, colorimetric, and potentiometric assays. Many of these techniques are generally considered to be inappropriate for routine enumeration of fecal coliforms in water samples. Only an ATP procedure, a colorimetric enzyme assay, and a ^{14}C-labeled substrate method are recommended by the APHA (1989). All these methods must be calibrated against standard curves obtained by viable count methods. This approach can lead to potentially large errors and nonlinear responses if environmentally stressed and substrate nonresponsive cells are present.

Berg and Fiksdal (1988) evaluated three fluorogenic substrates, 4-methylumbelliferyl-β-D-galactoside, 4-methylumbelliferyl-heptanoate, and 4-methylumbelliferyl-β-D-glucuronide for a rapid method to detect total and fecal coliforms based on enzyme activity. Samples of river water, potable water, or sewage effluents were filtered through 0.45 μm membrane filters, the filters incubated in a nutrient broth containing a fluorogenic substrate, lactose, sodium lauryl sulfate, and buffer at temperatures selective for each coliform group. Initial rates of substrate hydrolysis, measured over time by fluorescence, were related to inital coliform densities. Linearity of fluorescence over the first hour of incubation was very good. Of the substrates examined, 4-methylumbelliferyl-β-D-galactoside gave the best response time, providing a positive result in 15 min. Incorporating this substrate in M 7h FC agar (APHA 1989) provided a direct method for enumeration of fecal coliforms within 6 h.

Escherichia coli

This organism is considered the dominant fecal coliform in human and animal feces and is generally believed to be an indisputable indicator of fecal contamination from warm-blooded animals. Its thermotolerance facilitates its separation from most other members of the coliform group, using selective and differential methods. Replacement of the fecal coliform indicator with *E. coli* has been considered because of thermotolerant fecal coliforms whose presence in aquatic environments is not necessarily related to fecal contamination. *Klebsiella* spp. is a ubiquitous FC-positive organism that can be found in aquatic environments and is associated with plant materials (Bagley and Seidler 1977), industrial effluents (Huntley, Jones, and Cabelli 1976), and degraded water quality (Vlassoff 1977). The validity of the approved fecal coliform method has been questioned because of FC-positive klebsiellae and estuarine "fecal-coliform-mimicking" bacteria that proliferate in shellfish harvested from approved growing areas in the Gulf of Mexico and other areas during seasons of maximum water temperature (Miescier et al. 1985). Hood et al. (1983) reported *Klebsiella* spp. were abundant in Gulf-of-Mexico oysters and clams during the months from April through October. A similar phenomenon may also occur to a lesser extent in the water column. Another criticism focuses on the multiple-species composition of the fecal coliform group. It is reasonable to assume that the development of rapid and direct enumeration methodologies could be simplified if the target was a single species. For these and other reasons *E. coli* has been suggested as a more specific indicator of fecal contamination in estuarine waters.

However, we should ask if the information available is sufficiently compelling to support this conclusion. As noted for fecal coliforms, *E. coli* does not fulfill the requirements of an ideal indicator for various reasons despite the alleged exclusivity of its fecal origin. Besides its persistence and aftergrowth capabilities, its biomass is significantly impacted by the natural microbiota. In contrast, we would not anticipate that human enteric viral pathogens would enter the microbial food web and be removed through similar predator–prey or other microbially mediated interactions. This hypothesis implies a fundamental difference between viral and bacterial indicator removal mechanisms. A selective removal process of this kind may contribute to viral isolations without detecting indicator bacteria. Therefore, although *E. coli* may be a useful indicator downstream from definitive or large, stable point sources of fecal contamination, its value as a quantitative standard in shellfish-growing waters impacted by diffuse sources appears equivocal. This conclusion stands despite anticipated approaches to "fine-tune" the specificity, selectivity, and accuracy of traditional or "state-of-the-art" enumeration methods because the deficiencies noted owe their origin to ecological phenomena.

The APHA-approved MPN method using lauryl sulfate tryptose or lactose broth followed by EC broth is effective but suffers from the poor analytical precision because it is a statistical estimate of the true population density and is time-consuming. The most-probable-number-based A-1 method was developed by Andrews and Presnell (1972) and subsequently modified as A-1-M (Andrews et al. 1978) as a one-step enumeration procedure. The modified method incorporates an initial low temperature (35 ± 0.5°C) 3 h resuscitation period followed by incubation at (44.5 ± 0.2°C) in a waterbath for 21 ± 2 h (APHA 1984*a*). Cook (1981) developed a temperature control mechanism to automatically program an incubator for the required resuscitation and incubation periods at the appropriate temperatures. *E. coli* can be enumerated by direct count using an EPA-approved mTEC procedure (USEPA 1985). Water samples are filtered through 0.45 μm membrane filters, placed on mTEC medium, incubated for 2 h at 35°C to allow for resuscitation of stressed or injured cells, and incubated at 44.5°C. After 22 h, the filter is removed and placed on a pad saturated with urea. Target colonies are urease-positive and turn yellow to yellow–brown.

Nonapproved or Emergent Enumeration Methods for *Escherichia coli*

Rapid methods for detection of *E. coli* in water based on hydrolysis of various fluorogenic or chromogenic conjugated substrates have been developed in recent years and few have been evaluated in shellfish waters. A review of several methods through 1986 can be found in Hartman, Petzel, and Kaspar (1986). Several assays are based on the ability of *E. coli* to produce catabolic enzymes in response to particular substrates that are conjugated to a fluorogenic or chromogenic moiety through a *β*-D glycosidic linkage. Hydrolysis of the glycosidic bond releases the fluorogen or chromogen, which can be detected by fluorescence or spectrophotometry. Kilian and Bülow (1976) evaluated a variety of nitrophenyl-conjugated enzyme substrates for rapid detection of bacterial glycosidases in representative genera including *Escherichia* spp. A widely used fluorogenic substrate is 4-methylumbelliferyl-*β*-D-glucuronide (MUG), which has been shown to be hydrolyzed by *E. coli* and some *Salmonella* and *Shigella* spp. (Kilian and Bülow 1976). However, Dahlen and Linde (1973) found glucuronidase activity in some *Bacteroides* spp. and *Corynebacterium* spp. Basic assumptions concerning these methods include: (1) that the ability to hydrolyze the substrate is uniformly present in the target species; (2) that this ability can be expressed under a given set of cultural conditions; (3) that the interferences caused indirectly (overgrowth) or directly (false positives) by the presence of other organisms can be minimized to an acceptable level; and (4) that the sample matrix itself does not autofluorescence or contain enzymes that will hydrolyze the assay substrate. Reports suggest that only a small proportion of *E. coli* recovered from environmental samples are *β*-glucuronidase-negative and cannot hydrolyze this substrate (Freier and Hartman 1987; Adams et al. 1990). However, Chang, Brill, and Lum (1989) and Lum and Chang (1990) concluded the incidence of glucuronidase-negative strains is much higher than the aforementioned reports suggest and caution against use of MUG hydrolysis as a singular method to detect *E. coli*. MPN, plate, and microtiter-based methods employing 4-methylumbelliferyl-β-D-glucuronide (MUG) have been described by Feng and Hartman (1982). Freier and Hartman (1987) developed two membrane filtration media containing MUG and inhibitors to gram-positive bacteria for simultaneous recovery of both total coliforms and *E. coli* within 24 h. The method, which incorporates a 35°C incubation temperature, does not require a resuscitation step and effectively recovers the target organisms from sewage and surface freshwater. There are no reports of these media applied to shellfish waters. Ley et al. (1988) synthesized the glucuronide, indoxyl-β-D-glucuronide, as a less expensive analog to MUG and chromogenic compounds. Watkins et al. (1988) evaluated the chromogenic glucuronide BCIG (5-bromo-4-chloro-3-indoxyl-β-D-glucuronide) for the specific, differential identification of *E. coli* from wastewaters and shellfish using direct enumeration methods on agar media. Adams et al. (1990) described a colorimetric method for enumeration of *E. coli* based on hydrolysis of the conjugate *p*-nitrophenol-β-D-glucuronide in a liquid freshwater medium at 37°C and 44°C. The method is based on a graded response whereby the time required for spectrophotometric detection of the chromogen is inversely related to the *E. coli* cell density. Relationships between log cell density and time to detect the dye are established on the basis of a series of standard curves. Interference, demonstrated in the presence of high numbers of competing microorganisms such as *Klebsiella* spp. and *Enterobacter* spp., resulted in overestimations of the "true" cell densities. It is probable that sublethal stress caused by hyperosmolarity, temperature stress, sunlight, starvation, etc. would also cause departures from the predicted densities.

The advantages of direct counting methods compared to enumeration methods based on the cultivation of target cells are widely recognized (e.g., Daley 1979; Atlas 1982). Specific issues concerning indicator enumeration based on viable recovery methods have been reviewed by Roszak and Colwell (1987). Many studies demonstrating nonculturability have used *E. coli*, *Salmonella* spp. or *Vibrio* spp. as target organisms. A variety of direct counting procedures to distinguish viable from nonviable cells have been developed. Representative approaches include cell elongation (Kogure, Simidu, and Taga 1979), measurements of electron transport (Zimmerman, Iturriaga, and Becker-Birck 1978), autoradiography (Tabor and Neihof 1982), use of fluorescent antibodies (Desmonts et al. 1990), measuring the change in activity over time of ^{3}[H]-thymidine-labeled DNA recovered from cells exposed to seawater (Servais, Billen, and Vives-Rego 1985), and recombinant methods such as gene probes and PCR-based amplification of target sequences (Steffan and Atlas 1988). The APHA recognizes use of fluorescent antibodies as a valid method for direct detection of microorganisms (APHA 1992). Singh, Pyle, and McFeters (1989) demonstrated the advantages of combined automated image analysis with the direct viable counting method of Kogure, Simidu, and Taga (1979) for bacterial enumeration. Advantages include rapidity and improved counting efficiency. Although this method was used under ideal conditions, e.g., pure cultures of *E. coli* grown in laboratory media, the concentration of nalidixic acid used was observed to significantly affect direct viable counts, the effect generally being to decrease counts with increased antibiotic concentration. Furthermore, the exposure time and concentration of nalidixic acid required to maximize the counts of pure cultures of *E. coli*, *S. typhimurium*, *P. aeruginosa*, *Y. enterocolitica*, and *V. cholerae* varied over almost a tenfold range. This observation emphasizes the importance of prior knowledge of the effects of nalidixic acid concentration, assay incubation time, and substrate on enumeration of a target indicator or segment of a microbial community in natural samples by direct viable count.

Thus far, the methods using recombinant DNA/RNA techniques applied to environmental microbiology have included (1) direct detection through hybridization of the target nucleic acid sequence using naturally occurring organisms in an environmental sample or biological matrix, or (2) detection by colony hybridization after cultivation of the target microorganism on a particular medium. Although the first method has been limited by the amount of DNA or RNA required to hybridize with a probe, use of cell concentration techniques (Somerville et al. 1989) and new PCR methods (Atlas and Bej 1990; Bej et al. 1990) offers improved sensitivity. Atlas and Bej (1990) claim a sensitivity of 1–10 fg of genomic DNA, which is equivalent to only 1–5 *E. coli* cells in a 1–100 ml water sample. Issues of concern with the viable approach include those problems associated with viable recovery methods, e.g., interference and overgrowth of the target organisms, poor selectivity and specificity, sublethal stress, nonculturability, etc. Amy and Hiatt (1989) illustrated several of these problems using a DNA chromosomal gene probe against colony blots prepared from water samples containing indigenous microbiota and target *E. coli.* Optimum conditions for detection of target cells were the absence of competing microbiota and use of selective pressure, conditions that are probably unrealistic with natural samples. Knight et al. (1990) used a commercially available DNA probe for direct detection of *Salmonella* spp. from estuarine water samples. Sterivex filter units (Millipore Corporation, Bedford, MA) were used to concentrate and recover cells from water samples. Similar methods could be applied to recovery of indicator bacteria. The probe hybridization procedure, which can be accomplished in 2 days, detected *Salmonella* spp. DNA in direct extracts of cell concentrates and in samples negative for cultured cells following addition of medium to the filter unit. At this time, it is not possible to determine with sufficient accuracy the number of cells, viable or nonviable, corresponding to a given yield of probe-bound DNA. Sayler and Layton (1990) recently reviewed the environmental applications of nucleic acid probe techniques for detection of microorganisms.

Fecal Streptococci

The term "fecal streptococci" has been used to describe a group of taxonomically diverse streptococci that are gram-positive, catalase-negative, nonspore-forming, facultative anaerobes associated with the gastrointestinal tracts of humans and animals. Functionally, the fecal streptococci were variously divided into groups and subgroups on the basis of serology and physiological characteristics. Thus, the serologically defined group-D streptococci of the fecal streptococci (Clausen, Green, and Litsky 1977), was divided into an "enterococcal" subgroup (*S. faecalis*, *S. faecium*, *S. avium*, and *S. durans*) and a "nonenterococcal" subgroup (*S. bovis* and *S. equinus*). As noted already, the latter term is curiously inappropriate and has contributed to confusion regarding the meaning of the term "enterococci" (Deibel and Hartman 1984).

Recent studies have indicated that although the "nonenterococcal" species of the genus *Streptococcus* share the group-D antigen and a fecal habitat, these similarities do not reflect the underlying taxonomic relationships. Comparative analysis of oligonucleotides from 16S-rRNA of representative streptococci (Ludwig et al. 1985), and nucleic acid hybridization studies, have led to creation of a new genus, *Enterococcus* (Schleifer and Kilpper-Balz 1984), that includes members of the enterococcal group as well as other species not associated with humans. The genus *Enterococcus* now includes nine valid species and proposals for the inclusion of three newly discovered species have been made (Facklam and Collins 1989). *S. bovis* and *S. equinus*, were taxonomically distinct and were not considered to be valid members of this genus owing to differences in physiology and metabolic characteristics (Bergey's Manual of Systematic Bacteriology 1986).

Although far from being numerically dominant in human feces, constituting only about 0.1% or less of the gut microbiota (based on cell densities) (Holdeman, Good, and Moore 1976) this group has been considered as an important indicator of fecal pollution for a considerable period of time. This is attributed to its association with feces, the alleged differences in numbers of coliforms to numbers of fecal streptococci in human and animal feces (Geldreich and Kenner 1969), and observations suggesting that the enterococcal group does not manifest many of the negative characteristics associated with the coliform group, one of the most important being its reported inability to grow in seawater (Slanetz and Bartley 1965). As a group, the complex nutritional requirements of these organisms are generally considered to preclude extraenteral growth (Mundt 1982).

Use of the ratio of fecal coliforms to fecal streptococci has been accorded considerable importance in the literature and much discussion has concerned its validity as an index capable of revealing the nature of a pollution source, i.e., human versus animal fecal contamination (Geldreich and Kenner 1969; Feachem 1975). Values of the ratio equal to 4.0 or more are assumed to reflect human pollution; values less than 0.7 animal (Geldreich and Kenner 1969). This is because there are proportionately more fecal streptococci in the feces of animals than man. A number of investigators have presented data that critically question the validity of this concept in freshwater environments, including the accepted values of the ratio in feces (McFeters et al. 1974; Wheater, Mara, and Oragui 1979). Values of the ratio in feces from a variety of animals and man varied considerably and certain animal species exhibited mean values essentially similar to human sources (Kjellander 1960; Wheater, Mara, and Oragui 1979). Furthermore, the ratio did not remain stable in receiving waters and was difficult to interpret in systems with multiple sources of fecal pollution. Significantly less attention has focused on the survival characteristics of fecal streptococci in marine and estuarine waters (Slanetz and Bartley 1965; Vasconcelos and Swartz 1976; Lessard and Sieburth 1983). These studies have yielded contradictory results, so that a definitive conclusion regarding use of the ratio in these waters is not possible. However, in an estuarine environment, differential "dieoff" of fecal coliforms and fecal streptococci and the species comprising these groups, and the effects of biological and physical/chemical factors, will likely contribute to variations in the ratio suggesting that it would be difficult to deduce the origin of pollution based solely on its value. Furthermore, it is difficult to see how the recommended criteria for use of the fecal coliform/fecal streptococcus ratio (Geldreich and Kenner 1969), can be met in growing areas polluted by multiple, diffuse sources of fecal contamination. Finally, as will be noted, enumeration of fecal streptococci in estuarine waters is very method-sensitive and contributes a source of variation that must be minimized if the ratio is to be viewed as a meaningful parameter (Brodsky and Schiemann 1976). Attention to reduce variability is also important because these organisms occur at significantly lower densities than fecal coliforms. Fortunately, the newest edition of Standard Methods (APHA 1992) cautions against the use of the ratio to differentiate pollution sources.

Although certain genera (e.g., *E. faecalis*, *E. faecium*) of the fecal streptococci have been considered specific to feces from human or other warm-blooded animals, phenotypically similar strains and biotypes can be isolated from other environmental sources (Mundt 1982; Clausen et al. 1977; Beaudoin and Litsky 1981). Strains and biotypes of *E. faecalis* and *E. faecium* can be isolated from a variety of plant materials, reptiles, and insects (Mundt, Johnson, and Khatchikian 1958; Mundt 1963; Geldreich, Kenner, and

Kabler 1964; Geldreich and Kenner 1969). Moreover, Mundt, Coogin, and Johnson (1962) reported growth of *E. faecalis* biotypes on plants and vegetables. In comparison, the distribution and viability of *S. bovis* and *S. equinus* in extraenteral environments appears restricted, exhibiting significantly reduced survival in aquatic habitats (Slanetz and Bartley 1965; Geldreich and Kenner 1969; Wheater, Mara, and Oragui 1979). The usefulness of this group as an indicator of fecal pollution in shellfish-growing waters appears to depend on the development of specific enumeration methods that distinguish between strains of fecal versus nonfecal origin.

A large variety of media have been developed for the recovery of fecal streptococci (Brezenski 1973; Yoshpe-Purer 1989). Methods have been based on most-probable-number techniques, direct pour-plate or membrane filtration techniques, some of which incorporate resuscitative steps to minimize the sublethal stress. Comparisons of the various media and methods can be found in references such as Switzer and Evans (1974), Clausen, Green, and Litsky (1977), Pagel and Hardy (1980), Voltera et al. (1985), and Yosphe-Purer (1989). It is evident from this literature that there exist considerable differences in recovery efficiency, specificity and selectivity among the different methods, and that each method requires some degree of confirmatory testing. Use of this group as an indicator in marine and estuarine shellfish waters requires improved methods, specifically the selectivity and specificity of media, coupled with an improved understanding of factors affecting the survival of this group. Methods evaluation should be performed in shellfish-growing areas, with attention to identifying typical fecal streptococcal species recovered and those microorganisms indigenous to shellfish-growing waters that produce false positives. Essentially similar recommendations with respect to methods were articulated quite a few years ago by Kjellander (1960) and Hartman, Reinbold, and Saraswat (1966). Thus, until recently KF was considered to be the medium of choice in the 15th edition of Standard Methods (APHA 1985), although anecdotal reports by many investigators and most recently by Yosphe-Purer (1989) suggested that KF medium is not sufficiently selective to apply to marine waters. The 18th edition of Standard Methods (APHA 1992) now recommends the use of azide-dextrose broth with confirmation on Pfizer selective enterococcus agar (PSE) for MPN determinations and m Enterococcus agar for membrane filtration assay.

Other considerations aside, use of this group as a fecal indicator necessitates adoption of a common functional definition of fecal streptococci (as with fecal coliforms), that is based on a formalized and unique method for recovery of specific enterococcal and streptococcal species associated with feces. This must be viewed as a challenging but formidable task.

Enterococci

The term "enterococci," with its connotation of fecal origin, is "understood" by some workers to exclude the fecal streptococcal species (*S. equinus* and *S. bovis*) associated almost exclusively with animal gastrointestinal tracts. This usage is misleading because the term implies streptococci sharing a fecal habitat. Diebel and Hartman (1984) consider the terms group-D streptococci and enterococci the same. Creation of the genus *Enterococcus* spp. (Schliefer and Kilpper-Balz 1984) (whose constituents are "enterococci"), which contains several nonhuman species (Facklam and Collins 1989) but excludes *S. equinus* and *S. bovis*, does little to resolve this semantic confusion.

Functionally, the enterococci are defined on the basis of selective recovery obtained using one of several methods developed for this purpose (e.g., Slanetz and Bartley 1965; Isenberg et al. 1970; D'Aoust and Litsky 1975; Levin, Fischer, and Cabelli 1975). The range of species selected varies with each method, some being more selective for the classic "enterococcal" (*E. faecalis*, *E. faecium*) group organisms than others. In reviewing this literature the reader must be wary of the operational nature of the results and current taxonomic relationships.

The most recent use of enterococci in marine waters is by Cabelli et al. (1983). The mE procedure was used in polluted marine and estuarine waters to derive a statistical relationship between indicator density and the incidence of swimming-associated gastrointestinal disease. Cabelli et al. (1983) concluded that enterococcal densities were better predictors of health risk than fecal coliforms or *E. coli*. Prospective epidemiological studies to evaluate the enterococci and other indicators as predictors of enteric disease associated with consumption of raw shellfish were recently conducted and the results are being evaluated (A. Dufour, USEPA, personal communication). Enterococci are now recommended as the indicators of choice for recreational waters by the US Environmental Protection Agency (1986). Dufour (1980) has modified the mE procedure by substituting indoxyl-β-D-glucoside for esculin in the primary medium.

Validation of the enterococci as an indicator of public health risk in shellfish-growing areas will require improved recovery methods. The range of selectivity and specificity characteristics of methods for enumeration of enterococci has been mentioned. Current recovery methods require confirmatory testing of presumptive isolates (Ericksen and Dufour 1986), increasing the time and cost of analysis. Although the mE-based method has been applied to marine and estuarine waters (Cabelli et al. 1983), its utility in nonpoint-source-impacted shellfish-growing areas is undetermined. Similarly, enterococcal presence in stormwater runoff (Geldreich et al. 1968; Pagel and Hardy 1980) and differences concerning its persistence in estuarine and marine waters (Lessard and Sieburth 1983; Slanetz and Bartley 1965;

Vasconcelos and Swartz 1976) demonstrate a need for distribution and survival studies. Research requirements include: (1) improving methods for recovery of stressed cells; (2) improving specificity for various *Enterococci* spp. and *Streptococci* spp.; (3) reducing the incidence of false positives, which can vary with temperature, season, and geographic region; and (4) minimizing background growth. The relative occurrence and densities of nonfecal biotypes of *E. faecalis* and *E. faecium* in nonpoint source and stormwater runoff impacted areas should be determined. Rapid methods for the confirmation of selected enterococcal species, based on serological or biochemical characteristics, could significantly improve its candidacy as an indicator in shellfish-growing waters. Until recently, there have been few reports of direct methods for the enumeration of this group or species identification, e.g., fluorescent antibody-based methods (Abshire and Guthrie 1971; Pugsley and Evison 1975). Bosley, Facklam, and Grossman (1983) describe a useful and rapid (4 h) identification method to separate enterococci from group-D non-enterococci based on hydrolysis of L-pyrrolidonyl-β-naphthamide (PYR). A colony hybridization method, employing oligonucleotide probes synthesized for specific sequences of 23S rRNA of selected enterococci, was used to successfully detect and identify *E. faecalis*, *E. faecium*, and *E. avium* in mixed culture (Betzl, Ludwig, and Schleifer 1990).

Alternate Indicators for Water

Bacteroides fragilis Group

Although the obligate anaerobic bacteria are the dominant constituents of the human fecal microbiota (Moore and Holdeman 1974), attempts to utilize these organisms as indicators of fecal pollution have been limited. This may be attributed to the need for rigorous anaerobic cultural and enumeration procedures and the assumption that anaerobes do not persist under extra-enteral conditions.

Bacteroides spp. are nonsporulating obligately anaerobic gram-negative bacteria considered to be the dominant genus in human feces (Holdeman, Good, and Moore 1976; Salyers 1984). Like the bifidobacteria, *Bacteroides* spp. is an important candidate indicator of fecal pollution as its major habitat is restricted to the gastrointestinal tracts of humans and warm-blooded animals. As a group, *Bacteroides* may be more tolerant of oxygen than the bifidobacteria; some *Bacteroides* spp. produce catalase, and *B. fragilis* and *B. distasonis* synthesize superoxide dismutase (Salyers 1984).

Until recently, the term *Bacteroides fragilis* group (BFG), was used to describe the dominant bacteroides found in the human colon which included *B. fragilis* and its subspecies. The subspecies were subsequently accorded species rank on the basis of DNA homology studies (Salyers 1984) and dominant species now include *B. vulgatus*, *B. fragilis*, *B. thetaiotaomicron*, and *B. distasonis*. The term BFG still appears in the current literature and, although taxonomically incorrect, it has utility as a simple phrase referring to the dominant human colonic bacteroides species. *B. vulgatus*, *B. thetaiotaomicron*, and *B. distasonis* are numerically dominant (10^{10} cells/g dry weight feces) and *B. vulgatus* that can constitute $> 10\%$ (as cell count) of the culturable fecal microbiota (Holdeman, Good, and Moore 1976). *B. fragilis*, which is more aerotolerant than *B. vulgatus*, occurs at densities similar to *E. coli*. The observation that *Bacteroides* species other than those found in humans dominate in ruminants (Macy 1981) supports the use of BFG species as a human-specific indicator. Macy (1981) provides a summary of host-related species composition and methods for recovery and cultivation of nonpathogenic *Bacteroides* spp.

Although *Bacteroides* spp. and *B. fragilis*, specifically, are medically significant and can be isolated from a variety of clinical specimens and blood samples (Goldstein and Citron 1988), there is little information describing the occurrence of nonpathogenic bacteroides in natural waters. Post, Allen, and Reid (1967) first suggested use of the BFG group as an indicator of fecal pollution. Allsop and Stickler (1985) evaluated BFG as a fecal indicator based on its presence in sewage, various freshwater, and marine environments. Organisms belonging to this group were readily isolated downstream from waters impacted by sewage. Because *Bacteroides* spp. disappeared faster from these waters than *E. coli*, Allsop and Stickler (1985) concluded the ratio of BFG to *E. coli* counts reflects "aging" of a fecal pollutant and also the proximity to the source. The presence of BFG organisms in natural waters was attributed to human fecal pollution because analysis of feces from a variety of domestic animals and feral birds revealed that BFG species occur at much lower densities (i.e., 10^5–10^{10}-fold lower) in common domestic farm animals (cattle, horses, pigs, chickens, and sheep) (Allsop and Stickler 1985). Higher densities occur in domestic pets and seagulls but these were still lower than levels in human feces. A variety of obligate anaerobes including *Bacteroides* spp. were detected during the warmer months at a polluted river site (Daily et al. 1981). These observations, admittedly limited, support the potential use of BFG as a human-specific indicator in nonpoint-source-impacted shellfish-growing areas.

General comments made concerning *Bifidobacteria* spp. recovery methods also apply to this group. Perhaps more work on identification methods has occurred with *Bacteroides* spp. owing to its clinical importance as a pathogen and possible involvement in colon cancer. Allsop and Stickler (1984) developed an improved medium for recovery of BFG from natural waters. Penicillin is incorporated into the medium to inhibit clostridia and membrane filtration (0.22 μm) is used for sample concentration. Application of this medium to samples of sewage, and fresh and marine waters revealed interference by obligate and facultative anaerobes. Increasing the concentration of gentamycin improves the selectivity of the medium but decreases the recovery efficiency. A resuscitation period is incorporated to minimize the effects of sublethal injury. Serious consideration of the BFG group requires a detailed assessment of this recovery method in shellfish-growing waters, including its selectivity, validation of confirmatory steps required, as well as data describing the effects of season on BFG occurrence, and persistence in estuarine waters.

Anaerobic bacteria vary in their sensitivity toward oxygen (Loesch 1969). Loesch (1969) classified anaerobes into two groups: (1) those that do not exhibit growth on agar media if pO_2 levels exceed 0.5% were called strict anaerobes, and (2) those that could grow at pO_2 levels from 2% through 8% (mean 3%) were called moderate anaerobes. *B. fragilis* was found among the most oxygen-tolerant of the latter group. Cells exposed to the atmosphere on an agar surface survived for at least 8 h with only a small reduction in colony counts. This tolerance explains the detection of this organism in environmental waters but also suggests that modifications to sample storage and processing protocols that reduce oxygen exposure could improve recovery and enumeration efficiency. Walden and Hentges (1975) confirmed the moderate tolerance of *B. fragilis* to oxygen and demonstrated that oxygen, not Eh, was adverse to anaerobic growth because anaerobes grew well under conditions of positive, intermediate, and negative oxidation–reduction potentials in the absence of oxygen. Onderdonk et al. (1976) confirmed the independence of Eh on growth but observed that, under conditions of continuous culture, oxygen concentrations from 10% to 100% were bacteriostatic and elicited a decline in metabolic activity.

Chromosomal DNA probes have been developed to speciate *Bacteroides* as an alternative to conventional phenotypic methods. Roberts, Moncia, and Kenny (1987) and Morotomi, Ohno, and Mutai (1988) used whole-cell dot–blot assays that can correctly differentiate *Bacteroides* by species. Kuritza and Salyers (1985) used a cloned DNA probe to enumerate *B. vulgatus* in feces and the concentrations measured were similar to viable counts. Although the probe was highly specific for a segment of the *B. vulgatus* genome, its use for enumeration of cells in environmental samples is unlikely because of its low sensitivity (ca. 10^7–10^8 cells required). A sample concentration procedure (e.g., Somerville et al. 1989) that would eliminate the problems associated with particulates in estuarine water, i.e., clogged filters, and interfering overgrowth by the autochthonous microbiota (Amy and Hiatt 1989) will be needed. However, these probes should be valuable in supplanting or augmenting the specialized and labor-intensive biochemical identification of presumptive isolates. Use of *B. fragilis* bacteriophage as an alternate indicator is discussed in the section on viral indicators.

Bifidobacteria

Bifidobacterium spp. are obligately anaerobic, gram-positive, nonmotile bacteria that possess a characteristic pleomorphic and branching cell morphology. They are potentially important candidate indicators of fecal pollution as the habitat of many species appears restricted to the feces of adult humans and infants and major groups of warm-blooded animals (Levin 1977; Cabelli 1978*a*; Bergey's Manual of Systematic Bacteriology 1986). In humans, bifidobacteria are a major component of the intestinal microbiota occurring at densities greater than 10^{10} cells/g dry feces and some species may comprise more than 6% of the culturable microbiota (Holdeman, Good, and Moore 1976). A unique metabolic feature of this genus is that glucose is metabolized exclusively and characteristically through the fructose-6-phosphate shunt (bifid shunt) (Bezkorovainy 1989). The first reaction step is mediated by the enzyme fructose-6-phosphate phosphoenolketolase (F6PPK). This enzyme can be detected using a colorimetric assay and is a reliable phenotypic characteristic of the genus (Scardovi 1981; Bergey's Manual of Systematic Bacteriology 1986). Bifidobacteria can also utilize ammonium as a sole source of nitrogen. The sensitivity of these organisms to oxygen varies with species and strains, the presence of CO_2 allowing some species to tolerate exposure to air for several hours.

A potentially valuable property of this group relates to its ability to differentiate human from animal fecal pollution based on "human" and "animal" strains (Levin 1977). Tanaka and Mutai (1980) found that *B. adolescentis* and *B. longum* accounted for 74% of the bifidobacterial strains isolated from human feces. Levin and Resnick (1981) were unable to isolate bifidobacteria dominant in human feces (i.e., *B. longum, B. adolescentis*) from a variety of domestic and wild animals except swine. Mara and Oragui (1983) confirmed the low occurrence of these microorganisms in animal feces and developed a medium based on sorbitol fermentation, which selects for bifidobacteria derived from human sources. The ability of human-specific strains to ferment sorbitol can vary and not all sorbitol-fermenting bifidobacteria are derived from human feces (Bergey's Manual of Systematic Bacteriology 1986). Nevertheless, the possibility of differentiating human from animal fecal pollution should provide an impetus for the evaluation of this anaerobe.

Use of the bifidobacteria as an indicator of fecal contamination in shellfish-growing waters has been considered (Cabelli 1978*a*), but recent studies have focused primarily on its occurrence and recovery from freshwater environments (Mara and Oragui 1983; Carrillo, Estrada, and Hazen 1985; Munoa and Pares 1988). Information on the distribution of these bacteria, albeit limited to natural and sewage contaminated freshwater, suggests that its use as an indicator of very recent fecal pollution is valid as the organisms appear incapable of extraenteral growth.

Although obligate anaerobes, Gyllenberg, Niemela, and Sormunen (1960) reported that bifidobacteria as well as *E. coli* survive in freshwater. In contrast, Levin and Resnick (1981) observed that *B. longum* and *B. adolescentis*, exposed *in vitro* to fresh and marine water samples, are less persistent than *E. coli*. *B. adolescentis* populations decline considerably after 48 h of *in situ* exposure in tropical freshwaters (Carrillo, Estrada, and Hazen 1985). Under these same conditions *E. coli* densities remain constant or increase. Results of an *in vitro* experiment to compare survival of *B. adolescentis* in fresh and estuarine water (Kator and Rhodes 1988) confirmed these observations. *B. adolescentis* cells were incubated in flasks containing membrane filtered (0.2 µm) water at 6°C and 25°C. *B. adolescentis* persistence at 6°C was moderately better than *E. coli* but at higher temperatures was significantly worse. In contrast, Levin and Resnick (1981) observed considerably less persistence of bifidobacteria from sewage exposed *in vitro* to membrane-filtered seawater (32 psu). Survival was also independent of temperature (at temperatures of 4°C, 12°C, and 20°C), with approximately 15% of the initial density remaining after about 6 h. Individual species and strains varied in survival capability.

The generally low bifidobacterial densities found in receiving waters may be attributed in part to poor recovery owing to oxygen toxicity and sublethal stress, factors that are exacerbated by selective media and harsh recovery methods. Munoa and Pares (1988) demonstrated that the inability of *Bifidobacterium* spp. cells to produce colonies on a selective medium was caused by sublethal injury following exposure to seawater. To minimize the effect of sublethal stress on enumeration, a two-layer recovery procedure was designed that incorporated plating and incubation of the sample on resuscitative medium (reinforced clostridial agar). This was followed with an overlay of BIM-25, a selective medium developed to improve the poor selectivity of YN-6 medium with environmental samples (Levin and Resnick 1981). Unfortunately, *B. adolescentis* did not grow as well on BIM-25 as other *Bifidobacterium* spp. (Munoa and Pares 1988). Improving recovery methods for this group of microorganisms, especially human-specific species, should be an objective of future research efforts. Beerens (1990) described a modified Columbia (Pasteur Production, Bioservice, France) agar medium containing propionic acid that is selective and enhances recovery of bifidobacteria. This and other recovery media should be evaluated using resuscitative techniques and membrane filtration because of the need for sample concentration. Despite the recognition that the bifidobacteria are sensitive to oxygen, use of prereduced media or an oxygen-free environment for recovery manipulations has not been prescribed. Although Levin and Resnick (1981) suggest that oxygen toxicity is not a concern with recovery of bifidobacteria, sorbitol-fermenting strains are catalase negative, and data such as those of Munoa and Pares (1988) and Kator and Rhodes (1988) suggest factors that exacerbate sublethal stress should be minimized. Differences in the tolerance of bifidobacterial species or strains to oxygen (Bergey's Manual of Systematic Bacteriology 1986) may target the choice of an indicator bifidobacterial species. The role of particle association in protecting *Bifidobacterium* spp. and other anaerobes from oxygen in the environment is undetermined. *In situ* exposure studies are needed to measure the persistence and recovery of selected bifidobacterial species in estuarine waters and sediment.

If using indicators to identify sources of fecal pollution to nonpoint-impacted estuarine growing areas, the unequivocal fecal origin and limited survival properties of the bifidobacteria are potentially valuable because their detection implies immediate fresh sources. The appearance of bifidobacteria in feeder streams may be used to locate sources of fecal pollution and to differentiate human from nonhuman sources. Although a discrimination of vertebrate source will require studies to verify the restricted host range of the sorbitol-fermenting bifidobacteria species composition in feces of domestic animals, humans, and septic tank effluents, the ability to differentiate human from animal pollution could aid management strategies directed toward reduction of pollutant sources. Finally, various bacteriophages, including those lytic to *Bacteroides fragilis* (Jofre et al. 1986; Tartera and Jofre 1987), are being considered as human-specific indicators of fecal pollution. Similarly, if bacteriophages lytic to bifidobacteria occur in feces or sewage, these could be used as human-specific viral indicators of fecal contamination.

Clostridium perfringens

C. perfringens is an anaerobic, gram-positive, spore forming rod whose presence in receiving waters has been linked to contamination by feces and wastewaters (Cabelli 1977*a*; Bisson and Cabelli 1980). Bisson and Cabelli (1980) developed a membrane filtration method for *C. perfringens* recovery applicable to saline waters based on a highly selective medium (mCP). The composition of this medium was later modified to reduce the concentration of a costly component without compromising its selectivity (Armon and Payment 1988). Sartory (1986) compared the recovery of *C. perfringens* on mCP and egg-yolk-free tryptose-sulfite-cycloserine (TSC) agar as membrane-based tests. For a variety of sample types, which included polluted river water, egg-yolk-free TSC was found as selective as mCP and more efficient. Egg-yolk-free TSC was recommended because of its ease of preparation, lower cost, and availability of a simple confirmation scheme.

Bisson and Cabelli (1980) concluded that *C. perfringens* has value as an indicator of chlorination efficiency and the presence of unchlorinated sources of fecal contamination, and as a "conservation tracer" delineating the areal impact and transport of wastewater effluents. Moreover, detection of vegetative cells in the environment reflects fresh and untreated fecal matter because the persistence of vegetative cells is very short (Bisson and Cabelli 1980). Fujioka and Shizumura (1985) confirmed the utility of *C. perfringens* to detect wastewater discharge into freshwater streams. However, there is doubt if *C. perfringens* would be a useful indicator in nonpoint-impacted shellfish-growing areas because (1) it is so persistent that it may be difficult to index current pollution conditions; (2) it is widely distributed in soils and sediments; and (3) it is carried into growing areas from extraenteral sources by stormwater runoff and transport of suspended sediment (Matches and Liston 1974; Smith 1975).

Rhodococcus coprophilus

Rhodococcus coprophilus is an aerobic nocardioform actinomycete proposed as an indicator of domestic farm animal fecal pollution (Rowbotham and Cross 1977*b*; Mara and Oragui 1981; Oragui and Mara 1983). As a fecal indicator, this organism is unique because it is associated with the feces of domestic grazing farm animals but is not considered an active component of the rumen microbiota (Rowbotham and Cross 1977*b*). Rowbotham and Cross (1977*b*) noted that numbers (colony-forming units/g) of *R. coprophilus* in pasture grass and in dung collected from cattle fed this pasture grass were similar, implying no multiplication occurs during passage through the animals. In contrast, significant increases in densities of *R. coprophilus* that occur during incubation of fresh dung at a moderate temperature (ca. 20°C) with water, confirms the coprophilic habitat of this organism. Consequently, this organism is commonly found in pasture grass and soils grazed by cattle and other domestic farm animals. Studies have shown that *R. coprophilus* is absent from human feces but consistently found in feces of cattle, sheep, pigs, horses, donkeys, farm-raised poultry and sporadically in dog and seagull feces (Mara and Oragui 1981).

R. coprophilus can survive both dessication and temperatures of 2–4°C in dung for periods of up to 6 weeks (Rowbotham and Cross 1977*b*). Goodfellow and Williams (1983) concluded that the coccal stage of *R. coprophilus*, carried by runoff into either freshwater or saltwater habitats, does not grow but remains viable. *R. coprophilus* persists *in vitro* for 17 weeks in nonfiltered freshwater incubated at 5°C, 20°C, and 30°C, whereas *E. coli* and fecal streptococci disappeared within 5 weeks (Oragui and Mara 1983). These investigators suggested that these and other properties of the organism could be used (in conjunction with other bacterial indicators) to determine the temporal characteristics, location and source of fecal contamination to freshwater systems. Information concerning the distribution and persistence of this organism in estuarine waters is insufficient to evaluate its utility as an indicator in shellfish-growing waters. In an experiment using a recovery medium (MM3) devised by Mara and Oragui (1981), Kator and Rhodes (1988) compared its persistence to *E. coli* over a 30 day period in 0.2 µm filtered estuarine water incubated at 6°C and 25°C. Although *R. coprophilus* appeared to multiply more slowly than *E. coli*, its persistence in water over the range of salinities and temperatures tested appeared generally better than that of *E. coli*, and it manifested less initial aftergrowth at 25°C. Whether the observed density increases were due to growth or an artifact of the organism's fragmentable nocardioform morphology (see methodological considerations below) must be addressed in future studies. Similarly, the persistence of this organism in marine and estuarine sediments has not been examined, although it is readily isolated from marine and estuarine sediments (Goodfellow and Williams 1983; Attwell and Colwell 1984; Goodfellow and Haynes 1984; Kator and Rhodes 1989). Isolation of *R. coprophilus* in a small subestuary of the York River, Virginia, was more frequent and the densities higher in water and sediment adjacent to a livestock farm compared to locations lacking this activity (Kator and Rhodes 1989). Studies of *R. coprophilus* persistence in estuarine water and sediment, related particularly to seasonal temperature, are needed to assess its value as an index of current pollution conditions.

Disadvantages now associated with use of this indicator are primarily limitations inherent in the current enumeration procedure (Mara and Oragui 1981). Only small volumes (0.2 ml) of sample can be spread plated onto MM3 agar, currently the medium of choice. Direct spread plating limits detection to 1 cfu/ml (e.g., five replicate plates containing 0.2 ml each). A recovery procedure is needed for concentrating *R. coprophilus* from water using MM3 or other suitable medium for expressing characteristic colony morphology. The combination of membrane filtration (Pisano, Sommer, and Lopez 1986) with the novel selective method of Hirsch and Christensen (1983) could be evaluated for the enumeration of *R. coprophilus*. Hirsch and Christensen (1983) described a "selective" method for recovery of actinomycetes that eliminates bacterial contamination. This is based on the ability of actinomycetes to penetrate the "pores" of a membrane filter on a growth medium, whereas bacteria remain on the filter surface. After an incubation period the filter is discarded and only actinomycetes that penetrate the filter form colonies. Another limitation noted by Mara and Oragui (1981) is the long incubation period (17–18 days) required for development of characteristic stellate colonies having both substrate hyphae and bright orange central papillae (Rowbotham and Cross 1977*a*). The prolonged incubation is a consequence of (1) the low organic content of the medium, (2) the presence of inhibitors used to suppress bacterial growth, (3) the photochromogenic nature of color production, and (4) the time required to visibly develop the characteristic colony morphology. Biochemical confirmation of nocardioforms to the generic level is very time-consuming and an identification on the basis of colonial morphology alone is neither reliable nor valid (Goodfellow and Williams 1983; Bergey's Manual of Systematic Bacteriology 1986). Rapid and direct methods for species verification (and enumeration) are needed. An indirect method to detect or identify *R. coprophilus* using actinophage appears doubtful because of the broad cross-reactivities of actinophage to both *Nocardia* and *Rhodococcus* (Prauser 1984).

A precise enumeration of *R. coprophilus* is complicated by the inherent morphogenetic nature of nocardioforms, and particularly for *R. coprophilus*, which forms extensively branched hyphae and coccoid elements that are connected in chains of varying length. Thus, a given colony-forming unit in a water or sediment sample can originate from a variety of cell configurations. Rowbotham and Cross (1977*a*) observed that a typical colony enumerated from dung is derived from single or multiple coccoid elements, the latter being the most prevalent form. The complicating effect of *R. coprophilus* morphology on enumeration is neither a desirable indicator characteristic nor conducive to analytical precision.

Streptococcus bovis

S. bovis may be the dominant fecal streptococcus in warm-blooded animals (Kjellander 1960; Wheater et al. 1979). Use of *S. bovis* as a specific indicator of animal fecal pollution was first suggested by Cooper and Ramadan (1955). Since their report few investigators have evaluated its use as an indicator of animal fecal pollution in freshwater and none that we know of in marine and estuarine waters. Wheater, Mara, and Oragui (1979) claim that the proportion of *S. bovis* to total fecal streptococci was largest in ruminants but note the important role of diet and geographic location on this ratio. However, the association of *S. bovis* with ruminants is not unique as it has been found in the feces of dogs, cats, various birds, horses, and pigs (Kenner, Clark, and Kabler 1960; Clausen, Green, and Litsky 1977). Osawa and Mitsuoka (1990) isolated *S. bovis* biotype I (mannitol-fermenting) in the feces of koalas on a selective medium. Geldreich and Kenner (1969), Clausen, Green, and Litsky (1977), and Wheater, Mara, and Oragui (1979) were unable to isolate *S. bovis* from human feces. Oragui and Mara (1981) drew similar conclusions but did isolate *S. bovis* from sewage effluent. This contradiction was resolved when they located an abattoir that was discharging into the sewage treatment facility. However, other workers have isolated *S. bovis* from the feces of healthy humans (Kjellander 1960; Switzer and Evans 1974). Kjellander (1960) isolated *S. bovis* from about 15% of humans examined; Dalton et al. (1986) observed a fecal carriage rate in healthy humans of 10–16%. About 1% of the group-D streptococci recovered by Abshire (1977) from human feces during an evaluation of a new presumptive medium for this group were *S. bovis*. Osawa and Mitsuoka (1990) indicate that mannitol-fermenting *S. bovis* (biotype I) are commonly isolated from ruminants whereas mannitol nonfermenting strains (biotype II) are most frequent in human infections. Some workers suggest that these contrasting results are due in part to dietary variation and regional effects. Another explanation is that investigators used methods with different recovery efficiencies, selectivity and specificity characteristics. Based on the literature, the hypothesis that *S. bovis* is a unique indicator of animal fecal pollution remains equivocal.

Greater concurrence exists concerning the survival of *S. bovis* in natural waters compared to other enteric aerobic cocci. *S. bovis* mortality in freshwater is much greater than that of *E. faecalis* or *E. coli* (Geldreich and Kenner 1969; McFeters et al. 1974; Clausen et al. 1977; Wheater, Mara, and Oragui 1979). Results of an *in vitro* exposure experiment using an isolate of *S. bovis*, exposed to filtered feeder stream or estuarine water at 6°C and 25°C, confirmed its inability to persist, especially at the higher temperatures (Kator and Rhodes 1988). As stated earlier, poor survival may be viewed as a positive attribute because detection of *S. bovis* implies fecal pollution of very recent origin. Conversely, its poor survival violates the requirement that an ideal indicator be at least as persistent as enteric pathogens.

Although *S. bovis* can be isolated with other *Enterococcus* spp. and *Streptococcus* spp. on some of the media used for recovery of these groups (e.g., Switzer and Evans 1974), only one medium is specifically designed for its isolation (Oragui and Mara 1981). Littel and Hartman (1983) described a selective medium for the fecal streptococci that differentiates *S. bovis* from other streptococci and enterococci based on hydrolysis of a fluorogenic substrate and a colorimetric indicator of starch hydrolysis, amylose azure. The ability of *S. bovis* to ferment starch has been considered to be an important phenotypic characteristic separating it from *Enterococcus* spp. and other fecal streptococci. Membrane-bovis agar (m-BA) (Oragui and Mara 1981) is a selective medium for *S. bovis* whose specificity is based on the ability of *S. bovis* under anaerobic conditions to utilize NH_4^+ as the sole source of nitrogen and the absence of a requirement for exogenous vitamins. A resuscitation step to minimize the effect of temperature stress is incorporated. m-BA is more specific and more efficient for the recovery of *S. bovis* from freshwater and sewage than KF. Subsequently, Oragui and Mara (1984) described a modified m-BA medium (called mm-BA) containing less sodium azide because of reports noting its inhibitory effect on *S. bovis* strains from different geographical regions. Using m-BA, Oragui and Mara (1981) reported that more than 65% of the fecal streptococci in animal feces were *S. bovis.* Moreover, a proportion (10%) of typical isolates later confirming as *S. bovis* failed to ferment starch. Isolates lacking this characteristic could lead to an underestimation of *S. bovis* densities using the medium of Littel and Hartman (1983). A significant proportion of isolates we confirmed as *S. bovis* that were recovered on mm-BA from freshwater feeder streams and shellfish-growing waters in a subestuary of Chesapeake Bay did not hydrolyze starch (Kator and Rhodes 1991).

Our experience using this medium with samples from freshwater feeder streams and estuarine shellfish-growing areas revealed departures from selectivity and specificity characteristics reported by Oragui and Mara (1981, 1984). Thus, in the analysis of approximately 400 mm-BA isolates, a significant proportion (65.1%) of "typical yellow colonies," i.e., presumptive *S. bovis,* were confirmed as *E. faecium, S. salivarius* and non-group-D streptococci (Kator and Rhodes 1991). This figure is considerably higher than the 1.3% false-positive rate observed by Oragui and Mara (1981). Conversely, in terms of false negatives 8.9% of the atypical, nonyellow colonies were confirmed as *S. bovis. E. faecium* was frequently recovered on m-BA, a point subsequently noted by Mara and Oragui (1984). Although *S. bovis* has potential as a source-specific and time-sensitive indicator, and the mm-BA method has the advantage of sample concentration, use in shellfish waters requires an evaluation of the medium's recovery efficiency, specificity and selectivity over seasons and in different geographic regions. Also needed is the development and validation of rapid screening methods for confirmation of presumptive *S. bovis.*

Osawa and Mitsuoka (1990) reported a selective medium for recovery of *S. bovis* based on treatment of brain heart infusion agar with tannin and addition of colistin-oxolinic acid to inhibit enterobacteria in feces. Although the medium selectively recovers and differentiates *S. bovis* biotypes I and II present in fresh koala feces, it has not been evaluated for other animal feces, applicability to membrane concentration, and is not particularly effective recovering *S. bovis* in pure culture.

Bacteriophages

Occurrences of enteric viral disease attributed to shellfish consumption (Richards 1985) have stimulated research for an appropriate viral indicator. Inadequacies of the current bacterial standard have been further emphasized by the lack of parity in survival characteristics of fecal coliforms and enteric viruses (Feachem et al. 1983) and the detection of enterovirus in shellfish from approved growing waters (Goyal, Gerba, and Melnick 1979; Vaughn et al. 1979; Ellender et al. 1980). Richards (1985) has advocated the direct use of enterovirus as indicators, in particular poliovirus because it is easily cultivated and prevalent in sewage due to universal vaccination. Although cost, method, and time constraints currently preclude routine detection of enteric viruses, such analyses could be used to assess potential health risk in growing areas lacking identifiable sources of human fecal pollution and closed to direct harvesting because of elevated indicator densities.

Compared to the costs and limits of direct viral detection, bacteriophage indicators are appealing due to their resistance to disinfection (Keswick et al. 1985) and physical factors that eliminate bacterial indicators (Debartolomeis 1988), their ease of detection, low analytical cost, and short assay periods. Justification for use of coliphages as indicators of fecal and sewage pollution or as simulants of human enteric viruses in estuarine and marine receiving waters (Ayres 1977; Borrego et al. 1987, 1990; Cornax et al. 1991; O'Keefe and Green 1989; Grabow et al. 1984; Kott et al. 1974; Kott 1981; Scarpino 1975) has been reviewed by Gerba (1987) and IAWPRC Study Group on Health Related Water Microbiology (1991). Gerba (1987) expressed concern over the paucity of basic data describing the ratios of specific coliphages to viral pathogens and the occurrence, persistence, and seasonal stability of coliphages in shellfish-growing waters. The following studies illustrate the basis of his concern. Vaughn and Metcalf (1975) recovered enterovirus from shellfish-growing waters free of coliphages. Seeley and Primrose (1980) described a subpopulation of coliphages apparently capable of replication at temperatures found in freshwater environments. *In situ* replication of somatic coliphages in estuarine water has been reported (Vaughn and Metcalf 1975). It is evident that studies of coliphage ecology in marine and estuarine waters are necessary to evaluate its utility as an indicator.

Use of male-specific RNA coliphages as indicators of sewage pollution has been discussed (Havelaar and Hogeboom 1984; Furuse 1987; Havelaar and van Olphen 1989). The gastrointestinal tract of warm-blooded animals and domestic sewage are major habitats for these viruses (Furuse et al. 1978). FRNA coliphages share some properties with human enteroviruses, such as type of nucleic acid, structure and size, although these similarities do not necessarily translate to similar functional properties in saline environments or shellfish. FRNA phage may also be source-specific. Furuse et al. (1978) isolated and classified FRNA phages in domestic waste from various countries in South and East Asia into four major serological groups (groups I, II, III, and IV). The relative proportion of FRNA phages in domestic waste and sewage ranges from 10% to 90% of total coliphages (Furuse et al. 1978; Osawa et al. 1981*b*). Most FRNA phages from countries other than mainland Japan belonged to group III and those in mainland Japan domestic waste were group II. Dominance of phage groups in Korea was intermediate between Japan and Southeast Asia, i.e., groups II and III were equally prevalent (Osawa, Furuse, and Watanabe 1981*b*). Differences in climatic temperature and its effect on viral replication was hypothesized to be responsible for the distributions observed (Osawa, Furuse, and Watanabe 1981*b*; Snowdon and Cliver 1989). From a global perspective, fewer FRNA phages were detected in sewage from Central and South American countries than in Asian sources (Furuse et al. 1978). Additional studies (Osawa, Furuse, and Watanabe 1981*a*) showed that FRNA phages belonging to group I were only detected in feces or gastrointestinal contents of mammals (domestic farm and feral zoo animals) other than humans. FRNA phages isolated from pigs belonged to groups I and II, and those from humans groups II and III. Phages belonging to group III were exclusive to humans. However, in a study of domestic sewage from treatment plants in Japan (Furuse et al. 1981), FRNA phages belonging to groups I, II and III were found. Occurrence of group I phages, albeit at low frequencies compared to those in II and III, was attributed to inputs of sewage derived from animal sources such as slaughterhouse wastes. Basic studies concerning the ecology and occurrence of FRNA phages in sewage and feces in the United States are lacking. Poppell (1979) recovered FRNA coliphages at low frequency (2%) from human feces and consistently from all parts of municipal sewage collection systems in the Northeast United States. The validity of FRNA serological groupings to differentiate human and animal fecal contamination should be examined, and, if corroborated, might be useful in nonpoint-impacted growing areas.

Two host assay systems have recently been developed that are selective for F-specific phages. Previous coliphage assays required labor-intensive screening to identify RNase sensitive phages to avoid counting somatic and filamentous DNA phages. Debartolomeis (1988; Debartolomeis and Cabelli 1991) developed and applied a F-specific sensitive host strain to estuarine and marine waters. The assay uses a male *E. coli* host mutant (called Famp) which is also resistant to lysis by DNA somatic phages. RNase has to be incorporated in a parallel assay to distinguish between FRNA and FDNA filamentous coliphages. A second F-specific coliphage assay was developed by Havelaar and Hogeboom (1984) through the addition of an *E. coli* plasmid coding for sex pilus production to a F^- *Salmonella typhimurium* host strain (WG45). *S. typhimurium* was chosen as the host to avoid interference from somatic coliphages which are relatively abundant in sewage. Theoretically, plaques observed on this host would be attributed primarily to FRNA phage, because somatic salmonellae and FDNA filamentous phages occur at much lower densities in sewage. This assay has been applied to studies of feces and wastewaters (Havelaar, Hogeboom, and Pot 1984; Havelaar and Nieuwstad 1985; Havelaar et al. 1986) but not to marine or estuarine waters. Recently, Rhodes and Kator (1991) compared densities of fecal coliforms and phages lytic against F^+ *S. typhimurium* WG49 from samples collected along a salinity gradient in an estuary subject to nonpoint pollution. Verified FRNA phages were infrequently recovered from 100 ml samples of feeder streams or estuarine water with fecal coliform densities significantly above the growing-area standard (range = < 2–7900 FC 100 ml^{-1}). Moreover, higher levels of phages were detected in estuarine sediments compared to feeder stream sediments. Although fecal coliform densities in freshwater feeder stream sediments were one to four orders of magnitude larger than phage densities, densities of phages and fecal coliforms in estuarine sediments were similar. This pattern of phage distribution was inconsistent with the fecal coliform data which identified the feeder streams as sources of fecal pollution.

Our experience confirms the critical importance of the bacterial host strain to the coliphage assay (Sinton and Ching 1987; Havelaar and Hogeboom 1983; Seeley and Primrose 1982). Host strains vary as to the accessibility, specificity, and location of phage receptor sites. Evaluation of coliphage indicator systems must be conducted and verified using field samples collected to obtain seasonal coverage. Seeley and Primrose (1982) considered coliphages to be inappropriate indicators because of the likelihood of interfering phages. For example, Vaughn and Metcalf (1975) found the abundance of coliphages in shellfish-growing waters varied over a three-year period as a function of the *E. coli* host employed. In the study previously mentioned, Rhodes and Kator (1991) randomly picked plaques on WG49 for confirmation as male-specific phages. Of the total number of plaques produced on *S. typhimurium* WG49, 99% (293 of 294) were produced by RNase-resistant phages. Phages purified from these plaques were lytic against the female parent *S. typhimurium* WG45, were not lytic against a male strain of *E. coli*, and were lytic against environmental *Salmonella* spp. isolates. The susceptibility of the *S. typhimurium* WG49 host strain to MS2 was routinely retested and confirmed. As a result, not only is the sanitary significance of these data uncertain, but the application of this particular assay host to nonpoint-source-impacted growing areas may be inappropriate.

If FRNA phages are to be used as fecal indicators, more must be known about their ecology and fate. A possible limitation concerns their comparative occurrence in feces and sewage; FRNA phages are an infrequent component of human (and some animal) feces (Havelaar, Furuse, and Hogeboom 1986; Furuse 1987; Havelaar 1987) and occur at densities significantly higher in sewage. Their abundance in sewage treatment plants may be the result of extraenteral multiplication at ambient temperatures on hosts that formed pili at temperatures above 30°C (Havelaar, Furuse, and Hogeboom 1986; Havelaar 1987; Havelaar and Pot-Hogeboom 1988). However, Poppell (1979) calculated the low occurrence of FRNA phages found in humans was sufficient to account for the densities found in sewage because the numbers of phages contributed by those individuals was very high. Therefore, the extent to which FRNA phages can infect and lyse enteric bacteria in wastewater facilities, septic systems and growing-area environments is an issue of concern because such multiplication would vary the ratios of indicator to bacterial and viral pathogens. Thus, although FRNA (FDNA) phages may serve as indicators of wastewater and sewage contamination, their use as indicators of fecal contamination or a predictor of health risk in shellfish-growing waters requires analysis. Unless it can be demonstrated that male-specific phages occur frequently and at high densities in septic leachate, these phages may be a poor indicator of human fecal contamination in nonpoint-impacted shellfish-growing areas. On the other hand, FRNA phages may have potential use as an indicator of animal contamination because these can be very abundant in the feces of certain domestic farm animals (Havelaar and Pot-Hogeboom 1988; Havelaar, Furuse, and Hogeboom 1986).

Another issue of concern is the persistence of bacteriophages in estuarine environments and shellfish. Borrego and Romero (1985) found selected coliphages persisted for extended periods, i.e., hundreds of days in sterilized seawater, but their numbers decreased rapidly in natural nonfiltered seawater. Mitchell and Jannasch (1969) reported that filtered natural seawater (using either 0.45 or 0.22 μm membrane filters) is strongly antiviral toward coliphage ΦX-174, producing a decrease of viral titer from 10^{12}/ml to 10^{3}/ml in 6 days. To assess the validity of bacteriophages as indicators of viruses, studies are needed to examine the survival of target phages under various seasonal conditions in natural waters, sediments, and commercial species of shellfish.

Quantitative assessment of F-specific phages in estuarine waters may require processing sample volumes of 100 ml or larger and the use of concentration methods. Cornax et al. (1991) concluded (perhaps prematurely) that male-specific coliphages and *B. fragilis* bacteriophages are inappropriate indicators in seawater because of their low densities in sample volumes of only 0.1 to 1.0 ml assayed by double-layer-agar method. Grabow and Coubrough (1986) used a single-agar-layer method to assay 100 ml sample volumes, although this is time-consuming and costly. Methods for phage concentration have relied primarily on concentration techniques developed for human enteric viruses. However, there is no *a priori* basis to assume that these methods are appropriate or equally effective for bacteriophages (Borrego et al. 1991). Seeley and Primrose (1982) reviewed a variety of viral concentration procedures and considered their applicability to phage concentration. Positively charged filters have been used successfully to concentrate coliphages from potable and freshwater systems (Logan et al. 1980; Singh and Gerba 1983). Although positively charged filters are considered effective for concentrating enterovirus from estuarine waters (Kilgen and Cole 1983), Havelaar (1986) reported that they were ineffective for the recovery of coliphage from artificial seawater. Concentration methods for enteric viruses from marine waters have relied primarily on the use of electronegative microporous filters (Sobsey 1987). Debartolomeis (1988) evaluated a method for recovery and concentration of FRNA bacteriophages modified after the procedure of Purdy et al. (1984, 1985) for the recovery of *Bacillus* spp. bacteriophage from water samples. The procedure involves adsorption of bacteriophages to host cells added to a water sample, concentrating the infected cells by centrifugation or membrane filtration, and plaque assay of the concentrate by agar overlay. Debartolomeis (1988) obtained mean recoveries of 103.5% and 87.4%, respectively, for naturally occurring F-specific phages in seawater and river water when compared against the direct overlay method. Using the same approach with marine samples spiked with *Bacillus* spp. phages, Purdy et al. (1984, 1985) obtained anomalous results. Gerba et al. (1978) described a successful protocol for the concentration of poliovirus from seawater using pleated membrane filters in conjunction with aluminum chloride flocculation. Isbister et al. (1983) incorporated 2,3,5-triphenyltetrazolium chloride into a single-layer coliphage assay procedure to improve plaque visualization. Borrego et al. (1991) evaluated a variety of electronegative and chemically produced electropositive filters for the recovery of coliphages from nonsaline water containing diluted sewage effluent. A protracted "drop-by-drop" elution technique with 3% beef extract and positive pressure gave superior recovery of coliphages adsorbed to diatomaceous earth filters treated with cationic polymers. Application of these and other methods for the enumeration of F-specific and other phages requires further evaluation before being considered for routine use in shellfish-growing waters.

Sobsey et al. (1990) recently described a simple membrane-filter-based method for the concentration of bacteriophages from drinking and surface water. This method may have applicability to shellfish waters provided that particulate loading on the filter is not an uncontrolled source of variation to overall recovery. Briefly, $MgCl_2$ is added to a water sample, which is then vacuum-filtered through a 0.45 μm membrane filter, and the filter placed face down on an agar surface inoculated with the assay host. Adsorbed phages subsequently desorb to produce plaques in this agar layer and a tetrazolium dye is used to improve detection. Particulates were shown to reduce plaque numbers, possibly by inhibiting phage desorption, suggesting that this method will have to be carefully evaluated in estuarine waters which are generally characterized by high particulate loads. Methods for the recovery of coliphages from a variety of representative estuarine and marine sediments must be developed and rigorously verified. In a study of this type, Armon and Cabelli (1988) compared the effectiveness of various eluants to release f2 phage experimentally adsorbed to purified clay minerals and those in natural sediment. Efficient recovery of phages from shellfish will be needed as the numbers are likely to be low because only small volumes of homogenized samples can be accommodated by direct plating methods. Brodisch et al. (1986) enumerated coliphages in mussel homogenates by direct plaque assay in a medium augmented by antibiotics to suppress background. A similar approach should be evaluated for different species of commercial shellfish in this country.

Use of direct-pour plate assay methods for phage enumeration generally requires steps to reduce or suppress the growth of background microbiota, depending on sample type. Incorporation of antibiotics, decontamination with chloroform (Tartera and Jofre 1987; Osawa, Furuse, and Watanabe 1981), membrane filtration (Tartera and Jofre 1987), and selective media (Kennedy, Bitton, and Oblinger 1985) have been used for this purpose. Adams (1959) used chloroform to lyse background cells to optimize coliphage recovery. Osawa, Furuse, and Watanabe (1981) diluted sewage samples with 5 ml of PG medium and treated with chloroform (5%, v/v) to kill bacteria. Kennedy, Bitton, and Oblinger (1985) compared chloroform (5%, v/v) with other approaches including selective media to reduce the background for recovery of coliphages from sewage and lake waters. Following an exposure interval after mixing of 1 h to chloroform, samples were removed from the aqueous phase for phage enumeration. Chloroform eliminated the background but also reduced plaque numbers on most media examined and the use of selective media was recommended. Tartera and Jofre (1987) found that although chloroform did not inactivate *B. fragilis* phages, it also did not suppress the growth of anaerobic sporeformers, which sporulated under anaerobic conditions. Potassium sorbate (0.05%) and a lowered pH (5.7) were used to eliminate spore germination. Cornax et al. (1990) compared use of chloroform (20%, v/v), antibiotics, and membrane filtration for removal of the background during the recovery of *B. fragilis* bacteriophages from sewage and sewage-contaminated seawater. Membrane filters (0.45 μm and 0.22 pore nitrocellulose) were used either untreated or treated with 10 ml of 3% beef extract (pH 9.5). Recovery was statistically better in terms of phage numbers and suppression of background growth using 0.45 μm filters treated with beef extract. It was hypothesized that treatment with beef extract combined with the larger-pore-size membranes is the most effective, because most bacterial cells were retained and yet phages were able to pass unadsorbed through the nonbinding filter matrix. The combined use of membrane filtration with incorporation of antibiotics was recommended to increase overall assay selectivity.

In contrast to coliphage, bacteriophages active against the anaerobe *Bacteroides fragilis* demonstrate a high degree of host strain specificity and appear to lack activity against other species of *Bacteroides* spp. (Booth et al.

1979; Cooper et al. 1984; Keller and Traub 1974; Kory and Booth 1986; Tartera and Jofre 1987; Tartera et al. 1989). As a potential indicator of fecal contamination, *B. fragilis* phages were detected exclusively in human feces and sewage and appear to reflect the dominance of the *B. fragilis* host in human feces (Booth et al. 1979; Cooper et al. 1984; Salyers 1984; Tartera and Jofre 1987; Tartera et al. 1989). Phages lytic to the most efficient host strain examined, *B. fragilis* HSP40, were recovered only from environmental areas subjected to sewage and never detected in nonpolluted areas or those occupied exclusively by feral animals (Jofre et al. 1986; Tartera and Jofre 1987). These observations, coupled with the apparent inability of *B. fragilis* phage to multiply in freshwater, seawater, or sediment habitats (Jofre et al. 1986; Tartera, Lucena, and Jofre 1989), suggest this phage is a promising subject for verification as an indicator of human fecal pollution. The low isolation frequency of HSP40 bacteriophages in humans, i.e., $\leq 10\%$, suggests the need for sample concentration and an assessment of its occurrence in waters and shellfish in growing areas contaminated by nonpoint pollution.

The occurrence of phages active against alternate bacterial indicators such as *B. adolescentis* or *B. breve*, *S. bovis*, and *R. coprophilus* appears to have received scant attention. Phage assay systems could provide an alternative detection method to viable enumeration procedures, which are unsatisfactory for reasons previously discussed. Enumeration of phages active against human-specific sorbitol-fermenting bifidobacteria could provide an assay system similar to that proposed for *B. fragilis* (Jofre et al. 1986). As noted, cross-reactivities between *Nocardia* spp. and *Rhodococcus* spp. phages may preclude the development of a *Rhodococcus* assay system.

Although the validation of a viral indicator will depend upon establishing a statistically significant relationship between densities of a given phage and target enteric pathogen or risk of enteric disease, such information will at first be very difficult and expensive to obtain. Therefore, candidate viral phage indicators must be carefully evaluated before these studies are begun to understand their ecology and to confirm their target specificities with samples collected from regionally characteristic growing areas over a variety of seasonal conditions.

Enumeration of Indicators in Shellfish

The object of this section is not an exhaustive coverage of a very large and (sometimes redundant) literature concerning the methods of recovery of indicators from shellfish. Rather, its purpose is to highlight areas we believe warrant attention or are of current interest. Specifically, all methods should be considered with regard to factors that affect indicator recovery discussed in previous sections and the overall need to improve analytical precision.

Methods for Shellfish Preparation

Procedures for preparing shellfish for enumeration of microorganisms vary with the microorganism being sought and the detection method. The first step of the currently approved method (APHA 1985) for detection of coliform organisms is an initial breakup and dilution of a sample consisting of 10–12 whole animals to release and uniformly distribute microorganisms throughout a homogeneous suspension. However, Al-Jebouri and Trollope (1981) found that processing only parts of the digestive system (through excision of the "stomach and intestine") of mussels yielded improved sensitivity of bacterial numbers compared to the total animal homogenate method. In contrast, Metcalf et al. (1980) recovered more virus from whole shellfish than from selected tissues, which included hepatopancreas from oysters and hepatopancreas and siphons for clams. Mechanical breakup of the tissue is normally performed by homogenization (low-speed blending or high-speed homogenization) or stomaching. Stomaching, which seems to be preferred for many foods, has only been evaluated for shellfish by a few investigators and it has not been rigorously tested. Trollope (1984) compared counts of lactose-fermenting bacteria from mussels (*Mytilus edulis*) after stomaching and mechanical blending. Average counts from stomached tissue were about 2 times higher than mechanically blended tissue. Andrews et al. (1978), in a comparison of stomaching and blending using a variety of foods, found stomaching to recover higher aerobic plate counts (APC) from oysters. A smaller-volume model was judged to be more efficient than a larger-volume one for the recovery of APC in oysters. Stomaching is intrinsically a uniform and "clean" method because the sample is encased in a plastic bag, preventing aerosolization and heating because the sample is repetitively struck with a paddle. However, its effectiveness for release of naturally polluting bacteria and viruses from different species of edible shellfish requires evaluation. It has been claimed, for example, that one problem with stomaching shellfish is bag-puncturing because shucked shellfish occasionally contain large bits of shell debris (A. P. Dufour, personal communication).

Because of better precision, methods that utilize membrane filtration for direct estimates of target indicators in shellfish may be preferable to MPN-based or direct pour-plate methods. Direct pour-plate methods are limited by small plant volume, the potential for heat-shocking stressed cells, and background interference. The potential accuracy of direct counting methods using membrane filtration for epifluorescence microscopy or culturable counts is influenced by requirements to: (1) optimize release microorganisms from the food matrix, (2) minimize clogging of the filter by reducing the sizes of food particles, and (3) prevent adsorption of food on and within the filter. Food particles that remain on the filter are a potential substrate for competing organisms and may obscure colony development of target microorganisms. An approach to achieve these goals is to treat shellfish with hydrolytic enzymes. Hydrolysis must break down the food matrix without reducing the target indicator population, especially if sublethally stressed cells are present. Trypsin and other enzymes have been used to effectively digest a variety of food types (Entis, Brodsky, and Sharpe 1982) for microbiological analysis using the hydrophobic grid membrane filter (0.45 μm) method. Enzyme digestion has also been coupled with the direct epifluorescent filter technique (DEFT) for the estimation of total microbial numbers in meat and poultry (Shaw et al. 1987). Rodrigues and Kroll (1988) proposed the use of DEFT for rapid enumeration of coliforms by counting microcolonies growing on selective media. DEFT methods are amenable to automated counting by image analysis and use of fluorescent antibodies for added specificity. Although homogenized clams and oysters have been treated with trypsin for enumeration of *E. coli* and enterococci by membrane filtration (A. P. Dufour, personal communication, Grabow, de Villiers, and Prinsloo 1991), an evaluation of the usefulness of enzymatic digestion for the microbiological analysis of shellfish is needed.

Whatever methods used for release of microorganisms, different species of shellfish are unique in terms of their overall structure and organization, physiology, and accumulation strategies. Given the same natural exposure conditions, Manila clams (*Tapes japonica*) consistently accumulate coliforms and fecal coliforms at higher concentrations than oysters (*Crassostrea gigas*) (Vasconcelos et al. 1969). Viral infection experiments with *Crassostrea gigas* suggest that viruses are located intracellularly, reflecting a different mechanism of sequestering than indicator bacteria (Hay and Scott 1986). Power and Collins (1990) found similar uptake patterns of *E. coli* and an icosahedral coliphage in various tissues of the mussel *Mytilus edulis*. Oysters are considered to be more easily homogenized than hard clams. Metcalf, Eckerson, and Moulton (1980) observed significant differences in the recovery efficiency of spiked viruses as a function of shellfish species. Use of identical procedures for hard- and soft-shelled clams resulted in lower viral recoveries from the latter. Additional extraction steps were necessary to release virus from soft-shell clams. In view of the national interest in alternate indicators and direct counting methods, it seems prudent to optimize the recovery methods as a function of target indicator *and* shellfish species. Commercially important filter-feeding shellfish include oysters (*Crassostrea virginica*, *Crassostrea gigas, Crassostrea lurida*), hard-shell clams (*Mercenaria mercenaria, Mercenaria campechiensis*), soft-shell clams (*Mya arenaria*), Pacific little-neck clam (*Tapes japonica*), and mussels (*Mytulis edulis*).

Approved Methods for Indicators

Total Coliforms and Fecal Coliforms

One approved method for examining the bacterial quality of shellfish is the multiple-tube fermentation test (APHA 1985). Used in either three- or five-tube most-probable-number (MPN) configurations, the method offers poor precision, providing an estimate of the "true" population and large confidence intervals. The largest sample portion generally inoculated is 10 ml of a 1:10 diluted homogenate which is equal to 1 g of shellfish meats. Media used are the same as for water, lactose or lauryl sulfate tryptose broths for the presumptive test, followed by EC broth and brilliant-green bile broth (BGB). Total analytical time for presumptive and confirmed tests is 72 h for fecal and 96 h for total coliforms. Confirmation that fecal coliforms are *E. coli* takes 10 days with this method. The validity of the approved fecal coliform MPN test, and particularly its use to assess product quality in terms of the recommended market guideline of 230 FC/100 g meats, has been questioned for shellfish from the Gulf of Mexico. Counts exceeding this guideline were attributed to EC-positive klebsiellae and estuarine "fecal-coliform-mimicking" bacteria which proliferate during seasons of maximum water temperature (Miescier et al. 1985). Hood et al. (1983) reported that *Klebsiella* spp. were abundant in Gulf-of-Mexico oysters and clams during the months from April through October.

There is only one approved pour-plate method to detect and measure densities of fecal or elevated temperature coliforms in shellfish (hard and soft clams) meats (APHA 1985). It offers a short analysis time (24 h) compared to the MPN, the improved precision of a direct count, but is restricted to the shellfish mentioned and less sensitive in that only 1 g of shellfish meats can be processed per plate. In addition to the important issue of recovery of sublethally stressed cells (Yoovidhya and Fleet 1981), other problems associated with this method are reports of significant interferences caused by background growth with shellfish from the Gulf of Mexico and variations in dye lot (John Miescer, FDA, personal communication).

Nonapproved and Emergent Methods

A succinct discussion of rapid methods, media evaluated, and international methods for the recovery of fecal coliforms and *E. coli* from shellfish meats, is available in the Compendium of Methods for the Microbiological Examination of Foods (APHA 1984*b*). The following sections focus on recent methods not covered in the Compendium.

Coliforms

Detection methods based on hydrophobic grid-membrane filters offer significant advantages in enumeration of indicators in foods (Sharpe, Peterkin, and Malik 1979). These include larger numerical operating range, improved precision, sensitivity, reduced background interference, and removal of inhibitory substances and substrates that may support the growth of background species. These are very desirable characteristics and, in combination with appropriate media, could prove valuable with shellfish if a reliable and reproducible method can be developed for the initial filtration step. Entis and Boleszczuk (1990) evaluated a hydrophobic grid-membrane filter method for enumeration of total coliforms and *E. coli* within 24 h. The method compared favorably with a MUG-based 3-tube MPN. The method has been used for scallops and shrimp, digested with trypsin, but not for bivalve shellfish.

Escherichia coli

As noted previously, the fluorogen, 4-methylumbelliferyl-β-glucuronide (MUG), which has found application for enumeration of *E. coli* in water, has also been examined for similar use in foods and seafoods (Frampton and Restaivo 1993). The initial work of Feng and Hartman (1982) was followed by Alvarez (1984) and Moberg (1985). Moberg (1985) examined factors such as specificity, sensitivity, and optimum MUG concentration in LST (lauryl sulfate tryptose broth)–MUG to detect *E. coli* in a variety of food and dairy samples. The LST–MUG method had a lower false-positive rate and detected more *E. coli* in non-seafoods than the standard APHA method. False positives were attributed primarily to staphylococci. Coliforms such as *E. cloacae* and *K. pneumoniae* did not produce detectable fluorescence from MUG in LST. Alvarez (1984) compared recovery of *E. coli* from a variety of fresh and frozen seafoods (including bivalve shellfish). MUG was used directly in lactose broth, in violet red bile agar (VRBA) as an overlay, and in M-Endo broth with membrane filters. All assays were performed within 24 h. The high specificity of MUG for *E. coli* was verified by confirming fluorescent tubes or colonies in EC broth, streaking on EMB, and with traditional confirmatory tests. The lowest false-positive rates (EC-positive, no fluorescence) and highest recoveries were found using lactose-broth–MUG. The latter observation was attributed to the nonselectivity of lactose broth toward stressed cells, an observation mirroring earlier findings of Feng and Hartman (1982), who recovered heat-stressed cells on LST–MUG. Koburger and Miller (1985) applied MUG–LST to the enumeration of *E. coli* in oysters. They found that the incorporation of MUG into LST was impractical because oysters possess endogenous glucuronidase activity which yields false positives. Incorporation of MUG into EC broth eliminated this problem, although an additional 24–48 h is required. The 25 oyster samples tested yielded 127 gas-positive EC tubes. Of this total 103 tubes were fluorescence-positive and only one was not positive for *E. coli.* Twenty-four tubes were both fluorescence-negative and *E. coli*-negative. However, Chang, Brill, and Lum (1989) suggest that the incidence of glucuronidase-negative strains of *E. coli* is higher than commonly observed and may lead to underestimations of *E. coli* densities in shellfish or water samples. Frampton, Restaino, and Blaszko (1988) evaluated another glucuronidase substrate, 5-bromo-4-chloro-3-indolyl-β-D-glu-curonide (X-GLUC) as an alternative to MUG. This chromogenic substrate does not require illumination by near-UV to detect β-glucuronidase activity and does not diffuse into agar (as MUG) but remains within a colony. Disadvantages include high cost, poor intensity in liquid medium, and possible interference if additional color-based tests are performed.

Finally, as previously mentioned, *E. coli* has been enumerated from shellfish by combining trypsin digestion with membrane filtration (A. P. Dufour, personal communication, Grabow, de Villiers, and Prinsloo 1991). Because of the advantages of direct count, these methods should be evaluated with regard to reducing the potential deleterious effects of hydrolysis on the recovery of stressed cells. Recently, Grabow, de Villiers, and Schildhaver (1992) compared various direct plating methods for enumeration of fecal coliforms and *E. coli* in shellfish. A spread-plate procedure with mFC without rosolic acid and a preincubation period yielded the highest bacterial counts.

Recovery of other indicators such as the enterococci and bacteriophages will require the development and validation of rapid and accurate methods. Methods currently available for the recovery of bacteriophages from shellfish meats are tedious, with recovery efficiencies that may vary with species of shellfish. Direct plating of homogenates using the conventional double-layer-overlay procedure lacks sensitivity and is visually demanding to read. Clearly, methods for the recovery of viral indicators from shellfish meats must be improved in sensitivity, reliability, and ease of use if they are to become routine.

CHEMICAL INDICATORS

Coprostanol

Coprostanol (5β(H)-cholestan-3β-ol) is formed by bacterial reduction of cholesterol and is one of the principal sterols in the feces of man and other mammals (Murtaugh and Bunch 1967; Walker, Wun, and Litsky 1982). Unique anaerobic cholesterol-reducing bacteria, possibly *Eubacterium* spp., use cholesterol as an electron acceptor, reducing the 5,6-double bond of the molecule to form coprostanol (Sadzikowski, Sperry, and Wilkins 1977). Individuals vary as to the amount of coprostanol produced. Wilkins and Hackman (1974) reported that 25% of a human test group reduced less than 50% of fecal cholesterol to coprostanol. Coprostanol can be the dominant sterol in domestic wastes.

Unlike the microbial indicators, which may be subject to chemical, biological, and physical processes that influence their numbers and detection, coprostanol offers advantages of an abiotic marker of sewage pollution. The areal distribution of coprostanol has been used to delineate impacts of point sources such as sewage outfalls or dumping areas (Kanazawa and Teshima 1978; Hatcher et al. 1981; Boehm 1983; Pierce and Brown 1984; Brown and Wade 1984; Düreth et al. 1986; Holm and Windsor 1986) but has not been applied to the study of nonpoint pollution. Attempts to correlate densities of bacterial indicators to coprostanol concentrations in estuaries subject to known inputs have yielded conflicting results (Goodfellow et al. 1977; Churchland and Kan 1982; Yde et al. 1982). Dutka et al. (1988) applying a "battery of biochemical, microbiological and bioassay tests" to evaluate the water quality of the Saint John River and its basin, concluded that fecal sterols were not a useful indicator of fecal pollution.

Concentrations of coprostanol in estuarine (Holm and Windsor 1986) and marine (Kanazawa and Teshima 1978) waters decrease with increasing distance from sources. Brown and Wade (1984) demonstrated that coprostanol in sewage effluent is primarily associated with particulate matter, which may be deposited immediately proximate to an outfall or transported out of the area depending on the dynamics affecting the distribution of fine-grained sediments. Increases of coprostanol concentrations with depth in estuarine waters is attributed to settling of sewage-associated particulates and resuspension (Wade, Oertel, and Brown, 1983). The distribution of coprostanol concentrations in waters adjacent to the entrances of major estuaries results from small- and large-scale physical processes and the buoyancy characteristics of particulate matter. The hydrophobic nature of coprostanol causes it to be associated with particles whose transport subsequently affects its spatial distribution.

Serious consideration of coprostanol as an indicator in nonpoint-impacted shellfish-growing areas remains hampered by the lack of data describing the background levels in such environments and possible origins from nonfecal sources. Laboratory studies using estuarine water and sediment amended with radiolabeled cholesterol demonstrate its conversion to coprostanol by microorganisms (Teshima and Kanazawa 1978). Nishimura and Koyama (1977) suggested that stanols in recent sediments are derived from phytoplankton and sterol conversion. Under anaerobic conditions 5β-isomers of stanols are produced from autochthonous sedimentary organic matter

(Nishimura 1982). Tornabene et al. (1974), in a report describing sterols of the diatom, *Nitzschia alba*, noted that cholesterol has been identified in both blue-green and red algae. Cholesterol and other sterols have been detected in both surface waters (Matthews and Smith 1968) and sediments (Attaway and Parker 1970) from the Gulf of Mexico. Kanazawa and Teshima (1971) found that cholesterol was usually the dominant sterol in both suspended and dissolved fractions in Kagoshima Bay, Japan. Pocklington, Leonard, and Crewe (1987) concluded that coprostanol was an equivocal indicator of fecal pollution, identifying the marine biota as a probable source because patterns of coprostanol occurrence in particulate matter were conjunctive with those of various chemical indicators of primary production, including natural phytosterols.

Qualitative and quantitative aspects of coprostanol degradation in estuarine systems remain poorly understood. Coprostanol appears to persist, particularly under anaerobic conditions, in both lacustrine (Nishimura and Koyama 1977) and marine (Bartlett 1987) sediments. Coprostanol attributed to marine mammals (Venkatesan, Ruth and Kaplan 1986) dating back to 3500 B.C. has been found in Antarctic sediments remote from anthropogenic inputs. Recovery of 5β-stanols from late Pleistocene sediments has also been reported (Nishimura 1982).

Observations that demonstrate the longevity of coprostanol as a geochemical marker raise questions concerning its applicability as a quantitative indicator of health risk or as a marker for growing areas affected by varied and intermittent pollution sources such as storm or agricultural runoff. Eganhouse et al. (1988) noted that concentrations of coprostanol and linear alkyl benzenes (the anionic surfactants found in household laundry detergents) were rather variable in samples of sewage sludge. This was attributed to differences in concentrations of coprostanol in human feces and the differential effects of biosynthetic and degradative processes that can take place in feces and during waste processing. Another argument against the use of coprostanol is that although samples can be preserved for analysis at a late date, the analysis remains technically detailed, lengthy, and costly. Alternate approaches may result in shortened processing time. Wun et al. (1979) described a column adsorption method using XAD-1 resin (Rohm and Haas, Inc., Philadelphia) for the rapid extraction of both cholesterol and chlorophyll from water samples. Hoskin and Bandler (1987) described a rapid thin-layer chromatographic method for the detection of fecal contamination through the recovery of coprostanol from foods. Although this method would be of limited value for the analysis of water samples, it could possibly be adapted for use with shellfish meats. Current improvements in solid-phase packings, especially the silica-based bonded packings, could result in rapid and improved recovery procedures for the detection of cholesterol from shellfish waters or shellfish. Krahn et al. (1989) described a rapid, semi-automated high-performance liquid chromatographic (HPLC) method for separating coprostanol from interfering compounds in sediment extracts. Finally, Eganhouse et al. (1988) sounded a cautionary note concerning prior and future quantitative surveys of coprostanol; unless its co-eluting epimer, epicoprostanol, is clearly resolved, reported concentrations have been and will be overestimated. Overall, the preponderance of information suggests that the use of coprostanol as an indicator of fecal pollution in shellfish-harvesting waters should be viewed as very low priority, because of multiple sources and absence of data supporting its use in nonpoint-polluted areas.

Hydrolytic Enzymes

Measuring the activities of selected hydrolytic enzymes as indicators of fecal pollution has been proposed. Lenhard (1967, 1969) examined the relationship between urease activity and the presence of sewage or sewage contamination in bottom sediments. Skiba and Wainright (1982) evaluated the urease activity in unpolluted and sewage-polluted beach sands as an indicator of sewage contamination. Urease activity was relatively high in sands adjacent to a sewage outfall and decreased with distance from the source. Urea is a major nitrogenous compound in urine, untreated or partially treated sewage, and its release presumably elicits elevated urease activity in impacted sediments owing to its availability as a bacterial substrate. Urease may also be present in the sewage source. Urease activity can be quantified by measuring evolved NH_4^+ (Lenhard 1969) or loss of urea (Sadler 1989). Either procedure is relatively uncomplicated and rapid compared to the standard fecal coliform assay. Positive relationships between urease activity associated with known fecal contamination (Lenhard 1967) and *E. coli* densities (Sadler 1989) have been observed but the data base is very limited. Because the urease assay measures a soluble constituent, it is probable that its ability to reflect the behavior of free or particle-associated bacterial or viral pathogens in estuarine waters would be imperfect. Data are also needed that describe how quickly sediment urease levels change in response to changing levels of sewage contamination. Compared to coprostanol, urease might be a simpler and less labor-intensive alternative to trace established sewage plumes or discharges.

Long-Chain Alkylbenzenes, Fluorescent Whitening Agents and Sodium Tripolyphosphate

Certain chemicals found in commercial and domestic detergents and present in municipal wastewaters and domestic sewage have been identified as waste-specific markers or indicators of sewage contamination. Eganhouse (1986) reviewed the chemistry, occurrence, and fate of long-chain alkylbenzenes in various environments. Because of their chemical stability, he concluded that the use of these compounds would be to delimit areal and temporal impacts of plumes reflecting the discharge of municipal sewage. Close, Hodgson, and Tod (1989) detected septic contamination of groundwater by measuring fluorescent whitening agents and sodium tripolyphosphate. These compounds, which are present in domestic detergents, were detected in 17% of groundwater samples and significantly correlated with densities of fecal coliforms. The value of these indicator compounds to detect the presence of malfunctioning septic systems in a watershed or transport of septic effluents through subsurface infiltration to estuarine systems has not been evaluated. This will also require the determination of compound stability and the extent of degradation by microorganisms in natural systems.

PROTOZOA

Acanthamoeba

Free-living protozoans belonging to *Acanthamoeba*, a genus ubiquitous in soils and freshwater environments, have been detected in sewage-contaminated sediments from freshwater, estuarine and oceanic dump sites and outfalls (Daggett, Sawyer, and Nerad 1982; O'Malley et al. 1982; Sawyer, Visesvara, and Harke 1977; Sawyer 1980; Sawyer et al. 1987). Generally, the occurrence of *Acanthamoeba* spp. has been associated with the elevated levels of fecal bacteria, although sediments positive for *Acanthamoeba* spp., enteroviruses, coprostanol, PCBs, and heavy metals were sometimes negative for bacterial indicators (Sawyer et al. 1987). Sawyer et al. (1987) proposed *Acanthamoeba* spp. as a monitoring indicator based on the persistence of its cysts (for periods up to 2.5 years), the pathogenicity of some strains, and its association with sewage. These amoebae are widely distributed in nature. The significance of their association with fecally polluted marine samples requires further study to determine if *in situ Acanthamoeba* densities reflect the utilization of bacteria growing in dump-site sewage sludge or the direct input of populations carried in the sludge (Daggett, Sawyer, and Nerad 1982; Sawyer et al. 1982). Perhaps, the most likely answer will be a combination of both processes. Quantitative data regarding *Acanthamoeba* densities in human feces, sewage, agricultural runoff and nonpoint-impacted areas would be needed to assess its applicability as an indicator in shellfish waters. The complexity and cost of this undertaking, drawbacks associated with indicator organisms that form resistant cysts, the lengthy incubation periods, and the degree of expertise necessary to distinguish *Acanthamoeba* spp. from other genera of free-living amoebae are concerns that do not support its use as a broadly-based indicator of fecal pollution.

CONCLUDING COMMENTS

Some investigators (Berg 1978; Richards 1985) have proposed direct detection of microbial pathogens as an alternative to indicators, citing documented examples of instances where current indicators failed to predict pathogen presence or pathogens were detected in the absence of indicators. In our view, arguments to retain surrogate microbiological indicators (Cabelli 1977*a*; James 1979) remain valid because of the unpredictable occurrence of pathogens, variations in pathogen virulence, morbidity rates, and minimum infective dosages within a target population. Ultimately, of course, pathogens intrinsically lack the ability to predict risk. Arguments based on the lack of accepted methods for direct and accurate routine detection of pathogens are less compelling. Recent advances in biotechnology suggest that the feasibility of direct pathogen detection is now a reality. Experimental gene probes for hepatitis A virus (HAV) have been used to detect HAV in polluted estuarine waters (Jiang, Entes, and Metcalf 1987; Metcalf and Jiang 1988) and a gene probe for Norwalk agent is now being developed. A gene probe for *Salmonella* spp. (GENE-TRAK Systems, 31 New York Avenue, Framingham, MA 01701) is commercially available.

Emergent methods may revolutionize some aspects of microbiology that traditionally have been associated with frustrating drudgery, elaborate and repetitive procedures, i.e., enumeration and phenotypic characterization of isolates. After more than half a century of using essentially unchanged cultivation methods for the recovery of the coliform indicator group, new methods for rapid detection are becoming available. These may be based on traditional cultivation methods, direct enumeration procedures, target phenotypic characteristics such as constitutive enzymes or antigenic factors, or may use recombinant DNA/RNA technology such as PCR to target an oligonucleotide sequence with high specificity. We welcome these new methods, anticipating their promise of ultimate specificity and rapidity but feel compelled to offer the following cautionary remarks.

First, the best detection systems will not eliminate the need to understand the ecology of an indicator; its persistence characteristics and the relationship between environmental exposure and the assay. For example, allochthonous bacteria exposed to marine and estuarine environments lose their ability to grow on selective and conventional bacteriological media (Barcina et al. 1989; Roszak and Colwell 1987; Rhodes, Anderson, and Kator 1983) and important diagnostic characteristics may be altered or lost. False negatives or misidentification can be a consequence of enumeration or identification schemes based on the expression of a sensitive phenotypic characteristic. *E. coli* loses its ability to ferment lactose (Kasweck and Fliermans 1978), β-galactosidase activity was reduced in cells exposed to seawater (Anderson, Rhodes, and Kator 1979; Munro, Gauthier, and Laumond 1987). Loss or alteration of plasmids encoding for antibiotic and heavy-metal resistance occurs in various *Enterobacteriaceae* during long-term starvation (Chai 1983; Caldwell et al. 1989) and streptococcal (and enterococcal) isolates from animal and environmental sources do not consistently yield valid reactions in miniaturized biochemical testing systems (Molitoris et al. 1985; Rhodes and Kator, unpublished results). Chang, Brill, and Lum (1989) reported that a significant proportion of *E. coli* isolated from human fecal samples were β-D-glucuronidase-negative.

Finally, it is conceivable that advances in sensitivity and accuracy of analytical methods could lead to requests for growing-area closures based on improved detection of pathogens. Such action could produce an untenable situation for the shellfish industry and regulatory agencies. Detection of pathogens in approved shellfish-growing waters using new (or old) methods should not be construed as a *prima facie* evidence of health risk. Cabelli's (1978*a*, 1979) arguments that standards must be based on functional relationships, i.e., correlating rates of illness from prospective epidemiologic investigations with indicator densities, must be heeded.

3

Human Enteric Pathogenic Viruses

Lee-Ann Jaykus

University of North Carolina, School of Public Health, Department of Environmental Sciences and Engineering, Chapel Hill, NC 27599

Mary T. Hemard

Nicholls State University, Department of Biological Sciences, Thibodaux, LA 70310

Mark D. Sobsey

University of North Carolina, School of Public Health, Department of Environmental Sciences and Engineering, Chapel Hill, NC 27599

VIRAL DISEASES ASSOCIATED WITH WATER AND MOLLUSCAN SHELLFISH

It is essential to examine the role viruses play in the transmission of infectious disease via the food chain. Shellfish, which provide high-quality protein, have historically made a large contribution to the American diet. In recent years, there has been a marked increase in the amount of seafood consumed due to public awareness of the nutritional values of both finfish and shellfish. Therefore, the consumption of molluscan shellfish has also risen. Molluscan shellfish are consumed in various conditions ranging from raw to thoroughly cooked. It should be noted that enteric viruses can survive the thermal processes of cooking including stewing, frying, baking, and steaming (Digirolamo, Liston, and Matches 1970). The potential for viral disease transmission via shellfish highlights the importance of monitoring both the waters from which they are harvested and the shellfish themselves.

Cliver (1988) reported that viruses are an important cause of foodborne illnesses. The number of outbreaks of virus-associated gastroenteritis and hepatitis A caused by waterborne and foodborne infections is increasing (Gerba 1988; Cliver 1988). The illnesses caused by enteric viruses are rising, while those caused by bacteria appear to be decreasing (Richards 1985). Since 1900, there have been more than 11,500 cases of shellfish-associated disease in the United States and Canada (Verber 1983). Typhoid accounted for nearly one-half of all of the shellfish-associated disease outbreaks; however, it has not been involved in any major shellfish-borne outbreaks in the United States since the early 1950s (Verber 1983).

The viruses associated with foodborne infections in the United States originate in human intestines and are shed in the feces (Cliver 1988). Fecal contamination of food either directly or indirectly accounts for the spread of enteric viruses by these vehicles. While direct transmission involves person-to-person contact, indirect transmission by feces and sewage contamination includes the vehicles of water and shellfish (Cliver 1988).

Foodborne and waterborne outbreaks of viral disease are usually self-limiting (Cliver 1988). People who have acquired foodborne viral gastroenteritis do not infect others in large numbers, although secondary transmission can occur. However, the outbreak will usually subside, rather than spread through an entire community. Occasionally, explosive, communitywide shellfish-borne viral disease occurs under the right conditions.

A waterborne outbreak occurs where two or more people become ill with similar symptoms after drinking water from a source that is epidemiologically proven to be the source of illness, and a foodborne outbreak occurs when two or more people have similar illnesses after ingesting a common food, and the food is epidemiologically proven to be the source of illness (CDC 1900).

The risks from waterborne viruses have been documented both epidemiologically in terms of disease and by virus detection in water (Berg 1976; 1987; Cabelli et al. 1982; CDC 1990*c*; Clark et al. 1981). Gastrointestinal illness of presumed viral etiology has occurred in people swimming in waters that were considered safe on the basis of coliform bacteria standards (Sobsey 1987; Cabelli 1982). The Centers for Disease Control (1990) reported that municipal water supplies, wells, streams, commercial ice, lakes (Baron et al. 1982), and pools are waters that have been associated with outbreaks of viral gastroenteritis. Furthermore, the presence of enteric viruses in waters and sediments has been well documented (Cole et al. 1986; Goyal, Gerba, and Melnick 1978; LaBelle et al. 1980; Vaughn et al. 1970).

Norwalk virus and related agents collectively referred to as small, round, structured viruses (SRSVs) are responsible for 23% of the reported waterborne viral gastroenteritis outbreaks (Keswick et al. 1985; Baron et al. 1982; Kaplan et al. 1982*a*, *b*; Kappus et al. 1982; Koopman et al. 1980). Norwalk virus and its importance in waterborne outbreaks is attributed in part to its apparent resistance to chlorine (Keswick et al. 1985). Outbreaks of Norwalk virus and similar agents have occurred from municipal water systems (Kaplan et al. 1982*a*), swimming pools (Kappus et al. 1982), and lakes (CDC 1990*c*). Five thousand cases of gastroenteritis caused by a Norwalk-like agent were traced to well water used to produce commercial ice in September 1987, with another 339 cases associated with a well water source in June 1988 (CDC, MMWR March 1990). Norwalk virus and the Snow Mountain agent have been reported to cause both waterborne and foodborne gastroenteritis (Gerba 1988; Brondum et al. 1985).

Keswick et al. (1985) reported that in addition to Norwalk virus and other SRSVs, rotaviruses and hepatitis A virus are the most important causes of waterborne infections involving viruses. Rotaviruses are a documented cause of waterborne outbreaks of gastroenteritis (Gerba 1988). Transmission of hepatitis A virus by contamination of water supplies, food, and drink is also well documented (Grabow et al. 1983; Peterson et al. 1983).

Those enteric viruses which have been shown epidemiologically to be transmitted by shellfish include Norwalk, Snow Mountain agent, astroviruses, caliciviruses, other SRSVs, hepatitis A virus, and non-A, non-B hepatitis virus (HEV) (Gerba 1988). Numerous clam, mussel, and oyster-associated outbreaks of illness have been caused by enteric viruses (Morse et al. 1986; Liston 1990; Richards 1985; Sobsey 1987; CDC 1990; Gonzalez-Ayala 1990; Cliver 1983). Between the years 1976–1980, 50% of the Norwalk-confirmed foodborne outbreaks involved shellfish (CDC 1990*a*). Of the 20,000–30,000 cases of hepatitis A reported to the Centers for Disease Control yearly, only 140 of these are clearly associated with foodborne outbreaks (Cliver 1988). However, if all foodborne hepatitis A were included in the Centers for Disease Control outbreaks, they would be 0.5% of the total hepatitis A reported in the United States (Cliver 1988). It is important to remember that some of these outbreaks have occurred from shellfish that were harvested from waters meeting the coliform bacteria standards (Sobsey 1987), and from shellfish meeting the acceptable levels of fecal coliforms (Morse et al. 1986).

Between the years 1961–1980, 1394 cases of shellfish-borne hepatitis A occurred along the U.S. Gulf and East Coast (Verber 1983), suggesting that hepatitis has become in this period what typhoid was up to the 1950s. During the years 1970–1978, there were 50 reported outbreaks of shellfish-associated foodborne disease in the United States (Bryan 1980; Larkin and Hunt 1982). Richards (1985) observed that during the past 50 years, there have been over 100 outbreaks of shellfish-associated enteric virus illness reported, of which a substantial percentage were of viral etiology.

Of the cases of shellfish-borne illness reported to the United States Food and Drug Administration, Northeast Technical Services Unit and occurring in the years 1973 to May 1990, 433 were hepatitis A, 175 were Norwalk virus gastroenteritis, 11 were Snow Mountain agent gastroenteritis, one was non-A, non-B hepatitis, 4479 were unspecified hepatitis, and 5919 were illnesses of unknown etiologies (Rippey and Verber 1988; Rippey 1990). During the years 1973–1987, the Centers for Disease Control reported the following cases of shellfish-borne illness: 335 cases of hepatitis A, 42 cases of other viral origin, and 3524 cases of unknown etiologies (CDC 1980, 1982, 1989). These outbreaks involved oysters, clams, mussels, and scallops (Rippey and Verber 1988; Rippey 1990).

INDICATORS OF ENTERIC VIRUSES

At present, bacterial indicators are used for evaluation of shellfish harvesting waters and shellfish (APHA 1985; USFDA 1988*a*). However, numerous studies have shown that enteric viruses may be present in shellfishing waters, sediments, and shellfish even though the numbers of indicator bacteria are within acceptable levels (Wait et al. 1983; Gerba et al. 1980; Dufour 1984).

The increase in the numbers of reported outbreaks of shellfish-associated and waterborne illness caused by enteric viruses highlights the need for a reliable indicator that can accurately assess water and shellfish safety with respect to enteric viruses. An indicator of enteric virus presence is a substance or an agent that when present in a sample suggests that viruses are also there (Cliver 1988). There have been many indicator organisms proposed for shellfishing waters, and shellfish. However, no single indicator, index or reference organism to insure the virological safety of these waters and shellfish has been identified and accepted (Elliot and Colwell 1985). Shellfish and their growing waters are regulated based on numbers of total coliforms and fecal coliforms, which are assumed to have originated in human or animal feces. The National Shellfish Sanitation Program and its predecessors developed the presently used most probable number (MPN) analysis of total and fecal coliforms to certify shellfish harvesting waters (USFDA 1988*a*). The sanitary quality of shellfish is based on an allowable standard of 70 MPN total coliforms or 14 MPN fecal coliforms/100 ml of growing waters as a median or geometric mean concentration (USFDA 1988*a*). There are also wholesale market criteria that the shellfish meat must meet. The meats are allowed up to 500,000 total aerobic plate count/g of meat and 230 MPN fecal coliforms/100 g of meat (USFDA 1988*b*; Martin and Pitts). The acceptability of shellfish as a food commodity is not based on only bacterial criteria. A sanitary survey of shellfish-growing waters prior to approval for shellfish harvesting, relaying or depuration and periodically thereafter is required by the National Shellfish Sanitation Program (see the chapters on these topics).

The National Academy of Sciences (Committee on Evaluation of Safety of Fishery Products 1991) discussed three major problems with the current indicator:

1. Non-*E. coli* fecal coliforms and even nonsewage-related bacteria may predominate in the fecal coliform population analyzed by APHA approved methods.
2. The fecal coliform indicator for waters does not indicate the presence of nonsewage-related naturally occurring aquatic bacterial pathogens.
3. The fecal coliform indicator does not correlate with the presence of human enteric viruses, which may be the pathogens most commonly associated with sewage contamination of waters and seafood.

Lack of correlation between numbers of enteric viruses present in shellfish and their growing waters and sediments and numbers of fecal coliforms in shellfish-harvesting waters and shellfish has been established (Elliot and Colwell 1985; Richards 1985, 1987; Cole et al. 1986; Gerba 1988; Kilgen, Cole, and Hackney 1988; Dufour 1984; CDC 1990; Sobsey 1987).

SOURCES OF ENTERIC VIRUSES IN SHELLFISH

Molluscan shellfish filter large quantities of water in order to obtain their necessary oxygen and nutrients and, therefore, are capable of filtering out numerous microorganisms, some of which may be human pathogens (Galtsoff 1964; Portnoy et al. 1975). These pathogenic organisms may come from treated and untreated municipal wastewater and sludge discharges during heavy rainfall periods, raw or partially treated sewage from fishing and hunting camps and other housing facilities which may only have septic tanks or other on-site waste treatment and disposal systems and boat waste discharges. Viruses occur in large numbers in sewage, and have been shown to survive secondary sewage treatment (Gerba et al. 1978; Hejkal et al. 1979). Typical virus reductions by conventional primary and secondary wastewater treatment, including effluent chlorination, are 90–99%. Greater virus reductions would require more advanced treatment processes and high levels of disinfection.

Individuals who are infected, regardless of the presence or absence of clinical symptoms excrete as much as 1 million infectious viruses per gram of feces (Berg 1983). Concentrations of infectious viral particles in sewage generally average 1000 to 10,000/l but may be as high as 100,000/l. Infectious virus particles can survive for several months under certain environmental conditions. Human sewage, therefore, is considered the most important source of human pathogenic viruses contaminating the shellfish environment (Berg and Metcalf 1978).

The isolation of enteric virus from seawater collected several miles from a sewage outfall has been reported (Denis et al. 1977). Enteroviruses have also been detected in seawater 1500 m from a point of sewage discharge (Shuval et al. 1971). Viruses in oysters in estuarine water at a temperature below 7°C retain infectivity for up to 4 months (Lo, Gilbert, and Hetrick 1976). Viruses have been shown to remain stable in artificial seawater for 40 weeks at 4°C (Lo, Gilbert, and Hetrick 1976). Viruses may also survive for longer periods of time by attaching to particulate matter and accumulating in sediments (DeFlora, DeRenzi, and Badolati 1975).

The potential to contaminate seafood with enteric viruses during processing and preparation by contaminated water supplies and infected food handlers also exists. Infected workers or carriers of enteric viruses are capable of transmitting their viruses by the fecal–oral route if poor personal hygiene occurs (Cliver 1988; Matches and Abeyta 1983). Hepatitis A is the only enteric virus documented to have been transmitted in seafood by infected food handlers (Ahmed 1991) although such transmission of other enteric viruses is possible. There is no documentation of transmission of other human enteric viruses by seafood products contaminated during processing, distribution, or food handling (Bryan 1986; CDC 1989; Rippey and Verber 1988). Human enteric viruses do not replicate in seafood products; therefore, time and temperature are important only in terms of how long enteric viruses will persist or survive in shellfish.

FAMILIES OF HUMAN ENTERIC VIRUSES

Human enteric viruses are taxonomized by their morphologic and chemical characteristics (White and Fenner 1986). They are relatively species-specific and receptor-specific for certain cells. They are transmitted by the fecal–oral route, and are stable outside the host. They are able to survive the acidic condition of the stomach, proteolytic enzymes, and bile in the duodenum, and initiate primary infection in the intestines. Like all viruses, enteric viruses are inert in food systems and are active only inside the host cells. More than 120 enteric viruses can be found in human feces (Melnick, Gerba, and Wallis 1978). The families of enteric viruses include picornaviruses, reoviruses, adenoviruses, caliciviruses, and parvoviruses, and unclassified viruses such as astroviruses, coronavirus-like agents, Norwalk-like viruses, Snow Mountain agent, and other small round viruses.

Picornaviruses

The picornaviruses make up the largest family of viruses. There are about 200 host-specific picornaviruses which have been identified in man. At least 69 of these can infect the human enteric tract (Gerba 1988; White and Fenner

1986). Picornaviruses are nonenveloped, possess icosahedral capsids, are 25–30 nm in diameter, smooth and round in electron microscopic appearance, constructed from 60 protomers, and replicate in the cytoplasm. Each protomer is comprised of a single molecule of four polypeptides, VP 1, 2, 3, and 4, or 1D, 1B, 1C, and 1A, respectively. The picornavirus genome is a single-stranded linear molecule of positive polarity RNA consisting of about 7500 nucleotides with a molecular weight of about 2,500,000 Da. The molecule is polyadenylated at its 3′ end with the protein VPg covalently linked to its 5′ end.

Picornavirus replication takes place in the cytoplasm where virion RNA that has entered the cell serves as its own messenger RNA. Messenger RNA is translated into a single large polypeptide (polyprotein) that is enzymatically cleaved stepwise to produce various functional proteins, as well as the viron proteins and their precursors. Formation of protomers and their association with virion RNA produced from a replicative form leads to provirions that mature into infectious virions and are released by lysis of host cells. Several hundreds to thousands of infectious viruses are produced per infected cell. The picornavirus family has been subdivided into several genera, including the genus enterovirus. This genus is further subdivided into groups: polioviruses, groups A and B coxsackieviruses, echoviruses, and enteroviruses. The last group consists of the "newer" enteroviruses.

Poliovirus (PV1, PV2, and PV3)

There have been no reported cases of seafood-associated poliomyelitis in the United States, and there is virtually no risk of infection with wild-type PV from fecally contaminated seafood or from seafood processing and handling (Kilgen and Cole 1991). The three wild types of polioviruses, PV1, PV2, and PV3, are no longer endemic in this country, due to the widespread use of live, attenuated, oral polio vaccine. Vaccine strains of poliovirus are generally the most common enteroviruses found in molluscan shellfish and their growing waters (Kilgen and Cole 1991). The Sabin vaccine strains have even been suggested as a possible indicator of pathogenic enteric viruses in shellfish and estuarine waters (Cole et al. 1986; Richards 1985; Sobsey et al. 1980). However, the recent shift to the use of disposable infant diapers and their disposal in solid waste landfills suggest that the quantities of vaccine strains of polioviruses in sewage are likely to decline.

Symptoms of poliomyelitis include malaise, sore throat, and the possibility of headache and vomiting. Headache and vomiting indicate aseptic meningitis. One percent of these cases progress to secondary central nervous system (CNS) involvement resulting in muscle pain, stiffness, and rapid development of paralysis. Motor function usually improves for a few months. Any remaining paralysis is considered to be permanent. Poliovirus infections have an incubation period of 10–15 days with a range of 3 days to 1 month. Virus particles are shed abundantly in the feces during the incubation period. Once the virus is ingested it multiplies in the pharynx and small intestine, then spreads to regional lymph nodes resulting in viremia and dissemination of the virus throughout the body. If the CNS becomes involved, the virus is carried to the anterior horn cells of the spinal cord and motor cortex of the brain resulting in either spinal (most common) or bulbar poliomyelitis. Mortalities in the bulbar form are due to cardiac or respiratory motor failure (White and Fenner 1986).

Most infections from poliovirus, prior to the Salk and Sabin vaccines, were minor. The ratio of clinical illness to subclinical infection by polioviruses varies with age and decreases with increasing age. In children the ratio is between 1 : 100 and 1 : 1000 (White and Fenner 1986). Epidemics and sporadic cases of poliomyelitis still occur mostly in underdeveloped countries and an international polio eradication initiative is in progress. Before vaccination programs were developed, there were summer epidemics in temperate countries (White and Fenner 1986). However, the Salk inactivated polio vaccine (IPV) and the Sabin live oral polio vaccine (OPV) have virtually eliminated poliomyelitis from developed countries. Reported cases of poliomyelitis in this country are due to failure of OPV, inadequate attenuation of one of the virus types, or immunosuppression of the vaccine recipients. Imported infections of unimmunized immigrants and spread into populations of unimmunized groups such as the Amish community in 1978 have also occurred. Continued immunization programs are imperative to avoid reintroduction of the disease until global eradication is achieved. Outbreaks continue to occur in developing countries in some parts of the world due to inadequate vaccine programs (White and Fenner 1986; Kew et al. 1990).

Hepatitis A

In 1988, 50% of the reported hepatitis cases in the United States were caused by hepatitis A (CDC 1900*b*), which is spread mainly by the fecal–oral route. Transmission is typically person to person, but contaminated fomites, water and food also contribute to its spread (CDC 1990*b*). The virus particle is 27 nm in diameter and it superficially resembles the enteroviruses in size, morphology and composition (CDC 1990*b*). Until recently, HAV was classified as enterovirus 72, but is now in a separate genus.

Hepatitis A (HAV) is ingested and multiplies presumably in the intestinal epithelium. Subsequent infection of the parenchymal cells of the liver is through the bloodstream. The virus is found in the feces 1 week prior to the clinical signs and fecal excretion declines once symptoms of illness appear. It may also be found in the blood 1 week prior to the appearance of the main clinical sign of dark urine, and it disappears after serum transaminase levels reach their peak. The incubation period is normally 4 weeks with a range of 15 to 50 days (CDC 1990*b*).

The onset of hepatitis A is associated with the clinical symptoms of fever, malaise, anorexia, nausea, and lethargy. Symptoms also include dark urine, jaundice, and an enlarged, tender palpable liver. Severity of the disease is usually related to age. Infected children are generally asymptomatic, and as the age increases, so does the likelihood and severity of illness (Bryan 1986; CDC 1990*b*; Cliver 1988; White and Fenner 1986).

HAV is hyperendemic in developing countries, especially in areas that are overcrowded, have inadequate sanitation and poor hygiene. The infections seen in these situations generally occur in children and are subclinical. In more developed countries the disease is seen most often between the ages of 15 and 30 (Cliver 1988; White and Fenner 1986), with a fatality rate of 0.6% (CDC 1990*b*).

HAV infection results in permanent immunity. Therefore, a vaccine could eradicate the disease, as has been done for poliomyelitis in developed countries. There are killed and live vaccine trials under way in the United States and other countries. Some vaccines are licensed in Europe and mainland China. Passive immunization with gamma-globulin following exposure to HAV still is the main method of control and prevention if exposure is expected or has occurred recently (Cliver 1988; White and Fenner 1986).

Other Enteroviruses: Coxsackie, Echo, and Enteroviruses

Although most enterovirus infections are subclinical, they can cause several clinical syndromes including neurological disease resembling poliomyelitis, cardiac and muscular disease, exanthems, respiratory disease, ocular disease, and gastrointestinal disease (White and Fenner 1986). These enteroviruses grow in both the throat and the intestinal tract. They are shed for longer periods of time in the feces than in the respiratory secretions, and are capable of being disseminated via the bloodstream to target organs. None of the well-characterized enteroviruses have been shown to be responsible for seafood-associated enteric illness (CDC 1989; Cliver 1988; Richards 1985; Rippey and Verber 1988). However, virus particles resembling enteroviruses have been observed in stools of cases of shellfish-associated enteric illness (Appleton 1987).

Reoviruses–Rotaviruses

Rotaviruses, members of the Reovirus family, are one of the most important causes of infantile gastroenteritis in the world. Worldwide it is estimated that they cause 140 million cases of diarrhea a year, with 1 million of these resulting in death (CDC 1990*a*). The viruses are shed in large numbers in the feces, up to a trillion particles per gram of stool, for 3–4 days (CDC 1990*a*). The infectious dose of rotavirus for a child is 10 virus particles and the infectious dose for healthy young adults is also very low (Ward et al. 1986; CDC 1990*a*). Spread of the disease is mainly by the fecal route, although aerosol transmission also has been suggested. Nosocomial outbreaks in nurseries are very common (White and Fenner 1986). No seafood-associated cases of rotavirus infection have been documented in the United States; however, rotaviruses have been detected in estuarine water and sediment (Rao et al. 1986).

Morphologically, the rotaviruses have a smooth and round outline with two concentric icosahedral shells giving the appearance of a wheel. Virions are 70 nm in diameter and contain double-stranded RNA consisting of 11 different segments and have a molecular weight of 12,000,000. There are at least five human serotypes, but based on genomic differences, there are many strains. The incubation period for rotavirus infection is 1–3 days. The virus multiplies in the differentiated columnar epithelial cells of the small intestine. In neonates, older children, and adults, infection with rotavirus is usually asymptomatic. Clinical illness is most likely between the ages of 6 and 24 months. Symptoms include vomiting for 3 days and watery diarrhea for 3–8 days (CDC 1990*a*). Fever and abdominal pain can occur frequently. If untreated, this can result in severe dehydration and death especially in undeveloped tropical countries where child health may be compromised by other factors such as poor nutrition (Cliver 1988; White and Fenner 1986).

Rotaviruses are estimated to cause 3.5 million cases of diarrhea a year in the United States (CDC 1990*a*). They account for 35% of diarrheal hospital stays and result in 75–125 childhood deaths annually in this country (CDC 1990*a*). Rotavirus diarrhea affects almost all children by the time they are 4 years of age. In the United States, rotavirus activity is mainly in the cooler months of October to April.

Improving the nutrition and hygiene of third-world populations would be the greatest aid in control of rotavirus-related mortalities. There is no vaccine available; however, research to develop a live oral vaccine is in progress (Cliver 1988; White and Fenner 1986).

Adenoviruses

Adenoviruses, originally isolated from human adenoids, cause respiratory, ocular, and genitourinary infections. Recently, they have been documented as causative agents of gastroenteritis. Because they are shed fecally, they can be spread by the fecal–oral route, either by direct contact, fomites, contaminated food, and contaminated swimming waters. No seafood-associated cases of adenovirus illness have been documented in the United States.

Adenovirus particles are icosahedral, 80 nm in diameter, with two main types of morphologically distinct capsomers numbering 252 in total. Of these, 240 are hexons and 12 fibers. The genome consists of a single, linear molecule of double-stranded DNA containing 23–45 kilo base pairs with a molecular weight of 20,000,000–25,000,000. Adenoviruses replicate in the nucleus of the host cell.

In the case of enteric infections, the virus multiplies in the small intestines; most infections are subclinical. The incubation period is 3–10 days. Clinical signs of infection include diarrhea, vomiting and fever; sometimes respiratory symptoms occur. The illness may last 1–2 weeks.

Adenoviruses types 40 and 41 have been shown to cause 10% of infantile gastroenteritis, and are responsible for 5–20% of hospitalized diarrheal children in developing countries (CDC 1990*a*). Adenoviral diarrhea occurs mostly in children less than 2 years of age, although infection is possible at any age. Illness due to enteric adenoviruses are seen at all times of the year.

There is no vaccine for the enteric adenovirus types 40 and 41.

Astroviruses

Astroviruses have a star-shaped appearance and a diameter of 28 nm. These RNA viruses have been described in feces from infants and they are responsible for gastroenteritis in all age groups. Illness is most frequent in young children and the elderly. Outbreaks due to contaminated water and oysters have been reported in England (Kurtz and Lee 1987; Herrmann et al. 1991; CDC 1990*a*). The incubation period is 2–4 days and the illness usually lasts 1–4 days. Symptoms include vomiting, diarrhea, fever, and abdominal pain. There are at least five serotypes of human astroviruses and four of these have been grown in cell cultures. This has made possible the development of improved diagnostic methods based on enzyme immunoassay and nucleic acid hybridization (Moe et al. 1991).

Caliciviruses

Human caliciviruses are responsible for many cases of gastroenteritis. The Norwalk virus is now known to be a calicivirus on the basis of genomic characterization (Jiang et al. 1990). Like other caliciviruses, it is about 35 nm in diameter and contains a single capsid protein and single-stranded, plus-sense RNA. Outbreaks of viral gastroenteritis due to the Norwalk agent have been associated with swimming in fecally contaminated waters, ingestion of fecally contaminated food or drinking water, and consumption of uncooked or partially cooked shellfish harvested from fecally contaminated estuaries.

The incubation period for Norwalk virus gastroenteritis is 24–48 h with the illness lasting 12–60 hours (CDC 1990*a*). Clinical symptoms are nausea, vomiting, diarrhea, and abdominal cramps. Twenty-five to fifty percent of those infected also report headache, fever, chills, and myalgia. Death has occurred in some adults due to electrolyte imbalances (CDC 1990*a*). Immunity following an infection with Norwalk virus is only temporary, lasting approximately 1 year. This may be one of the reasons for the high attack rate of 50–90% (Cliver 1988). Some individuals are resistant to infection despite low humoral immunity.

Caliciviruses are slightly larger than the picornaviruses with the virion measuring 35–40 nm in diameter. They have an icosahedral capsid with 32 cup-shaped (calix) depressions giving the particle its unique appearance. The genome is single-stranded RNA of about 7.5–8 Kb with a molecular weight of about 2,600,000. These viruses are considered relatively resistant to heat and acid, although the resistance of the human caliciviruses is not well documented. Norwalk virus appears to be quite resistant to chlorine (Keswick et al. 1985).

Infection by caliciviruses other than Norwalk virus usually occurs by the age of 12, most often between the age of 3 months to 6 years. The incubation period is 1–3 days and the illness lasts approximately 4 days. Symptoms include vomiting and diarrhea, and sometimes respiratory infection and fever (CDC 1990*a*).

Unclassified Viruses

The unclassified viruses include the named but poorly characterized agents of gastroenteritis such as Snow Mountain agent and other small round viruses (SRVs), such as Taunton, Hawaii, Montgomery County, Paramatta, Cockle and Ditchling (Madeley 1989). Generally, the illnesses caused by these viruses are similar to Norwalk virus gastroenteritis.

NANB Enteral Hepatitis; Hepatitis E Virus

Enteral NANB hepatitis is transmitted mainly by sewage-contaminated water and by person-to-person contact. The etiological agent of enteral or epidemic NANB hepatitis is a calicivirus and is designated hepatitis E Virus or HEV (Reyes 1990). Virions, 27–34 nm in diameter, have been recovered from patients with enteric NANB hepatitis. These agents have been transmitted to primates and human volunteers having immunity to hepatitis A virus. Subsequently, HEV is recovered from their feces (White and Fenner 1986; Bradley, Krawczynski, and Kane 1991).

Enteral NANB hepatitis can be more severe than hepatitis A with a high incidence of cholestasis. The mean incubation period ranges from 15 to 40 days. The highest attack rate is in young adults, especially pregnant women in the third trimester. The overall mortality rate for NANB hepatitis is only slightly higher than for hepatitis A, but significant mortalities in excess of 10% can occur in pregnant woman. The incidence of chronic active hepatitis is also extremely low in NANB hepatitis (CDC 1990*b*; Overby et al. 1983; White and Fenner 1986).

Enteral NANB appears to be endemic in the Middle East and Africa (Overby et al. 1983). Cliver (1988) noted water-associated outbreaks in India, Africa, the USSR, and recently Mexico. HEV is not prevalent in the United States, but there is concern that it could enter from Mexico and other endemic areas. The sanitary measures discussed previously with respect to prevention and control of other enterically transmitted viral diseases apply to enteral NANB hepatitis. There is no vaccine available at present.

EPIDEMIOLOGY OF SHELLFISH-BORNE VIRAL DISEASE

Human disease transmission due to the consumption of virally contaminated molluskan shellfish was first recognized in 1956 with the report of a large outbreak of clam-associated hepatitis A (Roos 1956). Since then, epidemiological evidence linking viral disease transmission to shellfish consumption has continued to increase. Epidemiological evidence clearly indicates that shellfish can serve as effective vehicles for the transmission of viral hepatitis and gastroenteritis.

Viral disease transmission due to ingestion of contaminated shellfish has been documented for many species of bivalves, including the Eastern oyster (*Crassostrea virginica*) (Mason and McLean 1962; Portnoy et al. 1975; Mockowiak, Caraway, and Portnoy 1976; Guzewich and Morse 1986; Morse et al. 1986), the European oyster (*Ostrea edulis*) (Gerba and Goyal 1978), the Sydney rock oyster (*Crassostrea commercialis*) (Murphy et al. 1979), the Pacific oyster (*Crassostrea gigas*) (Gill et al. 1983), the Northern quahog or hard-shell clam (*Mercenaria mercenaria*) (Dougherty and Altman 1962; Guzewich and Morse 1986; Truman et al. 1987), the sand or soft shell clam (*Mya arenaria*) (Gerba and Goyal 1978), the English mussel (*Mytilus edulis*) (Dienstag et al. 1976; Mele et al. 1989), the cockle (*Cardium edul*) (Appleton and Pereira 1977), various Italian clams (Mele et al. 1989), and a Chinese clam (Halliday et al. 1991).

Although human enteric viruses are a major cause of shellfish-associated viral disease, only a few have been epidemiologically linked to consumption of contaminated shellfish. These include hepatitis A virus (HAV), non-A, non-B hepatitis virus, Norwalk virus, Snow Mountain agent, other small round structured viruses (SRSV), astroviruses, caliciviruses, and "featureless" viruses, such as picornaviruses and parvoviruses (Appleton 1987; DeLeon and Gerba 1990). Previous reviews of shellfish-borne viral epidemiology have been provided by Mosley (1967), Goldfield (1976), Gerba and Goyal (1978), Richards (1985), and DeLeon and Gerba (1990).

Hepatitis A

Infectious hepatitis caused by hepatitis A virus (HAV) is one of the most serious viral illnesses transmitted by the ingestion of contaminated shellfish (Richards 1985). The first documented outbreak was attributed to the consumption of raw clams in Sweden in 1955 (Roos 1956). Richards (1985) presents a summary of U.S. surveillance data on shellfish-borne hepatitis, while Appleton (1987) discusses European data. These shellfish-associated HAV outbreaks show two distinct patterns in terms of magnitude: large outbreaks and small episodes involving clusters of illness. Additional epidemiological evidence indicates a third pattern consisting of sporadic cases of HAV infection due to the consumption of contaminated shellfish. These may be more significant than outbreak-related cases (O'Mahoney et al. 1983). It is generally accepted that the reported outbreak and individual case data represent only a small proportion of total epidemic and sporadic shellfish-borne hepatitis A infection (Goldfield 1976).

Small clusters of infection are primarily reported as small outbreaks involving defined sources of exposure or through surveillance data summaries. Investigation of the source of the contaminated shellfish is rarely done in these cases. However, Dienstag (1976) reported a small outbreak in which seven members of a family of 14 developed acute viral hepatitis approximately one month after a reunion. The epidemiological investigation implicated incompletely cooked mussels as the vehicle of infection. Serological testing confirmed the etiological role of HAV. The oysters were obtained from Australian harvesting waters showing intermittent fecal contamination over the two years prior to the outbreak.

Five large outbreaks accounting for more than 1200 cases have been documented in the U.S. since 1955. Mason and McLean (1962) reported 84 cases of infectious hepatitis in individuals living or vacationing in Mississippi and Alabama during the first three months of 1961. Oysters were epidemiologically implicated as the source of infection. They were harvested from the mouth of the heavily contaminated Pascagoula River. Sewage pollution of oyster-growing waters due to an overtaxed sewage treatment plant and a blocked sewage line was the source of contamination. In 40% of the cases a nonspecific gastroenteric illness occurred 15–24 h after ingestion of the raw oysters.

Dougherty and Altman (1962) noted a significant increase in the reported cases of hepatitis A in New York and New Jersey in the early months of 1961. A total of 459 patients reported a history of consuming raw clams and in 87% (368/423) of those traced, the source of the clams was found to be New York's Raritan Bay. This bay was known to have unfavorable bacteriological surveys frequently and the bay waters received a significant amount of sewage from New Jersey and New York municipalities. Additionally, the harsh winter conditions had forced harvesting of clams in more contaminated, ice-free waters.

Two large outbreaks of hepatitis A occurring in 1963–1964 were attributed to the consumption of contaminated clams. The clams from a Philadelphia outbreak were traced to Atlantic County, New Jersey, but the source of contamination was not clearly identified (CDC 1964). Ruddy et al. (1969) investigated the other outbreak which consisted of 123 cases of viral hepatitis following ingestion of raw clams in Connecticut. Cases were six times as likely to have consumed raw clams than were controls. The clams were harvested off the coast of Rhode Island but the source of contamination was never determined.

A large epidemic of hepatitis A in Houston, Texas, and Calhoun, Georgia during the fall of 1973 was attributed to the consumption of raw oysters (Portnoy et al. 1975). In Houston, 263 persons were identified as having oyster-associated hepatitis, while in Calhoun, 15 cases were identified. An outbreak of oyster-associated hepatitis occurred in Louisiana simultaneously (Mackowiak, Caraway, and Portnoy 1976). The oysters incriminated in all three locations were traced to approved oyster-growing areas along the eastern Louisiana coast. Investigation into the source of contamination was inconclusive, but it was believed that the flooding of the Mississippi River in April and August of 1973 led to fecal pollution of oyster-growing areas. Shellfish harvesting was closed in this area until the end of summer, 1973, due to high coliform counts. Upon reopening and harvesting in September, cases of hepatitis occurred. It appeared that HAV was filtered, accumulated, and retained by the shellfish for as long as two months. These data provided epidemiological evidence that the coliform standard as an indicator of the purity of shellfish-growing areas may be inadequate with respect to viruses (Portnoy et al. 1975; Mackowiak, Charles, and Benjamin 1976).

Two European outbreaks also are notable. A 41-person outbreak was identified in England in 1978 and attributed to the consumption of boiled and steamed mussels (Bostock et al. 1979). The mussels from waters in Eire went to a cleansing station in Anglesey prior to consumption. The survival of viruses after purification indicated inadequate treatment and viral persistence. The source of contamination was not identified. An outbreak in Livorno, Italy, was identified because the annual incidence of acute viral hepatitis doubled between April 1984 and January 1985 (Mele et al. 1989). A case control study of 75 patients revealed that illness was strongly associated with the consumption of raw mussels and clams within six weeks of the onset of illness. When the cases reporting raw shellfish exposure were classified on the basis of frequency of consumption, there was a clear dose–response relationship. The shellfish were most likely harvested from sewage-contaminated local waters. Comprehensive control measures instituted after the onset of the outbreak resulted in a tenfold decrease in disease incidence from the pre-epidemic period.

A distinct seasonal pattern of shellfish-borne hepatitis has been observed (Craun 1974). A rise in incidence reportedly occurs in late fall, with peak incidence occurring from January to March, followed by a decline in late spring. Most of the reported outbreaks in the United States are consistent with this seasonal pattern. It is noteworthy that in many states the shellfishing season corresponds to this same time period.

The importance of contaminated shellfish as a vehicle of infection in sporadic cases of hepatitis A has been investigated. Koff et al. (1967) conducted a prospective, controlled epidemiological study designed to determine the modes of transmission of nonepidemic hepatitis among patients in ten Boston hospitals from March 1963 to March 1966. Ingestion of raw shellfish was significantly more frequent in patients (34/185) than in controls (10/185). Additionally, steamed-clam ingestion was significantly more frequent in patients (13/104) than in controls (2/104). Confounding due to simultaneous contact with jaundiced people was unlikely. There was little crossover between ingestion of steamed and raw shellfish. When considered together, ingestion of raw and steamed shellfish was as frequent a potential exposure to hepatitis as was contact with jaundiced persons.

Stille, Kunkel, and Nerger (1972) studied 425 patients with viral hepatitis from May 1968 to October 1971 at the Frankfurt University Clinics. Consumption of contaminated shellfish accounted for 20% of the cases of infectious hepatitis. More recently, a case-control study in southeastern England revealed a statistically significant association between infective jaundice and the consumption of cockles, with as many as 25% of hepatitis A cases attributable to the consumption of contaminated shellfish (O'Mahoney et al. 1983). The authors postulated that the role of shellfish in sporadic cases of hepatitis is more significant than outbreak-related cases.

The largest documented epidemic of shellfish-associated HAV occurred in 1988 in Shanghai, China. Nearly 300,000 cases of hepatitis A, representing an attack rate of 4,083 per 100,000 population, were associated with consumption of raw clams harvested from a harbor area receiving untreated domestic sewage from the adjacent community and fishing vessels. The hepatitis A epidemic was preceded by an outbreak of clam-associated infectious diarrhea (Halliday et al. 1991).

Coastal areas generally report a higher frequency of hepatitis A, which may be a consequence of shellfish consumption, although the evidence for this is not definitive. Dougherty and Altman (1962) calculated a hepatitis A attack rate of 26/100,000 for all adults in New Jersey, but noted that attack rates in coastal counties were higher.

Non-A Non-B Hepatitis

Transmission of non-A non-B hepatitis by consumption of contaminated shellfish has been documented (Caredda et al. 1981, 1985, 1986; Alter et al. 1982; Bamber et al. 1983; Torne et al. 1988). Caredda et al. (1981) studied the epidemiological pattern of non-A non-B hepatitis at a Milan clinic and attributed 17.8% of the cases to ingestion of contaminated shellfish. A Maryland study revealed that 12.8% of the non-A non-B hepatitis cases investigated were associated with raw shellfish ingestion with an adjusted odds ratio of 3.81. Ingestion of raw shellfish was the third most common risk factor after parenteral drug use and history of blood transfusion (Alter et al. 1982). Tabor (1985) suggested that non-A non-B hepatitis is caused by at least three distinct viruses, one of which is transmitted enterally. Recently, Caredda et al. (1985) noted that the clinical features of non-A non-B hepatitis associated with shellfish consumption were similar to those noted for epidemic waterborne non-A non-B hepatitis. It is likely that the non-A non-B hepatitis epidemiologically associated with shellfish is due to the recently characterized hepatitis E virus, a likely calicivirus.

Viral Gastroenteritis

A comprehensive listing of shellfish-associated gastroenteritis in the United States through 1985 is presented by Richards (1985). For most of these outbreaks, viral etiology could not be proven but was the most probable cause. European surveillance data are provided by Gunn and Rowlands (1969) and Appleton (1987). Research in the 1970s led to the identification of many viral agents responsible for nonbacterial gastroenteritis, and epidemiological evidence links consumption of contaminated shellfish to the transmission of Norwalk virus, Snow Mountain agent, small round viruses, astroviruses, caliciviruses, and parvoviruses (Appleton 1987; DeLeon and

Gerba 1990). Detection methods for enteric viruses are sophisticated, largely inaccessible, and incapable of detecting many of the important enteric viruses of epidemiological importance in foodborne and waterborne disease. Furthermore, the reporting and investigation of foodborne disease outbreaks are fraught with a variety of problems. Hence, the number of unreported and undocumented outbreaks and cases of shellfish-borne viral disease in the population far exceeds the number reported (Richards 1985).

Despite the logistic and technical impediments, a number of studies have shown shellfish consumption to be responsible for outbreaks of viral gastroenteritis. In December 1976 and January 1977, 33 incidents of food poisoning affecting 797 people throughout England were epidemiologically linked to the consumption of steamed cockles (Appleton and Pereira 1977). Small round virus-like particles were detected in a high proportion of fecal specimens in three of the outbreaks. The particles were similar in size, morphology, and buoyant density to particles implicated in outbreaks of "winter vomiting" disease and nonbacterial gastroenteritis. The cockles were gathered from waters polluted by sewage during a period of widespread gastroenteritis in the population. The authors concluded that the heat treatment applied to the cockles after harvesting was inadequate to destroy the viruses.

Appleton et al. (1981) investigated nine separate British outbreaks of gastroenteritis associated with the consumption of raw shellfish between 1976 and 1980. Of the fecal specimens examined, 78/90 (87%) revealed the presence of small round virus-like particles resembling those seen previously in outbreaks of winter vomiting disease and the 1976 cockle outbreak. Similar virus particles also were detected in oysters implicated in an outbreak.

The first documented outbreak of shellfish-associated Norwalk virus gastroenteritis affected over 2000 people in Australia during the summer of 1978 (Murphy et al. 1979). The attack rate was as high as 85%, and laboratory diagnosis as Norwalk virus etiology was confirmed by electron microscopic examination of fecal specimens and antibody response in paired serum samples. The oysters were harvested from the Georges River near Sydney, where increased sewage contamination was attributed to high levels of rainfall.

A related outbreak of oyster-associated gastroenteritis due to Norwalk virus occurred in Darwin, Australia in December 1978 (Murphy et al. 1979; Linco and Grohmann 1980). Electron microscopic examination of stool samples from patients showed two distinct parvo-like viruses present, one of which was identified serologically as Norwalk virus. One sample of dissected oysters showed 22–25 nm virus-like particles by electron microscopic examination. The oysters had been taken from the Georges River during the summer of 1978 and were held frozen until consumed six months later in Darwin.

Eyles, Davey, and Huntley (1981) investigated five cases of gastroenteritis that were traced to a single batch of bottled oysters which were probably harvested from the Georges River during the summer of 1978. Norwalk virus particles were demonstrated in fecal specimens of the patients. Echovirus type 8 and a reovirus were isolated from the oysters using standard tissue culture techniques. While it is unlikely that the viruses isolated caused the observed disease, this was the first reported occasion in which the food vehicle incriminated in an outbreak of viral disease was shown to be contaminated with human enteric viruses.

Following the Australian Norwalk virus outbreak, several programs were implemented to minimize the occurrence of further disease. Grohmann et al. (1981) used human volunteers to consume depurated oysters originating from Georges River and Brisbane waters. Over a period of 32 weeks, 52 volunteers became ill after consumption of depurated Georges River oysters. Sixty percent of these illnesses occurred in a two-week period preceded by heavy rainfall. Norwalk virus was confirmed by examination of feces or paired sera in 37% of the patients. This study provided epidemiological evidence that enteric viruses may persist even after depuration.

Richards (1985) reported an outbreak of gastroenteritis among Louisiana oyster eaters in 1982 involving 472 cases. Norwalk-like viruses were suspected because testing did not implicate bacteria as the causative organisms.

During the three-year period of 1981–1983, there were numerous cases of shellfish-associated gastroenteritis in New York. The large number of documented cases was attributable in part to an increased awareness of shellfish-borne disease (Richards 1985). The first outbreak occurred in June of 1981 and raw clams were identified as the vehicle of infection (Guzewich and Morse 1986). With implementation of increased surveillance, thousands of shellfish-related illnesses were reported throughout the state of New York between May and December 1982 (Centers for Disease Control 1985; Guzewich and Morse 1986; Morse et al. 1986). These included 103 well-documented outbreaks in which 1017 people became ill, with 813 cases due to clams and 204 due to oysters. Disease was associated with both raw and steamed shellfish. Ten individuals subsequently developed hepatitis A. Norwalk virus was implicated as the predominant etiological agent based on features of illness and serological testing. Norwalk virus was also identified by radioimmunoassay in 4/6 clam and oyster specimens from two of the outbreaks, and virus-like particles were observed by electron microscopy in some clam extracts. The shellfish originated from several northeastern states. The most likely source of fecal contamination was runoff associated with heavy spring rains. The epidemic nature of shellfish-related illnesses in 1982 prompted dealers to import depurated, English clams in 1983. These were responsible for 14 separate outbreaks of gastroenteritis involving 2000 shellfish eaters in New York and New Jersey (Richards 1985). In 1984 and 1985, additional outbreaks of shellfish-related viral gastroenteritis were reported in New York state (Guzewich and Morse 1986).

Two clambake-related outbreaks in Rochester, New York, provided the first documentation of shellfish-associated gastroenteritis attributable to the Snow Mountain agent (Truman et al. 1987). A total of 55% (83/151) persons who ate raw or baked hard-shell clams became ill. Those who ate clams were up to 30 times more likely to develop disease than those who did not were and a dose–response relationship was demonstrated between attack rate and number of clams consumed. By using paired serum specimens, it was confirmed that clam eaters were more likely to seroconvert to Snow Mountain agent (8/12) than controls (0/4). The clams were harvested in Buzzard's Bay, Massachusetts, during a period of high rainfall and increased runoff.

Another outbreak of viral gastroenteritis that occurred in London in 1983 implicated raw oysters as the vehicle (Gill et al. 1983). The attack rate in those eating oysters was 79%. One individual subsequently developed hepatitis A. Examination of fecal specimens by electron microscopy revealed small round Norwalk-like virus particles. Immune electron microscopy using paired sera indicated a fourfold increase in antibody titer to small round viruses. The source of contamination was not identified; however, the oysters had been depurated for 72 h prior to consumption, indicating viral persistence after controlled purification.

Two recent outbreaks of gastroenteritis have been attributed to astrovirus and a calicivirus. Consumption of oysters at a naval base dinner resulted in a gastroenteric illness occurring in two phases. The first phase was associated with a small round virus, and was followed four days later by a second phase characterized by diarrhea and shedding of large numbers of astroviruses (Caul 1987). A human calicivirus outbreak transmitted by contaminated oysters demonstrated a high attack rate in all age groups (Cubitt 1987).

Other Diseases

Mahoney et al. (1974) detected Australia antigen, a marker for hepatitis B virus, in clams contaminated by drainage of untreated sewage from a Maine coastal hospital. Likewise, Kater, Kim, and Davidson (1969) isolated Australia antigen from one of four soft shell clams taken from a polluted estuary. However, Koff et al. (1976) could not detect hepatitis B surface antigen in hard- and soft-shell clams from certified New England clambeds. While shellfish have not been found to cause hepatitis B infection in the U.S. (DeLeon and Gerba 1990), Caredda et al. (1981) reported that shellfish ingestion was associated with up to 23.9% of hepatitis B cases in a Milan, Italy clinic. This conflicting evidence indicates that the role of bivalve mollusks in the transmission of hepatitis B remains speculative. There are no other scientific data to document a shellfish route of transmission for hepatitis B.

Two case control studies have revealed a significant association between Creutzfeldt–Jakob disease and ingestion of raw oysters and clams (Bobowick et al. 1973; Davanipour et al. 1983). However, these were small studies evaluating multiple potential risk factors. A firm epidemiological link between this disease and raw shellfish consumption has not been established.

Isolation of Viruses from Shellfish

Virus isolations from shellfish were reported as early as 1965. Since then, there have been additional studies and published review articles (Gerba and Goyal 1978; DeLeon and Gerba 1990). Enteric virus isolations have been reported for a variety of bivalves including the Eastern oyster (*C. virginica*) (Metcalf and Stiles 1965, 1968; Metcalf, Vaughn, and Stiles 1972; Fugate, Cliver, and Hatch 1975; Vaughn and Metcalf 1975; Goyal, Gerba, and Melnick 1979; Vaughn et al. 1979*b*, 1980; Ellender et al. 1980; Sobsey et al. 1980*a*; Cole et al. 1986; Richards and Hetrick 1988), the hard-shell clam (*M. mercenaria*) (Liu et al. 1968; Harshbarger, Shang, and Otto 1977; Vaughn et al. 1979*b*, 1980; Wait et al. 1983), *Mercenaria californicus* (Gerba and Goyal 1978), and mussels such as *Mytilus galloprovincialis* (Bendinelli and Ruschi 1969) and *M. edulis* (Watson, Inglis, and Anderson 1980). Additionally, Goyal et al. (1984) have reported the isolation of enteroviruses from blue crabs at sludge dump site off the northern Atlantic coast of the United States.

The viruses of public health importance found in shellfish are those associated with the alimentary tract of humans and in some cases other homeothermic animals as well and whose characteristics permit survival and transmission by feces. The reported virus isolates include polioviruses (Petrilli and Crovari 1965; Metcalf and Stiles 1968; Metcalf, Vaughn, and Stiles 1972; Fugate, Cliver, and Hatch 1975; Goyal, Gerba, and Melnick 1979; Vaughn et al. 1979*b*, 1980; Ellender et al. 1980; Wait et al. 1983), echoviruses (Metcalf and Stiles 1965, 1968; Bellelli and Leogrande 1967; Bendinelli and Ruschi 1969; Fugate, Cliver, and Hatch 1975; Goyal, Gerba, and Melnick 1979; Vaughn et al. 1979*b*, 1980; Ellender et al. 1980; Wait et al. 1983), coxsackie viruses (Metcalf and Stiles 1965, 1968; Bendinelli and Ruschi 1969; Denis 1973; Vaughn et al. 1979*b*, 1980; Watson et al. 1980; Wait et al. 1983), reoviruses (Metcalf and Stiles 1968; Watson et al. 1980), adenovirus (Watson et al. 1980; Pietri et al. 1988), and various bacteriophages (Vaughn and Metcalf 1975; Harshbarger, Shang, and Otto 1977; Gerba and Goyal 1978). Other viruses that have been isolated but not identified unequivocally include hepatitis A virus and SRSVs resembling Norwalk virus or caliciviruses.

The virus most frequently recovered from shellfish is poliovirus, whose prevalence is explained by the active immunization with live oral polio vaccine in developed and developing nations. Children immunized with live, attenuated polio strains can shed virus for several days at levels of 10^3–10^7 infectious units per gram of feces (Larkin and Hunt 1982; Richards 1985). Metcalf and Stiles (1968) reported 78.1% (89/114) culturable virus isolates from oysters to be polio, with 83.1% of these being poliovirus type 1. More recent studies have identified 88.1% (58/67) to 95.6% (241/252) of viral isolates from oysters to be poliovirus (Goyal, Gerba, and Melnick 1979; Ellender et al. 1980). Virtually, all of these isolates were polio type 1. Likewise, Wait et al. (1983) found 73.7% (14/19) of viral isolates from clams were poliovirus type 1. While the high percentage of poliovirus isolates is predictable, the relative predominance of type 1 remains unexplained.

Early studies reporting virus isolation from shellfish document that shellfish served as vehicles of human enteric viruses. Most of these studies sought to detect viruses in shellfish obtained from waters known to be polluted with treated or untreated domestic sewage. The first field studies on the occurrence of enteric viruses in shellfish and their overlay waters were done in New Hampshire estuaries polluted with raw sewage (Metcalf and Stiles 1965, 1968). As many as 25–60% of the oyster pools examined were positive for enteroviruses, and viruses were detected in oysters up to four miles from the pollution source. Similar results were reported by Liu et al. (1968), who found that 33% of clams and 55% of oysters from a polluted Rhode Island estuary harbored enteric viruses. Further evidence of widespread enteric virus contamination of shellfish was provided by Metcalf, Vaughn, and Stiles (1972) who reported that 28.6% of the oyster samples from a prohibited area of Galveston Bay, Texas, were virally contaminated.

Early survey data are limited in precision and accuracy due to a lack of dependable, quantitative techniques for the isolation of viruses from shellfish and water (Gerba and Goyal 1978). Only limited numbers of shellfish could be sampled and testing was frequently done only on crude homogenates. Additionally, many human enteric viruses are nonculturable. Several investigators used analytical approaches other than cell cultures to identify viruses in shellfish. Denis (1973) detected coxsackie A16 virus in shellfish by the production of flaccid paralysis in suckling mice. Harshbarger, Shang, and Otto (1977) reported the presence of Chlamydia phage in hard clams by the examination of extracts using electron microscopy. Pietri et al. (1988) reported the detection of hepatitis A virus in mussels by immunochemical methods.

Improved methods for extracting and concentrating viruses from shellfish were developed in the late 1970s and this enabled more quantitative studies. The recognition of lengthy viral persistence in shellfish and discontinuous patterns of pollution in overlay water prompted the sampling of shellfish from approved, conditionally approved, and prohibited harvesting waters. The first such study was reported by Fugate, Cliver, and Hatch (1975) who sampled oysters off the coast of Louisiana and Texas for 12 months in 1974. Only two of 17 (11.8%) Texas samples were positive for virus, and both of these were taken from approved beds. Goyal, Gerba, and Melnick (1979) monitored oyster-harvesting waters of the Texas Gulf coast and found 2/10 (20.0%) and 12/30 (40.0%) samples positive for enteric viruses from opened and closed waters, respectively. Viral concentrations in oysters ranged from 6–224 plaque-forming units (pfu) per 100 g of oyster meat.

Other studies also report virus isolations from shellfish harvested from open areas. Oysters from two Mississippi reefs were monitored for viral contamination from January to December 1978 (Ellender et al. 1980). Enteric viruses were detected in 32/109 (29.4%) oyster pools, of which 2/22 (9.1%) were harvested from open waters, and 30/87 (34.5%) came from closed waters. The concentration of viruses in samples ranged from 1 to 31 pfu per 100 g of oyster meat. Vaughn et al. (1979*b*, 1980) reported the occurrence of enteric viruses in oysters and clams from the coast of New Jersey and Long Island. Clams taken from the Great South Bay yielded 4/12 (33.3%) virus positive samples, with 2/7 (28.6%) pools taken from closed waters and 2/5 (40.0%) pools taken from open areas. Oysters harvested from Oyster Bay yielded viruses in 5/16 (31.3%) samples, with 3/8 (37.5%) samples taken from closed areas and 2/8 (25.0%) from open waters. Viral titers ranged from 8 to 200 pfu per 100 g of shellfish. Wait et al. (1983) found 9/28 (32.1%) coastal North Carolina clam samples positive for enteric viruses, with 3/13 (23.1%) of these coming from open waters. Viral concentrations ranged from 0.8 to 7.0 most probable number of cytopathogenic units (MPNCU) per 100 g of shellfish. Both Vaughn et al. (1979*b*, 1980) and Wait et al. (1983) reported no significant differences in the occurrence of enteric viruses in shellfish taken from open versus closed beds.

Several investigators have failed to isolate enteric viruses from shellfish originating from open waters. Sobsey et al. (1980*a*) reported viral isolation in 3/47 (6.4%) oyster samples taken from North Carolina coastal waters. In this case, 0/22 samples from open waters contained enteric viruses. Cole et al. (1986) analyzed oysters from closed and open beds in Terrebonne Parish, Louisiana and found 3/41 (7.3%) samples positive for enteric viruses; all three positive samples were from closed areas. Viral titres were low, ranging from 1 to 2 pfu per 100 g. Richards and Hetrick (1988) reported viral isolation from 5/81 (6.2%) oyster samples taken from a five-state region. Viruses were recovered only from oysters harvested from prohibited areas, with isolation reported for 5/50 (10.0%) of such samples. Virus titers ranged from 21 to 196 pfu per 100 g of oyster tissue. Khalifa et al. (1986) has reported that no enteric viruses were detected in 85 samples of clams and oysters collected from legal and restricted beds in Massachusetts. However, the interpretation of such negative results is problematic because nondetection may be due to either the absence of viruses or to methodological limitations (Sobsey et al. 1980*a*; Khalifa, Werner, and Timperi 1986).

Enteric viruses have been found in shellfish harvested from open areas, although with less frequency than in those from prohibited areas. While isolation rates for shellfish taken from approved and prohibited beds both range from 0% to 40.0% those for approved shellfish generally fall below 20.0%, and those for prohibited shellfish frequently exceed 25%. In all cases, the levels of contamination are relatively low, ranging from 1 to approximately 200 pfu per 100 g of shellfish. While these virus concentrations are well below those in sewage, it must be remembered that as little as one infectious unit for laboratory hosts potentially can initiate human infection. Furthermore, only some human enteric viruses are culturable, so that actual virus levels are higher than those measured.

The detection of enteric viruses in shellfish harvested from approved waters has prompted investigation into the relationship between viral contamination of shellfish and the water of their habitat. Metcalf and Stiles (1965, 1968) studied both oysters and overlay water and found that the coexistence of virus in overlay water and oysters occurred in only 43% of the cases. Goyal, Gerba, and Melnick (1978) reported parallel isolations in only 50% of the cases. When viruses were isolated from both water and oyster samples, the types of viruses identified in the two samples frequently differed (Goyal, Gerba, and Melnick 1978). Other investigators have reported generally poor correlation between the presence of phage or virus in shellfish samples and concurrent samples of sediment and/or water (Vaughn and Metcalf 1975; Gerba et al. 1980; Vaughn et al. 1980). Cole et al. (1986) reported no parallel isolation of viruses from 41 samples of oysters and overlay water. Investigators have concluded that viral examination of overlay water does not adequately predict viral contamination of shellfish. Discrepancies between viral contamination of water and shellfish are attributable to discontinuous pollution patterns and viral persistence in shellfish after disappearance from overlay water (Metcalf and Stiles 1968; Goyal, Gerba, and Melnick 1978). It has been suggested that the only reliable way to determine the virological quality of shellfish is to test the shellfish meat itself (Metcalf and Stiles 1968).

While early studies report enterovirus recoveries from shellfish primarily during winter months (Metcalf and Stiles 1968), more recent survey data cites virus isolation at other times of the year as well. Vaughn and Metcalf (1975) isolated enteric viruses and coliphages from oysters and overlay water in higher numbers in the summer months. Ellender et al. (1980) reported higher viral titers from prohibited oysters in the summer months, although peak isolation rates were noted in January. Likewise, other investigators have reported viral isolation predominantly in late summer and early winter months (Vaughn et al. 1980; Wait et al. 1983; Cole et al. 1986). This phenomenon probably corresponds to the late summer and early fall period of peak enterovirus excretion by populations in temperate climates (Wait et al. 1983). Shellfish dormancy at cold temperatures may account for prolonged enterovirus persistence during winter months (Metcalf and Stiles 1968). Goyal, Gerba, and Melnick (1979) reported the isolation of different virus types from the same shellfish-harvesting site over different sampling times and concluded that this may be due to fluctuating patterns of enteric virus infection in the community.

Most virus isolation studies have been conducted on fresh, raw shellfish harvested from open and/or closed waters, but some data do exist for market samples. Bendinelli and Ruschi (1969) reported the isolation of echoviruses from Italian market mussels, and Denis (1973) found coxsackie A viruses in market samples of oysters and mussels in France. Fugate, Cliver, and Hatch (1975) isolated poliovirus in an oyster sample imported into the United States from Japan. However, examination of 30 retail samples of cooked, peeled prawns in the United States yielded no virus recovery (Eyles 1983) and unpublished FDA data suggested that less than 0.01% of market oysters tested were positive for culturable enteric viruses (Larkin and Hunt 1982). These data indicate that culturable enteric viruses may be present in market shellfish, but their incidence in the U.S. market samples is generally low.

It has been assumed that shellfish are passive carriers of human enteric viruses since viral replication has never been demonstrated (Metcalf and Stiles 1968). Liu, Seraichekas, and Murphy (1966*a*, *b*) observed that enteroviruses in the digestive tract of quahogs were neither intracellular nor adsorbed to cells. Chang et al. (1971) inoculated proflavine-labeled, light-sensitive viruses into northern quahogs and later examined them for light-sensitive progeny. The failure to detect such progeny provided convincing evidence of the absence of viral replication in contaminated shellfish.

RELATIONSHIPS BETWEEN VIRUSES IN SHELLFISH AND BACTERIOLOGICAL STANDARDS

Methods for detecting enteric viruses and pathogenic bacteria in water and shellfish are considered too unreliable, complex, expensive, and time-consuming to be used for routine monitoring and surveillance purpose. Consequently, the fecal coliform index is still used as a sanitary quality indicator for shellfish and their harvesting waters. This index is considered appropriate because fecal coliforms are normal inhabitants of the gastrointestinal tract of warm-blooded animals and are excreted in the feces in large numbers. Their presence is taken as evidence of recent fecal pollution (Gerba and Goyal 1978). There are U.S. standards for shellfish and shellfish-harvesting waters based on the enumeration of total and fecal coliforms. For water, the most probable number (MPN) of total coliforms must be a median or geometric mean that does not exceed 70 per 100 ml and no more than 10% of the samples can exceed an MPN of 230 per 100 ml for a 5-tube decimal dilution test. Shellfish meat can contain no more than 230 fecal coliforms in 100 g (Wood 1976). Such standards are used to classify shellfish-harvesting areas as closed, conditionally approved, or approved.

The adequacy of bacteriological standards to indicate the virological quality of shellfish has been questioned. Viral isolations from shellfish harvested in approved waters have continued to be reported (Fugate, Cliver, and Hatch 1975; Goyal, Gerba, and Melnick 1979; Vaughn et al. 1979*b*, 1980; Ellender et al. 1980; Wait et al. 1983). One of the largest documented outbreaks of shellfish-related hepatitis in the United States was due to oysters harvested from approved waters (Portnoy et al. 1975). Additionally, evidence suggests that coliform bacteria are more sensitive to disinfection, natural inactivation in the environment, and depuration than are viruses (Gerba and Goyal 1978).

When investigators have examined the relationships between viral contamination of shellfish and several bacteriological indicators, no consistent and statistically significant relationships have been found between virus presence in clams and oysters and a variety of bacteriological and physicochemical quality parameters for water and shellfish (Fugate, Cliver, and Hatch 1975; Goyal, Gerba, and Melnick 1978; Ellender et al. 1980; Gerba et al. 1980; Vaughn et al. 1980; Wait et al. 1983; Kilgen et al. 1984; Cole et al. 1986; Grabow et al. 1989). Bacteriological indicators examined included total and fecal coliforms in overlay water, as well as total coliforms, fecal coliforms, *E. coli*, fecal streptococci, *Vibrio parahemolyticus*, and aerobic plate count for shellfish. Goyal, Gerba, and Melnick (1978) were able to isolate viruses from oyster samples when the total coliform MPN was as low as 3/100 ml water. Ellender et al. (1980) reported viral isolation from clams at fecal coliform levels less than 20 per 100 g of shellfish. Only Richards and Hetrick (1988) have reported a correlation between virus and fecal coliform levels in shellfish. They consistently isolated viruses from clams and oysters when fecal coliform levels exceeded 130 per 100 g of shellfish.

The consensus among investigators is that the current bacteriological standards for determining the quality of shellfish and shellfish-growing waters do not reflect the occurrence of enteroviruses. The use of coliphages has been suggested as an alternate indicator of enteric viral contamination in shellfish (Kott et al. 1974). The successful use of coliphages as indicators requires that their presence accurately and consistently indicate human enteric viral pollution (Kott et al. 1974). Vaughn and Metcalf (1975) reported the results of concurrent examination of sewage effluent, sediment, shellfish, and water for coliphages and enteric viruses. The majority of enteric virus isolations (63%) were observed in samples yielding no coliphage activity. In controlled experiments, oysters were found to accumulate 5–30 times more coliphages than model enteric viruses. Shifts in dominant coliphage type occurred during the course of the study and evidence for coliphage replication in estuarine waters was presented. Kennedy et al. (1986) also reported wide ranges of coliphage recovery from fresh oysters and poor correlation between coliphage counts and *E. coli* levels. Vaughn and Metcalf (1975) recommended that their coliphage system not be used as an index of shellfish hygiene. Male-specific coliphages, which are fecally excreted by a small percentage of humans but are consistently found at high levels in sewage and conventionally treated effluents, have been suggested as viral indicators. This is because their source, size, and resistance to chlorine disinfection are similar to certain enteroviruses (IAWPRC 1991). Burkhardt et al. (1989) recently reported that male-specific coliphages were unable to replicate within whole clams, although replication in clam homogenates was observed.

Other investigators have proposed direct assay for pathogenic viruses in water and shellfish as a means to monitor shellfish hygienic quality (Gerba and Goyal 1978; Richards 1985). Candidate viruses have included rotaviruses and enteroviruses, specifically poliovirus (Richards 1985). In previous studies, poliovirus was frequently isolated from naturally contaminated shellfish and was usually the predominant virus type (Metcalf and Stiles 1968; Goyal, Gerba, and Melnick 1979; Ellender et al. 1980; Wait et al. 1983). Live, attenuated oral polio immunization programs provided a regular supply of these viruses to sewage. However, detection of polioviruses in sewage has been erratic (Katzenelsen and Kedmi 1979; Irving and Smith 1981) and their effectiveness as an indicator of the virological quality of shellfish has not been adequately evaluated. Furthermore, the widespread use of disposable baby diapers and their disposal as solid waste have further reduced the quantities of vaccine strain poliovirus in sewage.

VIRUS ACCUMULATION BY SHELLFISH

Studies have demonstrated that many commercially valuable species of shellfish rapidly accumulate viruses when exposed to contaminated water. Uptake has been documented for oysters (*C. virginica, C. gigas, C. glomerata, O. edulis, O. lurida*) (Hedstrom and Lyke 1964; Metcalf and Stiles 1965; Hoff, Jakabowski, and Beck 1965; Mitchell et al. 1966; Hamblet et al. 1969; Hoff and Becker 1969; DiGirolamo et al. 1975; Bedford et al. 1978; Jensen 1980; Landry et al. 1982, 1983; Scotti et al. 1983), clams (*M. mercenaria, Mya arenaria, Tapes japonica*) (Liu et al. 1966*a*, *b*; Liu, Seraichekas, and Murphy 1967*a*; Seraichekas et al. 1968; Hoff and Becker 1969; Canzonier 1971; Metcalf et al. 1979, 1980*c*; Landry et al. 1982, 1983), mussels (*M. edulis, M. galloprovincialis*) (Duff 1967; Gerba and Goyal 1978; Croci et al. 1984; Gaillot et al. 1988), cockles (*Cerastoderma edule*) (Millard et al. 1987), and Hawiian bivalves of the genus Pinnidae and Isognomonidae (Dow et al. 1984). Various crab species (*Pachygrapsus* and *Hemigrapsus* species, *Cancer magister*, *Cancer antennarius*, and *Callinectes sapidus*) (DiGirolamo et al. 1972*b*; Hejkal and Gerba 1980) and the Florida conch and sea hare (Gerba and Goyal 1978) have also been found to accumulate viruses. Reviews of virus accumulation by shellfish have been presented by Gerba and Goyal (1978), Richards (1988*a*), and DeLeon and Gerba (1990). In most studies, the viruses were not only taken up, they were significantly accumulated by the shellfish, often to a much higher level than in surrounding water. Reported accumulation or "bioconcentration" factors varied from less than onefold to over 1000-fold. The highest uptakes were 900-fold and 1500-fold as reported by Hoff and Becker (1969) and Canzonier (1971), respectively. Most accumulation ranged from tenfold to 100-fold (Liu et al. 1966*a*; Mitchell et al. 1966; Hamblet et al. 1969; Seraichekas et al. 1968; Metcalf et al. 1979, 1980). In some cases, virus accumulation did not exceed the exposure level (Hedstrom and Lyke 1964; Metcalf and Stiles 1965; Duff 1967; Scotti et al. 1983; Millard et al. 1987), which may have been due to suboptimal conditions for metabolic activity (Hamblet et al. 1969).

The type of laboratory aquarium used as the exposure system in uptake studies affects the level of accumulation. Stationary or static water systems generally resulted in lower uptakes than did flowing water systems (Liu et al. 1966*a*, *b*; Hamblet et al. 1969). It has been suggested that flowing seawater more closely simulates the natural environment of shellfish and is, thus, more conducive to feeding. Static systems may lead to suboptimal physiological conditions for dissolved oxygen, food availability, and metabolic waste dilution, which also influence shellfish feeding (Hamblet et al. 1969).

Maximum virus accumulation takes place within a few hours of exposure to contaminated water. Investigators have reported significant uptakes after 1–3 h (Hedstrom and Lyke 1964; Metcalf and Stiles 1965; Liu et al. 1966*a*). Mitchell et al. (1966) found that oysters concentrated viruses 27-fold over the amount in seawater after only 1 h of exposure. Others have found a slightly longer exposure of 4–16 h is necessary to maximize uptakes (Metcalf et al. 1979). A leveling off or the reaching of a plateau of virus titers in shellfish has been observed (Liu, Seraichekas, and Murphy 1966*a*; Mitchell et al. 1966; Gaillot et al. 1988). This can be explained as an equilibrium between accumulation and elimination, indicating that viral concentration by shellfish is a dynamic process (Mitchell et al. 1966). As the concentration of viruses in water decreases, the viral content of the shellfish also decreases (Hedstrom and Lyke 1964; Liu, Seraichekas, and Murphy 1966*a*; Metcalf and Stiles 1965; Gaillot et al. 1988).

The degree of viral uptake and accumulation by shellfish depends upon several factors, including exposure time, viral concentration in water, presence of particulate matter or turbidity in water, salinity and pH of water, temperature, interspecies differences, individual shellfish differences, type of virus, and food availability. Because uptake of viruses by shellfish is dependent on active feeding or related activities, any factor affecting the physiological vigor of the mollusk can influence accumulation.

The investigation of the relationship between the level of viral contamination of water and viral accumulation by shellfish has produced conflicting results. Early studies concluded that large amounts of virus in seawater were required to demonstrate virus in oysters (Hedstrom and Lyke 1964; Metcalf and Stiles 1965). More recent studies confirm that the greatest bioaccumulation of viruses in other shellfish species occurs when there are the greatest numbers of contaminating viruses in water (Bedford et al. 1979; Metcalf et al. 1979). However, these studies used very high viral titers in waters. Using lower titers, Gaillot et al. (1988) observed that maximum numbers of viruses were recovered from mussels when the viral concentration in water was highest (390 pfu/ml). There were fewer viruses in mussels when concentrations in water were lower (64–71 pfu/ml). However, considerable virus uptake by mussels was noted at low water levels (3–8 pfu/ml).

Liu, Seraichekas, and Murphy (1966*b*) found a comparatively higher virus uptake efficiency by quahogs when levels in seawater were lower, (10–45 pfu/ml) rather than higher (1500–4000 pfu/ml). At the lowest virus levels of 10 pfu/ml seawater, they observed as many as 9000 pfu/g virus in digestive tracts of quahogs. They concluded that viral uptake by shellfish appeared more efficient when surrounding water was contaminated with low levels of virus. Canzonier (1971) also found virus accumulation factors exceeding 1000-fold for hard-shell clams contaminated with S-13 coliphage at low levels of 1–8 pfu/ml in water.

The previously cited studies report virus accumulation by shellfish when levels in water far exceed the norm. Investigation of the efficiency of accumulation at more typical contamination levels was undertaken when sporadic isolation of enteric viruses from shellfish was reported at water levels as low as 0.009, 0.001, and 8.34×10^{-5} pfu/ml (Goyal, Gerba, and Melnick 1979; Metcalf et al. 1979; Vaughn et al. 1979*b*, 1980). Landry et al. (1982) investigated accumulation by clams and oysters at levels of 0.002–0.18 pfu/ml for both feces-associated and monodispersed polioviruses. These levels corresponded to those reported for lightly to moderately polluted waters. They concluded that viral accumulation by shellfish may not be efficient when water column concentrations are less than 0.01 pfu/ml. However, upon prolonged exposure to water levels of 0.1 pfu/ml, an equilibrium between accumulation and depuration of viruses in shellfish was reached, thereby achieving a steady state of viral carriage. The authors suggested that sediments may serve as a potential source of viruses for accumulation, but subsequent studies revealed low viral accumulation in the presence of artificially contaminated sediments (Landry et al. 1983). The role of virus contamination of sediments to virus accumulation by shellfish remains unclear.

Virus association with solids present in water increases the extent of viral uptake by shellfish. Hoff and Becker (1969) found that crude virus preparations which contained cellular debris were concentrated 40–60 times greater by clams and oysters than clarified virus suspensions. Metcalf et al. (1979, 1980*c*) investigated the bioaccumulation of poliovirus 2 by the soft-shell clam using stock virus, solids-associated virus, and feces-associated natural virus. More viruses were bioaccumulated when suspended solids were present, with maximum efficiency for feces-associated virus. The authors concluded that the efficiency of bioaccumulation depends in part on the type of the suspended solids present. Millard, Appleton, and Parry (1987) also found that incorporation of normal fecal extract into virally contaminated water resulted in greater viral uptakes by mussels. Investigators agree that the physical state in which viruses exist in natural waters is an important consideration in assessing viral uptake by shellfish (Hoff and Becker 1969; Millard, Appleton, and Parry 1987). Three theories have been put forth to explain the improved uptake of solids-associated virus by shellfish (Metcalf et al. 1979). The particulate matter may stimulate shellfish feeding or alternatively, solids-associated viruses may have a greater likelihood of being ingested and adsorbed to mucus sheaths. It is also possible that viruses are afforded greater protection from inactivation in water by the presence of solids, thereby increasing their persistence in the estuarine environment.

Despite the increased uptake of viruses associated with particles, excess turbidity or suspended particulate matter can inhibit viral uptake by shellfish. Hamblet et al. (1969) reported 18-fold uptake in oysters subjected to low turbidity (16–24 ppm) seawater but only fivefold accumulation from high turbidity (54–77 ppm) seawater. They hypothesized that the high turbidity clogged the gills and palps, thereby interfering with pumping, feeding, and filtration.

The importance of temperature was documented by Metcalf and Stiles (1968) who demonstrated that oysters lying in polluted water at temperatures below 7°C failed to accumulate viruses. This was attributed to the cessation of pumping and feeding activities at low temperatures. At temperatures exceeding 7°C, accumulation was routinely observed, although the rate may be temperature-dependent. Meinhold (1982) found that the maximum uptake of poliovirus by the Eastern oyster occurred in 5 h at 6°C, 2–3 h at 17°C, and 1–3 h at 28°C. Bedford et al. (1978) reported that the rate of accumulation of radioactively labelled reovirus by rock oysters was not appreciably affected by temperatures of 15–25°C. Hence, it is likely that virus uptake efficiency is influenced primarily at temperature extremes, especially at very low temperatures where uptake rates are slower.

Interspecies differences among shellfish in virus uptake rates have been noted by DiGirolamo et al. (1975), who found that viral accumulation occurred four times faster in Olympia than Pacific oysters. More important perhaps is the difference in virus accumulation by individual specimens of the same species. Seraichekas et al. (1968) noted individual variability in virus uptake by clams of tenfold to 100-fold, while Canzonier (1971) found bioaccumulation to range from 100–1500-fold in digestive glands of individual clams. Individual differences in virus uptake rates have also been reported for soft-shell clams and Pacific oysters (Metcalf et al. 1979; Scotti et al. 1983). Factors suggested as influencing the efficiency of individual viral uptakes include physiological condition of shellfish, hydrographical conditions, size and weight of the specimen, and overcrowding (Seraichekas et al. 1983; Metcalf et al. 1979; Scotti et al. 1983). However, one study reported that the weight of the oyster did not significantly affect the amount of virus accumulated (Scotti et al. 1983).

Differences in accumulation rates have been noted for different viruses (Duff 1967; Canzonier 1971; Bedford et al. 1978). It is hypothesized that this is due to differences in surface properties among virus types. DiGirolamo et al. (1977) reported that attachment of viruses to shellfish mucus was due to ionic bonding to the negatively charged sulfate radicals of mucus. The strongest positively charged virus particles should bind most efficiently. Surface characteristics also influence virus association with particulates and sediments which may in turn affect the likelihood of viral uptake by shellfish (Canzonier 1971).

It has been suggested that increases in ionic concentrations (salinity) or alteration of pH weaken the bond between viruses and shellfish mucus (DiGirolamo et al. 1975). They observed that decreasing salinity from 28 to 14 ppt increased viral accumulation by 10%. It was hypothesized that this phenomenon was a consequence of competition between cations and viral capsid coats for mucus anions.

Investigators have proposed other factors affecting viral uptake by shellfish but none have proven significant. Bedford et al. (1978) showed that reovirus uptake by rock oysters was not influenced by the presence of food material such as algae in the surrounding water. Limited comparative uptake data has indicated that bacteria are accumulated faster than viruses (Bedford et al. 1978).

Most viral uptake studies in shellfish have been done using strains of poliovirus or coliphages. Reovirus, Semliki Forest virus, cricket paralysis virus, rotavirus, and coxsackie B5 virus have also been used (Bedford et al. 1978; Scotti et al. 1983; Dow et al. 1984). Millard et al. (1987) investigated the accumulation of HAV by cockles. Filter-feeding mussels were exposed for 24–48 h to a concentration of 3.2×10^5 rfu/ml seawater supplemented with normal fecal extract. The virus was recovered from shellfish extracts at levels as high as 9.2×10^4 rfu/ml and was present primarily in digestive tract tissue.

In addition to mollusks, crustaceans also accumulate viruses. DiGirolamo et al. (1972*b*) found that poliovirus uptake by West Coast crabs varied from 28% to 63% within 12 and 48 h of exposure to contaminated water, respectively. Coliphage uptakes ranged from 58% to 73%. When allowed to feed on virally contaminated mussels, the crabs accumulated 74–94% of the virus present in the shellfish. Hejkal (1980) reported that accumulation of viruses by blue crabs began within 2 h of exposure to contaminated seawater, with final concentrations rarely exceeding those present in the water. Highest concentrations of virus were found in the hemolymph and digestive tract.

MECHANISMS OF VIRAL ACCUMULATION AND ELIMINATION

Most viruses accumulated by shellfish are found in the digestive tract, as was shown by Metcalf and Stiles (1965) for viruses in the American oyster. Increasing numbers of viruses in alimentary tract tissues, lower at the mouth and higher in the digestive gland itself, indicated that the virus was accepted as a food. Similar evidence has been presented for Northern quahogs with over 90% of the total accumulated viruses isolated from digestive diverticula (Liu, Seraichekas, and Murphy 1966*a*, *b*). Accumulation studies on West Coast oysters revealed 67–71% of viruses in the digestive area (DiGirolamo et al. 1975), and likewise in soft-shell clams only a small fraction of viruses were found in nonalimentary tract tissues (Metcalf et al. 1979). While the digestive tissues harbor most of the accumulated viruses, studies have repeatedly confirmed that viruses can also be found in the hemolymph (Liu, Seraichekas, and Murphy 1966*a*; DiGirolamo et al. 1975; Metcalf et al. 1979, 1980).

Most investigators believe that the primary means of viral uptake is adsorption to shellfish mucus which is then ingested (Liu, Seraichekas, and Murphy 1966*a*; DiGirolamo et al. 1977; Hay and Scotti 1986). The attachment of the virus to mucus is primarily ionic and involves the binding of viral particles to sulfate radicals on the mucopolysaccharide moiety of the shellfish mucus (DiGirolamo et al. 1977). Bedford et al. (1978) have hypothesized that surface properties are the principal factor governing the accumulation of viruses by filter-feeding marine bivalves. They noted that the rate of accumulation of the lipoprotein-enveloped SFV was less than that for the protein-coated reovirus. Duff (1967) concluded that the more rapid uptake of poliovirus over coxsackie A virus in mussels may be due to differences in surface charge that affects adsorption to mucus. Considerable variation in uptake and elimination of two phages by hard-shell clams also indicates the potential influence of viral charge and stability on the kinetics of uptake and elimination (Canzonier 1971).

After adsorption to shellfish mucus, viruses enter the digestive tract. While most viruses remain in the digestive tract, some diffuse to other tissues. DiGirolamo et al. (1975) observed that the virus titer in the body of contaminated oysters steadily increased to 16% after 48 h, indicating diffusion of viruses from the digestive region to other body tissues. Some virus removal is done through the phagocytic action of the hemolymph. Fries and Tripp (1970) have demonstrated that shellfish leukocytes can phagocytize a 60 nm DNA virus within 2 h of exposure. Phagocytes may transfer viruses to tissues far removed from the normal elimination process, wherein they can remain for long periods of time (Canzonier 1971). Some viruses may become sequestered in digestive tract tissue and become refractory to the mechanisms responsible for elimination (Canzonier 1971; Cook and Ellender 1986; Hay and Scotti 1986).

Shellfish are able to rid themselves of at least some of the accumulated viruses when placed in clean waters. By stimulating natural feeding, the shellfish ingest food and eliminate viruses and other microbes in feces and pseudofeces during the normal digestion and excretion process. This is the predominant mechanism occurring during viral depuration and relaying, and it is entirely dependent on the successful stimulation of feeding (Metcalf 1987). A second mechanism of viral elimination may be through physical inactivation (Canzonier 1971).

Virus depuration studies report widely varying rates and extents of viral elimination and in some cases frequent persistence of viruses even after lengthy periods of depuration (Hedstrom and Lyke 1964; Mitchell 1966; Seraichekas et al. 1968; Canzonier 1971; DiGirolamo et al. 1975; Scotti et al. 1983; Sobsey et al. 1988; Power and Collins 1989). Canzonier (1971) has proposed two phases of elimination of viruses by shellfish. In the first, free or solids-associated viruses in the digestive tract are excreted in feces and pseudofeces by stimulated feeding. Elimination of these particles should be rapid and complete. However, the small numbers of viral particles sequestered in other shellfish tissues are less easily dislodged and their removal is independent of feeding activity. Their reduction by depuration is predicted to be much less efficient and may account for viral persistence. Hay and Scotti (1986) presented supporting evidence when they found that, while most labelled virus was removed effectively by depuration, a small amount was still present after 64 h. These viruses appeared to be intracellularly absorbed and associated with connective and epithelial cells surrounding the digestive tract.

VIRUS ELIMINATION BY DEPURATION AND RELAYING

Depuration and relaying are self-purification methods that rely on the ability of bivalve mollusks to eliminate contaminating microorganisms from their digestive tracts through normal feeding, digestion, and excretion. In relaying, shellfish harvested from contaminated waters are transferred to clean waters, where they filter-feed for a determined period of time. This process is subject to natural environmental fluctuations. Depuration is a controlled purification which is accomplished by placing contaminated shellfish in tanks containing purified flowing seawater used either once-through or recirculated. The shellfish are maintained in these tanks for a few days. Because the environmental conditions during depuration are controlled and therefore more readily optimized, this process is usually given 2–3 days. However, relaying in natural waters involves more variable and potentially suboptimal environmental conditions. Hence, greater amounts of time are used, usually two or more weeks (Larkin and Hunt 1982; Richards 1988*a*).

Studies have been done to determine the effectiveness of virus removal by depuration and relaying, and several reviews are available (Gerba and Goyal 1978; Richards 1988*a*; and DeLeon and Gerba 1990). Studies on virus elimination or persistence have been reported for oysters (*C. virginica*, *C. gigas*, *O. edulis*, *O. lurida*) (Hedstrom and Lyke 1964; Mitchell et al. 1966; Metcalf and Stiles 1968; Hamblet et al. 1969; Hoff and Becker 1969; Metcalf, Slanetz, and Bantley 1973; DiGirolamo et al. 1975; Gerba and Goyal 1978; Sobsey et al. 1980*b*, 1988; Scotti et al. 1983; Cook and Ellender 1986), clams (*M. mercenaria*, *M. campechiensis*, *M. arenaria*, *T. japonica*) (Liu, Seraichekas, and Murphy 1967*a*, *b*; Seraichekas et al. 1968; Hoff and Becker 1969; Canzonier 1971; DiGirolamo et al. 1975; Gerba and Goyal 1978; Metcalf et al. 1979; Docs and Noss 1989), and mussels (*M. edulis*) (Crovari 1958; Power and Collins 1989). Most studies have used poliovirus strains, although various coliphages, coxsackieviruses, echoviruses, other picornaviruses, and HAV also have been used (Liu, Seraichekas, and Murphy 1967*a*; Metcalf and Stiles 1968; Canzonier 1971; Metcalf, Slanetz, and Bartley 1973; Gerba and Goyal 1978; Sobsey et al. 1980*b*, 1988; Scotti et al. 1983; Docs and Noss 1989).

Maximum rates of viral elimination occur during the first 24–48 h of depuration, regardless of initial level of contamination (Hedstrom and Lyke

1964; Mitchell et al. 1966; Hamblet et al. 1969; Seraichekas et al. 1969; Hoff and Becker 1969; Metcalf et al. 1979; Docs and Noss 1989). In Eastern oysters, 90–95% of polioviruses were eliminated within 6–8 h of depuration and 99.9% were eliminated within 24 h (Mitchell et al. 1966; Hamblet et al. 1969). Slightly slower rates for clams were reported by Seraichekas et al. (1969), with more than 90% reduction of their viruses after 48 h of depuration. Similarly, Sobsey et al. (1980*b*) found that poliovirus reduction in oysters was rapid and extensive, with less than 3% of the virus remaining after 3 days.

After the first 28–48 h, virus depletion rates are much slower (Hoff and Becker 1969; Metcalf et al. 1979). Some experiments report reduction of viruses to nondetectable levels after depuration for 48–96 h (Liu, Seraichekas, and Murphy 1967*a*; Hamblet et al. 1969; Hoff and Becker 1969). However, other reports document incomplete virus removal (Hedstrom and Lyke 1964; Mitchell et al. 1966; Seraichekas et al. 1968; Canzonier 1971; DiGirolamo et al. 1975; Metcalf et al. 1979, 1980; Scotti et al. 1983; Sobsey et al. 1988; Power and Collins 1989). Early studies reported viral persistence after depuration for 100 h (Hedstrom and Lyke 1964; Liu, Seraichekas, and Murphy 1967*a*; Hoff and Becker 1969). More recently, two studies demonstrated incomplete virus removal after 6 days of depuration (Canzonier 1971; Metcalf et al. 1979, 1980). Scotti et al. (1983) reported the greatest degree of persistence with incomplete depuration of contaminated oysters after 10 days.

Depuration occurs more rapidly and efficiently in a free flow versus stationary or static system (Mitchell et al. 1966; Liu, Seraichekas, and Murphy 1967*b*; DiGirolamo et al. 1975). When DiGirolamo et al. (1975) depurated West Coast oysters in both static and free-flow systems, the free-flow system yielded a 99% reduction in poliovirus titer after 72 h, whereas up to 21% of the virus remained after 120 h in the static system. Mitchell et al. (1966) suggested that a static system limits some essential features for shellfish activity including dissolved oxygen, food, and dilution of metabolic waste. Virus elimination is dependent on active feeding, and the free-flow system better simulates the natural environment, which promotes increased physiological vigor and hence, faster depuration rates (Mitchell et al. 1966).

As for virus uptake, several factors affect the efficiency of depuration, including type of shellfish, individual shellfish variability, level of contamination, virus type, temperature, salinity, turbidity, and particle association.

Depuration rates are affected by the type of shellfish. For example, under identical conditions, the Eastern oyster can eliminate poliovirus to undetectable levels in 24 h, while 48 h is required for the soft-shell calm (Hoff and Becker 1969). However, variability in the depuration efficiency of individual shellfish may be of greater importance. It has been observed that during the latter phases of depuration (48–72 h), while the majority of shellfish were free of viruses, a few still harbored low virus levels (Seraichekas et al. 1968; Metcalf et al. 1979, 1980). Scotti et al. (1983) found a 0–100-fold range in depuration levels for virus-contaminated Pacific oysters when assaying individual shellfish. Differences in individual depuration rates may reflect variations in viral uptake efficiency or physiological inactivity of the shellfish (Seraichekas et al. 1968; Scotti et al. 1983). Metcalf et al. (1980*c*) concluded that depuration and relaying of shellfish should reduce microbial contamination but cannot assure that all shellfish will be virus-free. Limited survey data on depurated and relayed shellfish have shown a general absence of enteric viruses (Eyles and Davey 1984; Cook and Ellender 1986). However, depurated shellfish have been epidemiologically linked to disease (Bostock et al. 1979; Grohmann et al. 1981; Richards 1985).

Depuration efficiency is highly dependent on the initial contamination level of the shellfish. Less time is required for virus elimination from lightly contaminated shellfish (Liu, Seraichekas, and Murphy 1967*b*; Metcalf and Stiles 1968; Canzonier 1971; Metcalf et al. 1979; Cook and Ellender 1986). Liu, Seraichekas, and Murphy (1967*b*) found that hard-shell clams contaminated with 10^2 pfu of polioviruses per gram of meat required 24 h to eliminate to nondetectable levels, whereas 72 h were required at 10^3 pfu contamination levels. No apparent difference in rate of elimination was seen. Mesquita (1988) exposed mussels to a coliphage at low water titers for long times and high titers for short times. Depuration patterns for both contamination schemes were similar; however, exposure to high titers for short times showed more rapid phage elimination. Both treatments, however, resulted in a prolonged retention of a portion of the initial phages.

Elimination rates of viruses in shellfish differ with virus type, which may be due to differences in virus surface charges. Viruses with a strong positive charge may attach more effectively to the negatively charged sulfate radical of shellfish mucus and, therefore, depurate more slowly (Duff 1967; DiGirolamo et al. 1975). Power and Collins (1989) reported that both the relative rates and patterns of elimination of coliphage A1-5a and poliovirus from depurated mussels differed. They suggested that comparability in sizes and shapes of these viruses indicates that elimination may not be a function of these characteristics alone. Sobsey et al. (1987, 1988) compared the depuration of HAV and poliovirus in experimentally contaminated oysters. HAV reduction was slow with up to 18% viral persistence after five days. Poliovirus persistence was less than 1% after five days. Other investigators have concurred that poliovirus depuration proceeds quickly with 80–99% removal in 48 h (Liu, Seraichekas, and Murphy 1967*a*; Hoff and Becker 1969; Metcalf et al. 1979; Davis 1986). After the initial decline, however, low levels of poliovirus may persist for up to six days (Hoff and Becker 1969). Because of virus-specific differences in depuration patterns, investigators caution the use of model viruses for the prediction of overall viral depuration efficiency in shellfish (Sobsey et al. 1988; Power and Collins 1989).

The efficiency of viral depuration is influenced by seawater temperature, with lower temperatures resulting in a slower rate of shellfish feeding and, hence, slower virus removal (Cook and Ellender 1986). There is a threshold temperature below which no depuration occurs, presumably due to shellfish inactivity. This temperature is estimated to be 7–10°C (Metcalf and Stiles 1968; Cook and Ellender 1986). Metcalf and Stiles (1968) found that oysters in water of less than 5°C showed virtually no depuration after four months. Liu, Seraichekas, and Murphy (1967*a*) found that complete depuration took 48 h at 20°C but 96 h at 5°C. This temperature effect may help explain why most outbreaks of shellfish-related hepatitis A occur in winter and spring. If shellfish are contaminated in the late fall at temperatures optimal for uptake and this is followed by a sudden temperature drop to 6°C or less, feeding will stop and viral contamination will persist until water temperatures rise (Liu, Seraichekas, and Murphy 1966*b*; Metcalf and Stiles 1968).

Above threshold temperatures, depuration rates may also be temperature-dependent, with more rapid depuration occurring at higher temperatures. Meinhold (1982) showed poliovirus depuration by Eastern oysters was greatest at 28°C. Sobsey et al. (1987, 1988) demonstrated the rate of poliovirus elimination in *C. virginica* was more efficient at 23°C than at 12°C or 17°C. However, HAV elimination by the same species was not extensive regardless of the temperature (Sobsey et al. 1987, 1988).

Dramatic changes in salinity affect depuration. Liu, Seraichekas, and Murphy (1967*b*) found that a 50–60% reduction in salinity stopped the process, while a 7.5% reduction did not significantly affect depuration rates. Sobsey et al. (1987) reported that at salinities of 8–28 ppt, there were no significant differences in poliovirus elimination by Eastern oysters. However, oysters depurated HAV more efficiently at salinities of 28 ppt than at 8 or 18 ppt (Sobsey et al. 1987).

Turbidity and particle association do not appear to appreciably affect depuration as long as the concentrations do not interfere with normal metabolic function. Hamblet et al. (1969) reported that oysters contaminated with poliovirus depurated equally well in low (8–21 ppm) and high (54–80 ppm) turbidity waters. Meinhold (1982) concluded that while particle association was a factor in viral uptake, it is not important for elimination. However, Hoff and Becker (1969) reported that crude cell-associated viruses persisted longer in Manila clams and Olympia oysters than did free, filtered viruses.

Other factors have been proposed as affecting viral elimination rates from shellfish but none have proven significant. The presence or absence of food (algae) had no effect on HAV or poliovirus elimination rates in oysters

(Sobsey et al. 1987). Power and Collins (1990) found that food availability affected depuration rates only in filtered seawater, in which case elimination was faster. There was an effect from spawning on poliovirus and coliphage elimination by mussels, but the investigators considered it to be negligible (Power and Collins 1989). Hedstrom and Lyke (1964) demonstrated a failure to disinfect viruses in oyster tissue by chlorination of overlaying waters.

The effectiveness of relaying depends on many of the same factors as depuration. Cook and Ellender (1986) found that the rate at which microorganisms are reduced is slower in relayed than in depurated oysters. This is attributable to natural fluctuations in temperature and salinity, differential concentration of enteric viruses, mechanical disturbance of the animals, and perhaps other environmental factors. Elimination proceeded faster at higher water temperatures, up to 31°C. At temperatures below 10°C, oyster activity and enteric microbe elimination were slowed (Cook and Ellender 1986). Sobsey et al. (1980*b*) studied the extent of enterovirus elimination in relayed oysters and found that during winter and spring months, virus elimination exceeded 99.9% in 30 days. Most rapid rates occurred at water temperatures of 20–25°C. During the summer and fall months, viral persistence up to 20% was observed after 30 days. Cook and Ellender (1986) reported that oysters with high levels of initial contamination frequently failed to eliminate all viruses after 45 days of relaying. Salinity of the water had little effect, provided it was sufficient to maintain metabolic function. Oysters under severe physiological stress failed to eliminate viruses as quickly as healthy oysters (Cook and Ellender 1986).

The relationship of viral elimination to bacterial elimination in shellfish has been investigated. While one study has reported similar patterns and rates of removal for viruses and bacteria (Hoff and Becker 1969), most studies found differences in reduction rates (Hedstrom and Lyke 1964; Canzonier 1971; Sobsey et al. 1980*b*; Scotti et al. 1983; Cook and Ellender 1986; Burkhardt et al. 1988; Power and Collins 1989, 1990). Generally, viral elimination rates are slower than those observed for bacteria. Canzonier (1971) found that elimination of *E. coli* from clams proceeded much more rapidly than did phage S-13. Sobsey et al. (1980) found that elimination of *E. coli* and enterococci in relayed oysters was much more extensive than was elimination of enteroviruses. Scotti et al. (1983) concluded that viral and bacterial depuration rates were unrelated for *C. gigas*, with much lower rates for the viruses.

Power and Collins (1989, 1990) reported that not only the relative rates but also the patterns of elimination of *E. coli*, coliphages, and polioviruses during depuration of mussels were significantly different. The mechanisms effecting elimination may vary from one microorganism to another and may depend on phenomenon such as differential tissue adsorption, interstitial sequestering, and spontaneous/hematocyte-induced inactivation (Canzonier 1971; Scotti et al. 1983; Power and Collins 1989, 1990). It is generally agreed that the lack of correlation between viral and bacterial reduction in shellfish means that bacterial elimination is not a reliable index of viral elimination during depuration and relaying (Canzonier 1971; Sobsey et al. 1980*b*; Scotti et al. 1983; Cook and Ellender 1986).

VIRAL PERSISTENCE IN PROCESSING AND STORAGE

Transmission of human viral disease by consumption of cooked shellfish has been documented. Epidemiological evidence clearly implicates incompletely cooked mussels as the cause of a small HAV outbreak (Dienstag et al. 1976). Boiled cockles have been responsible for various viral gastroenteritis and HAV outbreaks in Europe (Appleton and Pereira 1977; Bostock et al. 1979). A large outbreak in England in 1976 prompted implementation of double cooking which was still ineffective in preventing disease (Millard, Appleton, and Parry 1987). In the United States, viral gastroenteritis has been epidemiologically linked to the consumption of steamed and baked clams and oysters (Morse et al. 1986; Truman et al. 1987). A significant association between steamed-clam ingestion and non-epidemic hepatitis A has also been noted (Koff et al. 1967).

Bivalve mollusks present a particular risk because heat treatments are usually minimal in order to avoid excessive heating that results in unacceptable organoleptic changes. However, enteric viruses are quite heat-resistant. Limited numbers of transmission experiments in human volunteers have shown that HAV can withstand heating at 56°C for 30 min (Havens 1945). Koff and Sear (1967) observed that clam shells open within 1 min of steaming and at an internal temperature of 70°C or less. This heat treatment was insufficient to inactivate all infectious viruses. Peterson et al. (1978) reported that thermal treatment conditions of 140°F for 19 min failed to completely inactivate HAV in artificially contaminated oyster homogenates.

DiGirolamo et al. (1970) investigated the fate of poliovirus in contaminated oysters that were heat processed by stewing, frying, baking, and steaming. While different inactivation kinetics were noted for the four treatments, virus survival was 7–13% in all cases. Most impressive was a 7% survival of poliovirus after 30 min of steaming which resulted in a completely unpalatable product. The investigators demonstrated inadequate heat penetration of oyster tissues. Processing conditions of 70°C for 30 min were recommended to assure complete inactivation of poliovirus in shellfish.

In contrast to DiGirolamo et al. (1970), Millard, Appleton, and Parry (1987) found that heat inactivation of poliovirus and HAV in cockles processed by boiling or steaming provided adequate heat penetration, with internal temperatures reaching 91.5–94°C within 2 min. Boiling or steaming at 95°C resulted in complete inactivation of bioaccumulated HAV and poliovirus within 2 and 3 min, respectively. The authors recommended an internal temperature of 85–90° be maintained for 1 min to inactivate viruses in shellfish.

The discrepancies in study results may be due to variability in shellfish species, size, time of harvest, contamination level, initial temperature, and other factors (DeLeon and Gerba 1990). Even greater processing variability is expected at the consumer level, and therefore, it is difficult to recommend a standard heat treatment that will inactivate all enteric viruses in contaminated shellfish. Additionally, because increased heat treatment frequently results in undesirable organoleptic changes, consumer acceptance of such recommendations is questionable.

Shellfish other than bivalves may not pose as great a risk from enteric viruses because lengthy cooking is typical. When contaminated blue crabs (*Callinectes sapidus*) were boiled, 99.9% of poliovirus 1, rotavirus SA-11, and echovirus 1 were inactivated within 8 min (Hejkal and Gerba 1980).

Gamma irradiation has been found inadequate as a means of inactivating viruses in shellfish. Poliovirus-contaminated oysters exposed to high doses (400 krads) of gamma radiation demonstrated virus survival of 7–13% (DiGirolamo et al. 1972*a*). This dose also resulted in unacceptable organoleptic changes to the product. Further studies are needed to better define the ability of gamma irradiation to inactivate viruses in shellfish.

Studies have indicated that viruses can survive in shellfish for long periods at low temperatures. In the natural environment, Metcalf and Stiles (1967) investigated virus survival in the American oyster using three model enteroviruses. No reduction in virus titers was observed for 30 days when water temperatures were 1°C. Only when water temperatures rose above 8°C were the viral titers rapidly reduced. The same authors found that contaminated oysters held in water of 7°C or less retained coxsackie B virus and echovirus for up to four months without appreciable loss in titer (Metcalf and Stiles 1968). Canzonier (1971) reported that coliphage-contaminated clams retained 50–90% of the virus when water temperatures were 8–16°C. Viruses also can survive in shellfish for long periods of dry, cold storage. Metcalf and Stiles (1965) found that Coxsackie B3 virus persisted in the digestive gland and alimentary tract of contaminated oysters for up to 28 days at 5°C with no significant loss in titer. DiGirolamo et al. (1970) found that poliovirus 1 could survive for 30–90 days in whole and shucked Pacific and Olympia oysters held at 5°C. Reductions were 10%, 60%, and 87% after 5, 15, and 30 days of refrigerated storage, respectively. It was postulated that the tendency of the viruses to aggregate and their incorporation by ionic bonding to shellfish mucus may be the mechanisms responsible for viral persistence in chilled shellfish (DiGirolamo et al. 1970).

Tierney et al. (1982) investigated virus survival in refrigerated oysters at more realistic contamination levels than those used by DiGirolamo et al. (1970). Mean inocula were 3–3100 pfu of poliovirus per oyster. Generally, there was no reduction in virus numbers over 28 days of storage at 5°C, so the authors concluded that enteroviruses present in oysters at the time of harvest probably will persist until purchased and consumed.

Survival of viruses in edible West Coast crabs (*Cancer magister* and *Cancer antennarius*) stored at 8°C has been noted. After 120 h of refrigerated storage, 29–40% of the coliphage model virus still remained, despite extensive organoleptic decomposition of the product (DiGirolamo and Daley 1973).

DiGirolamo et al. (1970) reported that poliovirus was very stable in whole Pacific oysters held frozen (− 17.5°C) for up to 12 weeks. Loss of virus titer was gradual during the storage period; 91% of the virus was recoverable after two weeks, 40% after 6 weeks, and 10% after 12 weeks. Studies of viral persistence in frozen West Coast crabs using coliphage T4 as a model showed a gradual decrease in virus titer over 30 days at − 20°C. Survival rates after 20 and 30 days of frozen storage were 42–55% and 17–35%, respectively (DiGirolamo and Daley 1973).

METHODS FOR DETECTING ENTERIC VIRUSES IN SHELLFISH

A variety of methods have been investigated for recovering and concentrating enteric viruses from shellfish. Two general schemes have proven successful, designated as extraction–concentration and adsorption–elution–concentration. The goal of both approaches is to separate viruses from shellfish meats, provide a low-volume aqueous solution that is free of cytotoxic material, and recover most of the viruses present in the shellfish sample. An ideal method would be simple, fast, inexpensive, able to quantitatively recover all virus types, and be applicable to different shellfish species. None of the methods currently available fulfills all of these criteria.

Extraction procedures employ conditions facilitating viral separation from shellfish tissues. This is primarily done through the use of specialized reagents and adjustments to pH. The tissues are then removed by filtration and/or centrifugation. The resulting supernatants are frequently of high volume, turbid, and toxic to tissue culture. Consequently, they are further treated to remove interfering substances (clarification) and concentrate viruses (concentration). Tested treatments include polyelectrolyte flocculation, acid precipitation, ultracentrifugation, ultrafiltration, ether extraction, and hydroextraction. However, the poor quality of the supernatant samples to which these treatments are applied can compromise virus recoveries. Many of the extraction methods utilize several consecutive clarification and concentration steps in an effort to produce final extracts that can readily be assayed in tissue culture.

Alternative virus extraction procedures developed in the 1970s were adsorption–elution–concentration methods. Shellfish are initially homogenized under conditions promoting viral adsorption to tissues, usually by pH adjustment. The tissues are recovered by centrifugation and the adsorbed viruses eluted by suspension in solutions of appropriate salt concentration and pH. A second centrifugation removes the shellfish solids and provides a supernatant that can be further clarified and concentrated by acid precipitation, polyelectrolyte flocculation, ultrafiltration, ultracentrifugation, or freeze–thaw followed by centrifugation.

EXTRACTION AND EXTRACTION–CONCENTRATION METHODS

The early methods for recovering enteric viruses from shellfish were developed for clams, oysters, and mussels to investigate viral occurrence, uptake, and elimination. Shellfish samples were usually pooled (3–12 individual specimens) and homogenized with or without the addition of diluent. Appropriate diluents included peptone (Hoff and Becker 1969), phosphate-buffered water (Mitchell et al. 1967; Liu, Seraichekas, and Murphy 1967*a*), Hank's balanced salt solution (HBSS) (Liu, Seraichekas, and Murphy 1966*a, b*), and nutrient broth (Liu, Seraichekas, and Murphy 1967*a*; Hamblet et al. 1969). Some investigators did not use diluents (Metcalf and Stiles 1965; Bendinelli and Ruschi 1969). The homogenate was centrifuged at low speed to remove solids. The supernatant fluid was then assayed directly (Liu, Seraichekas, and Murphy 1966*a, b*) or further clarified by centrifugation (Liu, Seraichekas, and Murphy 1967*a*), ethyl ether extraction (Metcalf and Stiles 1965; Mitchell et al. 1966; Hamblet et al. 1969; Bendinelli and Ruschi 1969), or fluorocarbon extraction (Mitchell et al. 1966; Duff 1967). Duff (1967) modified the general procedure by homogenizing mussel pools in chilled fluorocarbon and performing only one centrifugation step. Metcalf and Stiles (1965) and Bendinelli and Ruschi (1969) further concentrated ether-extracted samples by ultracentrifugation, which resulted in a virus-containing pellet that could be resuspended in a small volume of appropriate diluent.

These methods have shortcomings which make their routine use in virus recovery from shellfish impractical. Furthermore, there is little quantitative evidence to indicate their recovery efficiencies. Large sample volumes preclude virus assay of the entire extract, and cytoxic components are frequently present in the final extracts. While Mitchell et al. (1966) reported no cytotoxicity for oyster extracts, Duff (1967) showed that undiluted mussel extracts were toxic to tissue culture, even at 1:10 dilutions. Homogenates obtained from clams were particularly cytotoxic (Konowalchuk and Speirs 1972).

Herrmann and Cliver (1968) presented a study evaluating the efficiency of early extraction procedures. Experimentally inoculated hard-shell clams were blended in glycine–NaOH (GN) buffer, magnesium chloride, agamma calf serum, and Freon TF. The addition of bentonite was optional. The homogenates were centrifuged and the supernatant was clarified by a second extraction with Freon TF. Extracts were further concentrated by ultracentrifugation in the presence of polyethylene glycol (PEG). The pellet was resuspended in 0.5 ml of phosphate-buffered saline (PBS) with 2% agamma chicken serum. Using coxsackievirus A9 as the model, this procedure gave losses of less than 25% at input levels exceeding 8×10^6 pfu/sample. The 50% sensitivity endpoint was in the range 3–4 pfu/25 g sample. Substantial cytotoxicity was removed by successive clarification and concentration.

Konowalchuk and Speirs (1972) investigated alternative clarification and concentration schemes in an effort to reduce cytotoxicity of oyster, mussel, and clam extracts. Homogenized, centrifuged extracts from oysters were the least toxic, and relative toxicities of mussels and clams were 75- and 200-fold, respectively. Ether treatment of extracts yielded a twofold reduction in cytotoxicity. Acidification of extracts to a pH of 3.0–3.5 resulted in precipitation of cytotoxic components which were then removed by centrifugation. Clam and oyster extracts treated by acidification had 1/125th and 1/8th, respectively, the toxicity of the original extract. Recoveries ranged from 50% to 60% of input virus, with virus loss being primarily associated with solids removal in the first centrifugation. Ultracentrifugation was used to further concentrate samples, resulting in a fourfold decrease in volume but a twofold increase in cytotoxicity.

Kostenbader and Cliver (1972) developed a method for extracting viruses from oysters that incorporated a polyelectrolyte typically used to flocculate solids in water and wastewater. Cat-Floc T (Calgon Corp, Pittsburgh, PA), a polydimethyldiallyl ammonium chloride, promotes cohesion of oyster solids and aids separation of viruses. It was the only polyelectrolyte sewage flocculant effective for shellfish clarification (Kostenbader and Cliver 1981). The procedure involves extraction of viruses in minced shellfish in the presence of GN buffer and 1% polyelectrolyte. After allowing sufficient time for flocculation, the solution is filtered successfully through cheesecloth, a glass fiber prefilter, and a 0.45 μm membrane filter. Concentration to a 5 ml volume is carried out by ultracentrifugation or ultrafiltration. Recoveries of 10–100% were obtained at virus input levels of 1.0×10^4, 3.0×10^3, and 20 pfu/sample. Further work gave recoveries of 95–100% at input levels of

100–1.0×10^4 pfu/sample, and 37–75% at input of only 1 pfu/sample (Kostenbader and Cliver 1973). Oysters inoculated with high levels of reovirus (> 10,000 pfu/g) gave 88% recoveries (Kostenbader and Cliver 1981). The species of oyster, type of enterovirus, and magnitude of initial virus inoculum had no effect on the efficiency of recovery (Kostenbader and Cliver 1972). Oyster and clam extracts concentrated by ultrafiltration produced significant cytotoxicity. This was counteracted by the addition of concentrated Eagle's medium and serum (CES) which converted the extract to a matrix that is 1 × Eagle's minimum essential medium (MEM) and 2% serum. Despite this treatment, concentrated clam extracts produced severe cytotoxicity that hindered interpretation of results (Kostenbader and Cliver 1973).

Watson et al. (1980) used a modification of the polyelectrolyte flocculation/filtration method to extract naturally occurring viruses from mussels. The volume of GN buffer and the concentration of Cat-Floc are increased, and a centrifugation step is substituted for the initial filtration through cheesecloth.

Sullivan and Peeler (1982) combined Freon® extraction, sonication, and polyelectrolyte flocculation to develop virus extraction procedures for individual oysters. The oyster is homogenized in a mixture of cold NaCl, nonfat dry milk, $MgCl_2$, and Freon TF, pH 9.0. The homogenate is shaken, sonicated, and centrifuged. The supernatant is then treated with activated carbon and Cat-Floc, mixed, and centrifuged to remove the floc. The resulting supernatant is assayed directly or filtered to remove residual cytotoxicity. When oyster samples were inoculated with low levels of virus, average recovery was 59%, with a 95% confidence interval of 52–66%.

Another extraction modification combines the polyelectrolyte flocculation and filtration of Kostenbader and Cliver (1972) with an acid precipitation step to concentrate viruses (Vaughn et al. 1979*a*; Landry, Vaughn, and Vicale 1980). These investigators found that acid precipitation alone resulted in significant virus loss to supernatant fluids. However, supplementation of clarified extracts with 6% beef extract resulted in only 3–6% virus loss to supernatant. One-hundred gram oyster samples are homogenized in GN buffer and Cat-Floc and stirred to facilitate flocculation. The homogenate is centrifuged to remove solids and the resulting supernatant is filtered. The filtrate is supplemented with 6% beef extract and adjusted to pH 3.5 to promote acid precipitation. The virus-containing precipitate is collected by centrifugation, dissolved in phosphate buffer, and adjusted to neutral pH. When compared to the adsorption–elution–concentration method of Sobsey et al. (1978), this procedure yielded higher poliovirus recoveries from oyster homogenates.

Richards et al. (1982) altered the procedure of Vaughn et al. (1979*a*) and Landry, Vaughn, and Vicale (1980) to optimize poliovirus recovery from artificially contaminated oysters. They recommended increasing the volume of buffer used in homogenization, adjusting homogenate pH to 9.5 prior to Cat-Floc addition, and elimination of the filtration step. When oyster homogenates were seeded with 3500–5840 pfu poliovirus per sample, recoveries ranged from 93% to 120%.

Richards and Goldmintz (1982) also evaluated the use of cross-flow filtration for recovery of poliovirus from oyster homogenates. The Acroflux system consists of a capsule containing a pleated 0.2 m acrylic polymer membrane filter of large surface area which is continuously rinsed by cross-flow action. Oyster extracts are recirculated through the Acroflux capsule which provides separation of the aqueous virus-containing buffer phase from the lipid phase. Mean virus recoveries were 39.1–46.1%.

Metcalf, Eckerson, and Moulton (1980 *a, b*) developed a single virus extraction procedure that was applicable to oysters, hard- and soft-shell clams. It represents compromises in methods which permit reasonably effective recoveries of several enteric viruses from all three shellfish species. This complex procedure entails homogenization of 100 g pools in phosphate-buffered saline (PBS) and 3% beef extract followed by pH adjustment to 8.0–9.5 and sonication. Sonication improved the release of virus. After centrifugation, viruses are concentrated by acid precipitation of the supernatant at pH 4.0 and low conductivity. The precipitate is recovered by centrifugation, resuspended in PBS, and treated with Cat-Floc to remove cytotoxic components. After another centrifugation, the remaining supernatant is concentrated by acid precipitation in the presence of 3% beef extract. The final precipitate is resuspended in a minimal volume of PBS, adjusted to neutral pH, and treated with antibiotics. The resulting shellfish extracts had virtually no cytotoxicity to tissue culture. At virus inputs of less than 100 pfu/sample, recoveries of 47–65%, 50–93%, and 40–80% were obtained for oysters, hard- and soft-shell clams, respectively. Average efficiency was 63%. Recoveries were dependent on virus type, generally exceeding 50% for the enteroviruses polio 2, echo 7, and coxsackie B1 in all species. Reovirus recoveries were less impressive, with a lower detection limit estimated as one of every three viruses present.

Tierney et al. (1980) reported a virus extraction procedure for oysters and clams that was a modification of existing methods for ground beef. The glass-wool hydroextraction method involves homogenization under conditions promoting virus separation from shellfish solids, followed by filtration to remove solids and hydroextraction to concentrate viruses. One-hundred gram shellfish samples are homogenized in Tricine buffer supplemented with $MgCl_2$ and DEAE-dextran, followed by pH adjustment to 9.0. The homogenate is stirred for 1 h at 37°C to monodisperse and liquify fat globules. Filtration through glass wool removes shellfish solids and the virus-containing filtrate is collected in a dialysis bag. Concentration is achieved by hydroextraction against PEG for 12–18 h. The concentrated sample is then treated with diatomaceous earth and centrifuged. Simultaneous Freon extraction is employed if cytotoxicity is suspected. The resulting supernatant is treated with antibiotics and diluted with Tricine prior to assay. Some cytotoxicity was noted even after Freon extraction and dilution. When evaluated in a comparative study of virus recovery methods, the glass-wool hydroextraction procedure detected all but one of the eight test viruses. Quantitative recoveries of approximately 50% were attained at inoculum levels of less than 1 pfu/g shellfish (Larkin and Metcalf 1980). This procedure has not been widely accepted due to significant sample toxicity and the cumbersome nature of the hydroextraction step.

Sullivan (1981) reported an alternate concentration scheme for the glass wool procedure which involves acid precipitation and resuspension of the precipitate in MEM-based medium, followed by Freon® extraction and sonication. After centrifuging to separate the Freon®, the virus-containing supernatant is detoxified using activated carbon followed by recentrifugation and assay of the final supernatant. When oysters were contaminated with an average of 42 pfu of coxsackievirus B2/100 g, recoveries ranged from 14% to 100% and averaged 23 pfu/sample (55%).

More recently, Sullivan (1983) has reported extraction procedures that can be used on large shellfish samples (100 g) and yield a low-volume final eluent (4 ml) which can be assayed in a single 150 cm tissue culture monolayer. The oyster sample is blended at pH 9.0 in a cold solution of Eagle's MEM supplemented with Hanks salt (MEMH), nonfat dry milk, $MgCl_2$, and Freon TF. The homogenate is centrifuged and the virus-containing supernatant diluted to less than 0.35% salts followed by acid precipitation at pH 4.8. The precipitate is harvested by centrifugation and eluted at pH 7.3 in 16 ml of cold 2×MEMH, 5% Cat-Floc, and $NaHCO_3$, followed by recentrifugation. The precipitation and elution are repeated a second time at a lower volume of 4 ml. When the procedure was tested on 35 oyster samples each containing 25 pfu of poliovirus 1, average viral recovery was 84%, with a 95% confidence interval of 75–93%. Later modifications of this procedure utilized PBS rather than MEMH, supplemented the initial acid precipitation with nonfat dry milk, and used only one precipitation–elution scheme to reduce cytotoxicity (Sullivan, Peeler, and Larkin 1985). When tested on oysters and clams at input levels of 35–79 pfu poliovirus per sample, this modification gave recoveries of 35–38%.

ADSORPTION–ELUTION–CONCENTRATION METHODS

The development of adsorption–elution–concentration schemes for the extraction of viruses from shellfish evolved from the observation that enteroviruses can be efficiently adsorbed to a variety of diverse materials by controlling pH and ionic conditions. Then, the adsorbed viruses can be eluted by changing pH and/or ionic concentrations. These principles have been widely used to concentrate viruses from water and wastewater (Sobsey, Wallis, and Melnick 1975). Sobsey, Wallis, and Melnick (1975) found that the association of poliovirus with oyster solids was dependent on pH and salt concentration. At pH 5.5 and salt concentration less than 1500 mg NaCl/l, 99% of the added polioviruses were found associated with oyster tissues. Adsorption of a model reovirus and adenovirus to oyster tissues also approached 99% at pH 5.0–5.5 (Sobsey, Carrick, and Jensen 1978; Carrick 1980). Similar results were obtained for all three viruses in hard-shell clams (Howard 1980). Finance et al. (1981) suggested that adsorption was also dependent on the type of shellfish and exchange surface. Johnson, Cooper, and Straube (1981) reported that a 62% optimal poliovirus adsorption to Japanese cockle tissue occurred at pH 4.5. A study on freshwater clams also found that maximum viral adsorption (99.97%) homogenates occurred at pH 4.5 (DeLeon et al. 1986).

Sobsey, Wallis, and Melnick (1975) evaluated the effect of pH and conductivity on virus elution from oyster solids and found that poliovirus readily elutes in a 0.05 M glycine–saline solution at pH 3.5 and 7.5. When elution efficiencies for poliovirus, reovirus, and adenovirus were compared, optimal elution always occurred at pH 7.5. However, elution efficiency was virus-dependent, with optima of 98–99%, 61–65%, and 84% for poliovirus, reovirus, and adenovirus, respectively (Sobsey, Carrick, and Jensen 1978; Carrick 1976). Howard (1980) reported similar results for hard-shell clams. Elution rates have been found to vary with shellfish species. Finance et al. (1981) reported that optimal poliovirus elution from the surface of mussel tissue occurred at the higher pH of 9.0. Johnson, Cooper, and Straube (1981) reported efficient elution at pH 9.5 for *T. japonica* contaminated with poliovirus; loss in pH 7.5 glycine–saline approached 89%. DeLeon et al. (1986) also used high pH (9.5) to successfully elute poliovirus from freshwater clam tissue with 83% efficiency.

The first efforts at concentration after adsorption and elution involved ultrafiltration of oyster extracts. Virus-containing supernatants of 150 ml volumes were filtered through 0.45 μm membrane filters followed by ultrafiltration to approximately 4 ml. Virus recovery in filtrates averaged 90% (Sobsey, Wallis, and Melnick 1975), the loss primarily occurring during the prefiltration step (Carrick 1980). Concentration of filtered supernatants by acid precipitation at pH 4.5 gave efficient recoveries of poliovirus, reovirus, and adenovirus, with no more than 5% loss to supernatant fluids (Sobsey, Carrick, and Jensen 1978; Carrick 1980). Howard (1980) reported similar results for hard-shell clams. When comparing acid precipitation and ultrafiltration concentration of oyster and clam extracts, there were no significant differences in final virus recoveries (Sobsey, Carrick, and Jensen 1978; Carrick 1980; Howard 1980). Finance et al. (1981) compared ultracentrifugation and ultrafiltration concentration methods and found them to be equally efficient with recovery rates exceeding 92%. Most investigators prefer acid precipitation to alternate procedures due to its ease and speed.

The principal mechanism proposed for virus adsorption to and elution from shellfish tissue involves electrostatic interactions between the virus, the adsorbent material, and the ions present in the solution (Carrick 1980). These electrostatic forces can be manipulated by adjustment of pH and salt concentration. The protein outer surfaces of many nonenveloped viruses such as enteroviruses exhibit an isoelectric point between 4.0 and 6.0. At this pH, viral adsorption to noncharged surfaces, such as shellfish tissue, would occur. During elution, a higher pH and the addition of negatively charged chloride ions cause the virus surface to become increasingly electronegative, which along with mechanical agitation leads to elution or repulsion of the virus from the shellfish tissue components. Electrostatic interactions may account for virus-specific differences in adsorption and elution efficiencies. However, the role of other factors, including hydrophobic interactions and differences in shellfish composition, is suspected since species differences in efficiencies have also been noted.

In an early adsorption–elution–concentration scheme, viral adsorption to oyster tissue is promoted by homogenization in distilled water at pH 5.5 and salt concentration less than 1500 mg/l (Sobsey, Wallis, and Melnick 1975). After centrifugation, the sedimented solids are washed in 0.05 M glycine of pH 5.0 and low conductivity, followed by a second centrifugation. The viruses are eluted from the precipitate by suspension in 150 ml of 0.05 M glycine-buffered saline and subsequent pH adjustment to 3.5. Centrifugation separates the shellfish solids and the resulting supernatant is clarified by filtration and further concentrated by ultrafiltration. The ultrafiltered concentrates are of low volume (4 ml) and nontoxic to tissue culture. Recovery experiments using oysters seeded with four different model enteroviruses yielded efficiencies of 24–97% at virus input levels of 34–4790 pfu/sample. Average recovery efficiency was 63%. Recovery was not dependent on virus type or inoculum level.

Modifications to the adsorption–elution–concentration scheme provided for recovery of different types of enteric viruses. The major changes included elimination of the washing step, establishment of an optimal diluent–tissue ratio of 1:7 for elution, and an increase in the elution pH from 3.5 to 7.5 (Carrick 1976; Sobsey, Carrick, and Jensen 1978). Acid precipitation at pH 4.5 is an effective alternative to concentration by ultrafiltration (Sobsey, Carrick, and Jensen 1978; Jensen 1980). Generally, acid precipitates can be resuspended in a small volume of phosphate buffer with subsequent pH adjustment to neutrality. The addition of 0.1% Cat-Floc to the extract results in precipitation of cytotoxic components, which are removed by centrifugation. The final supernatant has significantly less cytotoxicity, allowing for improved virus recoveries in cell culture (Wait, Kimball, and Sobsey 1981). Howard (1980) found this method to be effective in the recovery of enteric viruses from hard-shell clams, the only modification being initial virus adsorption at the lower pH of 4.5.

The modified adsorption–elution–concentration procedure of Sobsey et al. (1978) entails homogenization of 100 g samples of shellfish in 700 ml distilled water at pH 5.0 (oysters) or 4.5 (clams) and low salt concentration (< 2000 mg/l). After centrifugation, the solids-adsorbed viruses are eluted by resuspension in 700 ml of 0.05 M glycine–0.14 M saline with subsequent pH adjustment to 7.5. The shellfish solids are removed by centrifugation and filtration and the viruses in the filtrate are concentrated by acid precipitation at pH 4.5. A third centrifugation yields a virus-containing sediment that is resuspended in a small volume of phosphate buffer and adjusted to pH 7.5. Removal of additional cytotoxicity is achieved by treatment with 0.1% Cat-Floc, centrifugation, and recovery of the final supernatant. A recent review by Sobsey (1987) suggests the removal of the filtration step and the routine use of polyelectrolyte flocculation. When this method was tested on oysters experimentally inoculated with three model enteroviruses at 15–472 pfu/sample, recovery efficiencies ranged from 16% to 76%, with an average of 48% (Carrick 1976; Sobsey, Carrick, and Jensen 1978). For experimentally contaminated clams, recovery efficiencies averaged 63% at virus inputs of 85–1416 pfu/sample (Howard 1980). Recoveries were not affected by virus type or input level. Wait et al. (1983) used this method to recover naturally occurring enteric viruses from clams harvested off the North Carolina coast.

Ellender et al. (1980) further modified the Sobsey method for examination of naturally contaminated oyster samples. The changes include elimination of the intermediate filtration step and substitution of the phosphate-buffered diluent with glycine–saline for resuspension of the final acid precipitate. The resuspended precipitate is then centrifuged at high speed to remove extraneous solids. After an overnight freeze–thaw treatment, centrifugation is repeated, followed by membrane filtration of the supernatant prior to assay. When tested with seeded oyster samples, recoveries ranged from 61 to 73%. Baker (1978) also reported that successive freeze–thaw followed by centrifugation and filtration is an effective means by which to remove extraneous solids from shellfish extracts.

Finance et al. (1981) used a modification of the adsorption–elution–precipitation scheme to recover poliovirus from shellfish experimentally contaminated by submerging in a virus-containing solution. Using this method, the surface-bound viruses are eluted by submerging in demineralized water at pH 9.0 and high conductivity. Concentration is achieved by ultrafiltration or ultracentrifugation. When tested using small quantities of poliovirus (1–50 pfu/sample), recoveries approached 100%. However, the routine applicability of this protocol is questionable since most viruses present in shellfish are located in digestive tract tissues rather than surface-bound.

Johnson, Cooper, and Straube (1981) developed an adsorption–elution–concentration method to extract viruses from the Japanese cockle, *T. japonica*. Adsorption at pH 4.5 and low salt concentration is followed by centrifugation to recover shellfish solids. These are resuspended in glycine–saline buffer of pH 9.5, rather than the lower pH of 7.5 suggested by other investigators. Kostenbader and Cliver (1972) also used this higher pH to elute polioviruses from oyster tissue. Addition of a 0.5% Cat-Floc solution enables elution and removal of cytotoxicity to be done simultaneously. After centrifugation, the virus-containing supernatant is concentrated by acid precipitation at pH 4.0 in the presence of 6% beef extract. Another centrifugation yields a precipitate that is resuspended in a small volume of phosphate buffer, adjusted to physiological pH, and assayed for viruses. Recoveries averaged 73% and ranged from 42 to 104% when this method was applied to poliovirus in clam homogenates at levels of 90–660 pfu/100 g. When tested on clams contaminated by uptake from water, coxsackie B2 virus was recovered at a concentration of 26 pfu/100 g sample.

DeLeon et al. (1986) used a similar procedure to recover poliovirus from experimentally contaminated freshwater clams. In this case, adsorption of virus to clam homogenates is optimal at pH 4.5 in a 0.09 M glycine solution. After centrifugation, the viruses are eluted from clam tissue in a glycine–saline solution supplemented with 3% beef extract at pH 9.5. No concentration procedure was investigated. Freshwater clam extracts were not toxic to tissue cultures and an overall recovery of 83% was noted at input levels of 10^5–10^6 pfu/sample.

Sullivan et al. (1984) described a modified adsorption–elution–concentration scheme utilizing a composite of published techniques. These investigators found that viral adsorption and elution are temperature-dependent processes, with most efficient adsorption at 37°C and elution at 0°C. In the first step, 100 g oyster samples are processed by homogenizing in 150 ml of warm water at pH 4.8, followed by centrifugation to recover virus-adsorbed solids. The viruses are eluted from the pellet at pH 9.0 in a mixture of cold Eagle's MEM, nonfat dried milk, $MgCl_2$, and Freon TF and centrifuged again. The supernatant fluids are diluted in water, precipitated at pH 4.8, and centrifuged again to obtain the pellet, which is resuspended in a small volume (4–10 ml) of phosphate buffer. Cat-Floc is added to precipitate residual cytotoxicity and after a final centrifugation, the virus-containing supernatant is assayed. Extract volumes were low and not cytotoxic. When 30 oyster samples were experimentally seeded with approximately 95 pfu of poliovirus per sample, the average recovery was 55.4%, with a 95% confidence interval of 53.3–57.5%. This method was further evaluated in a collaborative study using six different samples and five independent laboratories. Recoveries varied from 35% to 55% at inoculum levels of 30–100 pfu/100 g; the coefficient of variation ranged from 34% to 50%. The limit of detectability was estimated to be 14 pfu/100 g sample when recovery was at 35% (Tierney et al. 1985). This procedure is recommended by the U. S. Food and Drug Administration (FDA) and appears in the FDAs Bacteriological Analytical Manual (Sullivan et al. 1984).

COMPARISON OF METHODS

Kostenbader and Cliver (1972) compared recoveries using the Freon extraction method of Herrmann and Cliver (1968) and their own polyelectrolyte flocculation and filtration procedure. The polyelectrolyte flocculation/filtration yielded 90–98% recovery as compared to 86–90% recoveries for Freon-extracted samples.

Vaughn et al. (1979*a*) and Landry et al. (1980) compared the relative efficiencies of the polyelectrolyte flocculation and filtration method of Kostenbader and Cliver (1972) against the adsorption–elution–concentration method of Sobsey, Carrick, and Jensen (1978). They reported two to five times more virus recovery in Cat-Floc-treated samples. When extraction was followed by acid precipitation in the presence of 6% beef extract, efficiencies were 66.5% for adsorption–elution–precipitation methods and 99.6% for polyelectrolyte flocculation–filtration.

Larkin and Metcalf (1980) conducted a cooperative study in which ten investigators analyzed oysters, hard- and soft-shell clams experimentally contaminated with polioviruses 1 and 2, coxsackievirus B3, echovirus 7, and/or echovirus 17. Homogenates were prepared in a single lab, split, and shipped to participants who applied methods of their choice. Variations in recovery efficiency ranged from 10% to 90% and depended on the virus type, shellfish species, and method. Polioviruses were consistently recovered by all methods; however, most investigators had difficulty isolating echoviruses. The glass-wool hydroextraction procedure (Tierney et al. 1980) and modifications of the adsorption–elution–precipitation procedure (Sobsey, Wallis, and Melnick 1975; Sobsey, Carrick, and Jensen 1978) detected the lowest levels (< 1 pfu/g) of virus in shellfish homogenates. Hard- and soft-shell clam extracts demonstrated considerable cytotoxicity. The investigators concluded that the methods evaluated were limited to detection of only some of the enteroviruses and in many cases to contamination levels exceeding 1 pfu/g.

Six techniques were evaluated by Cole et al. (1986) in the largest comparative study to date. Methods included modifications of those described by Ellender et al. (1980), Sobsey, Wallis, and Melnick (1975), Sobsey, Carrick, and Jensen (1978), and Richards et al. (1982). Three additional protocols were evaluated. The first of these was developed at Nicholls State University and was based on the procedure of Ellender et al. (1980) followed by filtration through electropositive filters for adsorption of negatively charged viruses. The viruses were eluted from the filters in a small volume of glycine–saline buffer supplemented with beef extract prior to assay. The second protocol was an adsorption–elution–precipitation procedure modified by the addition of Cat-Floc during the elution step and beef-extract supplementation before acid precipitation (Sobsey 1982). The last method involved a series of elution–precipitation steps using glycine buffer at high pH (9.5–11.5) and low pH (5.5) followed by freeze–thaw and Cat-Floc treatment to remove cytotoxic agents (Cole et al. 1986, personal communication).

The best recoveries were obtained using the methods of Ellender et al. (1980) and Sobsey (1982). A statistical model was developed which evaluated the six methods according to recovery rate, final extract volume, degree of cytotoxicity, and bacterial contamination. The model demonstrated that the method of Ellender et al. (1980) was superior for routine field use in the detection of enteric viruses in Louisiana oysters.

Most reported methods have been evaluated using polioviruses, coxsackieviruses, and echoviruses as models. Recent improvements in cell culture techniques have facilitated the evaluation of HAV recovery in shellfish samples. Two adsorption–elution–precipitation methods have been evaluated in oysters experimentally inoculated with wild type HM-175 extracted from primate fecal material (Sobsey 1985; Wait and Sobsey 1986). The first method was that described by Sobsey, Wallis, and Melnick (1975), Sobsey, Carrick, and Jensen (1978) supplemented with Cat-Floc treatment after acid precipitation. The second method was that recommended by the U.S. FDA (Sullivan et al. 1984). In trials with mean input of eight radioimmunofocus units (rfu) per 50 g sample, the Sobsey method recovered 48% of the seeded HAV, while the FDA procedure yielded a 22% recovery efficiency. At higher HAV input levels, the Sobsey method was always significantly more efficient at HAV recovery. When oysters were contaminated with high levels (10^3–10^6 rfu/sample) of cell-grown HAV, mean recoveries were 46%, 6%, and 2.1% for the Sobsey, FDA, and Richards et al. (1982) methods, respectively (Keating 1985; Keating and Sobsey 1985). Clearly, the Sobsey method was superior to the other methods tested for the recovery of HAV from oysters.

Results reported by Millard et al. (1987) on the evaluation of the methods of Sobsey et al. (1975, 1978), Johnson, Cooper, and Straube (1981), and Richards et al. (1982) for the recovery of HAV from experimentally contaminated cockles conflict with those reported by Sobsey and colleagues. In this case, the Sobsey method proved inconsistent and frequently failed to recover the virus. The investigators concluded that the method of Richards and colleagues proved most satisfactory for the recovery of HAV from cockles.

The recovery of rotaviruses from oysters was evaluated using two nonhuman agents as models (Werner and Sobsey 1983). The adsorption–elution–precipitation method followed by polyelectrolyte flocculation demonstrated average recoveries of 11% and 36% for Nebraska calf diarrhea virus (NCDV) and Simian rotavirus SA-11, respectively. Modifications to the protocol improved NCDV recovery to 32%, but wide variability remained.

RECOMMENDATIONS ON RECOVERY AND CONCENTRATION METHODS

Collaborative studies have yielded conflicting results and no one procedure is clearly superior in the recovery of enteric viruses from shellfish. Recovery efficiencies show wide variations and may be affected by shellfish species, virus type, and level of contamination. Methods have been described for oysters (*C. virginica*, *C. gigas*, and *O. edulis*) (Konowalchuk and Speirs 1972; Kostenbader and Cliver 1972, 1973; Sobsey, Wallis, and Melnick 1975, Sobsey, Carrick, and Jensen 1978; Vaughn et al. 1979*a*; Landry, Vaughn, and Vicale 1980; Ellender et al. 1980; Metcalf et al. 1980*a*, *b*; Tierney et al. 1980; Larkin and Metcalf 1980; Finance et al. 1981; Sullivan 1981, 1985; Richards et al. 1982; Sullivan and Peeler 1982; Sullivan et al. 1983; Sullivan, Peeler, and Larkin 1985; Tierney et al. 1985), clams (*M. mercenaria*, *M. arenaria*, *Corbicula fluminea*) (Herrmann and Cliver 1968; Konowalchuk and Speirs 1972; Kostenbader and Cliver 1973; Sobsey, Carrick, and Jensen 1978; Metcalf et al. 1980*a*, *b*; Tierney et al. 1980; Larkin and Metcalf 1980; Johnson, Cooper, and Straube 1981; Sullivan 1985; DeLeon et al. 1986), mussels (*M. edulis*) (Konowalchuk and Speirs 1972; Finance et al. 1981), and cockles (*T. japonica*) (Johnson, Cooper, and Straube 1981).

Cliver et al. (1983) argued that it is most reasonable to select a test method based on the shellfish species to be analyzed since most methods are not widely applicable to other species. They recommended three general methods as described by Landry et al. (1980) (Method A), Metcalf et al. (1980) (Method B), and Johnson, Cooper, and Straube (1981) (Method C). In testing *C. virginica*, *C. gigas*, or *O. edulis*, Method A is recommended. Method B is suggested for *M. mercenaria* and *M. edulis*, while Method C should be chosen for *T. japonica*. Alternative protocols were also suggested.

Sobsey (1982, 1984, 1987) has extensively reviewed the methods for isolation of enteric viruses from shellfish. Early recommendations included three adsorption–elution–concentration methods as described by Sobsey, Wallis, and Melnick (1975), Sobsey, Carrick, and Jensen (1978), Ellender et al. (1980), and Johnson, Cooper, and Straube (1981). Two extraction–concentration procedures were also recommended: those reported by Landry et al. (1980) and Metcalf et al. (1980). The glass-wool hydroextraction method of Tierney et al. (1980) was also suggested (Sobsey 1982, 1984). Recommended extraction–concentration methods include those of Vaughn et al. (1979), Richards et al. (1982), and Landry et al. (1984). The suggested adsorption–elution–concentration methods are those described by Sobsey et al. (1978) with modifications by Wait et al. (1981) and Sullivan, Peeler, and Larkin (1985). These methods are recommended because they are relatively simple, require minimal laboratory equipment and supplies, and have been used successfully to detect naturally occurring enteric viruses in field samples. The effectiveness of adsorption–elution–concentration Scheme B has been verified by collaborative testing (Tierney et al. 1985). Additionally, Scheme A adsorption–elution–concentration has demonstrated efficient recovery of HAV from contaminated oysters (Keating and Sobsey 1985; Sobsey 1985; Wait and Sobsey 1986).

METHODS FOR ISOLATION, ASSAY AND IDENTIFICATION OF ENTERIC VIRUSES

Virus assay of concentrates from shellfish, water, and sediments is typically based on the infectivity of viruses for susceptible, live laboratory hosts. A complete description of the assay procedures and the necessary equipment and materials for the recovery of viruses from environmental and shellfish concentrates is beyond the scope of this paper. Reviews and detailed instruction are provided by Lennette and Schmidt (1979), Gerba and Goyal (1982), Sobsey (1982, 1984, 1987), Smith and Gerba (1982), Cliver et al. (1983), Berg et al. (1984), Greenberg et al. (1985), Guttman-Bass (1987), and Richards (1987). This section will address some important considerations in adapting standard virological procedures to the analysis of concentrates from shellfish, water, and sediments. It will conclude with a brief description of some rapid and accurate viral diagnostic procedures being developed for use with environmental samples.

Sample Processing and Storage

Shellfish samples should be processed soon after collection, with recommendations that they be held no more than 2 h at 25°C or 48 h at 2–10°C. If processing cannot be done within this time frame, samples should be frozen at < -70°C. Final concentrates that have been treated with antibiotics may be stored under the same conditions with long-term storage recommended at $< -70°C$.

Decontamination of Concentrated Samples

Bacterial and mold contamination are frequently found in environmental concentrates and this can be controlled by the addition of antibiotics. Kanamycin–gentamycin, penicillin–streptomycin, nystatin B, and amphotericin B have been successfully used to control the growth of contaminating microorganisms. Antibiotic solutions are added immediately after extraction, followed by incubation at room temperature for 2 h or 37°C for 1 h. Subsequent freezing aids in the destruction of bacterial contaminants. The effectiveness of the treatment can be checked by spread-plating a small subsample of the extract on a general purpose agar and observing bacterial and mold growth after 24–48 h at 37°C. If contamination persists, chloroform or Freon extraction of the entire sample concentrate can be performed. Filtration is not recommended because of the potential for loss of solids-associated viruses.

Host Systems for Virus Isolation and Assay

The major host systems for detecting enteric virus infectivity are mammalian cell cultures of primate origin. Selection of cell culture type depends primarily on the viruses one wishes to detect, and the shellfish species matters little. Unfortunately, there is no single universal host system that will detect all or even a majority of the more than 100 different enteric viruses that can potentially contaminate shellfish and the estuarine environment. Some of the epidemiologically most important viruses such as Norwalk agent, hepatitis A virus, astroviruses, and human rotaviruses are difficult or impossible to recover using any of the currently available conventional host systems.

Primary or secondary human embryonic kidney and monkey kidney cell cultures are preferred host systems for many enteric viruses. However, few laboratories have the facilities and financial resources to maintain and utilize these host systems. Several convenient and sensitive alternative host systems have been suggested. Buffalo green monkey kidney (BGM), a continuous cell line derived from African green monkey kidney cells, is sensitive to many enteroviruses and reoviruses and is frequently used as a substitute for primary cells in the isolation of viruses from environmental concentrates (Dahling 1974). MA-104, a continuous rhesus monkey kidney-derived cell line, is sensitive to rotaviruses as well as many enteroviruses and, thus, provides complementary selectivity in virus isolation (Oglesbee, Wait, and

Meinhold 1981; Smith and Gerba 1982). The RD cell line is a continuous line of human rhabdomyosarcoma cells that has been shown to support the replication of certain group A coxsackieviruses as well as a wide variety of other enteroviruses (Schmidt et al. 1978; Morris and Waite 1980). Other cell lines useful for virus propagation and assay include BSC-1 monkey kidney cells, LLC-MK2 rhesus monkey kidney cells, MFK fetal rhesus monkey kidney cells, Vero African green monkey kidney cells, and HEp and HeLa human malignant epithelial cells (Smith and Gerba 1982). Numerous studies have compared the relative sensitivities of these cell lines in the isolation of enteric viruses from environmental samples, often with conflicting results (Schmidt et al. 1978; Schmidt, Ho, and Lennette 1976; Morris and Waite 1980; Watson, Inglis, and Anderson 1980; Oglesbee et al. 1981; Ridinger et al. 1982; Agbalika et al. 1984). Investigators agree that the use of at least two different host systems should enable the recovery of a wide spectrum of enteric viruses from shellfish, water, and sediment.

Assay Methods

There are two basic procedures for recovering and assaying viruses in cell culture: quantal methods (TCID 50 or MPN) and enumerative methods (plaque assay). Quantal assays rely on the development of cytopathic effects (CPE) in replicate cell cultures inoculated with different volumes or dilutions of environmental samples. The cultures are observed microscopically for CPE for up to 4 weeks and scored as positive or negative. The virus titer is estimated from the resulting dose–response data. The plaque assay is more precise than the quantal assay because relatively large numbers of individual infectious units are counted directly as discrete, localized areas of infection (plaques) in an agar-containing cell culture. Plaque assays can be done using cells in monolayer or cells suspended in agar. Early research recommended the quantal assay over the plaque assay, noting that it was more sensitive for detecting very low multiplicities of virus and was able to detect slow-growing and nonplaquing strains (Herrmann and Cliver 1968; Nupen 1970; Hill 1971; Schmidt et al. 1978). However, other evidence suggests that the agar cell suspension plaque assay is more efficient at virus recovery from environmental samples than is either the plaque assay in monolayer culture or the quantal assay in liquid culture (Morris and Waite 1980; Rao et al. 1984).

A potential problem of quantal assays is the inoculation of two or more viruses into the same cell culture. If one virus replicates much faster than others, the other viruses may not grow sufficiently to be detected. If two or more viruses grow, the culture may be scored as a single positive, causing an underestimate of the virus titer. Successful growth of two or more viruses in the quantal assay requires separation of the individual virus types by limiting dilution or subculture from discrete plaques to permit serological identification.

Sample Toxicity

A major problem in quantitative virus assay of concentrates from shellfish, water, and sediment samples is toxicity of the concentrates to tissue culture. Shellfish extracts are of particular concern, since they frequently contain dissolved and particulate shellfish tissue components. Although clarification procedures are intended to separate viruses from toxic matrix components, the treatment is not always effective. In quantal assays, sample toxicity may destroy all or most of the cells in the culture so that no virus isolations are obtained. In plaque assays, small particulate toxic material becomes localized by plaquing overlays, producing necrotic areas resembling virus plaques and constituting "false positives." Leong, Barrett, and Trussell (1978) reported that 43% of observed plaques on BGM cells were false positives. Fannin et al. (1978) were able to confirm less than 30% of the plaques on four different cell lines. In this case, the false positive rate was influenced by type of sample and host cell system. For this reason, all CPE-positive cultures must be confirmed as virus-positive by further passage into the cell culture. It is also recommended that all or some representative portion of plaques be confirmed by further passage.

A variety of measures can be taken in order to control cytotoxicity. Both the inoculum volume per cell culture and the adsorption time on monolayers may influence the degree of toxicity. These variables may be adjusted depending upon the level of toxicity of the sample. In addition, draining the inoculum from the monolayer after the adsorption period followed by rinsing with buffered saline, balanced salt solution, or cell culture medium may be helpful in controlling sample toxicity (Schmidt et al. 1978; Howard 1980).

Hejkal, Gerba, and Rao (1982) evaluated methods to reduce cytotoxicity of environmental concentrates and found filtration through positively charged depth filters reduced toxicity 1.2–4.4-fold without appreciable loss of virus. High-speed centrifugation (8500–35,000 × g for 1 h) was also effective. Richards and Weinheimer (1985) found that some brands of skim milk and beef extract can cause cytotoxicity as well as reduction in viral titer and plaque diameter. In this case, dialysis of samples prior to assay successfully removed cytotoxic components.

It is advisable to do a preliminary assay for viruses on a small volume of each concentrated sample in order to evaluate sample cytotoxicity and microbial contamination. A few replicate cell cultures are inoculated with different volumes of sample and incubated. The cultures are then evaluated for cell survival and quality, frequency of virus recovery, and approximate virus titer. Using the results of the preliminary assay, the remaining sample can be treated to reduce further cytotoxicity or eliminate microbial contamination. Furthermore, appropriate inoculum volumes can be chosen to minimize cytotoxicity and optimize viral quantitation.

Inoculation, Adsorption and Conditions

The most common inoculation procedure for monolayer cell cultures involves addition of a portion of the sample concentrate onto the monolayers of a series of cell cultures drained of their growth or maintenance medium. The inoculum is left on the monolayers for a period of one to several hours to promote viral adsorption. During this time, the inoculum is redistributed continuously by mechanical rocking or periodically (15–20 min intervals) by manual rocking. After the adsorption period, the inoculum may be removed and the cells washed, although this step is optional. Cells are then overlaid with liquid (quantal assay) or agar-containing (plaque assay) medium, incubated, and examined for virus-specific CPE or plaques.

Inoculation and adsorption conditions can significantly influence the efficiency of virus recovery from environmental concentrates. Important factors include inoculum volume, adsorption time, rocking conditions, and exposure conditions. Metcalf, Moulton, and Eckerson (1980*a*) investigated the inoculation and exposure conditions for shellfish concentrates in tissue culture. They found that optimal virus recoveries were obtained using inoculum levels of 1 ml or less per 45 cm^2 cell culture layer and an adsorption time of 2 h. These experiments were conducted under static conditions. The rate and extent of enterovirus adsorption was enhanced by continuous rocking (Richards et al. 1990; Richards and Weinheimer 1985). Tierney et al. (1980) found that comparable virus recoveries were obtained using 5 ml of inoculum when the adsorption period was increased from 1 to 5 h.

When more than one type of cell culture is employed, it is typical to divide samples into several equal portions and then inoculate small volumes of these portions into a series of cell cultures of each cell type. A disadvantage to this approach is that inoculated portions are exposed to only one cell type. If viruses in the inoculum will not grow in that cell type, they will not be detected. This common practice has been compared to the strategy of sequential addition of inocula representing the entire sample on two or more cell culture systems. Sequential adsorption results in more efficient recovery of viruses. Benton and Hurst (1989) found that plaque counts resulting from sequential adsorption were greater in 67% of the cases for four enterovirus stock cultures. Metcalf, Moulton, and Eckerson (1980*a*) found a recovery effectiveness of 88% for sequential adsorption versus 54% for the split sample method when testing shellfish extracts. Tierney et al. (1980) also noted that sequential adsorption may result in reduced cytotoxicity, since toxicity was apparently removed by the first cell layer while leaving the second cell layer intact.

Virus Recovery and Identification

Inoculated cultures are incubated and periodically examined for CPE or plaques for 1–3 weeks or until the cells begin to degenerate with age. Material from CPE-positive cultures or plaques is confirmed by further passage into fresh cell cultures. Material from CPE-negative cultures may be further passaged in a process called blind passage, which occasionally yields additional virus isolation. Confirmed virus isolates can be identified using conventional serological techniques (Lennette and Schmidt 1979; Berg et al. 1984; Greenberg et al. 1985).

Rapid and Sensitive Methods

Many of the viruses responsible for gastroenteritis and associated with environmental contamination do not grow appreciably in cell culture. Of particular concern is the hepatitis A virus. Cell culture-adapted strains have been developed by serial passage in tissue culture, but these strains generally do not produce notable CPE (Provost and Hilleman 1979; Daemer et al. 1981; Kojima et al. 1981; Gust et al. 1985). Quantitation of adapted strains of HAV can be done by immunofluorescence and radioimmunofocus assays (RIFA). The RIFA offers improved quantitation of HAV in cell culture by immune autoradiographic detection of foci (Lemon, Binn, and Marchwicki 1983). Sobsey and colleagues (Sobsey 1985; Sobsey et al. 1985; Sobsey, Oglesbee, and Wait 1985) has reported detection of cell culture-adapted HAV strain HM-175 in seeded water and oyster samples using this technique. More recently, Cromeans et al. (1987) have reported isolation of a rapidly replicating isolate of HAV strain HM-175 from persistently infected, serially passed cell cultures which induces a cytopathic effect and could be detected by conventional plaque assay. While the altered strain facilitates further investigation of detection methods for hepatitis A virus in environmental samples and model system studies on HAV behavior, it may not adequately model the behavior or provide for the detection of wild-type HAV strains.

Investigators have used a variety of substances to pretreat cell layers in an effort to enhance viral infectivity and improve recoveries from environmental concentrates. DeLeon and Gerba (1986) found that pretreatment of BGM monolayers with the proteolytic enzymes trypsin and ficin resulted in a threefold to fourfold increase in detection sensitivity for both laboratory strains and natural isolates of enteroviruses. Sobsey et al. (1983) found a twofold enhancement of hepatitis A virus infectivity in AGMK cells pretreated with pancreatin or trypsin. Nonidet P40 in low concentration was also a potent enhancer of HAV infectivity in AGMK cells (Wait, Sobsey, and Lemon 1985). Laboratory strains of enteroviruses, as well as viruses isolated from raw wastewater, were found to exhibit a onefold to sixfold enhancement in infectivity when BGM cell monolayers were pretreated with the cationic polymer polyethyleneimine (PEI) (Preston et al. 1988, 1989, 1990). PEI-treated cells adsorbed viruses much more effectively than did untreated cells. The efficacy of treatment of four continuous cell lines with 5-iododeoxyuridine (IDU) appeared to be virus and cell line-dependent, enabling some viruses to form plaques on normally nonpermissive cells, but being moderately inhibitory to others. Overall, IDU pretreatment resulted in an approximate twofold increase in plaque titers (Benton and Ward 1982; Hurst and Benton 1986). The results for mixed cell cultures have been disappointing, with cell line and virus-dependent synergistic and inhibitory effects noted (Benton and Hurst 1986; Hurst and Benton 1986).

Payment and Trudel (1985, 1987) have developed an enzyme immunoassay method based on immunoperoxidase for detecting a broad spectrum of cultivable human enteric viruses. This procedure involves inoculation of environmental concentrates onto suitable cell lines followed by detection of virus growth by an indirect immunoperoxidase method using human immune serum globulin (HSIG) as the source of viral antibodies. HSIG is an excellent broad-spectrum source of antibodies to human enteric viruses. When tested on water samples, this method gave virus titers 100–250 times higher than did parallel quantal assay for CPE (Payment and Trudel 1985). When wastewaters were evaluated, the immunoperoxidase method gave 10–50 times higher virus recoveries than did quantal assay for CPE (Payment and Trudel 1987). Even greater sensitivity could be attained after repassage into the same cell line. The improved sensitivity of this assay is attributed to its ability to detect viruses which grow in cell culture but fail to produce CPE.

The above improvements still rely on the ability of enteric viruses to replicate in cell culture. A number of direct immunochemical methods have been developed for the detection of noncultivable viruses in feces, tissues, and sera. The virus concentrations in these biological media are high and the tests are based on the detection of viruses as antigens rather than infectious agents. These methods are too insensitive to detect the low levels of viruses in shellfish. Additionally, sera are expensive and not always readily available. Immunofluorescence methods have been reported for the detection of poliovirus in water (Katzenelson 1976) and reoviruses in sewage (Ridinger et al. 1982). The detection limit of the latter was 10^4 reoviruses per liter of raw sewage. Dow et al. (1986) and Loh et al. (1981) recently developed a nitrocellulose enzyme immunoassay that was able to detect poliovirus and rotaviruses SA-11 and WA in mollusk tissue concentrates. The detection limit was high, estimated at 10^3–10^4 pfu poliovirus per 50 μl of extract. Pietri et al. (1988) recently reported failure to detect hapatitis A virus in sewage-contaminated oysters and mussels using classic cell culture, immune electron microscopy, and radioimmunoassay procedures.

Nucleic acid techniques may provide improved sensitivity and specificity in the detection of viruses in environmental concentrates. Gene probes and other nucleic acid techniques have recently been used for the detection of enteric viruses in these samples. Gene probe hybridization to genomic viral RNA has been reported to detect 2.5×10^5 physical particles of rotavirus SA-11 (Dimitrov et al. 1985), 10^4 physical particles of hepatitis A virus (Jiang et al. 1986), and 500–1000 pfu of rotavirus SA-11 and enteroviruses (DeLeon 1989; Shieh et al. 1990). Margolin et al. (1986) have reported the detection of 1–10 pfu of poliovirus using probes with a specific activity of 2×10^9 CPM/g of DNA. In general, RNA probes are five to eightfold more sensitive than DNA probes (Jiang, Estes, and Metcalf 1987; Shieh et al. 1990; Dubrou et al. 1990). Although most reports on detection limits are 100-fold less than a desired goal of 1–10 infectious units, several investigators have reported detection of enteric viruses in environmental samples. Margolin, Hewlett, and Gerba (1986), using a cDNA probe to poliovirus, detected enteroviruses in environmental samples, and these samples also were enterovirus positive by modified cell culture assay (DeLeon and Gerba 1985). In a later study, Margolin et al. (1989) used poliovirus and hepatitis A gene probes to detect enteroviruses in groundwater. Jiang et al. (1986) have used cDNA probes to detect seeded hepatitis A in estuarine samples with a sensitivity of 10^4 particles and in a subsequent study they detected naturally occurring hepatitis A viruses in estuarine samples using ssRNA probes (Jiang, Estes, and Metcalf 1987).

All virus detections in environmental samples by gene probes thus far reported have been made with radioactively labeled probes. Nonradioactive detection systems are approximately ten time less sensitive (Jiang et al. 1987) and, thus, not directly applicable for detecting the low number of viruses expected in environmental samples. However, the sensitivity of gene probes in environmental samples can be increased by the use of other molecular techniques in conjunction with gene probe hybridization. Recent advances in nucleic acid detection by gene probes coupled with target nucleic acid amplification techniques such as polymerase chain reaction (PCR) offer great sensitivity and detection specificity of enteric viruses (Shieh et al. 1990) and other microbial agents in the environment. Polymerase chain reaction (PCR) provides a 10^5–10^6 enrichment of a specific nucleic acid sequence and is capable of detecting a single copy of DNA (Ou et al. 1988; Saiki et al. 1988). Although *Thermus aquaticus* (Taq) polymerase cannot use RNA as a template, the PCR technique is adaptable to the amplification of RNA viruses by a simple reverse transcriptase reaction prior to polymerase chain reaction (RT–PCR). Recently, this approach has been used for the amplification of enteroviruses (Hyypia, Auvinen, and Maaronen 1989; Rotbart 1990; Chapman et al. 1990; Shieh et al. 1990), hepatitis A virus (Jansen et al. 1990) and rotaviruses (Gouvea et al. 1990). Major obstacles to the application of RT-PCR for enteric virus detection in environmental samples are the presence of sample components that interfere with RT-PCR enzymes and the need for substantial reductions of sample volumes to less than 100 μl. However, progress has been made in the application of these methods to sample concentrates from water (Schwab, DeLeon, and Sobsey 1993) and oysters (Jaykus, DeLeon, and Sobsey 1993).

4

Human-Associated Bacterial Pathogens

Cameron R. Hackney

Department of Food Science and Technology,
Virginia Polytechnic Institute and State University,
Blacksburg, VA 24061

Morris E. Potter

Division of Bacterial Diseases,
Center for Infectious Diseases, Centers for Disease Control,
1600 Clifton Road, NE, Atlanta, GA 30333

INTRODUCTION

This section reviews bacteria that cause disease in humans and the relationship of the diseases they cause to shellfish consumption. Recent detailed reviews have been published on each of these pathogens. Perhaps, the most comprehensive is the book *Foodborne Bacterial Pathogens* edited by Doyle (1989). This book gives detailed information on each pathogen, the illness it causes, its isolation and identification, and its control. Other recent reviews are included in the references. The objective of this review is to relate bacterial pathogens to safety issues of molluscan shellfish. Therefore, information from annual reports from the Centers for Disease Control and the Northeast Technical Services Unit of the Food and Drug Administration is included in each discussion. In addition, information on the isolation from shellfish and possible illness as reported in the literature is presented.

SALMONELLA TYPHI

General Overview

Salmonella typhi, the causative agent of typhoid fever, was first discovered by Eberth in 1880, and, therefore, it was called *Eberthella typhosa* for a number of years. The current genus name of *Salmonella* was assigned in honor of the American bacteriologist D. E. Salmon. In the generalized infections, termed enteric fevers, the organisms are spread throughout the body, and much of the symptomatology results from the endotoxin released from the bacterial cells. While any species of *Salmonella* may cause enteric fever, these infections are primarily caused by *Salmonella typhi*, *S. paratyphi* A, *S. paratyphi* B (*S. paratyphia* B (*S. schottmulleri*), or *S. paratyphi* C); by far; the most common infection in the United States is caused by *S. typhi* (Burrows 1985). The specific disease caused by *S. typhi* infection is typhoid fever. *S. typhi* is most often transmitted by food or water contaminated by feces or urine from infected humans. The food may be contaminated by careless handling after cooking, by inclusion of contaminated ingredients, by contaminated water, followed by inadequate cooking, or by using contaminated utensils for cooking or serving. When typhoid is transmitted from carriers, it is generally due to poor personal hygiene.

Organism

The salmonellae group, including *Salmonella typhi*, are facultatively anaerobic gram-negative rods belonging to the family *Enterobacteriaceae*. They are usually motile by means of peritrichous flagella. Studies of the genus resulted in the establishment of an identification procedure known as the Kauffman–White scheme, which arranges the organisms into serovars, or types, based on the presence of combinations of antigens making up the surface of the cell, or the composition of the antigens which make up the flagella providing motility to the organisms (Burrows 1985). Currently, there are over 2000 serovars or types recognized, as a result of testing with this system. *S. typhi*, *S. paratyphi* A, *S. paratyphi* C, and a few other strains have an additional antigen termed Vi. This antigen is similar to the O or cell antigens, but it is much more susceptible to heat. The Vi antigen is generally assumed to be associated with virulence of the organism. Also, the specific antibody to this antigen is thought to be protective.

Illness

S. typhi causes an acute systemic disease, typhoid fever, characterized by insidious onset of sustained fever, headache, central nervous system signs, malaise, anorexia, splenomegaly, and rose spots on the trunk. Diarrhea is uncommon, but stools are generally culture-positive for 2–3 months and, in the case of chronic carriers, can be intermittently positive for years (Feldman 1980).

Because persons differ in their susceptibility to infectious diseases, it is difficult to establish a minimum infective dose for any pathogenic microorganism. The Safe Drinking Water Committee of the National Academy of Science in 1977 noted that Hornick and coworkers had determined that when 14 persons were tested with doses of 1000 *S. typhosa* cells, none became infected. But when 116 persons were tested with 10,000 cell doses, 32 became infected.

When a susceptible human comes into contact with an infectious dose of the pathogen, there usually follows an incubation period of at least one to two weeks. Following this period, the symptoms gradually appear, including fever, chills, headache, constipation (in about 50% of cases), or diarrhea (in about 20% of cases), slowed heart rate, and muscle pain and soreness. The fever continues to rise stepwise, reaching a maximum 7–10 days after onset. The spleen and liver become enlarged. Peak fever levels may remain for a week or two. The white blood cell count is generally lower than normal and, for a short time, a rash may appear as discreet, rounded rose spots on the trunk. These may persist for 2–5 days, then fade.

Typhoid fever is a generalized infection spreading from the entry of organisms through the gastrointestinal tract. Experimental studies have determined that the organisms penetrate the intestinal epithelium although this tissue is fairly rapidly cleared of bacteria as they move into the mesenteric lymphatics, where they multiply. The bacteria move into the blood stream from the lymph. Intestinal symptoms result from inflammatory lesions produced in the mesenteric lymph tissues (Peyer's patches).

Complications of typhoid infection include intestinal hemorrhage, intestinal perforation, necrosis of lymphoid tissues, hepatitis, liver necrosis, ulcerated larynx, infection of the bone and joints, and inflammation of the lungs, gall bladder peritoneal lining, and other organs. The fatality rate of typhoid fever is 10% compared to $< 1\%$ for most forms of salmonellosis.

Association with Shellfish

Typhoid fever is worldwide in distribution, but varies widely in incidence, seasonality, and the vehicles associated with infection. Animal reservoirs of *S. typhi* are rare and cases almost always originate from an infected human. The mode of transmission is generally through food or water contaminated by feces. The historical association between shellfish and typhoid fever was identified early in the nineteenth century (see Chapter 1). During the first quarter of the twentieth century, 22 such outbreaks were reported in the United States (Shear and Gottlieb 1980). In 1925, a large outbreak of typhoid fever, associated with consumption of raw oysters, occurred, resulting in a 10% fatality rate and the establishment of the first national program on shellfish sanitation (Shear and Gottlieb 1980). However, it has been a number of years since raw molluscs were associated with an outbreak of typhoid fever in the United States (Bryan 1980; Rippey and Verber 1988). Since 1973, only one seafood-associated outbreak, involving 25 cases, was reported to CDC, and shellfish were not implicated as the vehicle. The last shellfish-associated typhoid outbreak reported in the NETSU data base occurred in 1954 (Rippey and Verber 1988).

Molluscs may acquire *S. typhi* from sewage-polluted water. However, in the United States, seafood-related typhoid fever is unlikely because there are few cases or carriers to discharge *S. typhi* into sewage or estuaries. The risk associated with imported seafood may vary, depending upon the rates of infection in the country of origin and the degree of pollution of the harvest waters. Recent sporadic cases have been reported in consumers of raw shellfish in the Mediterranean area (Torne et al. 1988) and, based on the incidence of typhoid fever in local populations, the risk associated with products imported from emerging nations would appear to be elevated (Parker 1984).

In addition to the risk posed by contaminated harvest waters, *S. typhi* can also be introduced to seafood by carriers during shucking, processing, or preparation. If contaminated foods are temperature-abused for several hours, the typhoid bacilli can propagate to an infectious dose. Available data suggest that at least 10^5 *S. typhi* in foods are necessary to cause disease (Blaser and Newman 1982). Therefore, only shellfish from grossly polluted waters and ready-to-eat seafood contaminated during preparation and then temperature-abused sufficiently to permit bacterial growth are likely to transmit typhoid fever.

Nishio, Nakamori, and Miyazaki (1981) examined the survival of *S. typhi* in oysters. The oysters were artificially contaminated by placing 60–80 cultured oysters in a 40 l tank containing UV-sterilized seawater inoculated with *S. typhi*. The oysters were exposed to the inoculum for 4 h at 15°C, then transferred to a holding tank for up to nine days. At the beginning of the storage the counts of *S. typhi* were as high as log 5 per 1000 g of oyster meat. A four log reduction was noted after 24 h; however, the counts did not decrease further and counts of 20–80 per 100 g were maintained throughout the storage period. In addition, these workers showed that freezing did not significantly reduce the population in the oysters over a 140 day storage at −20°C.

Methods for Detection

The methodology for the isolation and identification of salmonellae is discussed in Chapter 5. In general, the methods of isolation of *S. typhi* are like those of other members of the genus. Traditional methodology is a five-step process of pre-enrichment, selective enrichment, streaking onto selective media, biochemical identification and, finally, serological identification. A number of selective media, including Salmonella–Shigella agar, Hektoen enteric agar, and xylose lysine deoxycholate agar, have been developed to recover and distinguish *Salmonella* from other *Enterobacteriaceae*. Bismuth sulfite agar is highly selective for *S. typhi* (Difco 1984).

SHIGELLA

General Overview

Shigellosis is caused by invasion of the intestinal mucosa by *Shigella dysenteriae*, *flexneri*, *boydii*, and *sonnei* (IFT 1988). The normal habitat for shigellae is the intestinal tract of humans and nonhuman primates and, rarely, other animals; therefore, like typhoid fever, shigellosis usually originates with a human excreter. Human volunteer studies indicate that ingestion of fewer than 100 shigellae can cause disease, which is characterized by diarrhea, abdominal pain, fever, and vomiting (Morris 1986). Foodborne shigellosis is usually characterized by a high attack rate and a short incubation period, probably dose-related phenomena (Bryan 1979). Most outbreaks of foodborne shigellosis result from the contamination of raw or previously cooked foods during preparation by a carrier with poor personal hygiene. In view of *Shigella*'s ability to survive on seafood, the risk of seafood serving as a vehicle of shigellosis should parallel that of other foods. However, seafood-related outbreaks of shigellosis have been restricted primarily to raw shellfish.

Shigellae are readily killed by heat and low pH. However, under certain conditions they can survive for some time outside the host (IFT 1988).

Organism

The genus *Shigella* is grouped within the family *Enterobacteriaceae*. Shigellae are rod-shaped, gram-negative, nonmotile, facultatively anaerobic, catalase-positive (except for one serovar of *S. dysenteriae*), oxidase-negative, lactose-negative (except for a few strains), citrate-negative and H_2S-negative. With a few exceptions, shigellae ferment sugars without producing gas. The mol% G + C of their DNA is 49–53.

Virulent strains of *Shigella* are invasive, i.e., they invade the epithelial cells of the intestinal mucosa. Invasiveness can be measured *in vivo* by observing the keratoconjunctivitis in the guinea pig cornea (Sereny test) and/or *in vitro* by invasion of HeLa cells. Virulent strains of *Shigella* spp. carry a 120–140 MDa (megadalton) plasmid that is necessary for the expression of invasiveness.

Illness

Symptoms of the illness include abdominal pain; cramps; diarrhea; blood, pus, or mucus in stools; fever; vomiting; and tenesmus. The onset time is 12–50 h. The infective dose is very low and can be as few as ten cells, depending upon the age and underlying health status of the host. The disease is caused by virulent *Shigella* organisms attaching to and penetrating epithelial cells of the intestinal mucosa. After invasion, they multiply within the cells, and spread to contiguous epithelial cells, resulting in tissue destruction.

Diarrhea is accompanied by fever; nausea and, sometimes, toxemia, vomiting, cramps, and tenesmus. Pathogenesis sometimes occurs in young children with shigellosis. The illness is usually self-limited, lasting an average of 4–7 days. Most cases result in stools that contain blood, mucus, and pus (i.e., dysentery) resulting from a confluence of microabscess caused by the invasive organisms. Watery diarrhea and vomiting due to enterotoxins elaborated by *Shigella* may also occur. Bacteremia is uncommon. Human volunteer studies indicate that ingestion of fewer than 100 shigellae can cause disease (Morris 1986).

ASSOCIATION WITH SHELLFISH

Sewage pollution has been associated with outbreaks of shigellosis from shellfish. The number of cases is limited. The organism does not survive well and illness is most often the result of contamination by a handler. Cantoni et al. (1985) reported an outbreak of shigellosis from mussels (*Mytilus galloprovincialis*). The outbreak occurred in Milan in 1978 and approximately 100 people were affected. Studies of the mussels revealed the presence of *S. dysenteriae* and *S. boydii*. Taylor and Nakamura (1964) reported that *S. sonnei* and *S. flexneri* could survive at 25°C in clams for more than 50 days and in oysters for more than 30 days.

As in the case for the salmonellae, three surveillance systems for shigellosis exist at CDC. For the years 1978–87, an average of seven foodborne outbreaks affecting a total of 573 persons were reported each year to the foodborne disease surveillance system (CDC 1989). Seven outbreaks involving 137 cases were seafood-borne. Four of the seven outbreaks, involving 77 cases, were shellfish-associated. During the same period, an annual average of 14,460 and 18,498 total cases were reported through the laboratory-based *Shigella* Surveillance system and the MMWR, respectively. NETSU (Rippey 1991) reported four shellfish-associated outbreaks involving a total of 93 cases of shigellosis in the United States between 1894 and 1988. Nine persons were reported ill in Massachusetts in 1977, 11 in California and 26 in Arizona in 1979, and 47 in Texas in 1986. Between 1978 and 1987 NETSU reported 84 cases of shellfish-associated shigellosis (Rippey 1991).

In 1989 and 1990, additional cases of shigellosis from the consumption of oysters were confirmed or suspected (Rippey 1991). The cases were for the most part sporadic and only two outbreaks occurred. All the cases were reported in the state of Florida. The two outbreaks in which *Shigella* was suspected as the causative agent occurred in October 1989. In both incidences, four people became ill after consuming oysters. In the first outbreak, four of nine people became ill one day after eating the oysters. *S. sonnei* and/or *Vibrio parahaemolyticus* and *V. fluvialis* were suspected. In the other suspected outbreak, four of four people became ill, and *Shigella* or a *Vibrio* was suspected. (One case of *Shigellosis* in a two-year old girl was reported in November of the same year.) In 1990, four separate cases of seafood-borne shigellosis were reported. Three of the victims were female and the other was a male. Oysters were the vehicle in all the cases.

Methods of Detection

Media used for the isolation and detection of *Shigella* from suspect foods have been reviewed by Smith (1987). Speciation and identification of the suspect colonies from selective media is accomplished by the use of biochemical and serological techniques. DNA probes have been used with success to identify virulent shigellae, i.e., those shigellae strains that contain the 120 or 140 MDa virulence plasmid. These probes also react with enteroinvasive *E. coli* strains which contain a 140 MDa virulence plasmid.

Pre-enrichment

As with the isolation of salmonellae, pre-enrichment is often necessary to recover injured cells. The most widely used method is to blend 25 g of the food sample with 100 ml tryptic soy broth and incubate for 24 h at 35°C. Most procedures do not include a pre-enrichment step.

Enrichment

As with other enteric bacteria, it is usually necessary to enrich for the presence of *Shigella*. Two enrichment broths, such as GN (gram-negative) and selenite cystine broths, are most often used. If the presence of injured shigellae is suspected, a different procedure must be followed since it is probable that the enrichment broths GN and SC will prevent the detection of injured cells.

Selective Media

Aliquots of the enrichment broths are streaked, using a loop, onto selective media. At least two, and preferably three, selective media should be used: (a) a low-sensitivity medium such as MacConkey agar; (b) one of intermediate selectivity such as XLD agar; and (c) a medium of high selectivity such as SS agar. An agar of low selectivity allows the investigator to isolate fragile shigellae such as *S. dysenteriae*. XLD is an agar of intermediate selectivity and is suitable for the isolation of most *Shigella* strains.

Other Tests

The ability to bind Congo red form an agar medium has been shown to be associated with the virulence plasmid in *Shigella* species. Shigellae are gram-negative, nonmotile rods lacking urease. They do not utilize KCN, malonate, citrate, or acetate as sole carbon sources.

ENTEROPATHOGENIC *ESCHERICHIA COLI*

General Overview

E. coli is often thought of as an indicator of fecal pollution. In 1887, Escherich observed the ubiquity of what we now designate as *Escherichia coli* in human stools. Shardinger, in 1892, suggested that members of this species be used as an index of fecal pollution because they could be recovered more easily than *Salmonella* species (Kator and Rhodes 1991; Banwart 1989). Thus, in the earliest methods of water analysis for fecal contamination, the emphasis was on the recovery of *E. coli*. Its role as an indictor organism is discussed thoroughly in other chapters. Within this chapter, its role as a foodborne pathogen will be discussed.

Pathogenic strains fall into at least four categories: enterotoxigenic, enteropathogenic, enteroinvasive, and hemorrhagic (Medallion Laboratories 1987; Hitchins et al. 1992; Frank 1988). The first three are usually associated with human fecal contamination, whereas hemorrhagic strains are most often associated with farm animals. Therefore, hemorrhagic *E. coli* will be discussed separately in Chapter 5.

Enterotoxigenic (ETEC) strains can produce heat-stable (ST) and/or heat-labile toxins (LT) (Frank 1988; Hitchins et al. 1992). The LT toxin is functionally and antigenically similar to cholera toxin. These toxins stimulate adenylate cyclase, which increases the production of cyclic-AMP. Enterotoxigenic strains of *E. coli* have been implicated in infant diarrhea in developing countries, but most cases apparently do not have food vehicles. These strains are also associated with the travelers' diarrhea that visitors to underdeveloped countries often develop. The enterotoxigenic strains were first recognized in the early 1970s. Enterotoxigenic strains have been implicated in foodborne outbreaks involving raw shellfish in Japan, but not yet in the United States. Enterotoxigenic strains are usually associated with human fecal contamination.

The enteroinvasive strains produce dysentery similar to shigellosis. Like the enterotoxigenic strains, they are associated with human fecal contamination. No cases have implicated shellfish as the vehicle for the organism; however, the low infective dose required to cause infection makes such a transmission possible.

Enteropathogenic strains of *E. coli* are serotypes of the organism that mostly cause diarrhea in infants. The mechanism of illness is not known. The source of these strains is generally assumed to be an infected human. Foodborne illness by these strains has not been proven.

Organisms

E. coli is a gram-negative, non-spore-forming rod belonging to the family *Enterobacteriaceae*. Strains of *E. coli* are usually motile and seldom encapsulated. *E. coli* is present in the feces of warm-blooded animals and in the predominant facultative anaerobe in the normal human intestinal tract. *E. coli* is a mesophilic bacterium with a growth range of 10–47°C. It often finds its way into food through unsanitary conditions. It can grow quickly if the conditions are suitable.

Illnesses

Enterotoxigenic strains are a major cause of travelers' diarrhea and infant diarrhea in developing countries. The illness is characterized by watery diarrhea, nausea, cramps, and low-grade fever and malaise. Volunteer feeding studies indicate that a relatively large dose (100 million to 10 billion bacteria) of enterotoxigenic *E. coli* is probably necessary to establish colonization of the small intestine (Levine 1987; Hitchins et al. 1992). Infants may require fewer organisms for infection to be established. The onset time is usually approximately 24 h. During the acute phase of illness, large numbers of enterotoxigenic cells are excreted in the feces. The disease is usually self-limiting. In infants or debilitated elderly, appropriate electrolyte replacement therapy may be necessary. Enterotoxigenic strains are differentiated from nontoxigenic *E. coli* present in the bowel by a variety of *in vitro* immunochemical, tissue culture, or gene probe tests designed to detect either the toxins or genes that encode these toxins. The diagnosis can be completed in about three days. Both epidemiologic and volunteer studies have documented the pathogenic potential of *E. coli*, and further laboratory studies have demonstrated the role of plasmids in supplying critical virulence properties for most strains (Levine 1987). Enterotoxigenic strains elaborate heat-labile (LT) and/or heat-stable (ST) enterotoxins in the proximal small intestine. LT is closely related to cholera toxin and causes diarrhea by the same mechanism. ST appears to be mainly antiabsorptive and lacks the secretory activity of LT and cholera toxin (Gross 1983).

Enteroinvasive *E. coli* (EIEC) produce an illness known as bacillary dysentery. The EIEC strains responsible for this syndrome are closely related to *Shigella* spp. Following the ingestion of EIEC, the organisms invade the epithelial cells of the intestine, resulting in superficial intestinal lesions and a mild form of dysentery. Volunteer studies have documented that enteroinvasive *E. coli* strains produce disease by invading enterocytes, proliferating and, eventually, causing cell death, in the manner of shigellae (Hitchins et al. 1992). The illness is characterized by the appearance of blood and mucus in the stools of infected individuals. The infectious dose of EIEC is thought to be as few as ten organisms (same as *Shigella*). Dysentery due to EIEC usually occurs within 12–72 h following the ingestion of contaminated food. The illness is characterized by abdominal cramps, diarrhea, vomiting, fever, chills, and a generalized malaise. Dysentery caused by this organism is generally self-limiting, with no known complications.

Association with Shellfish

Much of the research on the isolation and incidence of pathogenic *E. coli* in shellfish has been done in Japan. Sato (1971) was perhaps the first person to report on the isolation of enteropathogenic *E. coli* from oysters. From July 1969 to February 1971, Sato examined 160 commercial samples of foods, including 66 ground pork samples, 34 chicken samples, and 60 oyster samples. Nineteen of the 60 oyster samples were positive for enteropathogenic *E. coli.* Other works published in Japan include an article by Kokubo (1978) that describes a study in which 405 oyster samples were examined for the presence of *E. coli* and a portion of the isolates were tested for pathogenicity. Only four pathogenic strains were isolated—one strain produced only heat-labile enterotoxin, while the other three strains produced only heat-stable enterotoxin. Ogawa et al. (1980) studied the incidence of enteropathogenic *E. coli* in seawater, oysters, river water, and sediment samples in Japan. In this extensive study, enteropathogenic *E. coli* serotypes were isolated from 14.4% of the seawater samples, 14% of the oyster samples, 15.3% of the river water samples and 3.7% of the sediment samples. The total *E. coli* levels and enteropathogenic *E. coli* levels were strongly related—as the numbers of *E. coli* increased, the frequency of isolation of the pathogenic strains also increased. Perez Martinez et al. investigated the incidence of enteropathogenic *E. coli* in raw oysters obtained from supermarkets in Mexico using inoculation of suckling mice to evaluate for toxin. Only 3.7% of the isolates produced heat-stable toxin. Stephen et al. reported the isolation of both enteropathogenic and enterotoxigenic *E. coli* from mussels in India.

Shellfish-Borne Illness Associated with Pathogenic *E. Coli*

Although most infections by *E. coli* appear to be related to the contamination of food or water, with some person-to-person amplification under unhygienic conditions (Gangarosa 1978; Gross 1983), there are currently no data to indicate that any seafood, including shellfish, is an important source for diarrheagenic *E. coli* infections in this country. Neither the CDC annual summaries (from 1973 to 1987) nor the NETSU data base (Rippey 1991) report any shellfish-borne illness associated with pathogenic *E. coli*. Pathogenic strains of *E. coli* (with the possible exception of enterohemorrhagic *E. coli*) are a much more important cause of diarrheal disease in the developing than in the developed world; thus, imported products may present a greater risk than domestic seafood.

Most of the reported illnesses attributed to enteropathogenic *E. coli* have occurred in countries other than the United States. In 1983, the annual report of the Tokyo Metropolitan Research Laboratory of Public Health reported on two outbreaks in oyster-associated food poisoning due to enterotoxigenic *E. coli* 027:H7. These outbreaks occurred in late 1980 and early 1981. In the first outbreak, 138 out of 333 people became ill with gastrointestinal symptoms after consuming raw oysters at a restaurant. The onset time averaged 33.5 h and the major symptoms were diarrhea, abdominal cramps, nausea, and chills. Enterotoxigenic *E. coli* was isolated from the stools. The organism produced heat-stable enterotoxin. In the second outbreak, also associated with oysters, 36 people became ill with symptoms similar to those described above. The same organism was isolated from the stools. This article was in Japanese and only the abstract was available in English.

Methods of Detection and Enumeration

American Public Health Association and U.S. Food and Drug Administration Methodology

The standard methodology of the APHA and FDA identification and enumeration of *E. coli*, which requires that isolates conform to the coliform and fecal coliform group definitions, is not adequate for identification of the species. In this procedure a sample is first inoculated into laurel sulfate tryptose medium followed by incubation at 35°C for up to 48 h. Tubes positive for gas production are subcultured into EC medium and incubated at 44.5°C for an additional 24 h. Tubes positive for gas production are then streaked onto eosine methylene blue medium for colony isolation. The *E. coli* isolates are identified by their IMViC (indole, methyl red, Voges-Proskauer and citrate reactions) pattern: + + − − (Type I) and − + − − (Type II) (APHA 1984; Hitchins et al. 1990). In addition, methodology is also critical. A relatively high incidence of Type II "*E. coli*" in samples may be explained partly by the fact that many cultures require 48 h to produce a detectable amount of indole. Therefore, additional tests are required for a proper identification of the species. To identify pathogenic strains of *E. coli*, the isolates must be further serotyped. This whole process is time-consuming and labor-intensive. It is important to note that pathogenic biotypes of *E. coli* frequently exhibit aberrant reactions with regard to fermentation of lactose, ability to grow at 45.5°C, and production of gas. Also (for enteroinvasive *E. coli*), both food and stool must be examined and contamination levels below the limits of sensitivity of the standard methodology may cause illness. It is estimated that the ingestion of as few as ten organisms may result in dysentery, and concentrations below $10^x - 10^y$ in food are likely to be negative.

Alternative and Rapid Methods

Rapid methods are to be thoroughly reviewed for a variety of microorganisms in a separate chapter. They will be only briefly described here. One of the most widely used and successful rapid methods for the detection and enumeration of *E. coli* is the MUG (4-methylumbelliferyl-β-D-glucuronide) method. The procedure, which requires the incorporation of MUG into LST broth (Hitchins et al. 1992), enumerates coliforms and *E. coli* at the same time; coliforms are enumerated by examining the tubes for gas production, while *E. coli* is detected by fluorescence. However, this procedure does not work for shellfish due to the presence of interfering endogenous glucuronidase in oysters (Koburger and Miller 1985). These workers recommended that the MUG be incorporated into EC medium, used in the second step of the standard isolation procedure to eliminate interference of the endogenous glucuronidase. Therefore, the recommended procedure for the detection of *E. coli* by the MUG method in shellfish takes 48 h using the most probable number procedure (Rippey, Chandler, and Watkins 1987). Generally, 50 μg MUG/ml is used for liquid media. After incubation, the tubes are held under a long-wave (365 nm) ultraviolet lamp in a darkened area. The presence of a bluish fluorescence indicates the hydrolysis of MUG by bacterial glucuronidase enzymes. The test is about 94% specific for *E. coli*—some *E. coli* strains, including enterohemorrhagic strains, do not produce GUD, while some shigellae (44%) and salmonellae (29%) do (Feng and Hartman 1983). GUD production by other *Enterobacteriaceae* is infrequent.

Recently, Holt, Hartman, and Kaspar (1989) developed an enzyme capture assay method for *E. coli* in oysters. Polyclonal antibodies were prepared against *E. coli* glucuronidase. The antibodies reacted only with GUD from *E. coli*, *E. vulneris*, and *Shigella sonnei*. They did not react against indigenous GUD from oysters. Using this procedure, *E. coli* could be identified in oysters in 24 h compared to 48 h for the LST–EC–MUG assay to Rippey, Chandler, and Watkins (1987).

In order to identify pathogenic strains of *E. coli*, the MUG positive tubes must still be streaked onto EMB and tested as described above. Therefore, no time is saved for positive identification. However, these procedures would be useful as a screening tool to determine that no *E. coli* is present.

MUG has also been added to solid media. The concentration recommended is approximately 100 μg/ml for agar medium that contain lactose. Despite published reports supporting the use of MUG in solid media, it appears that MUG does not work well when incorporated into a solid medium because the reagent diffuses into the medium and colonies are nearly impossible to count. Recently, the compound 5-bromo-4-chloro-3-indoly-β-D-glucuronide (BCIG) has been incorporated into media to specifically enumerate *E. coli* by plate count methods. When this compound is incorporated into media, black light is not required to count colonies. In our laboratory BCIG has been incorporated into tryptone bile agar in a repair detection system with excellent results. However, at certain times of the year, shellfish harvested from the Gulf coast may contain very high levels of non-*E. coli* fecal coliforms which may take it difficult to distinguish *E. coli* colonies (Paille et al. 1986). This method would allow rapid isolation of *E. coli* for the serological test for pathogenic strains.

Anderson and Baird-Parker (1975) developed a method for enumerating *E. coli* from foods. A cellulose acetate filter membrane is placed on the surface of selective tryptone bile agar (TBA) plates and the sample is then spread onto the membrane and the plates incubated at 44°C overnight. After incubation, 1–2 ml of indole reagent is placed in the lid, and the membrane filter is removed from the TBA and placed in the reagent. Indole-positive colonies stained pink within 5 min. This procedure can be modified to enumerate stressed cells by placing the filter first onto a nonselective media, such as nutrient agar or minerals-modified glutamate agar, for 4 h at 35°C followed by placing the filter on TBA as described above. The Anderson–Baird-Parker procedure did not work with oysters in Louisiana harvested during the summer because of the overgrowth of non-*E. coli* fecal coliforms (Kilgen, Cole, and Hackney 1988). Motes, McPhearson, and DePaola (1984) noted that the Anderson and Baird-Parker procedure gave lower counts compared to the standard APHA procedure.

Several other rapid methods have been developed and evaluated for the detection and enumeration of *E. coli* in shellfish. Speck, Ray, and Reed (1975) developed a repair detection for coliforms. McDonald et al. (1983) modified this procedure by incorporating agents to degrade hydrogen peroxide. Also, Hackney and Bissonnette (1978) modified the procedure of Speck, Ray, and Reed (1975) to enumerate fecal coliforms by elevating the incubation temperature. This procedure worked very well with shellfish and nearly all isolates were confirmed to be *E. coli.*

Motes et al. (1984) compared the Anderson and Baird-Parker method, the direct-count roll tube method of the United Kingdom, and the brilliant green lactose bile broth (BGLB) method of France including the standard five-tube MPN procedure of the U.S.

The BGLB method compared most favorably with the standard procedure; however, the shellfish species being evaluated influenced the methods. The roll tube method compared favorably with the standard U.S. procedure for mussels and water, but not with clams or oysters. A-1 medium has also been recommended to recover *E. coli* in shellfish (Andrew et al. 1975). Determining pathogenic *E. coli* by these methods would still require colonies to be isolated and serotyped, to identify known pathogenic types of *E. coli.* Sera for invasive serogroups 0124:K72, 028:K73, and 0112:K66 are found in *E. coli* polyvalents B, C, and D, respectively. Sera for all serotypes of *E. coli* involved in enteric illnesses are not commercially available.

Other Test for Pathogenic E. coli Strains

Y-1 Adrenal Cell Test for Heat-Labile (LT) Toxin

The heat-labile toxin stimulates the enzyme adenylate cyclase with the production of cyclic adenosine monophosphate. The toxin is closely related structurally and physiologically to *Vibrio cholerae* enterotoxin. The cyclic monophosphate is active in many systems. In this system it promotes the conversion of elongated fibroblast-like cells to round, refractile cells.

Sereny Test for Invasiveness of Enteroinvasive Strains

This examination should be limited to *Shigella*-like strains of *E. coli* that are members of *E. coli* 0 groups: 28, 112, 124, 143, 144 serogroups.

DNA Probe for Identifying the Heat-Labile Toxin Gene

This method will identify pathogenic strains encoding a heat-labile toxin (LT) gene. With the availability of a gene probe method, foods can be analyzed directly for the presence of enterotoxigenic *E. coli* and the analysis can be completed in about 3 days. Alternative methods, which involve enrichment and plating of samples for the isolation of *E. coli* and their subsequent confirmation as toxigenic strains by conventional toxin assays, may take at least 7 days.

STAPHYLOCOCCUS AUREUS

General Overview

Staphylococcus aureus is the second or third most commonly reported cause of foodborne diseases in the United States. The illness is an intoxication that results from consuming food in which the bacterium has grown and produced toxin. *S. aureus* is a common bacterium and it is associated with humans and other warm-blooded animals. The carrier rate in the human population is 6–60%, with an average of 25–30% of the population being positive for enterotoxin-producing strains. The main reservoir is the nose, but it also found on skin, hands, wounds, and boils. It can also be isolated from the air, dust, floors, and other environmental surfaces. It survives well in the environment.

Organism

S. aureus is a gram-positive spherical bacterium (coccus) which, on microscopic examination, appears in pairs, short chains or bunched, grape-like clusters. Some strains are capable of producing a highly heat-stable, protein toxin that causes illness in humans.

S. aureus is differentiated from the 20 other staphylococcal species by a combination of the following features: colonial morphology and pigmentation; production of coagulase, thermonuclease, acetoin, β-galactosidase, phosphatase, and alpha toxin (hemolysis); acid from mannitol, maltose, xylose, sucrose, and trehalose; novobiocin resistance; presence of ribitol teichoic acid, protein A, and clumping factor in the cell wall (Lancette and Tatini 1992).

Illness

The symptoms include nausea, vomiting, abdominal cramps, and diarrhea. The illness usually only lasts 1–2 days (Berdgoll 1979; Banwert 1989; Lancette and Tatini 1992). The onset of symptoms in staphylococcal food poisoning is usually rapid, averaging 2–4 h, but may range from 0.5–24 h. The onset time varies, depending upon individual susceptibility to the toxin, the amount of contaminated food eaten, the amount of toxin in the food ingested, and the general health of the victim. The most common symptoms are nausea, vomiting, retching, abdominal cramping, diarrhea, and prostration. All patients do not demonstrate every symptom associated with the illness. In more severe cases, headache, muscle cramping, and transient changes in blood pressure and pulse rate may occur. Recovery generally takes two days; however, it is not unusual for complete recovery to take three days, and sometimes longer, in severe cases. Death from staphylococcal food poisoning is very rare, although such cases have occurred among the elderly, infants, and severely debilitated persons. All people are believed to be susceptible to this type of bacterial intoxication; however, the intensity of symptoms may vary.

The Toxin

The toxin responsible for the illness is released during growth and has been well characterized. There are six enterotoxin types. They are single polypeptide chains with molecular weights between 27,000 and 34,000. The enterotoxins are heat-resistant and can withstand temperatures above boiling for long periods of time (Bergdoll 1979).

The conditions that affect the growth of *S. aureus* are often quite different from those for toxin production. In general, the relative ranges for toxin production are narrower than those for growth. For example, the temperature range for the growth of *S. aureus* is in the range 10–45°C (some strains can grow at 7°C), with an optimum at 35–37°C. Toxin production is maximum at the optimum, but very little toxin is produced at lower temperatures (< 20°C). The pH range for growth of the organism is 4.5–9.3, but enterotoxin production only occurs between 5.1 and 9. The optimum pH for both growth and toxin production is near natural.

The amount of toxin needed for illness varies, but values of 0.015–0.357 μg/kg body weight are most often reported. As little as 1 μg of enterotoxin A has been reported to cause illness. The *S. aureus* concentration in food needed to produce this much toxin is 100,000–1,000,000 cells per gram. The amount of enterotoxin produced will vary with the type of enterotoxin and the strain.

Association with Shellfish-Borne Illness

The CDC reported three seafood-associated outbreaks of *S. aureus* food poisoning between 1973 and 1987 (CDC 1989). Two outbreaks, involving 14 cases, were shellfish-associated. The other outbreak, involving three cases, was associated with other seafood products. The outbreaks were mostly attributed to infected food handlers (CDC 1989). The NETSU data base only reported five cases of shellfish-borne illness. This outbreak occurred in 1977 in Nevada. Neither the type of shellfish nor the location of harvest was given. Overall, it does not appear that shellfish are common factors for *S. aureus*, but may transmit disease if contaminated after harvest by a handler and temperature-abused to allow the organism to grow to high numbers.

Association with Shellfish

The isolation of *S. aureus* from molluscan shellfish has been reported. Balachandran and Surendran (1985) reported isolation from clams. Galassi et al. similarly reported its isolation in low levels in clams harvested from the Adriatic Sea. Saddik et al. reported its isolation from raw mussels sold in Egyptian markets.

Todorov (1968) examined mussels from sites near sewage outlets, bathing beaches, and clean waters, and isolated *S. aureus* from sites in all the three categories. The samples taken near sewage outlets were most often positive for the organism.

Sobsey et al. (1980) examined oysters from open and closed beds for their aerobic plate count (35°C), coliform count, fecal coliform count, *Vibrio parahaemolyticus* counts, *S. aureus* (coagulase-positive) counts, and for the presence of salmonellae, *Yersinia* and enteric viruses. *S. aureus* was isolated with greater frequency from oysters taken from shellfish beds closed to harvesting by state health officials. However, there was no correlation between the levels of fecal coliforms and *S. aureus* counts. The levels reported were quite low, averaging 0.9/g for oysters harvested from beds open to shellfish harvesting and 1.7/g for oysters harvested from closed beds. The range in counts was 0–10 and 0–20/g for oysters from open and closed beds, respectively.

Methods

Incriminated foods should be collected and examined for staphylococci. The presence of relatively large numbers of enterotoxigenic staphylococci is good circumstantial evidence that the food contains toxin. The most conclusive test is the linking of an illness with a specific food or, in cases where multiple vehicles exist, the detection of the toxin in the food sample(s). In cases where the food may have been treated to kill the staphylococci, as in pasteurization or heating, direct microscopic observation of the food may be an aid in the diagnosis. A number of serological methods for determining the enterotoxigenicity of *S. aureus* isolated from foods as well as methods for the separation and detection of toxins in foods have been developed and have been used successfully to aid in the diagnosis of the illness. Phage typing may also be useful when viable staphylococci can be isolated from the incriminated food, from victims, and from suspected carriers such as food handlers.

The confirmation procedure most frequently used to establish the identity of *S. aureus* is the coagulase test. Coagulase is a substance that clots the plasma of human and other animal species.

Frequently used staphylococcal isolation media which may restrict the growth of sublethally heated cells are mannitol salt agar, Vogel and Johnson agar, and egg yolk azide agar. On the other hand, injured cells are readily recovered on Baird-Parker medium (Lancette and Tatini 1992).

Low numbers of *S. aureus* in foods are determined by spreading 1 ml of sample suspension of the lowest dilution equally over three plates of Baird-Parker agar without egg yolk or Baird-Parker agar containing rabbit plasma–fibrinogen tellurite to which 0.5 ml of 20% sodium pyruvate was added. All black colonies surrounded by opaque fibrin halos (coagulase-positive) are counted. These plates are incubated at 65°C for 2 h and then overlayed with 10 ml of melted toluidine blue DNA agar, which is then solidified. After solidification, the plates are incubated at 37°C for 4 h. All colonies showing pink halos against a blue background are considered to be thermonuclease-positive (Lancette and Tatini 1992).

Detection of Enterotoxin

For detecting trace amounts of staphylococcal enterotoxin in foods incriminated in food poisoning, the toxin must be separated from food constituents by selective adsorption of the enterotoxin from an extract of the food onto ion exchange resins. Following concentrations, enterotoxin is identified by specific precipitation with antiserum (antienterotoxin). Then, physical and chemical procedures are used to selectively remove food constituents from the extract, leaving the concentrated enterotoxin(s) in solution. Rapid methods based on monoclonal antibodies (e.g., ELISA, reverse passive latex agglutination) have recently been developed for enterotoxins in food. These rapid methods can detect approximately 1.0 ng of toxin/g of food.

5

Animal-Associated and Terrestrial Bacterial Pathogens

Cameron R. Hackney

Department of Food Science and Technology,
Virginia Polytechnic Institute and State University,
Blacksburg, VA 24061

Morris E. Potter

Division of Bacterial Diseases,
Center for Infectious Diseases, Centers for Disease Control,
1600 Clifton Road, NE, Atlanta, GA 30333

ANIMAL-ASSOCIATED PATHOGENS

INTRODUCTION

This section will cover pathogenic bacteria whose primary host is either domestic and wild animals or soil. Most of the organisms discussed are recognized shellfish-borne pathogens; although *Yersinia* and *Yersinia*-like bacteria have been frequently isolated from molluscan shellfish, no cases of shellfish-borne illness from these bacteria have yet been reported. Similarly, *Listeria monocytogenes* is a concern of regulatory officials; however, no shellfish-borne cases have been documented. The purpose of this review is to relate the presence of these pathogens to shellfish-borne hazards. Recently, detailed reviews have been published on each of these pathogens. Perhaps the most comprehensive is the book *Foodborne Bacterial Pathogens* edited by M. Doyle (1989). This book gives detailed information on the pathogen, the illness caused by the organism, its isolation and identification, and its control. Other recent reviews are included in the references of each section. The objective of this review is to relate bacterial pathogens to safety issues of molluscan shellfish. Therefore, information from annual reports from the

Centers for Disease Control and the Northeast Technical Services Unit of the Food and Drug Administration is included in each discussion. In addition, information on isolation of these microorganisms from shellfish and possible illness as reported in the literature is given here.

SALMONELLAE

General Overview

The salmonellae group are facultatively anaerobic gram-negative rods included in the family *Enterobacteriaceae* (Flowers et al. 1992; IFT 1988; Kaufmann 1966). They are usually motile by means of peritrichous flagella. As mentioned in the Chapter on bacterial pathogens primarily associated with man, *S. typhi* was the first species of the genus discovered. Of the salmonellae associated with animal reservoirs, Gartner discovered *Salmonella enteritidis* as a rodent pathogen in 1888 (Burrows 1985). In the early years of the twentieth century, culture methods were developed to distinguish reactions among the different types of salmonellae. Over time the numbers of organisms included in the *Salmonella* group increased until, at the present, there are over 2000 recognized species of *Salmonella* (Davis et al. 1989). Lignieres was the first person to accurately describe the salmonellae group. The current nomenclature, "Salmonella" was assigned in honor of an American bacteriologist, D. E. Salmon (Davis et al. 1989).

Serological studies of the organisms resulted in the establishment of an identification procedure known as the Kauffmann–White scheme (Kauffmann 1966; Davis et al. 1989). This system arranges the organisms into serovars, or types based on the kinds and combinations of antigens making up the surface of the cell, or the composition of the antigens which make up the flagella that provide motility to the organisms (Burrows 1985). The Kaufmann–White scheme designates organisms by numbers and letters given to different O, Vi or H antigens (Burrows 1985; Davis et al. 1989). The Kauffmann–White designation is not a complete record of the antigenic nature of the organism. For example, lysogenization by certain phages may produce changes in the O antigenic characterization of certain salmonellas, but use of the formula for nomenclature does not require that changes in antigenicity confer different species or subspecies names on the organisms in question (Davis et al. 1989; Burrows 1985).

The names given to the salmonellae do not follow the usual rules of nomenclature for bacteria. At first the organisms were named by indicating the disease and the animal from which the organisms were isolated. Some of these names continue to be used, such as *S. choleraesuis* and *S. typhimurium*. More recently, this type of nomenclature has been abandoned because the older names imply more limited pathogenicity than is actually found. Currently, names are assigned indicating the city, state, or even the country where the organism was first isolated, i.e., *S. london*, *S. panama*, etc. In some cases new types are simply designated by the formulae utilized in the Kauffmann–White scheme. It is generally agreed that none of the recent methods of naming members of this genus are satisfactory and, in general, new serotypes should be designated by formula only (Burrows 1985). While there are over 2000 different serotypes of *Salmonella* sp. recognized, only about 200 of these are detected in the United States in any given year (Burrows 1985).

Salmonellae are all, to a greater or lesser degree, pathogenic to humans. These bacteria have very simple nutritional requirements, and are able to survive with at least minimal growth even in surface water supplies. In sewage, polluted sea, or fresh water *Salmonella* strains have been observed to survive for over four weeks, and in soil for six weeks or more (Flowers 1988; National Research Council 1977).

Raw foods, particularly those of animal origin, are the major vehicles for salmonellosis (IFT 1988; Flowers et al. 1992; Allred et al. 1967). The five most common food vehicles for *Salmonella* in the United States are beef, turkey, homemade ice cream (containing eggs), pork and chicken (Banwart 1989; Flowers et al. 1992). Turkey is the most common vehicle in Canada. However, many other foods have been involved in salmonellosis. For example, in 1985, the largest outbreak ever reported (18,000 cases) was traced to pasteurized milk produced in Illinois.

Illness

Salmonellosis, or salmonella gastroenteritis is a widespread bacterial infection which often begins with a sudden headache, abdominal distress, diarrhea, nausea, and sometimes vomiting (Hornick 1983). Although the infection may begin as only a gastroenteritis; on rare occasions, it may progress to an enteric fever which is manifested as a systemic infection (IFT 1988; Mandal and Mani 1976).

The food poisoning syndrome develops after ingestion of a food that contains a sufficient number of *Salmonella* cells to cause infection, usually between 100,000 and 100,000,000 cells, but sometimes fewer than 1000 cells (Blaser and Newman 1982; Hook 1985). The symptoms usually develop 12–14 h after ingestion of the food, although incubation times of greater than 24 h are not uncommon. Symptoms consist of mainly diarrhea along with nausea, vomiting, abdominal pain, headaches, and chills. The symptoms are often accompanied by prostration, muscle weakness, moderate fever, and drowsiness. Symptoms usually last only 2–3 days. In some cases, the stools may be bloody. Temperature elevations as high as 38–39°C are common, as are chills. Abdominal cramps occur in approximately two-thirds of the patients. In severe cases, particularly in children, fluid and electrolyte losses may be severe enough to cause clinically important dehydration. Toxic megacolon has been described as a rare complication (Mandal and Mani 1976).

Transient bacteremia occurs in less than 5% of adults with gastroenteritis, but at a somewhat higher rate in children and persons with major underlying diseases (Hook 1985; Hyams et al. 1980). Localized suppurative infections (bronchopneumonia, emphysema, endocarditis, pericarditis, arteritis, pyelonephritis, osteomyelitis, and arthritis) develop in about 10% of patients with bacteremia (Hook 1985; Hornick 1983). In the majority of uncomplicated cases, fever lasts for less than 2 days. Diarrhea usually lasts less than 7 days, but occasionally persists as long as several weeks (Hook 1985). In a clinico-epidemiologic study in the Pacific Northwest, mean length of illness due to infection by nontyphoidal *Salmonella* was 10.25 days (Seattle-King County Department of Public Health 1984). Illness is more severe in children, in the elderly, and in patients who have had a gastrectomy or gastroenterostomy, or who have achlorhydria, sickle-cell anemia, or other conditions that impair resistance to infection.

In foodborne outbreaks of salmonellosis, hospitalization rates have ranged from 0% to more than 50%. The Center for Disease Control (CDC) (1981) reported that 8% of 2356 cases in 51 outbreaks reported in 1978 were hospitalized. In the Pacific Northwest study, three (4.1% of 72 outpatients with positive cultures) were hospitalized (SKCDPH 1984). A case-fatality rate of 0.1% was observed for 16,172 patients in 261 foodborne outbreaks reported to CDC between 1972 and 1978.

Although there are differences by serotype, age-specific attack rates tend to peak between 2 and 4 months of age at approximately 300 cases per 100,000 population and decline rapidly thereafter. After age five, rates for all serotypes combined remain relatively constant at approximately ten cases per 100,000 per year (CDC 1981). In the Pacific Northwest study (SKCDPH 1984), 1.4 cases of salmonellosis per 100,000 population were hospitalized. Similar figures are obtained from reported cases and the percentages of hospitalization reported in studies of outbreaks. Based on the reported number of cases and estimated mortality rates, mortality ranges from 0.02 to 0.2 deaths per 100,000 population per year. Adjusting for deficiencies in the reporting system would make the estimated mortality rate higher.

Frequency of Illness

Salmonella species/serotypes other than *S. typhi* are among the most common causes of gastroenteritis in the United States. It is by far the most common cause of food poisoning reported to the CDC.

Salmonellosis has been a reportable disease in the United States since 1942, and data have been collected by CDC through its *Salmonella* surveillance system since 1963. The incidence of reported cases was approximately eight per 100,000 in 1963. In 1983, approximately 19 cases per 100,000 were reported (CDC 1984). It is likely, however, that reported cases represent only a small fraction of the total number of infections that occur (Banwart 1989). From data on outbreaks, it has been estimated that only one in 75 to 100 cases of salmonellosis in the United States is reported (Aserkoff, Schroeder, and Brachman 1970; Hauschild and Bryan 1980). Therefore, the incidence of actual infections ranges from 1500 to 2000 cases per 100,000 population. The overall reported incidence of salmonellosis in humans in the United States has increased steadily over the last 25 years.

Salmonellosis Associated with Seafoods

In the United States most outbreaks of salmonellosis are traced to contaminated products of terrestrial animals. However, vehicles for sporadic salmonellosis are rarely identified. While CDC and NETSU foodborne surveillance data indicate that seafood is a much less common vehicle for *Salmonella* than are other foods such as chicken and red meat; fish, and shellfish may be responsible for at least a small proportion of the total number of *Salmonella* cases that occur each year in the United States. However, current data are inadequate to make any attempt at estimating attributable risk.

When examining the importance of salmonellae contamination of seafood, it is useful to examine the overall incidence reported to CDC. CDC tracks disease incidence by several mechanisms, including the laboratory-based *Salmonella* Surveillance system and the epidemiology-based Morbidity and Mortality Weekly Report (MMWR). The surveillance tools used by these systems are not different, and their data do not agree. The annual foodborne disease incidence data for the 14 years period from 1973 to 1986 from the epidemiology-based-foodborne disease outbreak surveillance system documents an average of 55 foodborne outbreaks of nontyphoidal *Salmonella* infections affecting a total of 3944 persons each year. During the same time frame, only six seafood-borne outbreaks involving 147 cases were reported. Two of these outbreaks involving 40 cases were shellfish associated. Examining different surveillance systems, it is seen, during the 14 year period from 1973 to 1987, an annual average of 32,957 and 35,490 *Salmonella* cases were reported through the *Salmonella* Surveillance System and the Morbidity and Mortality Weekly Report (MMWR), respectively.

The NETSU data base, which attempts to document all occurrences of shellfish-borne disease outbreaks since 1894, reported only two shellfish associated outbreaks of confirmed nontyphoidal salmonellosis between 1894 and 1973 (Rippey 1991). A 100-case outbreak that occurred in Florida in 1947 was traced to contaminated oysters. The other outbreak occurred in New York in 1967 and involved in 22 cases. This outbreak was associated with oysters imported from England (Rippey and Verber 1988). No cases of *Salmonella* infections from shellfish were reported to NETSU between 1973 and 1988. However, several sporadic cases of salmonellosis associated with shellfish occurred in 1989 and 1990 (Rippey 1991). In September 1989, three cases of salmonellosis were associated with mussels harvested in Maine and consumed in Connecticut. *S. infantis* was isolated in two of the cases. In October and December 1989, oyster-associated cases were reported in Florida. In 1990, four separate oyster associated cases were reported in Florida (Rippey 1991).

With respect to other seafoods, outbreaks of salmonellosis associated with salmon prepared by a caterer have been reported (Cartwright and Evans 1988). The salmon was epidemiologically and microbiologically implicated, and multiple food-handling errors were identified. The origin of contamination of the fish was not determined, although raw chicken prepared at the same time as the salmon was also culture positive for *S. montevideo*, the outbreak serotype.

Outbreaks of salmonellosis have been associated with shellfish in other countries. For example, an outbreak of salmonellosis that involved 50 persons in Italy was associated with clams (*Venus verrucosa*) (Cantoni et al. 1985). The causative agents were *S. typhimurium* and *S. mbandaka*. The estimated count per clam was 400–800 cells, which implies that the infective dose was low. The NETSU data base on shellfish-associated outbreaks—Foreign Reports, did not report outbreaks due to *Salmonella* during the period from 1973 to 1990.

Association with Shellfish

Fraiser and Koburger (1984) examined various seafoods including clams and oysters from the east and west coasts of Florida, for the presence of *Salmonella*. The highest incidence of *Salmonella* was from clams harvested from the Gulf (west) coast of Florida. The shellfish were analyzed very quickly after harvest to improve recovery of salmonellae. In addition, individual animals were analyzed instead of using composite samples to increase the probability of isolating different serotypes of *Salmonella*. Eleven different serotypes of *Salmonella* were isolated, with as many as six serotypes being isolated from the same group of samples. The authors went on to theorize that salmonellae might be part of the free-living microflora of shellfish.

In this study, 43% of the clams tested were positive for *Salmonella.* This is one of the few studies that report the numbers of *Salmonella* present in the samples. In analysis of oyster meats the levels of *Salmonella* isolated was 2.2 per 100 g of tissue. Although this level of salmonellae would be unlikely to cause illness in most of the healthy adult consumers, retail food service establishments rarely screen customers to eliminate high-risk consumers, and relatively brief periods of temperature abuse could substantially increase the risk of illness for all consumers.

Other studies have reported the isolation of *S. pullorum* from butter clams of the Pacific northwest. Colburn et al. (1989) examined the microbiological quality of oyster (*C. gigas*) and water of live holding tanks at five different Seattle, Washington, markets—a water sample contained a nonpathogenic *Y. enterocolitica,* and *S. typhimurium* was isolated from one oyster sample.

In other countries salmonellae have frequently been isolated from mussels. For example, *Salmonella anatum* has been isolated form mussels in Italy (Carlotto and Bersani 1968). Reports from Germany also describe isolations from mussels (Havemeister 1969; Pietzsch 1970), and *S. braenderup* was isolated from market samples of mussels in India (Gopalakrishna and Varma 1987).

Relationship of the Presence of Salmonellae to Indicators

Andrews et al. (1974) examined whether coliforms were suitable as indicators of *Salmonella* in oysters and clams. During an 18 month period, 263 oyster and 96 clams samples were tested for coliforms, fecal coliforms, and the presence of *Salmonella. Salmonella* was isolated from 39 (14.8%) of the oyster and five (5.2%) of the clam samples. Although samples positive for *Salmonella tended to have high coliform and fecal coliform counts,* high numbers of indicators did not necessarily mean that the pathogen was present.

In later work the same group examined the comparative validity of members of the total and fecal coliforms groups for indicating the presence of *Salmonella* in the eastern oyster (*Crassostrea virginica*). In this study, 539 oyster samples and the corresponding harvest-water samples were analyzed. Occurrence of *Salmonella* in oyster meat more closely paralleled increases in fecal coliform counts than total coliform counts. In addition, more *Salmonella* were isolated from water meeting the total coliform standard than from water meeting the fecal coliform standard. *Salmonella* was not isolated from samples that met both the sanitary survey and fecal coliform standard. This study pointed out the importance of using the sanitary survey in conjunction with microbial analysis to ensure safety.

Andrews et al. (1976) studied the validity of members of the total coliforms and fecal coliform groups as indicators of the presence of *Salmonella* in hard (quahaug) clams. In this 2-year study, 214 samples of harvesting water and clams were used. Harvesting waters were classified as to whether they met either the total coliform standard of ≤ 70 coliforms per 100 ml or ≤ 14 fecal coliforms per 100 ml. The clams were further classified as to whether or not they met the market guideline of 230 fecal coliforms per 100 g of tissues. Inasmuch as none of the clams meeting the market guideline, or those harvested from waters meeting either of the standards, contained *Salmonella*, the investigators concluded that fecal coliforms were adequate indicators of shellfish safety with respect to *Salmonella*. *Salmonella* was isolated from some of the samples which exceeded the standards of the National Shellfish Sanitation Program.

Timoney and Abston (1984) studied the contamination and elimination of *E. coli* and *S. typhimurium* in the hard clam, *Mercenaria*. The bacteria were eliminated at similar rates; however, *E. coli* levels declined more rapidly than salmonellae. The organisms were eliminated from the clams by becoming associated (nonionicly bound) with feces and pseudofeces particulate matter. Most of the test organisms were eliminated between 6 and 24 h. This study indicates that *E. coli* is a good indicator with respect to salmonellae.

Hood et al. (1983) examined the relationship among fecal coliforms, *E. coli* and *Salmonella* species in freshly harvested Gulf of Mexico coast oysters and clams. *Salmonella* was only found in samples which exceeded the National Shellfish Sanitation Program's market guideline of 230 fecal coliforms per 100 g of the product. These investigators reported that low levels of fecal coliforms and *E. coli* were good indicators of the absence of *Salmonella*. However, high levels of these indicators did not necessarily indicate the presence of *Salmonella*.

Eyles and Davey (1984) observed that isolations of *Salmonella* from the Sidney Rock oyster was correlated to rainfall and to a lesser extent low salinity waters. The presence of salmonellae in this study was related to high *E. coli* counts.

Thomas and Jones (1971) examined the relationship between *E. coli* and salmonellae in mussels and determined a positive relationship between levels of the indicator and presence of the pathogen.

Elimination of Salmonellae and *E. coli* from Shellfish

It is interesting that there are species differences in depuration rates of species of *Salmonella*. For example, in a comparison of depuration rates for *S. typhimurium*, *S. montevideo* in Gulf oysters, Cook and Ellender (1986) found that *S. montevideo* persisted longer than *S. typhimurium*.

Matev et al. compared the depuration of *S. typhimurium* and *S. enteritidis* to that of *E. coli* and *Staphylococcus aureus* in artificially contaminated Black Sea mussels. *E. coli* was recovered for 6 days compared to 4 days for either *Salmonella* serotype and 2 days for *S. aureus*.

Rowse and Fleet (1982) observed that both *Salmonella* and *E. coli* survived in oyster feces and could be released in the overlying waters. Survival depended on water temperature. In later studies, Rowse and Fleet (1984) studied the effects of water temperature and salinity on the depuration of *S. charity* and *E. coli* form the Sidney Rock oyster. In this study, the organisms were eliminated at similar rates. For this species of oyster, elimination was most rapid at 18–22°C and salinities of 3.2–4.7%. Higher or lower temperatures and lower salinities slowed depuration.

Control

The microbial quality of harvest waters does not appear to be a good predictor of nontyphoidal *Salmonella* contamination, as oysters removed from closed and open beds had the same level of contamination (4%), and no correlation was observed between the presence of *E. coli* or fecal coliforms and *Salmonella* in fin fish or shellfish (Sobsey et al. 1980; D'Aoust, Gelimas, and Maishment 1980). Salmonellae do not appear to grow well in oysters during storage, and are not selectively retained during depuration or relaying, so contaminated oysters can be purified (Son and Fleet 1980). As with other pathogens, the risk of infection can also be minimized by assuring that shellfish are cooked and handled properly.

Antimicrobial use appears widespread in aquaculture at both therapeutic and subtherapeutic levels (Brown 1989). The possible development of antibiotic resistant strains of bacteria, such as *Salmonella*, resulting from these feeding practices, and the subsequent transfer of such resistant bacteria on seafood products to humans are of concern. High levels of resistance to tetracycline compounds, some of it transferable, have been documented in *Aeromonas* strains isolated from farmed catfish and their environment (DePaola, Hopkins, and McPhearson 1988).

DiGirolamo et al. (1970) examined the effect of freezing on the survival of salmonellae and *E. coli* in the pacific oyster. In this study salmonellae were more sensitive to freezing than *E. coli*.

Standard and/or Existing Methods for Detection in Growing Waters and Molluscan Shellfish

The examination of water and shellfish samples for *Salmonella* generally requires the use of methods different from those used in clinical laboratories because of low numbers and the possibility of injured cells. The standard methods usually employ four steps; preenrichment, enrichment, colony isolation, and identification.

It is generally accepted that preenrichment of processed foods in a nonselective broth medium facilitates detection of sublethally injured *Salmonella*. Lactose broth is the medium most often used for preenrichment of seafood products, including shellfish. This seems odd, since most *Salmonella* do not utilize lactose. North (1961) explained this phenomenon by concluding that when competing lactose-utilizing bacteria are present in a food sample, they utilize the lactose and the resulting drop in pH generates a bacteriostatic effect on competing microflora.

In the analysis of foods for *Salmonella*, the food is usually placed in the preenrichment broth and usually incubated for 18–24 h at 35–37°C and then a portion is subcultured to one or more selective enrichment broths. During preenrichment, injured cells repair and their numbers increase.

Selective enrichment media contain various selective ingredients which allow the proliferation of *Salmonella*, while inhibiting the growth of competing non-*Salmonella*, microorganisms. The two most widely used enrichments are selenite broth and tetrathionate broth. Selenite broth is often modified by the addition of cysteine (SC) or brilliant green dye to increase selectivity. In general, while selenite broths are widely used for the detection of foodborne *Salmonellae*, various modifications of tetrathionate broth have generally found wider applications. Use of two or more enrichment media provides greater recovery of *Salmonella* from different food products. Therefore, it is a good idea to use both types of broth (Flowers et al. 1992).

Incubation temperature markedly affects productivity of selective enrichment broths by suppressing growth of competing microorganisms but not *Salmonella*. Although selective enrichment media are frequently incubated at 35–37°C, several studies have shown increased recovery of *Salmonella* following incubation of selective enrichments at 43°C. Hackney et al. (1990) reported that increasing the incubation of selenite cysteine broth to 41–42°C greatly improved the recovery of salmonella from poultry samples. In this work, 50 samples, naturally, were preenriched in lactose broth at 35°C, followed by selective enrichment in selenite cysteine incubated at either 37°C or 42°C. Recovery was significantly greater at the higher incubation temperature. In addition, work in our laboratory has shown that increased incubation temperature will also improve the recovery from seafood samples, including shellfish (unpublished data).

Miller and Koburger (1984) examined methods of recovering salmonellae from naturally contaminated oysters. They compared pre-enrichment in lactose broth followed by selective enrichment in either tetrathionate or selenite broth incubated at either 35°C, 41°C, or 43°C to direct enrichment under the same conditions. In their studies direct enrichment at elevated temperatures gave the greater number of salmonellae-positive samples. Similar results were obtained in artificially contaminated samples.

Selective plating media that have been used for the isolation of *Salmonella* include: brilliant green (BG), BG sulfa, bismuth sulfite (BS), *Salmonella shigella*, MacConkey's, deoxycholate citrate, Hektoen enteric (HE), xylose lysine deoxycholate citrate (XLD), Shanson's, Rappold–Bolderdijk modified lysine iron, lysine–iron–cystine, MSRV, and novobiocin–brilliant-green–glucose (APHA 1986; USFDA 1984). None of these media are ideal for all situations. In general, it is recommended that two or more agar media be used for isolation of colonies. Plates are usually incubated 18 to 24 h. BS agar plates should be observed after 24 h of incubation, and any suspect *Salmonella* colonies should be tested biochemically and serologically.

Presumptive identification of typical *Salmonella* can be obtained from a minimal number of biochemical tests that may include urease, KCN, and indole. Biochemical tests including urease, lysine decarboxylase, growth in KCN broth, and production of indole are generally sufficient for presumptive identification of *Salmonella.*

Commercial Biochemical Multitest System or Kits

Numerous biochemical tests are routinely used to differentiate members of the family *Enterobacteriaceae* and to characterize presumptive *Salmonella* isolates. Commercially available multitest systems or diagnostic kits are convenient and generally provide reliable biochemical test results. The API 20E, Minitek, and Enterotube received AOAC approval in 1978, and the Micro-ID received first action approval in 1988.

Rapid Methods

Rapid methods for isolation and identification of *Salmonella* is covered in greater detail in a separate chapter on rapid methods. Several companies have developed methods for the rapid detection of *Salmonella* in food. A variety of techniques have been introduced, all of which appear to have functional sensitivity limit of 10^4 per sample. All of these rapid methods are screening methods, which means that presumptively positive results must be confirmed, whereas negative results are considered definitive.

DNA Hybridization (GENE-TRAK™)

This method is a screening test for presence of *Salmonella* in foods. Presumptive positive assays must be confirmed by streaking onto appropriate plating media and suspect colonies identified as described above. Genetic relationships between bacteria can be measured by DNA–DNA hybridization. In this technique, gene probes for unique nucleotide sequences in target organisms have been developed. The performance of a DNA hybridization assay for *Salmonella* in foods compares very favorably with conventional methodology.

A collaborative study among 11 laboratories indicated that the DNA probe method was significantly better than the standard culture method for the detection of *Salmonella* in turkey. No significant difference was observed between the methods for the other five foods (Flowers et al. 1987). The results of this study and earlier comparative studies led to an AOAC approval of the DNA probe method for detection of salmonellae in foods.

For most foods, the DNA probe method includes 14–24 h preenrichment, a 6 h selective enrichment, and 12–16 h postenrichment. After inoculation of the postenrichment GN-broth with 0.1 ml of enrichment culture, selective enrichments are incubated for 12–16 h.

The first DNA probes developed used radioactive labels, however, the use of radiation limited its use in some laboratories. Recently DNA probe methods have been developed that are calorimetric (nonradioactive) and employee different DNA probes than those used in the isotopic method. The calorimetric assay probes hybridize the probes (DNA) to target rRNA. Following preenrichment, selective enrichment, and postenrichment of test samples, bacteria are lysed and labeled *Salmonella*-specific. DNA probes are added and the solution is incubated to allow hybridization to occur if target rRNA is present. The calorimetric DNA assay as described can be completed in 2.5 h, following approximately 44 h of culture enrichments.

Fluorescent Antibody Method

The fluorescent antibody (FA) technique was the first immunological method proposed for the detection of salmonellae. The method was extensively evaluated and eventually approved by AOAC in 1975 (APHA 1986).

Enzyme Immunoassay

Several enzyme immunoassay (EIA) have been proposed for detection of salmonellae in foods. The first EIA method approved by AOAC was the "BioEnzaBead" method (Organon Teknika, Durham, NC, formerly Litton Bionetics). Comparative studies on the productivity of this method based on the reaction of two antibodies, MOPC 467 and 6H4, with *Salmonella* antigens were conducted by D'Auost (1981), Eckner et al. (1987), and Flowers et al. (1989). No significant differences between the AOAC culture method and EIA-confirmed are observed.

The Q-TROL™ *Salmonella assay* (Dynatech Laboratories, Inc., Chantilly, VA) utilized monoclonal antibodies and a fluorogenic reagent. It has AOAC approval. The extensive evaluations of the EIA methods clearly indicate that this technology is an attractive alternative to conventional culture methods for detection of salmonellae in foods.

CAMPYLOBACTER

General Overview

Campylobacteriosis is one of the most common causes of bacterial foodborne disease in industrialized countries, including the United States (Seattle-King County Department of Public Health 1984; Totten 1987; Franco 1988; Stern et al. 1992). Only recently has its importance been realized because methodology to detect the organism in food and feces was not available before 1972 (Doyle 1981). Since the first outbreak of foodborne campylobacteriosis was reported to FDSS in 1978, State health departments have reported numerous additional outbreaks, affecting thousands of people. Nationwide laboratory-based *Campylobacter* surveillance was initiated in 1982. By 1986, 39 states were enrolled and reported 10,066 cases.

Campylobacter species, which were once thought to be primarily important as pathogens of domestic animals, were placed in the genus *Vibrio* until 1974 because of their shape (Stern et al. 1992; Doyle 1981). The organism now known as *C. jejuni* was grouped with *V. fetus.* The genus *Campylobacter* currently consists of at least 18 species, subspecies, and biovars, with 17 names officially recognized by the International Committee on Systematic Bacteriology (Franco 1988; Stern et al. 1992).

Human illness is associated with three species of *Campylobacter*, *C. jejuni*, *C. coli*, and *C. laridis.* These organisms are carried in the intestinal tract of animals and, therefore, may contaminate foods of animal origin. In addition, fecal contamination of harvesting waters may allow shellfish to be a vehicle for the pathogens (Rippey 1991). *C. jejuni* is recognized as a leading cause of acute bacterial gastroenteritis. It is recognized as both a food and waterborne pathogen. Foodborne illness is usually associated with the consumption of products of animal origin. In addition, *C. coli* and *C. laridis* are also recognized causes of gastroenteritis, but less frequently than *C. jejuni.* These three species are collectively referred to as the *C. jejuni* group.

Although *C. jejuni* is not carried by healthy individuals in the United States or Europe, it is often isolated from healthy mammalian and avian species (Franco 1988). It also has been isolated from nonchlorinated water sources such as streams and ponds.

The Organism

Campylobacter are curved, spiral gram-negative rods that are nonspore-forming and microaerophilic (Simbert 1984). Campylobacters grow between 25°C and 43°C, are motile, oxidase positive and do not ferment or oxidize carbohydrates (Simbert 1984; Franco 1988). The campylobacters can be broadly placed into two groups on the basis of the catalase test (Patton et al. 1989; Barrett, Patton, and Morris 1988). The catalase-positive campylobacters are most frequently associated with human disease. Campylobacters are generally inactive in many conventional biochemical tests and identification is based on only a few morphological and biochemical features.

The Illness

C. jejuni infection may cause diarrhea, which may be watery or sticky and can contain blood (usually occult) and fecal leukocytes (white cells) (Blaser 1981; Black et al. 1988). The diarrhea is usually accompanied by malaise, fever, and abdominal pain. Less commonly reported signs and symptoms include nausea, tenesmus, bloody diarrhea, joint pain, and vomiting. Onset of diarrhea occurs 2–7 days after exposure. The diarrhea generally lasts 5–10 days, and resolution of other signs usually occurs by 2 weeks past onset. Relapses are not uncommon (about 25% of cases). Most infections are self-limiting and are not treated with antibiotics. However, treatment with erythromycin does reduce the length of time that infected individuals shed the bacteria in their feces (Mandal, DeMol, and Butzler 1984).

Clinically, campylobacteriosis can present as a water diarrhea, a dysentery-like syndrome mimicking inflammatory bowel disease, or as an extra-intestinal infection. Although uncommon, extra-intestinal infections including meningitis, cholecystitis, and urinary tract infections have been reported. The relative importance of host factors vs. strain differences in the clinical expression of infection is uncertain.

The pathogenic mechanisms of *C. jejuni* are still being studied so it is difficult to differentiate pathogenic from nonpathogenic strains. Three probable mechanism of pathogenesis are observed with *Campylobacter*: invasion, translocation, and toxin production (Johnson and Lior 1984; Ruiz-Palacios et al. 1983; McCardell et al. 1984; Lee et al. 1986).

Intestinal epithelial invasion appears to occur in the 20–30% of cases of campylobacteriosis with bloody diarrhea and mucus and cellular exudate in their stools (Blaser and Reller 1981; Franco 1988; Walker, Wun, and Litsky 1982). Infection most commonly involves the terminal ileum and colon.

Translocation is characterized by penetration of the intestinal mucosa with minimal surface damage, and proliferation in the lamina propria and mesenteric lymph nodes. The mesenteric adenitis that appears to cause most of the unnecessary appendectomies in patients with *Campylobacter* infections is clinical evidence that *C. jejuni* also can use this mechanism. Thus *C. jejuni* is clearly capable of being invasive. Overt colitis and mesenteric adenitis are common clinical presentations, and many patients have blood, mucus, and leukocytes in their stools.

The profuse watery diarrhea that frequently results from *Campylobacter* enteritis, especially in children in developing nations, provides clinical evidence that this mechanism can be used by *C. jejuni* (Johnson and Lior 1984). *C. jejuni* isolates recovered from patients with watery diarrhea produce an enterotoxin that appears functionally and immunologically similar to cholera toxin and *E. coli* LT (Ruiz-Palacios et al. 1983). *Campylobacter* enterotoxin induces fluid accumulation in both rat and rabbit ileal loops, causes elongation of CHO cells and rounding of Y-1 mouse adrenal cells, and can be blocked by anticholera-toxin antibody (Klipstein and Engert 1984; Johnson and Lior 1984). Although these studies confirm enterotoxin production by *C. jejuni*, strains isolated from asymptomatic carriers and patients with watery diarrhea did not differ in their ability to produce the toxin.

Kiehlbauch et al. demonstrated that *C. jejuni* survives within monocytes for as long as a week, which could explain, in part, the fever and other systemic signs that occur in absence of bacteremia. Bacteremia is rarely documented among cases of campylobacteriosis. Engulfed *C. jejuni* survived longer than free bacteria in the study by Kiehlbauch et al., but not as long as the more typical intracellular pathogens such as Listeria.

Complications of infection by *C. jejuni* are uncommon. Abdominal pain mimicking appendicitis can lead to unnecessary appendectomy. About 1% of patients suffer from reactive (aseptic) arthritis arising 7–10 days post onset, and this can persist for several weeks to months. Reiter's syndrome also has been reported to follow acute campylobacteriosis. Guillain–Barre syndrome occurs, but is less common than reactive arthritis. Death attributable to *Campylobacter* infection in the United States is estimated to occur at a rate of approximately 120–360 per year.

The infective dose of *C. jejuni* is considered to be small. Human feeding studies suggest that about 400–500 bacteria may cause illness in some individuals, while in others, greater numbers are required (Black et al. 1988; Robinson 1981).

Complications are relatively rare, but infections have been associated with reactive arthritis, hemolytic uremic syndrome, and infections of nearly any organ following septicemia (Stern et al. 1992). The estimated case/fatality ratio for all *C. jejuni* infections is 0.1, or one death per 1000 cases. Fatalities are rare in healthy individuals and usually occur in cancer patients or in the otherwise debilitated. Although anyone can have a *C. jejuni* infection, children under five years and young adults (15–29) are more frequently afflicted than other age groups.

Association with Foods

Known reservoirs of *C. jejuni* include poultry, cattle, and pets. Many healthy chickens are intestinal carriers of *C. jejuni* and these bacteria frequently contaminate raw chicken. Surveys have shown that 20–100% of retail chickens are contaminated. Raw milk is also a source of human infections; *jejuni* are often carried by healthy cattle and by flies on farms. Other meat animals also frequently harbor the bacterium (Doyle 1981; Banwart 1989). Nonchlorinated water may also be a source of infections. However, properly cooking chicken, pasteurizing milk, and chlorinating drinking water kills the bacteria.

Campylobacteriosis Associated with Shellfish Consumption

NETSU reported one domestic shellfish-associated outbreak of campylobacteriosis between 1894 and 1988, an outbreak of 16 cases due to contaminated hard clams that occurred in New Jersey in 1980 (Chapter 8 of this Book). In addition, *Campylobacter* was suspected in several outbreaks reported to the NETSU where the etiological agent was listed as unknown (Rippey and Verber 1988). The 1991 update of the NETSU data base reported several outbreaks and cases of *Campylobacter* associated with shellfish with onset of illness in 1989–1991. Most of these illnesses were reported in the state of Florida. In a 1989 outbreak, two people became ill three days after consuming oysters; *C. jejuni*, *Vibrio parahaemolyticus*, and *V. vulnificus* were all isolated from ill persons. In another incident, a single case of confirmed *Campylobacter* infection was reported in Lee County, Florida. In December 1989, four separate cases of confirmed *Campylobacter* associated with shellfish were reported (Rippey 1991). In 1990, six cases of *Campylobacteriosis* were reported in Florida—five of the incidents involved oysters and clams were implicated in the other case. The age of the victims ranged from 23 to 70 years. In addition, another case of *Campylobacter* illness from oysters was suspected in Alabama in 1989 (Rippey 1991).

While the risk of *Campylobacter* infection associated with eating raw shellfish appears low, there are inadequate data to make any assessment of real attributable risk. Estimates of the relative risk of eating various types of seafood, as compared with other food items, involve many estimates and a great deal of uncertainty.

Isolation from Shellfish

Arumugaswamy and Proudford (1987) reported the isolation of *C. jejuni* and *C. coli* from Sidney Rock oysters. These investigators were able to detect these organisms in 17 of 79 samples. This work is interesting, because the Sidney Rock oyster is usually harvested from water of fairly high salinity. *Campylobacter* species are reported to be very sensitive to environmental conditions; however, in the Sidney Rock oyster, survival was reported during refrigeration and freezing. Arumugaswamy et al. (1988) allowed the oysters to feed in waters containing approximately 10,000 cells of *C. jejuni* and *C. coli* per ml. The oysters were then subjected to depuration. They were depurated for the 48 h period usually allowed for depuration systems. These investigators also investigated survival of the organisms during storage as shellstock at 20°C and 30°C, on the half shell during refrigeration, shucked and bottled, stored, refrigerated, and frozen. Survival was better at the lower temperature and in the shucked product. The organisms survive for months during frozen storage at −20–24°C. The organisms failed to multiply during room temperature storage, but survived for 2–9 days. At 3°C or 10°C the organism survived 8–14 days growth.

Another *Campylobacter* species linked to illness in humans is *C. laridis*. This organism has been isolated from mussels (Owen et al. 1988); however, shellfish-associated disease in humans due to infection by *C. laridis* has not been documented.

Control

C. jejuni is a somewhat fragile microorganism and is easily destroyed by acids, sanitizers, and drying. It does not grow at refrigeration temperatures. *C. jejuni* temperature range for growth is 32–45°C. Control of *C. jejuni* is facilitated by the organism's lack of resistance to environmental factors (Franco 1988). It is sensitive to drying, oxygen, low pH, and heat. Its sensitivity to NaCl limits the importance of marine species as vehicles of campylobacteriosis (Doyle and Roman 1982). However, modified atmosphere packaging and careful attention to refrigeration may increase the risk of seafood-borne campylobacteriosis.

The optimum temperature for growth is 42°C to 43°C, and thermal inactivation occurs at 48°C. Heat injury and repair of *C. jejuni* occurs at 46°C, and the existence of an injured, viable but nonculturable form of the organism has been reported. *C. jejuni* survives best in foods held at refrigeration temperature.

C. jejuni is very sensitive to sodium chloride. The optimum concentration for recovery or enumeration of *C. jejuni* is 0.5% NaCl. Although *C. jejuni* dies more rapidly in the presence of sodium chloride, the organism can persist for weeks in refrigerated (4°C) foods with 6.5% salt.

Isolation Procedures

Campylobacters are both microaerophilic and capnophilic, i.e., the organism grows best under conditions of reduced atmospheric concentrations of oxygen and under high carbon dioxide concentrations. The optimum atmospheric composition for growth of *C. jejuni* is 5% oxygen, 10% carbon dioxide, and 85% nitrogen.

In order to enhance recovery the shellfish sample should be stored in an oxygen-free environment (100% N_2). The addition of 0.01% sodium bisulfite is suggested. Sodium bisulfite is a reducing agent that will help control exposure to oxygen.

Only small numbers of campylobacters are usually present in foods and food samples require selective enrichment. One of the most widely used enrichments is the Doyle and Roman (1982) enrichment broth, which consists of brucella broth plus added supplements. The procedure can detect as few as two cells of *C. jejuni* per gram of inoculated meat with a sensitivity of 96% (Doyle and Roman 1982). Selective agar media used for isolating *C. jejuni* from foods include Campy-BAP, the modified CCDA-Preston blood-free medium, the "Butzler" formulation, and the "Skirrow" formulation.

C. jejuni is catalase positive, oxidase positive, will reduce nitrate and will not ferment or oxidize glucose. *C. jejuni* produces detectable H_2S with the lead acetate test, whereas H_2S production is negative with TSI agar. The bacterium will grow in the presence of 1% glycine but will not grow in 3.5% NaCl. *C. jejuni* will grow at 42°C but will not grow at 25°C.

Serotyping can be obtained on request from the *Campylobacter* Reference Laboratory, Centers for Disease Control, for strains from outbreaks and for approved studies.

EDWARDSIELLA TARDA

General Overview

E. tarda causes opportunistic infections in humans, and appears to cause foodborne and waterborne diarrheal disease as well, although the latter is less certain (Farmer 1981). Healthy individuals rarely have stool cultures positive for *E. tarda* or carry antibodies against it, whereas isolation and seropositivity rates from some populations with diarrhea are much higher. In Zaire, there appears to be a relationship between *E. tarda* and *Entamoeba histolytica*, and it is possible that under certain conditions *E. tarda* plays the role of copathogen.

Edwardsiella tarda, like other members of the family *Enterobacteriaceae*, are small straight rods about 1 × 2–3 μm long (Farmer and McWhorter 1984). It grows well, but slowly, on commonly used enteric media.

The mode of action of *E. tarda* as an enteric pathogen is unclear (Farmer 1981). Environmental strains of *E. tarda* may generally be nonpathogenic, but sometimes become pathogens by acquiring virulence factors such as invasiveness or toxigenicity encoded by plasmids or phages from other *Enterobacteriaceae.*

Association with Foods

E. tarda is frequently isolated from animals, including fish and other aquatic animals, and their environment (Farmer 1981). Freshwater species of fish in Zaire and Viet Nam have high rates (50%) of intestinal carriage of *E. tarda*, corresponding to elevated rates of positive stool cultures from patients with diarrhea in those and other tropical countries (VanDamme and VanDepitte 1980). Although most isolates are from healthy animals, *E. tarda* causes emphysematous putrefactive disease in channel catfish (Meyer and Bullock 1973; Farmer 1981). The reservoir of *E. tarda* is the intestinal tract of animals, and human wound infections result from contact with *E. tarda* in the environment. It is rarely isolated from humans in developed nations.

CDC has no specific surveillance system for *E. tarda* infections; state health departments have not reported any foodborne outbreaks of such infections to FDSS (CDC 1983, 1985, 1989).

E. tarda has been suspected but not proven as a cause of shellfish-borne illness (Rippey and Verba 1988; Rippey 1991). It was isolated with other pathogens from shellfish implicated in human illness; therefore, the cause of the illness was not determined.

CLOSTRIDIUM PERFRINGENS

General Overview

Clostridium perfringens has been associated with human disease, mostly gas gangrene, for almost a century (Hobbs 1979; Labbe 1989). However, it was not until the 1940s that it was first associated with food poisoning (Labbe 1989). *Clostridium perfringens* food poisoning is associated with proteinaceous food-products due, in part, to the large number of growth factors required for its growth. *Clostridium perfringens* food poisoning is one of the most commonly reported types of human foodborne illness (Labbe and Harmon 1992).

Organism

C. perfringens is a gram-positive spore forming bacterium. Strains of the organism are usually divided into five toxin types (A–E), based on the production of four extracellular lethal toxins (α, β, ε, ι). All strains produce α-toxin (Labbe 1989). Types B–E produce one or more of the other lethal toxins. Virtually all food poisoning episodes are caused by type A; however, type C has been implicated in rare instances. Only type A strains are associated both with the intestinal tract of animals and soil. All other types are reported to be obligate parasites and while they can be isolated from animals and occasionally man, they are not isolated from soil (Hobbs 1979; Labbe 1989).

The food poisoning strains of type A produce relatively heat-resistant spores that will survive boiling for more than one hour; and heat-sensitive strains, whose spores do not survive boiling for 10 min (Labbe 1989). The food poisoning strains are usually nonhemolytic or produce only a small zone of hazy hemolysis on blood agar. In contrast, "classical" strains of *C. perfringens* produce spores that are killed at 10 min at 100°C and produce double zone hemolysis on blood agar (Labbe and Harmon 1992; IFT 1988).

Illness

The symptoms of *C. perfringens* food poisoning include severe abdominal cramps and a pronounced diarrhea. Nausea and vomiting are rare, and headache and fever are usually absent. The onset of symptoms is usually 8–12 h after ingestion of the food and the illness usually does not persist for more than 24 h (Hatheway et al. 1980; Banwart 1989). A heat-labile enterotoxin produced only by sporulating cells induces the major symptom of diarrhea in perfringens poisoning (Banwart 1989). The enterotoxin appears to be released *in vivo* in the intestine by the sporulating organism. The enterotoxin can be produced in food during sporulation, but it has not been proven that illness has resulted from preformed toxin in foods (Craven 1980).

Association with Foods

C. perfringens is common in raw meats, poultry, dehydrated soups, and certain other foods or food ingredients. Thus, its mere presence is inadequate to identify the suspected food. In food poisoning outbreaks, demonstration of

$> 10^5$ organisms per gram in an epidemiologically implicated food supports a diagnosis of perfringens poisoning when substantiated by clinical evidence.

The bacterium has exacting growth requirements, requiring 13 amino acid and 6 vitamins. Foods of animal origin are more likely to fulfill the growth requirements by *C. perfringens.* Meat and poultry products account for most of the reported illness, with seafood products only accounting for approximately 2% of the reported outbreaks (Banwart 1989).

Most of the outbreaks of *C. perfringens* food poisoning have been associated with food from food service establishments. Cooking of foods contaminated with *C. perfringens* will kill vegetative cells of the organism, but the spores will survive. Cooking tends to lower the oxidation/reduction potential of foods and heat-shocks the spores into activation, creating ideal conditions for growth of the organisms. Temperature abuse of the food after cooking allows the organism to grow to high numbers. *C. perfringens* grows very quickly, with a generation time of as short as 8.5 min reported in some foods (Willardsen et al. 1979). Growth can occur at temperatures as high as 50–52.3°C; thus, if warming trays in food service establishments are not kept at proper temperatures growth can occur. Time–temperature abuse of cooked products is usually a critical factor in food poisonings of *C. perfringens* origin. The number of organisms normally found in foods is usually low compared to the high number required to induce illness. The critical number needed to induce illness has been estimated at between 1×10^6 and 5×10^8 (Labbe 1988; Labbe and Harmon 1992).

C. perfringens is considered to be part of the normal microflora of soil, and the source of foodborne *C. perfringens* can be soil, dust, water, spices, or the food itself. Type A is the strain mostly associated with food poisoning. Virtually all soils examined have contained type A *C. perfringens* at levels of 10^3–10^4 per gram (Smith). It is also present in the intestinal contents of most animals at levels of 10^3–10^5. This level is usually attained in human infants after six months of age (Labbe 1988). This bacterium also has been isolated from the intestinal contents and surface of fish, but at considerably lower concentrations than those observed in mammals. In addition, it is often isolated at low levels from shellfish and has been suggested as an indicator of fecal pollution.

Association with Shellfish

The presence of *C. perfringens* in shellfish has been documented worldwide. Burrow (1974) reported that 56% of mussel samples were positive for the organism. Inal, Hildebrandt, and Siems (1974) also reported the isolation of *C. perfringens* from mussels in Turkey. Ayres (1975) reported the organism's isolation from a number of shellfish, including the European flat oyster, mussels, and hard clams. Fruin (1978) reported that most of the *C. perfringens* isolated from foods including clams were type A. Saito (1990) reported a high incidence of *C. perfringens* in oysters in Japan; 12% of the isolates from oysters were positive for enterotoxin production, compared to 6% of the isolates from food handlers, 2% of isolates from dogs, and 10% of water isolates.

Tia Son and Fleet (1980) observed that oysters (*Crassostrea commercialis*) were commonly contaminated with low levels of *C. perfringens* and *Bacillus cereus*. These organisms could be removed by depuration or relaying to clean water at rates similar to that of enteric bacteria such as *Escherichia coli*. *C. perfringens* in artificially contaminated oysters rapidly died off during refrigerated storage, whereas *B. cereus* populations remained stable.

Association with Seafood-Borne Illness

Examination of CDC data (CDC 1983, 1985, 1989) shows that *C. perfringens* accounted for 5.4% of the outbreaks and 16.6% of the cases of illness associated with fish over the 15 year period from 1973–1987. Additionally, it was responsible for 4.3% of the outbreaks and 3.1% of cases associated with shellfish during the same period. Since the illness associated with *C. perfringens* is usually mild, the number of cases are probably much higher than that reported to CDC.

The NETSU data base did not report any cases of shellfish-borne illness since 1894 (Rippey 1991). Since the CDC data base includes crustaceans in its classification of shellfish, it is possible that the shellfish-borne illness caused by *C. perfringens* reported by CDC may not have involved molluscan shellfish.

Analysis of the reported incidence of seafood-borne illness caused by *C. perfringens* indicates that *C. perfringens* is of little importance as a seafood-borne pathogen. It may be of greater importance as an indicator of pollution than as a pathogen. Madden, Buller, and McDowell (1986) recommended *C. perfringens* be the indicator of choice for depuration systems. Because the spores survive well in the environment, *C. perfringens* is far more likely to be present in polluted shellfish than *Escherichia coli*, and *C. perfringens* depurated from shellfish at similar rates. In addition, enumeration of the *C. perfringens* is easier than *E. coli*.

Control

The attributes of *C. perfringens* that contribute most to the organism's ability to cause foodborne illness are its ability to grow at high temperatures and the survivability of its spores. Therefore, it is important that cooked food be kept hot or cooled quickly. *C. perfringens* does not grow at refrigeration temperatures and will often be reduced in numbers at cold temperatures (Labbe and Hormon 1992). *C. perfringens* does not grow at pH values below 5.0 or above 9.0, but grows rapidly at values in between. The minimum water activity for growth is 0.95. It is not inhibited by 4% NaCl but is inhibited by 5–6% NaCl. Nitrite can be used for inhibition in cured meat systems. The spores are killed by normal canning temperatures but usually will survive cooking, drying, and freezing with little decrease. The vegetative cells are sensitive to these treatments (Craven 1980).

Spores of different strains of *C. perfringens* may vary widely in their heat resistance. Some may withstand 100°C for several hours, whereas others are inactivated by a few minutes or less at the same temperature. In most environments, heat-sensitive strains outnumber heat-resistant strains. Both heat-resistant and heat-sensitive strains may cause food poisoning.

Methods of Detection/Isolation

C. perfringens cells may lose viability if suspect foods are frozen or held under prolonged refrigeration before analysis, thereby making it difficult to incriminate the organism in food poisoning outbreaks. In such cases, gram-stained smears of food remnants or homogenates should be examined for the presence of large-celled bacilli typical of *C. perfringens*. The α-toxin method may be used to estimate nonviable populations of *C. perfringens*. Foods responsible for food poisoning outbreaks usually contain a sufficiently large number of cells to be readily detectable by direct microscopic examination.

Several solid media have been devised for quantitation of *C. perfringens* including neomycin blood agar, sulfite polymyxin sulfadiazine (SPS) agar, tryptone sulfite neomycin (TSN) agar, Shahidi Ferguson perfringens (SFP) agar, D-cycloserine blood agar, oleandomycin polymyxin sulfadiazine perfringens (OPSP) agar, tryptose sulfite cycloserine (TSC) agar, egg yolk-free tryptose sulfite cycloserine (EY-free TSC) agar, and trypticase-soy-sheep blood (TSB) agar (APHA 1985). TSC agar, or its modified form, EY-free TSC agar, has been documented as the most useful of the media for quantitative recovery of *C. perfringens*, with adequate suppression of the growth of practically all facultative anaerobes (APHA 1985; Easterbrook and West 1987; Labbe and Harmon 1992). The incorporation of D-cycloserine in TSC and EY-free TSC agar effectively inhibits growth of most enterococci. Presumptive sulfite-reducing colonies of *C. perfringens* from the selective differential media often are confirmed by their nonmotility, their ability to reduce nitrate to nitrite and their ability to liquify gelatin within 44 h and produce acid from raffinose within 3 days (APHA 1985).

Abeyta, Wekell, and Peeler (1985) compared the use of iron milk medium, [using the most probable number (MPN) procedure], and direct plating on tryptose sulfite cycloserine agar (TSC) and Shahidi-Ferguson-Perfringens agar. There were no significant differences among the procedures for enumeration of *C. perfringens* in clams, oysters, and turkey. Iron milk medium is less expensive than the direct plating methods; however, MPN procedures are less precise than direct plating procedures for *C. perfringens* (Easterbrook and West 1987).

Detection of *C. Perfringens* Enterotoxin

A number of serological assays have been reported for the rapid detection of enterotoxin. The most rapid methods available are the enzyme-linked immunosorbent assays (ELISA; 2, 22, 23) and a reversed passive latex agglutination assay (RPLA; 12) marketed by Oxoid U.S.A., Inc., Columbia, MD. The RPLA method is easier to perform than the ELISA (Labbe and Harmon 1992).

YERSINIA ENTEROCOLITICA

General Overview

Three species of *Yersinia* are pathogenic to humans: *Yersinia pestis*, *Yersinia pseudotuberculosis*, and *Yersinia enterocolitica*. *Y. enterocolitica* will be discussed in this section. Recently, several new species have been separated from *Y. enterocolitica*, as this species was originally defined, to distinguish what have been called "*Yersinia enterocolitica-like*" bacteria or "environmental" strains. Many of these new *Yersinia* species were previously distinguished as specific biotypes of *Y. enterocolitica*. The data indicates that none of these new species are pathogenic for humans (Schiemann and Wauters 1992).

Y. enterocolitica is not uncommon in the environment or in foods and, as a psychrotroph, can grow at refrigerator temperatures (Archer and Young 1988). Because most strains appear to be nonpathogenic, identifying pathogens by testing isolates for calcium dependence, autoagglutination, Congo Red binding, or other traits associated with pathogenicity are important aspects of control programs (Schiemann and Wauters 1992).

Organism

Yersinia enterocolitica is a gram-negative coccobacillus to rod-shaped bacterium 0.5–0.8 × 1–3 μm. It is nonmotile at 37°C, but motile with peritrichous flagella when grown at temperatures below 30°C. Sucrose, but not lactose, is fermented.

Yersinia is classified in the family *Enterobacteriaceae* and demonstrates the characteristics that define this family: gram-negative bacilli that are usually nitrate reductase positive, fermentative, oxidase negative, and facultative with respect to oxygen requirements. *Yersinia* is usually urease positive, and *Y. enterocolitica* is more biochemically active at 25°C than at 35–37°C. For example, a positive Voges–Proskauer test is observed only at the lower temperature (Doyle 1989; Schiemann and Wauters 1992).

Illness

Yersiniosis generally presents as an acute enteric disease manifested by watery diarrhea (in young children), mesenteric lymphadenitis mimicking appendicitis (in older children), or enterocolitis, with accompanying fever, headache, pharyngitis, anorexia, and vomiting. Erythema nodosum, Reiter's syndrome, and iritis are among the possible complications of yersiniosis. The incubation period can be from 4 to 10 days, complicating source tracing (Vantrappen et al. 1982).

Factors conferring pathogenicity on *Y. enterocolitica* are unknown, although it is likely that some interaction between plasmids and serotype-determining chromosomal genes that is involved (Archer and Young 1988; Miller and Falkow 1988). A chromosomal gene codes for a 103 KDa protein (invasin) that interacts with mammalian cells to cause active endocytosis (Isberg et al. 1987). In addition, pathogenic serotypes all contain a 40–50 MDa plasmid (Morawa et al. 1987). *Y. enterocolitica* produces a heat-stable enterotoxin similar to that of toxigenic *E. coli*, but only at temperatures below 25°C, so its clinical relevance is uncertain (Vantrappen et al. 1982).

Association with Foods

Y. enterocolitica is found worldwide in feces, food, water, and the environment (Lewis and Chattopadhyay 1986). It is commonly found in specimens from swine slaughterhouses, and consumption of raw sausage made from tonsil-containing pork scraps has been associated with disease (Tauxe et al. 1987). It has also been isolated from market meats, vacuum-packed beef, mussels, oysters, and ice cream (Morris and Feeley 1976). The organism grows well at 4°C, increasing its potential for foodborne illness spread by consumption of refrigerated foods.

The epidemiology of yersiniosis is quite different in North America, with the exception of eastern Canada, from that in many other countries. Yersiniosis in the United States has been characterized by periodic foodborne outbreaks, whereas it is endemic in eastern Canada, European countries, and Japan. Where yersiniosis is endemic, serotype 9:3 is most common, followed by 0:9 in incidence, both of which can be found as normal throat flora in slaughter-age pigs, on fresh porcine tongues, and in ground pork.

There is no specific surveillance system in place at CDC for yersiniosis. During the 14 years 1973–1986, State health departments did not report any outbreaks of yersiniosis associated with seafood. Likewise, NETSU did not report any shellfish-associated outbreaks of yersiniosis between 1894 and 1990 (Rippey 1991).

Three of the four documented foodborne outbreaks in the United States involved serotype 9:8, which has been the most common serotype associated with human illness in the United States and western Canada. Human carriers were the likely sources of the organism in two of these outbreaks.

Association with Shellfish

Y. enterocolitica has been isolated from finfish and shellfish, and is fairly common in oysters (Lee 1977). Lee (1977) was one of the first to report the isolation of *Y. enterocolitica* from oysters. He reported the isolation of atypical strains which are now recognized as a separate species. In a study of finfish and shellfish from Puget Sound, 3.8% of fresh seafood sampled in retail markets were positive for *Y. enterocolitica* (Abeyta 1983).

Colburn et al. (1989) examined the microbiological quality of oysters (*C. gigas*) and water of live holding tanks at five different Seattle, Washington markets. One oyster sample contained a nonpathogenic *Y. enterocolitica*. In addition, *Yersinia* species were isolated from a tank at two different sampling times.

Peixotto et al. (1979) studied the presence and growth of *Y. enterocolitica* in oysters (*C. virginica*), shrimp, and crab. These workers were able to recover the organism in six of 45 oysters samples. The isolates differed from clinical isolates and they may have been nonpathogenic. Counts increased in raw oysters during 14–21 days of storage at 0–2°C. Storage at higher temperatures, 5–7°C, resulted in growth between 2 and 10 days.

Gizzarelli et al. (1985) examined the behavior of *Y. enterocolitica* in artificially contaminated mussels *Mytilus galloprovincialis*. When either 10^2 or 10^4 *Y. enterocolitica* per ml of tank water was added, the levels of *Y. enterocolitica* remained constant in the water. In the mussels, the levels initially rose but later paralleled that of the water. The mussels were removed from the water and stored at 4°C for four days. The counts of *Y. enterocolitica* increased by one log in most samples during storage. In separate depuration studies it was shown that *E. coli* was removed far more quickly from the mussels than *Y. enterocolitica*. It took 24 h of depuration to remove *Y. enterocolitica* (at the level of 100/g), whereas *E. coli* was gone after just 6 h.

Methods of Detection/Isolation

Yersinia spp. grow well on blood agar and bile-containing media, such as MacConkey agar, and recovery from otherwise uncontaminated material such as blood is not difficult. Because of overgrowth by more rapidly growing microorganisms, cold enrichment in buffered saline for 2–3 weeks is sometimes necessary to isolate small numbers of *Yersinia* from contaminated specimens. The type of food being cultured should influence the enrichment method chosen (Peixotto et al. 1979).

The most efficient procedures for recovery of *Y. enterocolitica* from foods have incorporated at least one and often two enrichment steps before plating onto selective-differential agar media. "Cold enrichment" has commonly used 4°C as the incubation temperature. Some work suggests that enrichment at 10°C for 3 days or at 15°C for 2 days is effective.

Many highly selective enteric plating media used for isolation of *Yersinia* tend to be inhibitory at temperatures of 35°C to 37°C, but are effective at 22–25°C.

Yersinia is more resistant to high pH than other gram-negative bacteria. This property has been exploited for isolation of *Yersinia* by treating enrichment broths for a very short period to alkali (KOH) before plating them onto agar media. If alkali treatment is used, it should be in addition to secondary enrichment and not as a substitute for this step.

HEMORRHAGIC *ESCHERICHIA COLI*

General Overview

An indepth discussion of pathogenic *E. coli* is given in the chapter on bacterial pathogens whose primary hosts are humans. General methods for isolating the organisms are discussed and will not be repeated here.

The hemorrhagic strain (0157 : H7) causes a hemorrhagic colitis-bloody diarrhea. *E. coli* serotype 0157 : H7 strains are not invasive, but produce large quantities of one or more related, potent toxins that cause severe damage to the lining of the intestine. These toxins, verotoxin (VT), shiga-like toxin are closely related or identical to the toxin produced by *Shigella dysenteriae*. Hemorrhagic *E. coli* is most often associated with cattle, and the majority of cases are traced to raw or undercooked ground beef; rare roast beef. Unpasteurized milk has also been implicated (Doyle and Schoeni 1987; Kirshman et al. 1987; Remis et al. 1984; Hitchins et al. 1992).

Illness

Enterohemorrhagic *E. coli* causes hemorrhagic colitis, hemolytic uremic syndrome, and thrombotic thrombocytopenic purpura (MacDonald et al. 1988). Diarrhea tends to be bloody and abdominal cramps are severe. The illness is characterized by severe cramping (abdominal pain) and diarrhea which is initially watery but becomes grossly bloody. Occasionally vomiting occurs. Fever is either low-grade or absent. The illness is usually self-limited and lasts for an average of 8 days. Some individuals exhibit watery diarrhea only (Levine 1987). The infective dose is unknown. However, from a compilation of outbreak data, including the organism's ability to be passed person-to-person in the day-care setting and nursing homes, the dose may be similar to that of *Shigella* spp. (ten organisms). Some victims, particularly the very young, have developed the hemolytic uremic syndrome (HUS), characterized by renal failure and hemolytic anemia. In the elderly HUS plus two other symptoms, fever, and neurologic symptoms, constitutes thrombotic thrombocytopenic purpura (TTP). This can have a mortality rate in the elderly as high as 50% (Levine et al. 1978; Levine 1987).

Hemorrhagic colitis infections are not often reported, but this is probably not reflective of the true frequency. In the Pacific Northwest, *E. coli* 0157 : H7 is thought to be second only to *Salmonella* as a cause of bacterial diarrhea. Due to the unmistakable symptoms of profuse, visible blood in severe cases, those victims probably seek medical attention, but less severe cases are probably more numerous.

Association with Shellfish and Shellfish-Borne Illness

Unlike other pathogenic strains of *E. coli*, *E. coli* 0157 : H7, seems to principally have a bovine reservoir. Therefore, it would only be expected to be a seafood-borne health hazard when harvest waters are polluted by run off from cattle production units. Research is needed to determine its survival in the estuarine environment. No seafood-associated cases have ever been reported to CDC or NETSU (Rippey 1991; CDC 1989).

Methods of Isolation

Hemorrhagic colitis is diagnosed by isolation of *E. coli* of serotype 0157 : H7 or other verotoxin producing *E. coli* from diarrheal stools. Alternatively, the stools can be tested directly for the presence of verotoxin. Confirmation can be obtained by isolation of *E. coli* of the same serotype from the incriminated food.

E. coli 0157:H7 will form colonies on agar media that are selective for *E. coli.* However, the high temperature growth procedure normally performed to eliminate background organisms before plating cannot be used because of the inability of these organisms to grow at temperatures of 44.0–45.5°C that support the growth of most *E. coli.* The use of DNA probes to detect genes encoding for the production of verotoxins (VT1 and VT2) is the most sensitive method devised (Hitchins et al. 1992).

LISTERIA MONOCYTOGENES

General Overview

Listeria has been recognized as an infective agent for humans and other animals since the early 20th century, and reports of illness resembling listeriosis are described in the literature as early as 1891 (Donnelly 1988). *Listeria monocytogenes* was first definitively described by Murray et al. in 1926 (cited in Donnelly 1988) in connection with an epizootic among laboratory-raised guinea pigs and rabbits. Prior to the 1980s, listeriosis was primarily a veterinary concern, where it was associated with abortions and encephalitis in sheep and cattle. It was recognized that many of the animal listeriosis cases were foodborne. While cases of human listeriosis appeared in the literature in the 1960s, it was not until the 1980s that the organism became recognized as a foodborne pathogen. Only the hemolytic species of the genus have documented virulence. These include *L. monocytogenes, L. seeligeri,* and *L. ivanovii*; in as much as only *L. monocytogenes* is the only species consistently pathogenic, the discussion in this chapter will be restricted to *L. monocytogenes.*

Organism

The genus species name *Listeria monocytogenes* was first proposed in 1940. Before that names included *Bacillus hepatis, Bacterium monocytogenes,* and *Listerella hepatolitica. Listeria* are usually uniformly shaped gram-positive, nonspore-forming, rod-shaped bacteria approximately 0.4–0.5 × 0.5–2.0 μm. In older cultures the organism may appear larger and somewhat curved (Seeliger and Jones 1986). Motility is produced by a few peritrichous flagella at temperatures of 20–25°C, but motility is reduced or absent at 37°C. The cells show characteristic tumbling motility when grown at 20–25°C. *L. monocytogenes* colonies produce a narrow zone of *β*-hemolysis on blood agar, caused by a soluble hemolysin. The production of *β*-hemolysis on blood agar plates has been closely linked with pathogenicity (Gaillard et al. 1986). Often the reactions are weak and may not extend beyond the margin of the colony at 24h. Colonies appear somewhat small (0.0–1.5 mm), low convex, and dew-shaped. In addition, the colonies have a characteristic blue green appearance when observed by oblique lighting (Seeliger and Jones 1986). They grow best at temperatures of 30–37°C, but will grow over a range of temperatures from 1°C to 45°C. *L. monocytogenes* does require preformed growth factors including biotin, riboflavin, thiamine, isoleucine, leucine, and valine. In addition, growth is stimulated by arginine, histidine, methionine, and tryptophan (Seelinger and Jones 1986). The organism grows over a pH range of 5.0–9.6, but also survives in food products with pH levels outside these parameters (Connor, Brackett, and Beuchat 1986). Growth occurs in 10% NaCl, and *Listeria* survive in 25.5% salt at 4°C (Mitscherlich and Marth 1984).

The organism is facultatively anaerobic, but growth is best under microaerophilic conditions with 5–10% CO_2. The bacterium grows well anaerobically and aerobically. In general, the isolation of pathogenic *Listeria* spp. from clinical materials is made without difficulty, and the bacteria grow readily in a variety of common laboratory media. Isolation of *Listeria* from nonsterile specimens requires enrichment followed by selective plating (Donnelly et al. 1992).

Both the somatic (O) and flagellar (H) antigens are used to determine antigenic types of *L. monocytogenes.* The species is divided into 16 serotypes 1/2a, 1/2b, 1/2c, 3a, 3b, 3c, 4a, 4ab, 4b, 4d, 4e, 7, 4f, 4g, 5, and 6 (Seeliger and Hohne 1987; Pearson and Marth 1990). Serotypes 4b and 1/2a have caused the majority of the listeriosis cases, 59% and 18%, respectively. A bacteriophage typing system has been developed for some of the strains, but cannot be used for all strains.

Clinical Features

L. monocytogenes can infect humans and animals in a number of routes, including oral (food- and waterborne), ocular, cutaneous, respiratory, or urogenital. This review will concentrate on the infection by the oral route. Infection by the organism can exhibit a number of clinical features depending upon the physiological state of the host and the virulence of the ingested strain.

Host factors to a large extent determine the severity of the illness (Mascola et al. 1988; Beninger, Savoia, and Davis 1988; Gould et al. 1986; Donnelly et al. 1992). Host factors include the age, underlying illness, state of the immune system, and pregnancy. In compromised individuals, illnesses that may be only a minor inconvenience to a healthy person can be life threatening. Also, the number of organisms needed to cause illnesses in compromised individuals is often far less than that required for healthy individuals (Archer and

Young 1988). Host factors will be of increasing concern to food processors in the future as a greater portion of the population will be more susceptible to infection by foodborne pathogens and more likely to suffer severe consequences of infection (Archer and Young 1988; Davis et al. 1989). *L. monocytogenes* most often causes illness in four subsets of the population: immunocompromised individuals, pregnant women, fetuses and young children, and the elderly (Mascola et al. 1988; Beninger, Savoia, and Davis 1988; Gould et al. 1986; Read et al. 1985; Seelinger and Finger 1976). While these populations are at the greatest risk, *L. monocytogenes* can also infect and cause disease in persons who do not have recognized predisposing factors.

L. monocytogenes is a facultative intracellular parasite. In foodborne infections, the bacterium enters the body through the intestine. It has a variable incubation period before onset of symptoms, ranging from one day to one month or longer. The ingested cells enter the body through the ileal villi and are subsequently phagocytized by macrophages (Marth 1988). *Listeria* induces its uptake by mammalian cells, and from its intracellular location causes disease. One recognized virulence mechanism of *L. monocytogenes* is a hemolysin, listeriolysin, which promotes intracellular survival (Kvenburg 1991; Marth 1988; Pearson and Marth 1990). Listeriolysin inhibits macrophage-mediated presentation of antigen to T cells, which interferes with CMI against *Listeria* (Cluff and Ziegler 1987). If the illness proceeds to more severe disease, the clinical spectrum can include meningitis, septicemia, and perinatal disease (Nieman and Lorber 1980; Lamont, Postelthwaite, and MacGowan 1988; Newman, Waycott, and Cooney 1979). Other manifestations include brain abscesses, cranial nerve palsies, diffuse encephalitis, endocarditis, pneumonia, hepatitis, and arthritis (Numan et al. 1979). Changes occurring in the blood include accelerated sedimentation rate and granulocytic leukocytosis with predominant granulocytosis (an increase in the number of cells within the cytoplasm that contain granules usually with an increase in white blood cell count). In addition, spinal fluid appears turbid with the appearance of monocytes, and pressure is elevated (Seelinger and Jones 1986; Ciesielski, Swaminathan, and Broome 1987). Treatment usually consists of antibiotics such as ampicillin, penicillin, or erythromycin.

Neonatal listeriosis occurs in two forms (Lamont, Postelthwaite, and MacGowan 1988). The late-onset disease appears between days 3 and 35 of life usually as meningitis; maternal symptoms are usually absent and infection of the neonate generally occurs during passage down a contaminated birth canal. The more common early-onset disease results from intrauterine infection, appears during the first 5 days of life, and is characterized by septicemia, pneumonia, diarrhea, and CNS signs. Typically, the mother reports a febrile flu-like illness shortly before delivery. Common symptoms of the mother include fever, chills, backache, headache, and discolored urine.

Meningitis can occur in the pregnant host but is very uncommon. The organism can be isolated from blood, umbilical cord blood, lochia (discharge from uterus and vagina following delivery of a baby), tissue obtained by curettage, vaginal mucus, urine, and placental tissue (Seelinger and Jones 1986; Lennon et al. 1984).

Epidemiology

It has been difficult to estimate the extent of listeriosis in the United States. Before 1986 listeriosis was not reportable, and now its reporting system is laboratory-based and passive. Outbreaks of listeriosis have been traced to cole slaw, soft cheeses, and milk. In 1986 and 1987, CDC and six health departments conducted active surveillance for culture-confirmed sporadic listeriosis in Los Angeles County, Missouri, New Jersey, Oklahoma, Tennessee, and Washington (Broome et al.). The study identified 229 cases from the 34 million persons living in the study areas, providing minimum estimates of 1600 cases of massive listeriosis and 400 deaths per year. A case-control study performed in conjunction with the active surveillance project identified consumption of raw hot dogs and undercooked chicken as risk factors for acquiring listeriosis; questions on seafood consumption were not included on the questionnaire (Schwartz et al. 1988). The CDC study was modified slightly and repeated in 1988–1990 (Schuchat et al. 1991). The more recent study supported the estimated rates for invasive listeriosis, and confirmed the importance of soft cheeses and *L. monocytogenes*; although questions about seafood consumption were asked, no seafood products were identified as risk factors in this study.

Distribution

Listeria monocytogenes is widely distributed. The microorganism is not easily inactivated by environmental influences, such as sunlight or freezing and thawing. *L. monocytogenes* has been isolated from numerous environmental sources such as silage, soil, decaying vegetation, sewage, damp earth, straw, and feces (Donnelly et al. 1992). *L. monocytogenes* is saprophytic and lives in close association with soil and plants. *L. monocytogenes* survival in soil and on vegetation does not seem to depend on continual reintroduction, but rather from *Listeria*'s ability to exist as a free-living organism in these environments.

Listeria is found in association with a wide variety of fish, birds, and domestic and wild mammals (Gray and Killinger 1966; Gray 1958; Brackett 1988). In humans, incidence rates of intestinal carriage have ranged from 0.6% to 70% in various populations (Kampelmacher and van Noorle Jansen

1969, 1972; Ralovich 1986). The carriage rate varies among animal species and may somewhat depend on their feeding habits. For example, the carriage rate is higher in sea gulls that fed on garbage compared to ones that did not.

Association with Foods

Contaminated food is becoming increasingly well recognized as an important vehicle of *L. monocytogenes* (McLauchlin 1987; Ciesielski, Swaminathan, and Broome 1987). Prevalence in selected food products as reported by Doyle (1989) is: poultry (15% to 53%), dry sausage (22%), sausage (4.1%), frozen chopped beef (9.6%), and lamb (14%). These rates were reported from studies in several countries. In the United States, a study conducted by the FDA showed that the incidence of *L. monocytogenes* in foods is shrimp (5.1%), cooked lamb (4.9%), cheddar cheese (1.4%), Swiss cheese (0%), butter (0%), ice cream (3.8%), and novelty ice cream (8.9%) (Doyle 1989). An FDA study of the incidence of *L. monocytogenes* in various seafoods, including shrimp, crabmeat, lobster tail, langostinos, scallops, surimi-based imitation seafoods and cold smoked fish, found 26% of the samples positive for *L. monocytogenes* (Pearson and Marth 1990). FDA found *L. monocytogenes* in seven of eight categories of seafood cultured in fiscal year 1988 (FDA Pathogen Surveillance Program).

In the 1980s dairy products received the most attention with respect to *Listeria.* Although only two well documented foodborne outbreaks have been associated with dairy products in the U.S., the organism has been isolated from a number of dairy products, including Brie soft ripened cheese, Camembert cheese, Mozzarella cheese, chocolate milk, and ice cream (Rosenow and Marth 1987). A nationwide survey performed in 1986 showed that the incidence of *L. monocytogenes* in raw milk varied from 0% to 7%. The overall incidence of *L. monocytogenes* in raw milk was estimated to be 4.2%. Twenty-five of the 27 strains (93%) isolated from the raw milk were pathogenic to humans. The concentration of *L. monocytogenes* in raw milk has been shown to be low; probably < 1.0 cell/ml (Lovett, Francis, and Hunt 1987).

L. monocytogenes can be isolated from a number of animals which makes it likely that poultry and red meats are contaminated with the bacterium. *L. monocytogenes* was isolated from 57% of fresh and frozen poultry that were sampled. It has also been reported that 15% of oven-ready poultry samples contained *L. monocytogenes* (Brackett 1988). In another study, 70% of ground beef, 43% of pork sausage, and 48% of poultry were found to be contaminated with the organism (Lee and McClain 1986; McLauchlin 1987). *L. monocytogenes* is less often isolated from fruits and vegetables. It has been identified in fresh lettuce, celery, tomatoes, and cabbage (Beuchat et al. 1986).

Association with Molluscan Shellfish

Listeria monocytogenes has only rarely been isolated from molluscan shellfish. While no confirmed cases of listeriosis have implicated molluscan shellfish in the United States or other countries, an outbreak of perinatal listeriosis was tentatively attributed to consumption of raw finfish and shellfish in New Zealand. In this study 22 cases occurred in an eleven month period during 1980. While the cause of the epidemic was not determined, the authors stated that shellfish and raw fish consumption may have played a part. This conclusion was based on examination of dietary habits of the mothers. The greatest number of cases and highest rectal carriage rate was in women of Pacific Island Polynesian descent. This group tended to consume more shellfish and raw fish (Lennon et al. 1984). Overall reports of isolation of *L. monocytogenes* from molluscan shellfish are rare.

Isolation, Identification, and Enumeration

There are three general types of methods presently being used to analyze foods for *L. monocytogenes*. The first, and least used of these, is direct plating of a food suspension onto a selective solid medium. Among the best media for direct plating (Cassidy et al. 1988*a*, *b*) are LiCl-phenylethanol-moxalactam agar (LPM; Lee and McClain 1986), Modified Vogel–Johnson Agar (MVJ; Buchanan, Stahl, and Acher 1987), and Despierres agar (Donnelly et al. 1992). Problems associated with direct plating include low levels present in foods and the possible presence of injured cells. Injured *L. monocytogenes* not detected by conventional enrichment procedures may be resuscitated and grow at refrigeration temperatures and present a consumer health problem. The level of *L. monocytogenes* in foods in most cases is less than 100 per gram (Lovett 1988). This means that enrichment is generally necessary to accurately detect the organism. Almost all of the methods developed for isolation of *L. monocytogenes* use one or both of two distinct characteristics of the bacterium for enrichment: the ability of *Listeria* to grow at refrigeration temperatures and its resistance to many antibiotics. The most frequently used methods to isolate the organism is a cold enrichment. A 4°C cold enrichment was first used by Gray and Killinger in 1948 (Bryan 1979). Cold enrichment is typically used as a pre-enrichment step. During cold enrichment at 4°C, the psychrotrophic *L. monocytogenes* multiplies to detectable numbers. The major disadvantage of cold enrichment is the time needed—it may take from a few weeks up to several months to isolate the organism.

Current enrichment procedures use antibiotics to inhibit competitors along with higher incubation temperatures and take only a few days. A number of enrichment procedures have been developed. Doyle (1989) took advantage of the microaerophilic nature of *L. monocytogenes* for developing a scheme for isolating the organism from foods. No cold enrichment procedure was utilized. Instead, antibiotics and the microaerophilic environment were used to further select for *L. monocytogenes.*

Isolation media often used after enrichment of *L. monocytogenes* include Modified McBride *Listeria* (MML) agar, McBride *Listeria* (ML) agar, and Martin's *Listeria* agar (MaL) (Lovett 1988). Lithium chloride polyphenylethanol moxalactam (LPM) agar has been recommended for isolation of *L. monocytogenes* from meat and poultry products, and is extremely selective for the organism (Lee and McClain 1986). More recently, Oxford and modified Oxford agar have been recommended for the isolation of *Listeria* from food, meat, and other products. One of the problems with most isolation media is overgrowth of *Staphylococcus* spp. Cultural similarities to *Staphylococcus* spp. have been demonstrated by the abundance of new *Listeria* isolation media containing selective and differential agents designed to recover staphylococci (Buchanan, Stahl, and Archer 1987).

A commonly used method for the isolation of *L. monocytogenes* in both the U.S. and in Europe is a procedure developed by the Food and Drug Administration (FDA) (Lovett, Francis, and Hunt 1987, 1991). The US FDA method has been successfully used for isolation of *Listeria* in seafoods, dairy products, and other ready to eat foods.

The United States Department of Agriculture (USDA) is using a different procedure to detect *L. monocytogenes* in meat products. This procedure was developed because older US FDA methods did have the sensitivity needed to detect *Listeria* in heavily contaminated ground beef samples (McClain and Lee 1988). The method developed at the USDA and later modified involves a two-stage enrichment procedure. The primary enrichment broth contains nalidixic acid and acriflavine for selectivity. After incubation of the primary enrichment broth, the culture is transferred into a second enrichment broth, developed by Fraser and Sperber (1988). This medium is a modification of the original USDA secondary enrichment broth, containing lithium chloride and ferric ammonium citrate. Esculin hydrolysis, indicative of the presumptive presence of *Listeria,* is noted by observing blackening in the broth. Colonies are streaked onto modified Oxford agar. Modified Oxford agar contains mannitol to help distinguish between colonies of *Staphylococcus* and *Listeria* (Difco 1991). The former are able to ferment mannitol with the formation of acid, whereas *Listeria* does not.

Presumptive identification of *Listeria* focuses upon colony morphology, Gram reaction, tumbling motility, catalase reaction and for *L. monocytogenes,* β-hemolysis. *Listeria* are catalase-positive and oxidase-negative. They ferment carbohydrates to acid but do not produce gas. They hydrolyze esculin and sodium hippurate and are methyl red positive. Ammonia is produced from arginine. *Listeria* give negative reactions for hydrogen sulfide production, indole, nitrate reductase, gelatin liquefaction, and starch and urea hydrolysis (Lovett 1987).

Rapid and Alternative Methods

All of the isolation methods described involve lengthy procedures to complete confirmation of the organism. To shorten identification time, rapid methodology has been developed, including flow cytometry, an enzyme linked immunosorbant assay (ELISA) and a β-hemolysin gene probe.

Donnelly, Baigeut, and Briggs (1988) developed a method of flow cytometry. Flow cytometry characterizes a bacterial cell population by morphology, nucleic acid profile, and surface antigens as the cell passes through a laser beam. The species must be labeled with polyvalent antibodies for cell surface antigens and propidium iodide for cellular DNA (Donnelly et al. 1988; Pearson and Marth 1990). Unfortunately, current methods involving flow cytometry of *L. monocytogenes* give many false positive and false negative results.

Farber and Speirs (1987) described the use of ELISA (enzyme-linked immunosorbent assay) to detect *Listeria* in raw milk. This procedure involved the use of monoclonal antibodies specific for *Listeria* flagellar antigens. In addition, a monoclonal antibody procedure specific for *L. monocytogenes*, *L. innocua*, and *L. welshimeri* has been developed. All strains of *L. monocytogenes* were positive using this monoclonal antibody assay (Siragusa and Johnson 1990). This technique using antibodies against *L. monocytogenes* holds great promise for a variety of foods, but further laboratory and field research needs to be conducted before this methodology can be applied on an industrial scale. A rapid commercial assay is *Listeria*-tek, available from Organon-Teknica (Durham, NC). *Listeria*-tek, is an ELISA-based assay which employs two monoclonal antibodies specific to *Listeria* (Butman et al. 1988). Total assay time required to perform this test is 40 h.

All virulent strains of *L. monocytogenes* have been found to contain a β-hemolysin. In 1987, Datta et al. developed a β-hemolysin gene probe. The DNA probe contains a 500 base pair strand complementary to the β-hemolysin gene and has been used primarily for research purposes and for naturally contaminated raw milk samples. Current concerns with the gene probe are that contaminated product must have >10 *L. monocytogenes*/g and not all strains of *L. monocytogenes* react with the probe. The first commercially available gene probe from GENE-TRAK Systems (Framingham, MA) is a nucleic acid hybridization assay based on detection of unique *Listeria* 16S RNA sequences (Klinger et al. 1988).

TERRESTRIAL-ASSOCIATED BACTERIAL PATHOGENS

BACILLUS CEREUS

General Overview

Bacillus cereus is a ubiquitous microorganism, easily isolated from dust, soil, and near-shore sediment. It can also be isolated from the feces of 14–15% of people (Kramer and Gilbert 1989; Harmon et al. 1992).

B. cereus is a gram-positive, facultatively aerobic sporeforming rod, occurring in short to long chains (Breed et al. 1957). The cells are large and the spores do not swell the sporangium. Spores are usually abundant after 24 h of growth but are mostly absent after 48 h of growth. The organism sporulates freely in most nonselective well-aerated media (Breed et al. 1957). *B. cereus* is gelantinolytic and proteolytic. These and other biochemical characteristics are used to differentiate and confirm the presence of *Bacillus* spp. although these characteristics are shared with *B. cereus* var. mycoides, *B. thuringiensis* and *B. anthracis.* Differentiation of *B. cereus* from related species depends on motility (most *B. cereus* are motile), presence of toxin crystals (*B. thuringiensis*), hemolytic activity (*B. cereus* and others are beta hemolytic while *B. anthracis* is usually nonhemolytic) and rhizoid growth which is (characteristic of *B. cereus* var. mycoides).

There are two distinct types of *B. cereus* food poisoning (diarrheal and emetic), each caused by a distinct metabolite. The diarrheal type of illness is caused by a large molecular weight heat liable protein, while the vomiting (emetic) type of illness is believed to be caused by a low molecular weight, heat-stable peptide.

The symptoms of *B. cereus* diarrheal food poisoning are similar to those observed in *Clostridium perfringens* food poisoning. The onset of watery diarrhea, abdominal cramps, and pain occurs 6–15 h following consumption of contaminated food. Nausea may accompany diarrhea, but vomiting rarely occurs. Symptoms persist for 24 h in most instances.

The emetic type of food poisoning is characterized by nausea and vomiting within 0.5–6 h after consumption of contaminated foods. Occasionally, abdominal cramps and/or diarrhea may also occur. Duration of symptoms is generally less than 24 h. The symptoms of this type of food poisoning parallel those caused by *Staphylococcus aureus* foodborne intoxications (Harmon et al. 1992).

The type most likely associated with shellfish is the diarrheal type. The emetic type has almost exclusively been associated with rice and starchy products. The presence of large numbers of *B. cereus* (greater than 10^6 organisms/g) in a food is indicative of active growth and proliferation of the organism, and is consistent with a potential hazard to health. These high numbers could be reached during prolonged storage out of water.

Kawano et al. isolated *B. cereus* from oysters in Japan. This group noted that enterotoxigenic *B. cereus* are widely distributed in nature. Vasconcelos and Lee (1972) isolated *Bacillus* species from the Western oyster (*Crassostrea gigas*). Thi Son and Fleet (1980) observed that oysters (*Crassostrea commercialis* were commonly contaminated with low levels of *C. perfringens* and *Bacillus cereus*. These organisms could be removed by depuration or relaying to clean water. Their depuration rates were similar to that of enteric bacteria such as *Escherichia coli*. They further observe that *C. perfringens* in artificially contaminated oysters rapidly died off during storage, whereas counts of *B. cereus* remained stable during refrigeration.

6

Association of *Vibrionaceae*, Natural Toxins, and Parasites with Fecal Indicators

John Liston

University of Washington,
Institute for Food Science and Technology,
3707 Brooklyn Avenue, NE, Seattle, WA 98195

VIBRIONACEAE

Nature and Occurrence

This family includes genera of straight or curved, rod-shaped, gram-negative bacteria, typically characterized by fermentative utilization of glucose and production of cytochrome oxidase (Baumann and Schubert 1984). These characteristics differentiate the family from the oxidase-negative *Enterobacteriaceae* on the one hand and the nonfermentative *Pseudomonadaceae* on the other. Three genera of the *Vibrionaceae*, *Vibrio*, *Plesiomonas* and *Aeromonas*, are commonly associated with aquatic environments, and species in each genus have been implicated in shellfish-borne human disease. The genus most frequently associated with disease is *Vibrio*, while only a single species in each of the other genera have been suggested to be a cause of shellfish-borne disease.

Vibrio

Bacteria in this group are curved, actively motile rods widely distributed in both marine and fresh waters. The genus includes both psychrotrophic and mesophilic species, but only the mesophiles are significant as a cause of human disease (Baumann, Furniss, and Lee 1984). The three species most prominently identified as causing shellfish-associated illnesses are *V. cholerae*,

V. parahaemolyticus and *V. vulnificus*. Other species implicated less frequently in such diseases include *V. mimicus*, *V. fluvialis*, *V. hollisiae* and *V. furnissi*. *V. cholerae* and *V. mimicus* have a low Na requirement but all the other species are mildly halophilic. *V. cholerae* is separated into two groups on the basis of the O1 somatic antigen. Strains possessing the O1 antigen typically cause cholera, and shellfish have been implicated as a vehicle in cholera outbreaks (DePaola 1981; Morris and Black 1985; Madden, McCardell, and Morris 1989). *V. cholerae* non-O1 is frequently identified as a cause of gastroenteritis of varying severity, associated with the consumption of raw shellfish (Morris and Black 1985; Janda et al. 1988). Another *Vibrio* associated with shellfish-related gastroenteritis is *V. parahaemolyticus*, which is probably the most widely distributed of all the pathogenic vibrios in the marine environment (Nolan et al. 1983; Morris and Black 1985; Janda et al. 1988). *V. vulnificus* is a particular invasive organism occurring in shellfish which causes a frequently lethal disease, primary septicaemia, in certain susceptible individuals (cirrhotics, hematochromatosis, immunodeficient) who have eaten raw oysters containing large numbers of bacteria of this species (Tacket, Brenner, and Blake 1984; West 1989). The other vibrios listed have been implicated in gastroenteritis of varying severity but usually not lethal in outcome, associated with the consumption of (usually) raw shellfish (Morris et al. 1982; Shandera et al. 1983; Brenner et al. 1983; Spellman et al. 1986).

All the *Vibrio* species implicated in shellfish-related disease appear naturally in the inshore and estuarine marine environments. This is true even in the case of *V. cholerae* O1, which was generally believed to be derived from human feces but is now recognized also to exist at a low level in estuarine waters of USA and, more sporadically, in living shellfish (Colwell, Kaper, and Joseph 1977; Feacham 1981; Feacham, Miller, and Drasar 1981; DePaola 1981; Kaper et al. 1982; Klontz et al. 1987; Kaysner et al. 1987). Strains of *V. cholera* non-O1 appear to be widely distributed in estuarine waters and shellfish and are ubiquitous in coastal waters of USA (Twedt et al. 1981; Hood et al. 1983; DePaola et al. 1983; Kaysner et al. 1990). Of all the *Vibrio* species considered here, *V. parahaemolyticus* is the most frequently isolated one and is consistently recovered from shellfish when water temperatures are above 15°C (Twedt 1989; DePaola et al. 1990). *V. vulnificus* has been shown to be frequently present in shellfish, particularly oysters, harvested from warm waters but the organism is also isolated from shellfish in northern regions during summer months (Roberts et al. 1982; Oliver, Warner, and Cleland 1983; Kaysner et al. 1987; Tilton and Ryan 1987). The other *Vibrio* species listed earlier have been isolated less frequently, but all have been associated with shellfish (Davis et al. 1981; Barbay et al. 1984; Morris and Black 1985; West 1989; Kaysner et al. 1990).

The most important determinant of frequency and the level of occurrence of mesophilic vibrios in shellfish and inshore waters is temperature. There are numerous reports in the literature concerning the difficulty of isolating vibrios when environmental temperatures are below 15°C, and seasonal variations in the numbers of vibrios in shellfish directly related to ambient water temperature have been described (Baross and Liston 1970; Tilton and Ryan 1987; West 1989). The cycle of the rise and fall of *Vibrio* populations has been linked to corresponding patterns of occurrence of chitinous zooplankters (Kaneko and Colwell 1973). Some reports have suggested a relationship between the numbers of mesophilic vibrios in water and the concentration of suspended organic materials and this has been misinterpreted as a relationship between *Vibrio* occurrence and sewage contamination in a few cases. However, the relationship is clearly nutritional in nature (Liston and Baross 1973; Singleton et al. 1982; Miller, Drasar, and Feacham 1984). Another determinant that determines the range of occurrence of particular *Vibrio* species is salinity, although as a group they may be isolated throughout most estuaries (Hood et al. 1983; West 1989). One factor that complicates the determination of the occurrence of vibrios in the aquatic environment is the recent discovery that these organisms may convert to a dormant nonculturable form when exposed to adverse nutrient conditions (Xu et al. 1983; Roszak and Colwell 1987).

Plesiomonas

This genus contains only one species, *P. shigelloides*, a gram-negative rod-shaped organism widespread in nature and found commonly in fresh waters and, less frequently, in inshore seawater (Schubert 1984). This organism appears to be an opportunistic pathogen and may be a cause of foodborne human gastroenteritis, although this is disputed by some scientists. The evidence for its role as an enteric pathogen is largely epidemiological in nature and human volunteer studies have been inconclusive (Herrington et al. 1987). Symptoms in the reported cases include primarily diarrhea, nausea, cramps and blood-streaked stools, which, together with other evidence of damage to the bowel wall, led Holmberg et al. (1986) to conclude the organism is enteroinvasive. Enteric disease due to *P. shigelloides* in USA has been strongly associated with the consumption of raw or undercooked oysters (Rutala et al. 1982; Kain and Kelly 1989). The oysters are reported to have been mainly from warm water regions (Holmberg et al. 1986).

P. shigelloides is widespread in nature and, although considered primarily as an aquatic organism, has been isolated from asymptomatic humans and both cold-blooded and warm-blooded animals (Miller and Koburger 1985). The organism is most common in surface fresh waters (Arai and Ikejima 1980) but has been reported from seawater (Cabelli 1978; Van Damme and

Vandepitte 1980). However, Zakhariev (1971) reported *P. shigelloides* dies out in 22–25 h in seawater. The organism has been isolated from oysters and clams (Rutala et al. 1982; Miller and Koburger 1985).

The organism shows a marked seasonal incidence, indicating the strong influence of temperature on its occurrence in nature, which has been likened to the behavior of vibrios (Tsukamoto et al. 1978). The observation that *P. shigelloides* dies out in seawater may indicate that salinity could be another factor influencing its incidence in marine environments.

Aeromonas

These gram-negative motile rod-shaped bacteria have long been recognized as aquatic bacteria having a particular association with fresh waters and fish (Rippey and Cabelli 1979; Popoff 1984; Williams and LaRock 1985). One species, *A. hydrophila*, is well recognized as a pathogen of fish (Shotts et al. 1972; Hazen et al. 1978) and an opportunistic pathogen in humans (von Graevenitz and Mensch 1968; Davis, Kane, and Garagusi 1978). *A. hydrophila* has been implicated as a cause of foodborne gastroenteritis in human patients (Rahman and Willoughby 1980; Buchanan and Palumbo 1985). Recently, *A. hydrophila* has been reported as a cause of diarrhetic disease associated with the consumption of oysters (Caspar 1982; Herrington 1984; Abeyta et al. 1986).

There is strong evidence that *A. hydrophila* is prevalent in estuarine waters and it is considered by most authors as an indigenous inhabitant of this environment (Hazen et al. 1978; Rippey and Cabelli 1979; Kaper, Lockman and Colwell 1981). The organism has been shown to be present in oysters and the overlying waters, both in growing areas and holding tanks (Abeyta et al. 1989; Colburn et al. 1989). The primary factor determining the incidence of *A. hydrophila* in nature is temperature, but there is some evidence that salinity may also play a role since the highest numbers have been reported during warm periods and in or near rivers (Cavari, Allen, and Colwell 1981; Williams and LaRock 1985).

Association with Fecal Indicators

Vibrio

The importance of the mesophilic vibrios as agents of shellfish-associated disease has stimulated many investigations of a potential relationship between their occurrence and the counts of fecal indicator organisms in shellfish and shellfish-growing waters. Such investigations have been aimed at the question of the origin of the organisms or the potential utility of indicator counts as an index of *Vibrio* levels. In fact, as noted above, the evidence that vibrios are indigenous to the estuaries is very strong and no useful correlation has been established between vibrios and fecal indicators. In one of the earlier studies, Thompson, Vanderzant, and Ray (1976) tested a large number of water, sediment and oyster samples from Galveston Bay for a relationship between *V. parahaemolyticus* occurrence and various environmental factors. They found no statistically significant correlation between fecal indicators and this *Vibrio*. Similar results were reported for *V. cholerae* in Chesapeake Bay by Kaper et al. (1979). A few years later, Roberts et al. (1982) reported an extensive study of the Louisiana Gulf environment, in which they found *V. cholerae, V. parahaemolyticus, V. fluvialis*, and *V. vulnificus* to be ubiquitous, but they too could detect no consistent relationship between *Vibrio* occurrence and fecal coliform counts. A study by Blake et al. (1982) again reported no significant correlation between the occurrence of *V. cholerae, V. parahaemolyticus*, and *V. vulnificus* and standard fecal coliform MPN counts in either seawater or oysters. They concluded that the vibrios are "part of the normal ecosystem and not of human fecal origin". A similar observation was made by Hood et al. (1983) for water, sediment and oyster samples from Appalachicola and Tampa Bay. Later reporting on a study of pathogenic bacteria from Naples Bay, Florida, Peterson, Yokel, and Lim (1984) noted that *V. cholerae* was isolated from stations with both high and low fecal coliform counts. Reporting results from an extensive study of *Vibrio* occurrence in Arcachon Bay (Gironde, France), Marchand (1984) indicated no correlation between the incidence of fecal coliforms and vibrios. More recently, Kaysner et al. (1987) reported on a significant study of the incidence of *V. cholerae* in shellfish, water and sediment taken from 261 sites along the entire West Coast of USA. Once again, there was no correlation between the occurrence of *V. cholerae* and fecal coliform counts. A report of a more recent study by the same group in Grays Harbor, Washington—a major oyster-growing area—showed no relationship between the incidence of *V. cholerae, V. parahaemolyticus, V. mimicus*, and *V. fluvialis* and fecal coliform counts (Kaysner et al. 1990). A recent report of a nationwide study involving water and oyster samples from the Atlantic, Gulf and Pacific coasts again failed to establish any significant correlation between fecal coliform counts and the incidence of *V. parahaemolyticus* (DePaola et al. 1990).

The evidence is, thus, very strong that the pathogenic vibrios are natural inhabitants of marine inshore environments and may be expected to occur in molluscan shellfish in numbers which are largely determined by ambient temperature and, to a lesser extent, by the salinity of the water in which they are growing. There is no substantive evidence that they are primarily of fecal origin, even though, in particular cases, *V. cholerae* O1 might be isolated from sewage (Roberts et al. 1982). Therefore, the use of counts of fecal indicator organisms to evaluate the likely occurrence of pathogenic vibrios in shellfish or shellfish-growing water is not feasible.

Plesiomonas

There are no published studies of the relationship between the occurrence of *P. shigelloides* in seawater or in shellfish and the counts of fecal indicator organisms. The organism has been repeatedly isolated from human feces (Holmberg and Farmer 1984) and from sewage (Zakhariev 1971). This would suggest that when water is contaminated with sewage, *P. shigelloides* and fecal indicator organisms might both be present. Nevertheless, the strong evidence that *P. shigelloides* is a natural inhabitant of aquatic environments (Miller and Koburger 1986) suggests that counts of indicator organisms cannot be relied upon to signal the presence or absence of this potential pathogen in water or aquatic animals.

Aeromonas

Where comparisons have been made, no correlation has been established between the incidence of *A. hydrophila* in shellfish or shellfish-growing waters and fecal indicator organisms count (Abeyta et al. 1989). This is not surprising in view of the ubiquitous occurrence of the organism in aquatic environments and the importance of environmental factors in influencing their numbers (Kaper, Lockman, and Colwell 1981). The organism is clearly not of fecal origin.

NATURAL TOXINS

Nature and Occurrence

Shellfish become toxic to humans as a result of feeding on toxigenic microalgae. Toxins from the algae may be accumulated in the viscera and other tissues of molluscs and these are transferred to humans when the animals are eaten. The toxins are not destroyed by cooking or processing and are indetectable organoleptically by consumers. They are normally detected by animal testing, although some chemical methods of analysis are now available (Shumway 1990). Three types of shellfish poisoning have been reported in USA and these are paralytic shellfish poisoning (PSP), neurotoxic shellfish poisoning (NSP) (Taylor 1985; Yasumoto 1990) and amnesic shellfish poisoning (ASP), originally reported in Canada (Todd 1990). PSP is caused by dinoflagellates of the genus Alexandrium (formerly *Gonyaulax*) in North America (most commonly, *A. catenella* and *A. tamarensis*), which produce a family of neurotoxins known as Saxitoxins (Hall et al. 1990). NSP is caused by brevetoxins produced by the dinoflagellate *Gymnodinium breve* (Trainer et al. 1990). Both these dinoflagellates typically produce the so-called red tides, that are discolorations of seawater due to blooms of the algae which occur somewhat unpredictably. Alexandrium blooms typically occur between April and October along the Pacific coasts from southern California to Alaska and in the Northeast from the Canadian maritime provinces to Long Island. Blooms normally do not last more than a few weeks and most shellfish clear themselves of toxins shortly after the bloom dies down. NSP is mostly confined to Florida and the adjacent states, but shows no seasonal periodicity of occurrence and is less common than PSP. Brevetoxic shellfish may take somewhat longer to eliminate all toxin. ASP is caused by domoic acid produced by the diatom *Nitzschia pungens* which occurs in cold waters (Bates et al. 1989). Again, the blooms are usually of relatively short duration but high toxin levels persisted for several months during the recent (1991–92) incident in Washington state. All three are dangerous toxins which can cause death in high doses and lasting neurological damage in the case of ASP.

There is a specific surveillance system in place in Maine, Massachusetts, and the four Pacific states involving the testing of shellfish from commercial and recreational growing areas for toxicity by a mouse bioassay during spring and summer (Shumway, Sherman-Caswell, and Hurst 1988; Nishitani and Chew 1988). Recently, states have begun testing for domoic acid also. Surveillance for *Gymnodinium* is dependent on observations of the very obvious red tide phenomenon, associated fish kills and reports of irritant gases. Analyses of water for the presence of increasing numbers of the toxigenic algae are not used as a surveillance technique in USA, although this is done in other countries (Corrales and Gomes 1990). Blooms of Alexandrium are initiated by the development of reproductive forms from cysts under conditions which have not been fully specified but include ambient water 14°C or above and nonturbulent sea conditions. *Gymnodinium* enter a rapid reproductive phase also when water conditions are suitable and in neither case is there any evidence of terrestrial or human influences. Conditions conducive to increases of *Nitzschia* populations also seem to be related to natural variations in water conditions, unrelated to human activity.

Association with Fecal Indicators

All the agents of toxic conditions in shellfish are photosynthetic organisms that occur naturally in the pristine marine environment (Taylor 1990). There is no evidence of any significant relationship between terrestrial influences, including runoff or sewage contamination and the development of blooms. When blooms occur, they result in toxic shellfish in areas of low fecal count and in those of high fecal contamination. For this reason, surveillance for PSP or NSP conditions is completely independent of the testing for fecal contamination.

PARASITES

Nature and Occurrence

There do not appear to be any well-authenticated cases of human illness related to naturally occurring parasites found in molluscan shellfish. There is one poorly documented report of illness due to Microsporidium from oysters, but this has not been generally accepted by experts in the field. The most likely candidates as parasites infecting humans, to be found in molluscs, are *Giardia* spp. and *Cryptosporidium* spp. Protozoan parasites of the genera Giardia and Cryptosporidium are a common cause of human disease and are consistently reported to be associated in nature with water (Craun 1984; Ongerth and Stibbs 1987; Casemore 1990). The species most commonly implicated in human disease are *Giardia lamblia* and *Cryptosporidium parvum*, but other species may also cause illness. The symptoms are typical of diarrhetic disease, including diarrhea, abdominal pain, anorexia and nausea, and the illness is self-resolving after a few days in most cases, although immunocompromised individuals may experience longer lasting and more serious symptoms. Both Giardiasis and Cryptosporidiosis can be foodborne diseases (Barnard and Jackson 1984; Casemore et al. 1986; Pederson, Carter, and Hadler 1988).

Giardia is spread by resistant cysts and *Cryptosporidium* by oocysts, and both have been shown to persist in fresh water for long periods of time (Casemore 1990). Unfortunately, no information is available in the literature concerning the survival of the parasites in seawater. The organisms are identified in food and water by microscopic procedures or by induction of disease in test animals, but, recently, immunological tests have been developed (deRegnier et al. 1989; Stibbs et al. 1988; Casemore 1990).

Association with Fecal Indicators

Giardia and *Cryptosporidium* cysts occur in detectable numbers in both raw and treated sewage (Rose, Darbin, and Gerba 1988) and, therefore, might occur concurrently with fecal indicator bacteria in some water samples. However, those organisms are so generally distributed in natural waters that it would be difficult to draw meaningful inferences from a common occurrence. Of course, the whole issue is moot until there is substantial evidence of actual parasite infection by one of these organisms or any other (e.g., *Entamoeba histolytica*) parasite where the vehicle is molluscan shellfish.

7

Methods to Recover Injured Indicators and Pathogens

Cameron R. Hackney

Department of Food Science and Technology,
Virginia Polytechnic Institute and State University,
Blacksburg, VA 24061

INTRODUCTION

When microorganisms are exposed to stress, they may become injured. It is probable that any species of bacteria may suffer sublethal injury when exposed to adverse conditions. Stresses in the marine environment that cause injury include fluctuations in temperature, changes in pH or ionic conditions, exposure to light (ultraviolet or visible), exposure to chemical residues (chlorine, ozone, etc.), exposure to oxygen and combinations of stresses. In addition, starvation or nutrient depletion can cause cells to be stressed and injured. A cell is considered injured when it grows in a nonselective medium but fails to grow in a media selective for that organism, or if it requires more pre-formed growth nutrients than normal (Ray 1989). Two types of injury are observed, structural and metabolic. Structural injury is observed when microorganisms can grow in a nonselective medium but fail to grow in a medium normally selective for that microorganism (Ray 1979). Physical damage to the cell membrane and other components is observed when structural injury occurs (Ray 1989). Metabolic injury is observed when a microorganism can form colonies on a nutritionally complete medium, but fails to grow on a minimal medium or a defined medium. In reality, the two types of injury rarely occur separately. The extent of damage to the cells differentiates between structural and metabolic damage (Ray 1989). Almost all injured cells have damage to the cell structure (i.e., the cell membrane), an extended lag phase, and a sensitivity to selective agents to which uninjured cells are resistant. When injury is more extensive, metabolic systems are damaged. These injured cells retain their capacity to repair in a suitable environment, such as found in foods or in a complex or nonselective media (Hurst 1980; Ray 1979).

Under certain conditions of metabolic injury such as starvation, bacterial cells may enter into a viable but non-culturable state. In this state the cells exhibit metabolic activity as observed with viable organisms, but cannot form colonies on any type of media. The cells are reduced in both size and metabolic activity. They can be detected using viable stains. (This type of injury is discussed in another chapter in this report.)

The extent of cell death and injury is dependent on the strain and types of bacteria, their phase of growth and initial concentration. Growth conditions, including medium composition, and growth temperature will influence the degree of injury. In general, gram-negative bacteria are more sensitive to stress than gram-positive bacteria. Actively growing bacteria in the log phases of growth are more sensitive than cells in the stationary phase of growth.

The susceptibility of microorganisms to injury is important to the shellfish microbiologist. When cells of indicator bacteria are injured and, therefore, not enumerated with selective media, there is a possibility of underestimating their numbers (Ray 1989). In addition, pathogens, if present, may not be detected in the injured state. Yet, these pathogens may still be infective and cause illness if consumed. It is very important to consider the effects of injury in the development of a rapid methodology for the recovery of microorganisms.

Rapid methods have been developed for many of the pathogens of importance in shellfish, including salmonellae, *Listeria*, *Campylobacter*, and others. These methods include gene probes (both ELISA and radio immunoassay), fluorescent antibody tests, impedance procedures and many others (see the chapter on rapid methods). These procedures have in common a minimum limit of detection, usually in the range 10,000–1,000,000 cells per ml. Since the levels of a specific pathogen in foods and water are usually very low, the numbers must be increased through enrichment. If the cells are injured they will not grow in selective enrichment media. Indicators and enteric bacterial pathogens in shellfish and their growing waters are usually present in low numbers and cells are often debilitated or injured. When a cell of a bacterial pathogen of fecal origin is exposed to the marine environment, a number of factors affect its survival. These include water temperature, exposure to disinfectants during sewage treatment, exposure to sunlight, pH, metal concentration in the marine environment, nutrient level of the environment, salt concentration and so on. The population of cells generally decreases in number over time. The cells usually enter an injured state, which gradually worsens, until the death of the cell. The survival rate will vary, depending upon the microorganism and the environment. In many cases the cells are metabolically injured due to nutrient depletion. These cells can recover if placed in the proper environment. Pre-enrichment is used to provide this environment for recovery. The type of pre-enrichment is very important. The cells must be provided with an environment that will encourage repair. As mentioned earlier, the population of cells must be increased if the cells are to be detected by either conventional or rapid methods. If the cells cannot be repaired and allowed to grow, the organism of concern will not be detected, no matter how easy or rapid the method. Injured cells have the ability to *repair* themselves if given the proper environment. Repaired cells respond to selective environments in the same manner as uninjured cells. This repair occurs during an extended lag phase of growth. The lag in cell division allows the microorganism to adjust to the new environment. During the lag phase, cells are not dividing, but they are physiologically active. Cell division occurs at the end of the lag phase.

The type of enrichment will vary, depending upon the type and extent of injury. For example, structurally injured cells may repair if placed in a rich nonselective medium, whereas a metabolically injured cell may not recover because the medium was "too rich". Instead, these cells may recover better in a more dilute solution of nutrients. Therefore, the type of enrichment is very important and must be geared to the specific pathogen and the expected injury.

CAUSES OF INJURY OF INDICATOR BACTERIA IN MARINE ENVIRONMENTS

In general, enteric indicator bacteria are not well adapted to survive in marine waters or in shellfish. They are exposed to a variety of chemical, biological, and physical stresses that may cause injury and death. More specifically, the stresses include sublethal concentrations of disinfectant used in wastewater treatment, exposure to light, fluctuations in pH, temperature, and osmotic concentrations. Other factors, including influencing recovery, include damage by oxygen, metals, etc.

Chemical Stresses

Chemical stresses that may cause injury to indicator bacteria in harvesting waters and shellfish include exposure to disinfectants and metals. Chlorine is the most widely used chemical disinfectant for wastewater treatment. While chlorine is lethal, at a low concentration (0.5 ppm) it will cause injury to the indicator population (McFeters 1989). Since chlorine is quickly inactivated by organic matter, it is easy to envision that a large part of the indicator population would be exposed to these low levels. Chlorine injuries have an extended large phase and, of course, will not grow on selective media (McFeters 1989).

Chlorine injury can reduce virulence of pathogenic bacteria (LeChevallier et al. 1985). Walsh and Bissonnette (1983) demonstrated that chlorine damaged adhesions of enterotoxigenic *E. coli.* The injured bacteria can repair themselves. Singh, Yeager, and McFeters (1986) showed that chlorine-injured enterotoxigenic *E. coli* were able to repair themselves in mouse ileal loops and resume toxin production. Chlorine injury of *E. coli* does not necessarily reduce survival in the aquatic environment. Walsh and Bissonnette (1989) examine the survival of uninjured and chlorine-injured enterotoxigenic *E. coli* in water. Chlorine injury did not reduce the survival time in their aquatic system.

Wastewater disinfectants such as ozone or chlorine dioxide may also cause cell injury. Chlorine dioxide is not inactivated by organic matter. Its ability to cause injury in bacterial populations has not been investigated.

Metals

Marine waters may contain various concentrations of metals, depending upon the location, pollution source, and natural environment. McFeters (1989) reported that in freshwater systems the presence of copper, zinc, lead, and nickel could be related to bacterial cell injury. Metals may cause metabolic damage to bacteria. Seawater may contain these metals and it would be expected that they would contribute to indicator cell injury.

Physical Stresses

Physical stress encounters by indicator bacteria exposed to marine environments include exposure to light (visible and ultraviolet), fluctuations in pH, temperature, osmotic pressure, etc.

Light Stress

Intestinal bacteria are not exposed to light in their natural environment. However, exposure to the aquatic environment will also lead to exposure to sunlight. The effects of solar radiation on the survival of indicator bacteria have been studied by several investigators (McCambridge and McMeekin 1981; Kapuscinski and Mitchell 1981; Fujioka et al. 1981). McCambridge and McMeekin (1981) examined the effect of solar radiation on the survival of bacteria of fecal and plant origin. Bacteria studies included *Klebsiella pneumonia, E. coli, Salmonella typhimurium, Enterobacter aerogenes, Erwinia herbicola,* and *Streptococcus faecium.* The survival was compared to that of *E. coli. K. pneumonia* was more susceptible to the lethal effects of sunlight, whereas all the other bacteria tested were more resistant than *E. coli.* It is interesting to note that *K. pneumonia* was more sensitive to the effects of sunlight than *E. coli.* Pialle et al. (1986) reported that fecal coliform-positive *K. pneumonia* could be present in very high levels in Gulf coast oysters harvested during the summer from approved waters. While there was a strong correlation between *E. coli* levels in water and oysters, there was no relationship between *K. pneumonia* levels in surface water samples and shellfish. Sensitivity to sunlight may help explain this phenomenon.

Fujioka et al. (1981) noted that the effects of sunlight on indicator bacteria were more pronounced in seawater compared to freshwater. In their studies the effect of sunlight was very dramatic on survival. This group diluted raw sewage 1 : 1000 in seawater and stored the cultures in sunlight or in the dark. Fecal coliforms and fecal streptococci were enumerated using membrane filtration and selective media. Both the groups survive several days when stored in the dark. However, in sunlight a one log reduction (90%) was observed in only 30–90 min for *E. coli* and 60–180 min for fecal streptococci. Since the investigators used selective media, only the uninjured cells were recovered and it was not possible to determine if the cells were dead or injured.

Kapuscinski and Mitchell (1981) clearly showed that solar radiation induced sublethal metabolic injury to *E. coli.* The injury was detected by differential plating on complete and minimal media. The addition of catalase or pyruvate to the minimal medium reversed the injury, indicating that the catalase system was affected by solar radiation. In earlier work, it was reported that catalase could be photo-inactivated (Kapuscinski and Mitchell 1981).

D'Aoust et al. (1974) observed that visible light (above 400 nm) was detrimental to several enzyme systems of *E. coli.* Respiration was inhibited by the light. Krinsky (1976) extensively reviewed the cellular damage to microorganisms initiated by visible light. In this review it was noted that, perhaps, the main mechanism of damage to the cell was related to the photo-induced formation of oxygen radicals and singlet oxygen. The consequences of oxygen damage will be discussed in a later section.

Near-ultraviolet (UV) light has been shown to cause genetic damage in *E. coli* K12 (Webb and Brown 1982; Moraes and Tyrell 1983). Moss and Smith noted that near-UV light also caused metabolic injury and that recovery was enhanced on media containing reduced nutrient levels.

Changes in Osmotic Pressure and Salt as Stress

Salt is commonly used as a preservative. The exposure of *E. coli* to sodium chloride results in damage to the cell membrane, with leakage of cellular constituents (Ray 1989). The addition of magnesium to the incubation medium suppressed cell leakage.

Toxicity of Oxygen

Oxygen can be toxic to microorganisms. Toxicity is not due to the reactivity of oxygen but to its reactive intermediates, hydrogen peroxide, superoxide radicals, and hydroxyl radicals. These compounds can be produced in response to light. In addition, respiring microorganisms produce the superoxide anion (Fridovich 1975). Aerobic microorganisms have mechanisms to overcome oxygen toxicity, including the enzyme systems superoxide dismutase (SOD) and catalase (Fridovich 1975). The physiological function of SOD is to provide a defense against oxygen toxicity by catalyzing the dismutation of superoxide radicals.

Any reaction which generates superoxide radicals will also produce hydrogen peroxide by this dismutation (Fridovich 1976). Hydrogen peroxide is the most stable intermediate in the reduction of oxygen to water (Fridovich 1975, 1976). Hydrogen peroxide is toxic to microorganisms. Oxygen-metabolizing organisms contain the enzyme catalase (Hassan and Fridovich 1979). Catalase serves to degrade hydrogen peroxide into water and oxygen. It is probable that hydrogen peroxide is toxic because it reacts with superoxide radicals by the Haber–Weiss reaction to yield hydroxyl radicals. Hydrogen peroxide can also react with the ferrous ion to yield the hydroxyl radical.

The hydroxyl radical is one of the most potent oxidants known. *E. coli* responds to increased concentrations of the superoxide radicals and hydrogen peroxide by increasing the synthesis of superoxide dismutase and catalase (Hassan and Fridovich 1979). When cells become injured, this defense mechanism is also injured, enhancing the lethality of hydrogen peroxide, superoxide radicals, and hydroxyl radicals and interfering with the ability of the cells to recover after sublethal stress. The catalase system of bacteria can be inactivated by a number of stresses, including heat, light (photo-inactivation), freezing, acid and many other stresses. For this reason injured cells of facultative anaerobic bacteria are often recovered in greater numbers under anaerobic conditions.

Fluctuations in pH and Temperature as Stress

The effect of acidic conditions on cell injury is well documented. Hackney and Bissonnette (1978) showed that acid mine water was very injurious to *E. coli* and *Enterobacter aerogenes*, with several log differences observed between counts on selective and nonselective media. Of course, very acid waters are not usually encountered in the marine environment. In our laboratory we have observed that during the time of heavy rains, the pH of surface coastal waters of Louisiana could fluctuate from near 8 to as low as 5. The pH quickly rose after the rains ended. It is not known if these fluctuations caused injury to the fecal indicators present (unpublished data).

The pH of the meat of shellfish such as oysters decreases during spoilage, going from 6.5 to 6.8 to as low as 5.2. This decrease may cause some injury to the indicator bacteria. It is well documented that acid will cause injury in other food systems.

Reductions in pH can be particularly injurious to cells (Hackney and Bissonnette 1978; Roth and Keenan 1971). The amount of injury and death is dependent on the type and concentration of the acid. In general, organic acids are more lethal than inorganic acids at similar concentrations. Increases in temperatures accelerate the amount of death and injury caused by the acids (Roth and Keenan 1971). Similarly, Blankenship (1981) found that the rates of injury and death of *Salmonella barely* were dependent on acid concentration and temperature. In this study, the more concentrated an acetic acid solution, the greater the rates of death and injury. Blankenship (1981) determined that a reduction in the activity of the enzymes lactate dehydrogenase and glucose-6-phosphate dehydrogenase resulted from the denaturation of proteins in the acid-injured cells. *S. barely* also exhibited RNA damage and required amino acids and peptones for recovery. In another study, acid-injured *E. coli* K-12 showed an increased sensitivity to selective agents in violet-red bile agar (VRBA) (Prybylski and Witter 1979).

Fluctuations in temperature can also lead to cell injury. In the intestinal tract the temperature is constant. In the marine environment, temperatures fluctuate widely. It has long been documented that freezing and heat will injure bacterial cells (Ray 1989). Cold shock (sudden lowering of temperature) has been shown to be a cause of injury (Leder 1972; Cowman and Speck 1969). Rhodes, Anderson, and Kator (1983) examined the stress index of *E. coli* cells exposed to seawater during various seasons. The stress index was related to recovery efficiencies. The greatest stress was observed at temperatures below 10°C. This group observed that total adenylates and ATP levels decreased at low temperatures, although the energy charge remained constant. The amount of injury, as measured by differences in counts on selective and nonselective media, was greatest at low temperatures.

Biological Stress

Starvation

It has long been documented that starvation is a significant stress for indicator bacteria. Enteric bacteria are heterotrophs, which are adapted to receiving high levels of nutrients and are stressed by the nutrient limitation usually encountered in aquatic systems (McFeters 1989). Enteric bacteria respond to nutrient limitation in different ways. Many of the indicator bacteria respond by carrying on a maintenance level of metabolism. Nutrient-limited indicators will decrease in size and require more pre-formed growth factors, including amino acids and vitamins.

REPAIR

Injured cells have the ability to repair themselves if given the proper environment (Hurst 1980; Hurst and Hughes 1981; Przybylski and Witter 1979). This repair occurs during the lag phase of growth. For this reason the lag phase of growth of injured cells is usually greatly extended. The lag in cell division allows the microorganism to adjust to the new environment. During the lag phase, cells are not dividing, but they are physiologically active. Cell size increases and protoplasm and enzymes are synthesized. Cell division occurs at the end of the lag phase. Repaired cells respond to selective environments in the same manner as uninjured cells (Bissonnette et al. 1975; Busta 1976; Flowers and Ordal 1979). Injured cells are sensitive to selective agents and will die if not given the conditions needed for repair. The conditions needed for repair are dependent on the type and degree of injury. For example, some freeze-injured cells will repair in phosphate buffer as opposed to distilled water. In other types of injury, certain amino acids or nucleic acids are needed for repair (Hurst et al. 1973; Payne 1978). In another example, metabolically injured cells, which will not form colonies on a minimal medium, will grow if trypticase peptone is added to this same medium (Ray 1989).

DETECTION OF INJURED CELLS

Injured cells are detected indirectly through differences in counts on selective and nonselective media. Within a stressed population, cells may be uninjured, injured, or dead. The uninjured cells can be enumerated on selective media, whereas both injured and uninjured cells are detected on nonselective media. The number of injured cells is determined by subtraction. The formula used to determine the percentage of the population that is injured is

$$\% \text{ injury} = \frac{\text{CFU nonselective} - \text{CFU selective}}{\text{CFU nonselective}} \times 100$$

It is important to note that several factors affect the determination of the population, including the nature of the selective medium (i.e. type of selected agents) and composition. In addition, under certain conditions, nonselective media may not recover all injured cells. Photooxidation may cause hydrogen peroxide to be developed in the medium, which can be lethal to injured cells. McDonald, Hackney, and Ray (1983) showed that the addition of compounds that degrade or block the formation of hydrogen peroxide in media will greatly increase the recovery in freeze- and heat-injured cells. In general, the longer the time between media preparation and use, the greater the inhibition of injured cells. Storage conditions of prepared media will also affect its ability to recover injured cells. Prepared media should be stored in the dark.

Formation of Hydrogen Peroxide in Media

Since the late nineteenth century, the formation of peroxide in bacteriological media has been recognized. Autoxidation is the mechanism responsible for the formation of hydrogen peroxide in the presence of water and oxygen (Barry et al. 1956). The reduction of oxygen to water occurs through a series of single electron transfers generating hydrogen peroxide, superoxide radicals, and hydroxyl radicals as intermediates (Fridovich 1976). Barry et al. (1956) noted that when citrate was autoclaved with manganese, hydrogen peroxide was formed. The peroxide formation was catalyzed by the manganese. The presence of reducing sugars or hydrazine derivatives, in place of citrate, resulted in similar levels of peroxide. Carlson et al. (1978) demonstrated a relationship between phosphate concentration and autoxidation of media. Heating a phosphate and glucose solution with a pH of greater than 6.5 to 120°C for at least 5 min resulted in the accumulation of hydrogen peroxide. The rate of autoxidation increased with increased heating time, increased pH, and increased concentrations of glucose and phosphate. Hydrogen peroxide also accumulated when other reducing sugars and phosphate were heated together. Metal ions were not necessary for the autoxidation. In addition, these investigators reported that the addition of SOD to an autoclaved solution of glucose in phosphate buffer decreased the rate of oxygen consumption. This decrease indicates that the superoxide radical formed in the autoxidation serves to generate further oxidation of the media. In a later study Carlsson et al. (1979) demonstrated that hydrogen peroxide is formed from cysteine in the presence of oxygen. Metal ions were required for this reaction.

Korycka-Dahl and Richardson (1978) reported that superoxide anions could be generated in the presence of light, certain amino acids (cysteine, methionine, histidine, tyrosine, and tryptophan) and riboflavin, all common ingredients in complex media.

Hoffman et al. (1978) reported that hydrogen peroxide and superoxide radicals were formed in sterile Brucella broth in concentrations that prevented the growth of *Campylobacter fetus* subsp. *jejuni*. These oxygen derivatives were present regardless of whether the medium was stored in light or dark.

Supplementation of media with compounds that degrade hydrogen peroxide or scavenge free radicals has been studied (Baird-Parker and Davenport 1965; Scott et al. 1973; Rayman, Aris, and El Derea 1978). Baird-Parker and Davenport (1965) were perhaps the first group to supplement a medium with pyruvate to improve the recovery of a microorganism. The medium they developed was for the recovery of *Streptococcus aureus*. Flowers and Ordal (1979) observed that, of the standard methods for the selective enumeration of *S. aureus*, only the Baird-Parker medium was effective in enumerating stressed cells. The catalase system of *S. aureus* is easily destroyed by heat and the pyruvate in the medium helps the injured cells to recover.

Rayman, Aris, and El Derea (1978) showed that the addition of 1% sodium pyruvate to trypticase soy agar increases the enumeration of heat-stressed *Salmonella senftenberg*. Increased counts were also observed when ferrous sulfate and potassium permanganate were added to the medium. These compounds degrade hydrogen peroxide. This group proved that pyruvate was not acting as a nutrient because, when other metabolic substrates were added to the trypticase soy agar, no increase in counts was found. Therefore, they concluded that pyruvate acted as an exogenous decomposer of hydrogen peroxide and not as a substrate for the cells.

Exposure of *E. coli* to solar radiation results in sublethal metabolic injury (Kapuscinski and Mitchell 1981). In this study the injured cells could grow on a complete medium but failed to produce colonies on a minimal medium. The addition of pyruvate to the minimal medium resulted in counts which were not significantly different from counts on the complete medium. When metabolic substrates, such as malic acid and succinate, were added to the minimal medium, the counts did not increase. However, when catalase was added to the medium, counts on the minimal medium and the complete medium were the same. This indicated that the pyruvate acted to degrade hydrogen peroxide and not as a growth nutrient.

Harmon and Kautter (1976) observed an increased enumeration of *Clostridium perfringens* when media are treated with catalase. This inhibition may be a result of the formation of hydrogen peroxide in stored media. Of course, the use of pre-reduced media will also increase the enumeration of anaerobes.

The oxygen tolerance of *Campylobacter fetus* subsp. *jejuni* is enhanced by the addition of pyruvate, sodium metabisulfate, norepinephrine, ferrous sulfate, catalase or superoxide dismutase, or a combination of ferrous sulfate, sodium metabisulfite, and sodium pyruvate (Hoffman et al. 1978). These compounds act by scavenging superoxide radicals (or degrading hydrogen peroxide) which occur spontaneously in the media.

McDonald et al. (1983) examined the response of freeze- and heat-injured *E. coli* to 13 compounds that degrade or block the formation of hydrogen peroxide. Nearly all agents tested an increased recovery on injured cells on both selective and nonselective media. 3,3′-Thiodipropionic acid and sodium pyruvate were the most effective. The authors recommended that thiodipropionic acid would be better in selective procedures because it is not used as a nutrient and will not give false positives.

ENUMERATION OF INJURED CELLS

Methods which allow the repair of injured cells have been developed. *Salmonella* is enriched in nonselective lactose broth before being exposed to a selective environment. It is now generally recommended that foods be pre-enriched in nonselective medium before selective enrichment, when the possibility of injured cells exists. Pre-enrichment in nonselective liquid medium has been adapted for the enumeration of injured cells. For example, Ray, Hawkins, and Hackney (1977) showed that the recovery of *Vibrio parahaemolyticus* was greatly enhanced if the samples were pre-enriched in tripticase soy broth (3% NaCl) before enrichment in selective glucose salt teepol broth. The pre-enrichment greatly enhanced the recovery in both water samples and shellfish samples. This procedure could be used in an MPN procedure to enumerate the organism. It is also possible to enumerate injured salmonellae in this fashion. In other examples of using pre-enrichment procedures, Lanz et al. (1976) developed a method for the selective enumeration of stressed *E. coli.* Capsules containing the selective agents brilliant green and oxgall were added to modified lactose broth. Gradual diffusion of the selective compounds converted the nonselective environment into a selective environment. Repair of the cells occurred in the nonselective environment. The brilliant green and oxgall selected for coliforms.

REPAIR DETECTION SYSTEMS

Methods have been developed which allow the repair of injured coliforms in a solid media before the introduction of selective agents (Speck, Ray, and Reed 1975). Injury is observed when a higher population can be recovered on nonselective medium compared to a selective medium. Because fecal bacteria are never present in pure populations, the use of selective media is mandated for their recovery and enumeration. Methods have been developed to recover injured bacteria. These include repair detection procedures. In these procedures the bacteria are first plated in a nonselective medium. After a period of recovery, the nonselective medium is overlaid with a selective medium. Inhibitors in the selective medium will diffuse into the nonselective medium, creating the desired selective environment for the recovery of a particular bacterium. Speck, Ray, and Reed (1975) developed this procedure for the recovery of injured coliform. Hackney and Bissonnette (1978) adapted this method for the enumeration of fecal coliforms and enterococci. Anderson and Baird-Parker (1975) (later modified by Holbrook, Anderson, and Baird-Parker 1980) developed a method for enumerating *E. coli* from foods. A cellulose acetate filter membrane is placed on the surface of selective tryptone bile agar (TBA) plates and the sample is then spread onto the membrane and the plates incubated at 44°C overnight. After incubation, 1–2 ml of indole reagent is placed in the lid, and the membrane filter is removed from the TBA and placed in the reagent. Indole-positive colonies stained pink within 5 min. This procedure can be modified to enumerate stressed cells by placing the filter first onto a nonselective media, such as nutrient agar or minerals-modified glutamate agar, for 4 h at 35°C followed by placing the filter on TBA, as described above. The Anderson–Baird-Parker procedure did not work with oysters in Louisiana, because non-*E. coli* fecal coliforms were present in orders of magnitude higher than *E. coli* and masked its recovery (Kilgen, Cole, and Hackney 1988).

Repair detection procedures have been used successfully to recover coliforms and fecal coliforms and other aerobic or facultative anaerobic bacteria. In our laboratory we have applied this procedure to recover anaerobes. Anaerobes are very sensitive to oxygen and are easily injured by exposure to environmental conditions. For recovery to be maximized, injured cells must be recovered. The best procedure for the recovery of anaerobes is the roll tube technique. While this technique has been in the literature for a long time, it was perfected by the VPI anaerobe laboratory. This procedure employs a set of conditions that ensure that the system is kept oxygen-free. This system is usually far better than other methods used to recover anaerobes, such as the gas-pack jar. We developed a repair detection system for *Bifidobacterium* using the roll tube system (unpublished data). The media chosen was that reported by Mara and Oragui (1983) and a repair detection procedure was developed to recover injured cells. For the nonselective medium, the same medium was used; however, the inhibitory agents nalidixic acid, kanamycin sulphate, and polymyxin B were omitted. The selective medium overlay contained 1.7 times the concentration of inhibitors of the original formulation. Since the diffusion of components is fairly rapid in agar systems, the concentrations will equilibrate to that of the original medium. Optimum media proportions were determined and found to be 7 ml of nonselective medium overlaid with 10 ml of the selective medium. The optimum recovery time is between 2 and 4 h.

8

Rapid Methods for Detection and Enumeration of Indicators

Marianne Walch

Center for Marine Biotechnology, University of Maryland System, 600 East Lombard Street, Baltimore, MD 21200

Ishrat Husein and Rita R. Colwell

Department of Microbiology, University of Maryland, College Park, MD 20742

INTRODUCTION

Two standard methods are normally used for the enumeration of sanitary indicator bacteria from water (APHA 1985). The multiple-tube fermentation (MTF) method gives a most probable number estimate of total coliforms after growth in liquid media. The membrane filter (MF) technique enumerates total coliforms on the surface of agar by providing a count of colony-forming units (CFU) per 100 ml of filtered sample. Both the MF and MTF methods have been extensively studied and are approved for regulatory monitoring purposes (Anonymous 1975; Bissonnette et al. 1977; Evans et al. 1981; Jacobs et al. 1986; McFeters, Cameron, and LeChevallier 1982; Morgan et al. 1989; Shipe and Cameron 1954). Both methods require confirmed and completed tests after the initial observation of a positive reaction on the primary enrichment/selection medium. This can add an additional 24–72 h in order to obtain the final result. In neither of the methods is an isolate identified with species, or are any members of the fecal coliform group differentiated from the total coliform group (Evans et al. 1981; Geldreich et al. 1972; Jacobs et al. 1986; LeChevallier, Cameron, and McFeters 1983; Morgan, Gubbins, and Morgan 1965).

Methods to detect the presence of enteric pathogens in aqueous environments also tend to require lengthy enrichment and confirmation procedures. The low numbers of these organisms that are generally present even in contaminated systems and the fact that many become stressed and nonculturable in low-nutrient aqueous systems (Roszak and Colwell 1987*a*) make sensitive detection of pathogenic bacteria a difficult task.

The need for more rapid methods for water quality measurement was recognized by the introduction in 1980 of section 919, "Rapid Detection Methods," in *Standard Methods for the Examination of Water and Wastewater* (APHA 1980). Detection of indicator organisms or pathogens in less than one working day would permit decisions important to public health protection to be made more quickly. Rapid analysis of shellfish and other perishable food items would allow faster release of products to the marketplace and timely identification of hygiene problems. If these methods also offer greater specificity, then the shellfish industry (and the consumer) would potentially benefit from fewer false alarms that lead to unnecessary closing of fishing areas. Furthermore, methods that incorporate more tests into a single medium, in addition to being more rapid, would potentially reduce the cost of water quality monitoring.

The search for rapid methods of bacterial detection and identification has been going on in recent decades. The impetus for, and development of, rapid techniques has come largely from the clinical side of microbiology. Clinical microbiology laboratories in the United States generate tens of millions of dollars per week, providing an enormous market for efficient and cost-effective biotechnology (Salzman and Gregg 1984). Clinical microbiology has been largely retrospective in providing identification and susceptibility data to be used in the treatment of infectious diseases, because complete microbial identification may take days or weeks after receipt of a specimen. Methods are required that allow laboratories to respond with at least a tentative identification of infecting organisms before physicians begin antimicrobial therapy.

Tools have been developed in recent years that make microbial identification and enumeration increasingly rapid. Unfortunately for water quality monitoring and management, a majority of these methods were created specifically for clinical applications, and data on their use or modification for environmental samples are sparse or nonexistent. Progress, even in environmental microbiology, has been rapid, however, and a number of new techniques and approaches are now, or soon will be, available for environmental samples. During the last few years, many of the older conventional methods such as plate counting and MPN analysis have been modified, and new methods, such as gene probes and *in situ* DNA/DNA hybridization, have been developed.

TABLE 8-1. Methods for Detection, Enumeration, and Monitoring of Microorganisms in Environmental Samples

Method	Advantages	Disadvantages
Plating/culturing	Fairly sensitive; selective, colorimetric media available; operationally feasible and easy; relatively inexpensive	Requires culture of organisms; detects phenotypic expression; limited by media, temperature, etc. long incubation may be needed; will not detect dormant or non-culturable organisms
Immunofluorescent microscopy	High specificity; high sensitivity when sample is concentrated; direct detection of nonculturable organisms; direct viable count (DVC) differentiates living and dead cells; can be automated with image analysis	Requires concentration of sample for sensitivity; no differentiation between living and dead cells unless DVC is used; relatively expensive if automated
DNA fingerprinting	Very high specificity; useful for epidemiological analyses	Complex and time consuming; relatively expensive; best use currently requires previous culture of organism
Gene probes	Highly specific; permits detection of genotype; detection of nonculturable organisms; recent methods allow easy extraction of nucleic acids from water; PCR allows very high sensitivity	Low sensitivity unless large volumes of water are concentrated; relatively expensive; methods for environmental samples still in development

Table 8-1 summarizes the commonly used and newly developed techniques for detection, identification, and enumeration of microorganisms in environmental samples. They include culture methods employing selective media and colorimetric reactions, direct and viable counting methods, and gene probe technologies. Many of these methods have been reviewed in detail elsewhere (Colwell et al. 1988; Herbert 1990). This chapter will review rapid methods that have become available in recent years which have application to the detection and/or enumeration of bacterial sanitary indicators and waterborne enteric pathogens.

ENRICHMENT AND ISOLATION MEDIA FOR RAPID DETECTION

The use of selective and differential media for enrichment and identification of bacteria are standard methods in microbiology. They have been, and still are, useful for the detection and enumeration of specific populations of microorganisms from the environment. Despite growing evidence of the prevalence and pathogenic potential of viable but nonculturable bacteria in aquatic ecosystems (Roszak and Colwell 1987*b*; Xu et al. 1982), public health criteria for the microbiological quality of water and shellfish are still based on conventional culture methods (APHA 1985). This, combined with the reluctance of many water quality laboratories to incorporate into their protocols some of the more sophisticated and potentially more accurate, but unfamiliar, enumeration technologies such as immunofluorescence and gene probe methods, has fueled a continuing search for more rapid, accurate, and cost-effective assays based on culture methodologies.

The detection, isolation, and enumeration of coliforms and *Escherichia coli* by conventional methods generally requires an enrichment step, using liquid media or membrane filters incubated on solid media, in each case followed by subculture and additional testing of positive cultures or colonies (APHA 1985). Detection of *E. coli* has been based primarily on three characteristics in culture: the ability to grow at 44°C, the production of acid and gas from lactose, and production of indole from tryptophan (APHA 1985). Ever since examination of waters for fecal indicators began near the turn of the century, investigators have sought to modify and improve the standard methods (Colwell et al. 1991).

Many of the enteric bacteria are pathogenic and, therefore, of significant interest to medical microbiologists, who have developed schemes for their isolation and differentiation (Edwards and Ewing 1972; Ewing 1972). Hence, a number of methods for rapid identification of gram-negative rods, particularly the *Enterobacteriaceae*, have been described in connection with routine clinical diagnostic work (e.g., Borchardt 1968; Closs and Digranes 1971; Elston et al. 1971; Johnstone and Brown 1970; Lassen 1975; Rhoden et al. 1973; Roberts 1971; Rosner 1970; Smith et al. 1971, 1972; Tomfohrde et al. 1973). Lassen (1975), for instance, described a three-tube method, combined with a dichotomous key, which, it was claimed, would identify virtually all of the *Enterobacteriaceae*, as well as a number of other frequently encountered gram-negative rods, at the genus or species level within 20 h following primary isolation from stool cultures. Such methods were developed for identification of cultures obtained from clinical specimens, and not for environmental detection and quantitation. Evaluation of media and methods for environmental monitoring using samples and cultures obtained from the environment in which they will be used is of critical importance. Yoshe-Purer (1989), for example, demonstrated that KF Streptococcus agar, which is considered one of the best media for membrane filter enumeration of fecal streptococci in water and is recommended for this use in Standard Methods (APHA 1985), is not suitable for seawater because many indigenous marine bacteria grow well on it and produce false positive reactions. Similarly, Olson

(1978) reported a modification of the standard MPN method which made it more suitable for use in seawater.

A number of reports appeared in the 1970s of rapid methods designed to improve the detection and enumeration of *E. coli* and fecal coliforms. Andrews and Presnell (1972) described a 24 h assay for *E. coli* in seawater that combined a new "A-1" medium with direct incubation at 44°C. The method was evaluated several years later for recovery of *E. coli* from oysters and clams and was demonstrated to give results comparable to those of lengthier standard methods (Andrews et al. 1975). Qadri, Buckle, and Edwards (1974) also described 24 h methods for detection and quantitation of both fecal coliforms and *E. coli* type I in shellfish. They found that brief preincubation of samples in MacConkey broth at 37°C before transferring tubes to 44°C eliminated heat shock effects. Reasoner et al. (1979) developed a 7 h membrane filter test for fecal coliforms and found it suitable for examination of surface waters and unchlorinated sewage. This test was evaluated some years later to establish its usefulness for emergency and disaster situations, which require extremely fast decisions to protect public health (Barnes et al. 1989).

Rapid culture methods for detection of *E. coli* in foods were also developed. The method of Anderson and Baird-Parker (1975) employed a membrane filter technique with tryptone–bile agar to detect indole production as an alternative to laborious MPN counts. A rapid enrichment serology method was described by Cranston and Matthews (1973).

Many reports describing radioisotope detection methods have appeared over the years. Interest in use of radiolabeled substrates as a way of detecting sanitary indicator bacteria in water dates to the mid-1950s when Levin and coworkers described the use of ^{14}C-lactose to detect fermentation by coliforms (Levin, Harrison, and Hess 1956, 1957; Levin et al. 1956). Later investigators used ^{14}C-formate to detect the production of hydrogen and $^{14}CO_2$ via the coliform formic hydrogen lyase system (Korsh et al. 1968; Levin, Harrison, and Hess 1957; Levin et al. 1959; Levin, Strauss, and Hess 1961; Scott et al. 1964*a*, *b*). More recently, ^{14}C-labeled glucose (Korsh et al. 1971), lactose (Bachrach and Bachrach 1974) and mannitol (Reasoner and Geldreich 1979, 1989) have been used. The approach of detecting radiolabeled metabolic products has only very recently been tested for use in marine waters and shellfish beds (Reasoner and Geldreich 1989). It appears to offer a promise as a rapid method of estimating water quality, but problems associated with using radioisotopes probably will limit its acceptance as a routine monitoring tool.

Direct plating procedures offer advantages of speed and simplicity over the conventional MPN analyses, which are laborious, less accurate, and require large numbers of replicates. Havelaar and During (1988) reported that a direct plating procedure gave better recovery of *E. coli* than the membrane filter method, using 0.1% sodium lauryl sulfate agar. Calabrese and Bissonnette (1990) incorporated catalase and pyruvate into an MF agar medium to degrade hydrogen peroxide, to which injured bacteria become increasingly sensitive. This led to improved recovery of chlorine-injured bacteria. Beneficial effects of hydrogen peroxide degradation in coliform media have also been reported by others (Calabrese and Bissonnette 1990*b*; Lee and Hartman 1989; McDonald, Hackney, and Ray 1983). Massad and Oliver (1987) described a cellobiose–polymyxin B-colistin (CPC) agar which offers significant advantages over many other media for primary isolation of vibrios. The medium is highly selective for *V. cholerae* and *V. vulnificus*, and each organism produces differently colored colonies. It shows promise for environmental monitoring of estuarine water and shellfish, although no field studies had yet been performed.

A new agar plate medium for *Salmonella* was recently developed by Rambach (1990) which allows rapid visual differentiation of *Salmonella* spp. from other *Enterobacteriaceae*. Colonies of *Salmonella* spp., which produce acid from propylene glycol, appear bright red on the medium. *E. coli*, which uses lactose (detected by a chromogenic indicator of β-galactosidase), grows as blue colonies. Species such as *Citrobacter* which are positive for both reactions appear purple. Those negative for both are colorless.

Hydrophobic grid membrane filter (HGMF) plating methods have been used successfully for rapid isolation of *E. coli* from food (Doyle and Schoeni 1987; Szabo, Todd, and Jean 1986). Todd et al. (1988) combined the HGMF method with an enzyme-labeled monoclonal antibody for 24 h enumeration of pathogenic *E. coli* 0157 in foods associated with disease outbreaks. The performance of a HGMF protocol for overnight enumeration of *Vibrio parahaemolyticus* was found to be superior to that of other methods for detection of this organism in sensitivity and specificity, as well as efficiency and ease of use (DePaola et al. 1988; Entis and Boleszczuk 1983).

Some of the most useful of the rapid culture methods for enteric bacteria feature composite media, which contain selective agents, reagents for growth, and chromogenic or fluorogenic indicators in a single tube. Berlutti et al. (1986) described a single-tube broth medium for detection of *Salmonella* subgroup I which combines ONPG hydrolysis, urease production, and indole production assays. Blue, yellow, or green color reactions in the tube differentiate different patterns of expression of these enzymes. The authors screened 2412 *Enterobacteriaceae* cultures and found nearly 100% agreement with the standard procedures. Another novel composite medium was developed for differentiation of the pathogen *E. coli* serotype 0157:H7 from other strains of *E. coli* in human feces (Farmer and Davis 1985). The procedure takes advantage of two unique properties of 0157:H7: its flagellar antigen H7 and its inability to ferment sorbitol after overnight incubation. A sorbitol-based medium was used, to which was added 1 ml/l of *E. coli* H7 antiserum. Strains of 0157:H7 gave a characteristic pattern in the medium; they did not ferment the sorbitol and were immobilized in the semisolid medium because of the reaction of their flagella with the antibodies.

By far the most notable of the composite media developed for detection and enumeration of enteric bacteria are those which incorporate chromogenic substrates for assays of glucuronidase and β-galactosidase activities. In 1976, Killian and Bulow (1976), in a survey of bacterial glycosidases, found that β-glucuronidase activity was confined to the genera *Escherichia*, *Shigella*, and *Salmonella*. Over 95% of *E. coli* strains and 40% of *Shigella* strains have been shown to produce β-glucuronidase (Hansen and Yourassowsky 1984; Killian and Bulow 1976). Based on this enzyme characteristic, procedures have been developed for the specific detection of *E. coli* from water (Feng and Hartman 1982; Frampton, Restaino, and Blaszko 1988; Weiss and Humber 1988), food (Moberg 1985; Robison 1984), and clinical specimens (Edberg and Trepeta 1983; Trepeta and Edberg 1984).

Various substrates have been incorporated into the media to detect β-glucuronidase activity. Most are fluorogenic methylumbelliferyl compounds (Berg and Fiksdal 1988; Feng and Hartman 1982). The most commonly used is 4-methylumbelliferyl-β-D-glucuronide (MUG). This compound is broken down by β-glucuronidase to release 4-methylumbelliferone, which fluoresces under long-wave ultraviolet light. Presence of fluorescence is considered to be a positive test for *E. coli*.

Chang et al. (1989) recently reported that more than 30% of *E. coli* in human fecal samples are β-D-glucuronidase negative and that tests based on activity of this enzyme may fail to detect a significant proportion of these bacteria in polluted water. High proportions of β-D-glucuronidase negative *E. coli*, however, have not been found in the U.S. Environmental Protection Agency equivalency tests of the Colilert test system (Bej et al. 1990).

Berg and Fiksdal (1988) tested three fluorogenic methylumbelliferyl substrates for rapid detection of fecal coliforms in drinking water and demonstrated that a direct membrane filter method incorporating MUG into agar allowed detection of as few as one fecal coliform per 100 ml within 6 h. Kasper and Tartera (1990) developed a glucuronidase test for *E. coli* which uses microtiter plates filled with coliform growth medium containing MUG. Water samples are added to the wells at various dilutions, and the MPN of *E. coli* is calculated from the number of fluorescent wells after incubation. An analysis of marine waters using this method showed very good correlation between the MUG results and those of standard methods. A rapid method for colorimetric enumeration of *E. coli* based on β-D-glucuronidase activity was described by Adams et al. (1990). Color development was followed spectrophotometrically and could be detected in only a few hours. This technique is too labor intensive for routine monitoring, but it is amenable to automation (Adams et al. 1990). Robison (1984) evaluated the MUG assay for detection of *E. coli* in foods and also found agreement of results with standard enumeration methods. Rippey, Chandler, and Watkins (1987) used a fluorogenic assay for enumeration of *E. coli* in shellfish. A modification of their method was developed and evaluated in both water and shellfish samples by Watkins et al. (1988). O'Brien and Colwell (1985) incorporated 4-methyl-umbelliferyl-conjugated substrate into taurocholate–tellurite–gelatin agar to differentiate *Vibrio* species.

A modification of the MUG method which incorporates simultaneous detection of both total coliforms and *E. coli* into a single composite medium has been extensively tested in recent years and is now marketed commercially as the Autoanalysis Colilert System (Access Medical Systems) (Edberg and Edberg 1988; Stratman 1988). The Colilert test is based on the ability of coliforms to produce β-galactosidase, which hydrolyzes and cleaves the substrate *o*-nitro-β-D-galactopyranoside (ONPG), releasing *o*-nitrophenyl, which is yellow. The Colilert test is a novel departure from the classical coliform culture methods that are based on lactose fermentation. It uses ONPG (for total coliforms) and MUG (for *E. coli*) both for essential nutrients and as an indicator system (yellow color and fluorescence). Simultaneous detection and enumeration of both groups of bacteria is obtained within 24 h. The commercial Colilert kit is described further below. Environmental studies have shown it to be simple, efficient, and reliable (Covert et al. 1989; Edberg and Edberg 1988; Edberg et al. 1988, 1989, 1990; Lewis and Mak 1989).

KITS AND MODULES FOR RAPID IDENTIFICATION

A number of manufacturers have developed kits for the identification of bacteria, of which a few types are in widespread use. Most of these commercial products, however, were developed for clinical use rather than for identification of environmental isolates. Problems often arise when such kits are used for the identification of aquatic bacteria (Austin 1988). The aquatic environment contains a wider range of taxa than are encountered in clinical diagnostic laboratories. The data bases of clinical kits and modules seldom cover environmental strains, and extensive data banks are not available for comparative purposes. Furthermore, isolates from environmental or industrial sources live under different selective pressures and, therefore, may be expected to give aberrant reactions that make them difficult to identify with these tests (Palmieri, Cartio, and Meyer 1988). Incubation times and temperatures specified in the kits are also seldom appropriate for aquatic microorganisms. Furthermore, plasmid DNA may alter profiles, leading to misidentification (Austin 1988).

Commercial kits which focus on the biochemical responses of the bacteria dominate the market. Serological kits are also available. Commercially available products for measuring biochemical reactions generally feature rows of microtubules or paper strips impregnated with freeze–dried test substrates. These are rehydrated when inoculated with bacterial suspensions, and, after incubation, the results are recorded as color changes, often following the addition of reagents. The patterns of reactions are then compared with data bases supplied by the manufacturer for that kit.

Much of the effort in development of commercial kits has focused on the *Enterobacteriaceae*, with the development of systems such as the Micro-ID (General Diagnostics, Morris Plains, NJ) (Aldridge et al. 1978; Berry and Badal 1979; Edberg et al. 1979), the Enteric-Tek (Bruckner, Clark, and Martin 1982), the Enterotube system (Grunberg et al. 1969), the API 20E (Analytab Products, Plainview, NJ) (Applebaum et al. 1982; Gooch and Hill 1982; McLaughlin et al. 1984) and the API Rapid E (Izard et al. 1984; Keville and Doern 1984; Murray, Gauthier, and Niles 1984; Overman et al. 1985*b*). Automated systems, such as the Autobac-IDX (General Diagnostics) (Barry et al. 1982), the AutoMicrobic System (Vitek Systems, Inc., Hazelwood, MO) (Johnson and Brinkley 1982), the autoSCAN-4 System (American Micro-Scan, Inc., Mahwah, N.J.) (Rhoden 1985), and the Avantage and Quantum II systems (Abbott Diagnostics, Irving, Tx.) (Jorgensen et al. 1984; Murray, Gauthier, and Niles 1984), have also been introduced for the identification of members of the *Enterobacteriaceae*. Rapid, convenient identification systems have also been developed for *Streptococcus* species, including fecal *streptococci* and *enterococci*. These include the API-20S (Analytab Products) (Applebaum et al. 1984*b*; Facklam et al. 1985; Keville and Doern 1982), the Rapid Strep system (Analytab Products) (Applebaum et al. 1984*a*; Facklam, Rhoden, and Smith 1984; Ruoff and Kunz 1983), the gram-positive identification (GPI) system (Vitek Systems) (Applebaum et al. 1984*b*; Facklam et al. 1985; Ruoff et al. 1982), and the RapID STR system (Innovative Diagnostic Systems, Inc., Atlanta, GA) (Applebaum et al. 1986; You and Facklam 1986).

Two of the biochemical identification kits have gained the most attention from aquatic microbiologists (Austin 1988). These are the API-20E "rapid" identification system for gram-negative fermentative rods, and the API-ZYME system, designed for characterization of enzymatic activities (both by Analytab Products, Plainview, New York). The API-20E system originally was designed for use on clinically important bacteria of the *Enterobacteriaceae*. It contains 20 biochemical reactions, including β-galactosidase, arginine dihydrolase, lysine and ornithine decarboxylases, utilization of Simmons citrate, H_2S production, urease, tryptophan deaminase, indole production, the Voges–Proskauer reaction, gelatinase, oxidase, and acid production from glucose, mannitol, inositol, sorbitol, rhamnose, saccharose, melibiose, amygdalin, and arabinose. The results are coded numerically into a 7-digit code for comparison with a data matrix.

Overman et al. (1985*a*) used the API 20E system to identify members of the family *Vibrionaceae*. It was found to be the most accurate of three different rapid identification systems that they tested. The API 20B test panel was used by Torsvik et al. (1990) to study the phenotypic diversity of bacterial strains isolated from soil. Palmieri et al. (1988) compared the API 20E to other systems for use in identifying isolates from cosmetics and pharmaceuticals. A panel of 4 h supplemental biochemical tests was developed by Edinger, Migneault, and Nolte (1985) to complete the identification of certain bacterial strains tentatively identified to genus by the API 20E. Haldane et al. (1986) added ornithine and lysine decarboxylase tests to the API 20E panel to improve screening for *E. coli* 0157:H7.

The API-ZYME test kit is a semiquantitative micromethod system designed for rapid detection of enzyme activities in bacteria, including acid and alkaline phosphatase, esterase, lipase, leucine aminopeptidase, trypsin, phosphohydrolase, β-galactosidase, and N-acetyl-β-glucosaminidase (Humble, King, and Phillips 1977). It consists of a plastic-covered strip with a series of microcupules containing dehydrated chromogenic substrates. The Api-ZYME kits have been used successfully to characterize a wide range of gram-positive and gram-negative bacteria in clinical laboratories (D'Amato et al. 1978; Nolte, Hollick, and Robertson 1982; Slots 1981; Tharagonnet et al. 1977). Austin (1988) provides a table of API-ZYME profiles for gram-positive bacteria from aquatic sediments. Straube et al. (1990) modified the system to characterize enzymatic activities of barophilic deep sea bacteria under *in situ* conditions of temperature and high pressure.

Kits designed for detection and enumeration of bacteria from the environment are very few. One system, in particular, has attracted considerable attention as a simple, rapid, and inexpensive technique for simultaneous enumeration and detection of both total coliforms and *E. coli* from water—the Autoanalysis Colilert system from Access Medical Systems, Inc. (Branford, CT). Colilert is derived from earlier β-glucuronidase-based substrate systems for *E. coli* identification, which employed a single inoculation for simultaneous detection of β-glucuronidase, ONPG, and indole (Edberg and Kontnick 1986). The Colilert test uses ONPG and MUG in a single dehydrated medium which is sold prepackaged in individual MPN culture tubes. Ten milliliters of water sample are added to the tubes, and the tubes are incubated for 24–28 h at 35°C. A presence–absence test is also available in which 100 ml of sample are added to a bottle containing defined substrate. In both the MPN and the presence–absence tests the medium is supplied in a ready-to-use powder form, with nothing to prepare or sterilize.

A national evaluation of the Colilert System to determine its equivalency to U.S. Environmental Protection Agency-approved Standard Methods (APHA 1985) was sponsored by the EPA and the American Water Works Association Research Foundation (Edberg et al. 1988, 1989). The test was evaluated for distribution system waters from ten geographic areas, representing a broad range of sources, both surface and subterranean. In both quantitative and presence–absence modes, no statistical differences were found between the results of Standard Methods and Colilert (Edberg et al. 1988, 1989). The test was sensitive, detecting down to one coliform or one *E. coli* per 100 ml and, in general, was not subject to false positives or negatives caused by noncoliform heterotrophic bacteria. On the basis of these studies and other EPA investigations, the Colilert system has been proposed as an alternative test procedure for analysis of total coliforms from drinking water (Edberg et al. 1990). The test has already been implemented as the routine analytical procedure for total coliforms for the distribution system monitoring program in Calgary, Alberta (Lewis and Mak 1989). The kit has not yet been widely tested for samples other than drinking water supplies (Covert et al. 1989). Its efficacy for marine and estuarine waters and shellfish meats remains to be established.

IMMUNOLOGICAL METHODS

The major drawback of most of the methods currently available for detecting and enumerating microorganisms in the environment is that they require culturing of the organisms to be monitored. Culture methods can be useful for detection and monitoring of microorganisms in the environment, but they also have significant disadvantages. Isolation and culture of microorganisms collected from natural samples such as water or shellfish are difficult if the organisms are easily overgrown by other bacteria in the sample, are fastidious in nutrient requirements, or have entered a viable but nonculturable state (Colwell et al. 1985; Roszak and Colwell 1987*b*; Xu et al. 1982).

Direct methods for observation and enumeration, on the other hand, provide relatively greater efficiency of detection. When direct methods are coupled with highly specific labeled antibodies, it is possible to target specific microorganisms very precisely without the need to culture them first (Wright 1991). Modifications of the procedures, in many cases, also allow assessment of viability and metabolic activity (Byrd and Colwell 1991; Kogure, Simidu, and Taga 1979).

The reaction between antibodies and their corresponding antigens are highly specific and can, therefore, be used as specific probes for the target antigen (e.g., a species-specific cell-surface protein). The use of antibodies as probes to detect microorganisms in the environment is based upon the ability to detect an immunologically distinct marker (Byrd and Colwell 1991; Wright 1991). Microbial antigens have different degrees of immunological specificity, ranging from type-specific antigens which occur among one, two, or three strains of a species to group- or species-specific antigens which are shared by larger numbers of related microorganisms.

Antiserum obtained from an animal immunized with a particular microorganism or antigen contains antibodies against several to many different portions of the antigen complex and is known as a polyclonal antibody probe. Polyclonal antibodies have long been used for identification and tracing of microorganisms. They are considered stable probes because they consist of mixtures of antibodies which recognize different antigens. Loss of expression of a single antigen generally does not result in failure of the polyclonal antiserum to react with a microorganism. However, such antisera may cross react with related organisms, and this sometimes presents problems in using these probes to identify specific strains of a microorganism.

Recent advances in production technology have made monoclonal antibody probes options for routine assays. Monoclonal antibody probes result from the fusion of an individual antibody-secreting cell and a malignant cell to produce an immortal hybrid cell line which secretes a single type of antibody that recognizes one, and only one, antigen (Goding 1986). Antibodies which react with antigens specific for a microbial strain or product can be selected and used individually or mixed together to produce a defined polyclonal antiserum.

Perspectives on choosing monoclonal or polyclonal antibody probes and limitations of both types of antisera have been reviewed by Wright (1991). Specificity is the key question when dealing with antibodies. The possibility of cross-reaction with other organisms must always be considered and controlled by use of highly specific monoclonal or well-adsorbed polyclonal antibodies. The cost of monoclonal antibodies is relatively higher, but this may be balanced by their much greater specificity. Services for providing both polyclonal and monoclonal antibodies are available from many universities and commercial companies.

Fluorescent Antibody Methods

Microscopic methods for direct enumeration of bacteria in environmental samples began to be developed in the 1950s, when water samples were filtered and stained with bright-field stains such as methylene blue (Erlich 1955; Jannasch 1953; Jannasch and Jones 1959). Simple staining and bright-field microscopy, however, do not easily allow differentiation of bacteria from detritus (Francisco, Mah, and Rabin 1973). Detection and direct counting of microorganisms in the environment has been significantly enhanced by the use of the fluorescent microscope. Cells attached to surfaces or collected on polycarbonate filters are stained with fluorochromes such as Acridine orange (3,6-tetramethyl diaminoacridine), which fluoresces orange or green upon intercalation with cellular nucleic acid (Daley and Hobbie, 1975; Francisco, Mah, and Rabin 1973; Hobbie, Daley, and Jasper 1977; Yamabe 1973). Acridine orange staining procedures have become routine for enumeration of microorganisms in water and food (Byrd and Colwell 1991; Hobbie, Daley, and Jasper 1977; Pettipher 1983). DAPI (4′,6′-diamidino-2-phenylindole), a DNA-specific stain which fluoresces bright blue when bound to DNA, is another fluorochrome which is widely used for counting aquatic microorganisms in eutrophic and seston-rich water samples (King and Parker 1988; Porter and Feig 1980). Because of the significant contrast achieved with fluorescence microscopy, small, starved or dormant bacterial cells, which are common in low-nutrient marine waters (Ferguson and Rublee 1976; Watson et al. 1977), can be easily detected.

Direct microscopic counts of samples collected from the natural environment and directly stained with fluorochromes do not reliably distinguish viable from nonviable cells (Yeh et al. 1987). Modifications of these techniques have become available which do allow estimation of the proportion of total cells which are growing, metabolizing, or respiring. These have been reviewed by Roszak and Colwell (1987*b*) and Byrd and Colwell (1991) and include the direct viable count (DVC) method to detect nutrient-responsive cells (Brayton et al. 1987; Brayton and Colwell 1987; Kogure, Simidu, and Taga 1979; Peele and Colwell 1981; Roszak and Colwell 1987*a*), microautoradiography to detect uptake of isotopically labeled substrates (Brock and Brock 1968; Fliermans and Schmidt 1975; Hoppe 1976; Meyer-Reil 1978; Pearl 1974; Peroni and Lavarello 1975; Ramsay 1974; Roszak and Colwell 1987*a*; Tabor and Neihof 1982*b*; Wright 1978), and reduction of 2-(*p*-iodophenyl)-3-(*p*-nitrophenyl)-5-phenyl tetrazolium chloride (INT) to detect respiring organisms (Dutton, Bitten, and Koopman 1983; King and Parker 1988; Maki and Remsen 1981; Tabor and Neihof 1982*a*; Zimmerman, Itturriaga, and Becker-Birck 1978).

The fluorochromes just described are useful for enumerating total cells or total viable cells in a sample, but are unable to distinguish different species or groups of bacteria. Probably, the most powerful of the presently available tools for monitoring specific microorganisms in aquatic environments is immunofluorescence (Goding 1986; Knapp, Holubar, and Wicks 1978; Turgeon 1990; Wright 1991), a technique in which antibodies to species- or strain-specific cell surface antigens are labeled with a fluorescent dye, usually fluoroscein isothiocyanate (FITC). Besides being very specific, the method can also be extremely sensitive when combined with filtration to concentrate a sample.

Immunofluorescent detection of viable but nonculturable *Escherichia coli* and *Vibrio cholerae* in aquatic microcosms (Brayton et al. 1987; Huq et al. 1990; Xu et al. 1982) demonstrates the usefulness of this technique in direct detection of microorganisms in aquatic and marine ecosystems. Enumeration of specific populations of organisms (and particularly of allochthonous indicators and pathogens) in these environments is complicated by the nonculturable stage, which many microorganisms undergo when placed in stressful environments. Immunofluorescence microscopy can enumerate these populations without the need for culturing the samples. Those organisms which can only be cultured from the environment with difficulty—e.g., the *Bacteroides fragilis* group of anaerobic bacteria (Fiksdal and Berg 1987)—can also be detected easily and rapidly.

Fluorescent antibody labeling has been used to examine a variety of environmental samples directly to detect specific organisms. Byrd and Colwell (1991) provide a summary of organisms presently being detected in the environment and clinical specimens by this procedure. Immunofluorescence has been used to trace bacteria in marine and estuarine systems (Grimes and Colwell 1983; Ward and Carlucci 1985). Indicator bacteria and enteric pathogens detected and enumerated directly using fluorescent antibodies include *Escherichia coli* (Abshire 1976; Ferrier 1986; Pugsley and Evison 1974; Xu et al. 1982), *Bacteroides fragilis* (Fiksdal and Berg 1987), *Campylobacter* spp. (Chan et al. 1983; Pugsley and Evison 1974), *Salmonella* spp. (Thomason et al. 1975), *Shigella* spp. (Tsunematsu 1977), *Streptococcus faecalis* (Abshire and Guthrie 1971), and *Vibrio cholerae* (Brayton et al. 1987; Huq et al. 1990; Tamplin et al. 1990).

A simple and particularly useful rapid method has been developed which allows not only enumeration of bacteria of a given species, but also assessment of their viability (Brayton and Colwell 1987). It combines fluorescent antibody (FA) staining with the direct viable count (DVC) method of Kogure, Simidu, and Taga (1979) and Peele and Colwell (1981). The DVC technique allows direct estimation of viable bacteria in water by preincubating samples with low concentrations of yeast extract and nalidixic acid. Nalidixic acid prevents cell division in gram-negative bacteria by inhibiting DNA replication and, subsequently, cross-wall formation (Deitz, Cook, and Goss 1966; Goss et al. 1964, 1965; Kantor and Deering 1968). This results in formation of elongated, metabolically active cells in the presence of the nutrient, which can easily be detected by fluorescent staining. The DVC method has been used in a variety of environmental studies (Grimes, Singleton, and Colwell 1984; Maki and Remsen 1981; Peele and Colwell 1981; Roszak, Grimes, and Colwell 1984; Xu et al. 1982).

Combination of immunofluorescence with the direct viable count (FA-DVC) has proven to be a useful detection procedure for specific bacterial species in the aquatic environment. It has been used successfully to enumerate viable but nonculturable cells of *Campylobacter pylori* in water (Shahamat et al. 1989). The method also has been used extensively in the field to detect and enumerate viable *Vibrio cholerae* O1 (Brayton et al. 1987; Brayton and Colwell 1987; Huq et al. 1990; Tamplin et al. 1990; Xu et al. 1984). A recent study of *V. cholerae* O1 in Bangladesh waters (Huq et al. 1990) compared results of enumeration by FA-DVC (using a monoclonal antibody probe specific to the A antigen of the O1 lipopolysaccharide) to those obtained by conventional culture methods. Of a total 876 samples collected in the study, *V. cholerae* O1 was detected in 563 (or 64.27%) by FA-DVC, while only three (or 0.34%) of the samples were positive for *V. cholerae* by conventional culturing methods. The importance of the FA-DVC technique is emphasized by the demonstration that these viable but nonculturable forms of *V. cholerae* O1 are still capable of causing disease. The same monoclonal antibody probe has also been used to study the association of *V. cholerae* with phytoplankton blooms (Tamplin et al. 1990).

The major disadvantage of the FA-DVC method is the subjective nature of visually differentiating elongated cells. The problem can be overcome by use of a particle counter (Zelibor, Tamplin, and Colwell 1987) or an image analyzer (Singh and McFeters 1990). By incorporating an automated approach in the FA-DVC, the overall ability to detect rapidly the viability of a specific organism in the environment is enhanced.

Enzyme Immunoassays

Enzyme immunoassay (EIA) methods use enzyme-labeled antibody conjugate to detect the presence and quantity of antigen in a sample (Turgeon 1990). Reaction of specific antibodies with the antigen (or cells) is detected by incubating with an enzyme-labeled conjugate which binds to the antigen–antibody complex. The enzyme activity is measured spectrophotometrically after the addition of the specific chromogenic substrate. A variety of enzymes are employed in enzyme immunoassay. The most commonly used are horseradish peroxidase and alkaline phosphatase.

The most widely used EIA is the enzyme-linked immunosorbent assay (ELISA). The ELISA has many variations, and has proven particularly useful for detection of pathogens in clinical specimens, because it is specific, objective, and rapid. Many commercial kits are available for detection of viral antigens and of specific antibodies in blood serum.

ELISA has been used to detect and enumerate viral pathogens in water and wastewater. Nasser and Metcalf (1987), for example, developed an amplified ELISA for the detection of hepatitis A virus in estuarine samples. Guttman-Bass et al. (1987) compared several ELISA techniques for detecting rotaviruses in water. A report of specific ELISAs for detection of Norwalk virus and Snow Mountain Agent has also appeared (Madore et al. 1986).

A few studies have been published in which bacteria were detected in environmental samples by ELISA. Both direct and indirect (using antibody to capture the antigen) ELISA techniques have been used to enumerate rhizobia in peat and soil (Nambair and Anjaiah 1985; Renwick and Jones 1985). ELISA also has been used to identify rhizobial strain in root nodule material (Fuhrman and Wollum 1985; Wright, Foster and Bennett 1986). A competitive ELISA technique was developed by Kemp et al. (1988) for enumeration of specific methanogens in fermentor and sludge samples. Enterotoxigenic *E. coli* were detected from primary cultures using an antibody to the heat-stable enterotoxin (Ronnberg, Soderlund, and Wadstrom 1985).

The dot–blot immunoassay is a qualitative variation of the ELISA which uses a precipitating dye substrate to detect antigen bound to nitrocellulose membranes (Wright 1991). The dot–blot method has found growing use for identification of bacterial colonies growing on agar. Antigen from the colonies is blotted onto a nitrocellulose membrane, and the membrane is subjected to ELISA, resulting in the precipitation of colored dots where positive colonies were located. This technique has been used to identify strains of rhizobia (Ayanaba, Neiland, and Zablotowicz 1986; Olsen and Rice 1989), to detect root colonization by mycorrhizal fungi (Wright and Morton 1989), to detect CFA antigens from enterotoxic *E. coli* (Lopez-Vidal and Svennerholm 1986), and to identify *Staphylococcus aureus* strains that produce enterotoxin B (Peterkin and Sharpe 1984).

Use of ELISA for environmental detection and enumeration may require sample concentration, leading, perhaps, to interference from other species in the concentrate. Generally, it is reported that cell densities of 10^3–10^4 ml are required for detection (Kemp, Archer, and Morgan 1988; Nambair and Anjaiah 1985; Renwick and Jones 1985). At present, the best use of both the ELISA and dot blot assays for environmental bacterial detection is a highly specific and rapid confirming test for strains isolated on primary enrichment media.

RESTRICTION PATTERN FRAGMENT ANALYSIS

Because every species has a unique nucleotide sequence that distinguishes it from every other species, each organism's genetic composition essentially is a fingerprint that can be used for its identification. Restriction endonuclease analysis of bacterial plasmid and chromosomal DNA has been shown to be useful in detecting genetic variation and identifying species and strains. It allows rapid differentiation between closely related strains.

Plasmid fingerprinting is used in many public health and clinical laboratories to understand the epidemiology of bacterial disease outbreaks (Mayer 1988). Plasmids are autonomous, self-replicating pieces of extrachromosomal DNA, which can be transferred between bacteria of the same or different genera. They can encode a variety of nonessential functions, such as toxin production or antibiotic resistance (Falkow 1975; Hardy 1981). Plasmid profiles are relatively easy and inexpensive to perform. DNA is extracted from the cells, and plasmid DNA (which are circular molecules) is separated from the linear chromosomal DNA. The plasmid DNA is cut by restriction endonucleases, which cleave DNA molecules at specific sequences, and patterns of fragmentation are determined by agarose gel electrophoresis. Plasmids that have identical sequences produce the same set of fragments (Mayer 1988).

The first reported use of plasmid profile characterization was applied to enteric bacteria, including diarrheagenic *E. coli*, *Shigella dysenteriae*, *Salmonella* spp., and *Streptococcus faecalis* (Meyers et al. 1976). Plasmid profiles subsequently have been used to study disease outbreaks caused by a number of enteric bacteria, including *Salmonella* (Brunner et al. 1983; Goldstein et al. 1986; Gotuzzo et al. 1987; Hadfield, Monson, and Wachsmuth 1985; Holmberg et al. 1984; O'Brien et al. 1982; Riley and Cohen 1982; Riley et al. 1983; Tacket et al. 1984; Taylor et al. 1982), *Shigella* (Crosa et al. 1977, Jamieson et al. 1979; Prado et al. 1987; Tacket et al. 1985), *Enterobacter* (John, Sherbaugh, and Bannister 1982; Muytjens et al. 1983), *Yersinia enterocolitica* (Mingrone et al. 1987) and *Campylobacter jejuni* (Tenover et al. 1984). The virulence of enterotoxigenic *E. coli* depends upon toxins and adhesive factors, both of which are plasmid encoded (Mazaitis et al. 1981; Murray and Tsao 1987). Plasmid profile patterns have been used to study outbreaks of disease caused by this organism and to study transfer of virulence plasmids to other enterobacteria (Murray et al. 1982; Murray and Tsao 1987; Wilshaw, Smith, and Anderson 1979; Yamamoto et al. 1984).

Plasmid fingerprinting is not useful for bacteria that contain no plasmids or contain only one large plasmid. A related technique that has proven to be useful in strain identification is restriction endonuclease analysis of chromosomal DNA. In this method, whole-cell DNA is extracted and digested with restriction enzymes, and the fragments are separated by agarose gel electrophoresis. The fragment patterns produced, visualized by ethidium bromide staining, often are distinct enough to enable rapid differentiation between species or strains. Comparison can be made between the banding patterns of unknown isolates and those of known reference strains (Drahos et al. 1985). This technique has been used successfully for typing strains of *Campylobacter*

(Bradbury et al. 1984; Langenberg et al. 1986), *Bacteroides* spp. (Bradbury et al. 1985), diarrhea-inducing *Clostridium difficile* (Kuijper et al. 1987), and *Salmonella* (Whiley et al. 1988). A related method involving fingerprinting of hypervariable sequences in *E. coli* has also been reported (Huey and Hall 1989).

NUCLEIC ACID HYBRIDIZATION METHODS

The unique base sequences of the nucleic acids of organisms and the extreme specificity with which these sequences hybridize to complimentary sequences have been exploited to produce a new set of powerful, rapid, and reliable techniques for detection and identification of microorganisms. Techniques such as DNA–DNA hybridization, DNA–RNA hybridization, DNA and RNA sequencing, DNA fingerprinting, and the use of gene probes for detection of specific sequences have been employed widely in the clinical setting and, more recently, in the field of microbial ecology. The tools of molecular genetics offer new and very powerful ways to detect and study organisms in their natural environment without the need for culturing organisms of interest.

The current surge of interest in nucleic acid hybridization probes and their widespread application has generated several reviews of the subject. A review by Tenover (1988) focuses on the use of probes in the clinical setting but is a comprehensive treatment useful to those who wish to use probes for detection of microorganisms in the environment. Reviews by Ogram and Sayler (1988), Hazen and Jimenez (1988), Holben and Tiedje (1988), and Knight et al. (1991) are aimed at the use of probes in environmental microbiology.

This section will review techniques of using, constructing, and using gene probes for environmental samples, with particular emphasis on applications of these technologies to the detection and enumeration of indicator bacteria and pathogens in the aquatic environment.

Extraction of Nucleic Acids from Water and Environmental Samples

Theoretically, the genetic material of every organism in the microbial community should be represented in DNA extracted from environmental samples. With probes and hybridization techniques of sufficient specificity and sensitivity, every species present in the original community can potentially be detected. The success of gene probe methods for detection and identification of bacteria in a given environment, thus, depends largely upon the ability to recover representative samples of nucleic acid from that environment. Methods for extracting and purifying DNA and RNA from laboratory cultures of microorganisms are highly developed (Sambrook, Fritsch, and Maniatis 1989), but techniques for obtaining pure, intact nucleic acids directly from environmental samples are still in their infancy. The development of such methods has been fueled in recent years by increasing interest in rapidly and accurately monitoring the fate and distribution of genetically engineered organisms accidentally or deliberately released into the environment (Atlas and Sayler 1988; Colwell, Levin, and Palleroni 1991; Knight et al. 1991; Morgan et al. 1989).

Two approaches have been developed for extraction of bacterial nucleic acids from soil and sediment. One is direct recovery of DNA by *in situ* lysis of the cells, followed by extraction of the liberated DNA from the sample (Ogram, Sayler, and Barkay 1987; Porteous and Armstrong 1990). This method gives high yield but may include extracellular DNA as well as material from soil organisms other than bacteria. Problems may also be encountered from contamination of the DNA by humic acids and other compounds which inhibit the efficiency of restriction enzyme digestion and hybridization. The second approach is to recover the bacteria from the soil before lysis using differential centrifugation (Holben et al. 1988; Torsvik 1980). This method removes soil material and contaminants and produces DNA that is less fragmented and more uniform in size (facilitating subsequent manipulations) and whose origin is known. All bacteria in the sample may not be recovered by this technique, but Holben et al. (1988) claim that the material recovered is representative of the community. Pace et al. (1986) and Olsen et al. (1986) have described a procedure for extracting nucleic acids from attached microbial populations.

Recovery of high yields of DNA or RNA from mixed populations of freshwater or marine planktonic microorganisms presents unique problems. Biomass in soil is usually high enough to get adequate recovery of nucleic acids from less than 100 g of sample (Ogram and Sayler 1988). Bacteria in aquatic systems, however, are often very small ($< 0.6\ \mu m$) and very dilute (ca. 10^6/ml), making nonselective collection of a sufficient number of cells and quantitative DNA extraction difficult (Fuhrman et al. 1988). Extraction and purification of nucleic acids from most water samples requires concentration, often of a 1000 or more liters, before lysis to get adequate yield in a volume that is easy to handle (Ogram and Sayler 1988). Extraction of cellular DNA without concentration yields insufficient material for cloning or hybridization (Paul and Myers 1982) and also includes extracellular DNA present in the water (D'Amato et al. 1978; Paul et al. 1988).

Several methods for concentrating bacteria in water samples have been tried, including continuous centrifugation (Ogram and Sayler 1988) and various filtration methods (Fuhrman et al. 1988; Morgan et al. 1989; Ogram and Sayler 1988; Paul, Jeffrey, and DeFlaun 1987; Somerville et al. 1989).

Continuous centrifugation is a gentle, but time-consuming method of concentration. The cells recovered may not be representative of the community, because the small cells typical of marine planktonic bacteria are much more difficult to sediment than larger cells (Fuhrman et al. 1988; Ogram and Sayler 1988). Most investigators have used the approach of concentrating cells on micropore filters. Filtration collects bacteria quickly and nonselectively. Small-pore filters have the problem of frequent clogging by particulates when used with natural aquatic samples. Fuhrman et al. (1988) used prefiltration through larger-pore filters, but a significant portion of the bacteria may be lost during this step (Somerville et al. 1989). Delong et al. (1988) used tangential flow filtration to concentrate over 1600 l of oligotrophic seawater down to a few milliliters. Another approach is to use filters with large surface area (Fuhrman et al. 1988) or high capacity (Somerville et al. 1989).

A simple, rapid method for extracting nucleic acids from aquatic samples has recently been reported by Somerville et al. (1989). Their system uses a high-capacity, cylindrically mounted filter through which water is pumped using a peristaltic pump. The filter is housed in a chamber with a volume of 2 ml. Once the water is passed through the filter, a lysis buffer can be added into the chamber, and the cells can be lysed directly on the filter without further manipulation. The lysate is drawn off the housing and both DNA and RNA can be purified. The advantage of this method is that significant volumes can be concentrated and the entire process is carried out in the disposable filter unit. The use of a portable pump and the ability to stabilize quickly and freeze the filter units for later lysis and extraction make the system highly amenable to field use. The application of this system for detection of *Salmonella* in estuarine water using a DNA hybridization probe has been reported (Knight et al. 1990*a*).

Construction of Gene Probes

Nucleic acid probes (or gene probes) are molecules of single-stranded DNA or RNA which have been labeled, either radioactively or chemically, so that they can be detected in a mixture of renatured nucleic acids. The probes are incubated under appropriate conditions with denatured DNA or RNA (e.g., the total cellular DNA extracted from a water sample) and hybridize specifically with regions of the target nucleic acid which are complimentary. Excess, unhybridized probe is selectively removed, and probe—target hybrids are detected using methods appropriate to the particular labeling strategy.

Probes may be constructed to suit the specific purpose of the analysis for which they are used (Ogram and Sayler 1988). Function-specific probes allow rapid determination of the potential of a microbial community for performing a given function, such as biodegradation of a pollutant or production of an enterotoxin, in the environment. Gene probes have also been used to identify and isolate microorganisms with specific catabolic genotypes in much less time than is required for conventional enrichment culture techniques (Ogram and Sayler 1988). Species- or group-specific probes constructed from parts of genes coding for ribosomal RNA are proving useful for the rapid identification and enumeration of bacterial species in environmental and clinical samples.

The usefulness of hybridization techniques is largely determined by the probe that is used. Probe specificity is a key factor, and it is important that probes be based on specific genes that are well characterized (Ogram and Sayler 1988). When large, uncharacterized plasmids, for example, are used as probes, unwanted, nonspecific homology with sample nucleic acids is likely. Similarly, when specific genes are cloned into a vector for use as a probe, it must be ensured that no homology exists between the sample and the vector itself by including a control probe of the vector alone in the protocol (Ambinder et al. 1986). It is also possible for a probe to be too specific. For example, in the case of function-specific probes, genes used as a probe may not be strongly conserved across genera, and the probe may fail to detect the function in species other than the one from which the probe came. In such a situation it may be necessary to include more than one gene encoding the function from a range of organisms (Ogram and Sayler 1988).

The basic strategy in probe construction is to isolate a unique base sequence in the target nucleic acid that is specific to the organism or function of interest, to reproduce that sequence in large quantities, and to attach a label or reporter molecule to the sequence so that it can be detected in hybridization reactions. The choice of probe material should be determined by the availability and source of the material for labeling, the degree to which the nucleic acid can be labeled, and the particular hybridization strategy to be used (McInnes and Symons 1989*a*). Nucleic acid probes can be grouped into three major categories, based upon strategies used to construct them: total genomic DNA; cloned restriction fragments; and synthetic oligodeoxynucleotides (Knight et al. 1991).

Total genomic DNA hybridization, the technique of using total genomic DNA of an organism as a hybridization probe has most often been used in bacterial systematics to determine relationships among closely related bacteria. The technique has not been used routinely to identify or detect microorganisms, because it is time-consuming and not well suited to large-scale screening. Another major disadvantage is the likelihood of cross-reaction arising from homology between genes highly conserved between species (Grimont et al. 1985). Sometimes, however, the base composition of an organism is so unique that use of whole-cell DNA as a probe is justified. Species of *Bacteroides* (Morotomi, Ohno, and Mutai 1988), *Campylobacter* (Esaki et al. 1988; Totten et al. 1985), and *Vibrio* have been detected successfully using total genomic DNA probes.

The approach most widely used is construction of gene probes from restriction fragments of DNA cloned into plasmid vectors. Recombinant DNA technology has allowed great flexibility in nucleic acid probe preparation. Essentially, any nucleic acid can now be cloned into a phage or bacterial vector and produced in large quantities. Details of methods for isolation, preparation, and cloning of nucleic acid for probes may be found in many of the available molecular biology methods manuals (Sambrook, Fritsch, and Maniatis 1989).

Basically, two methods are available for constructing probes using cloned DNA: shotgun cloning and targeted cloning (Knight et al. 1991). In the shotgun approach, random restriction fragments from a genomic DNA library are cloned, and the cloned fragments are tested for specific hybridization with the target nucleic acid. The first commercially available probe for *Salmonella* was constructed in this manner (Fitts et al. 1983; Fitts 1985, 1986). The advantage of this approach is that it can be much less time-consuming than the targeted approach, which involves identification and characterization of specific clones. A disadvantage is that it may be difficult or impossible to find a clone of the desired specificity, since size of the cloned fragments is determined by chance. A clone containing highly specific sequence may also contain less-specific sequence and result in unwanted homology. Nevertheless, a number of probes have been successfully developed using this approach (Atwood et al. 1988; Kuritza et al. 1986*a*, *b*; Salyers, Lynn, and Gardner 1983).

In contrast to the shotgun strategy for developing probes from genomic libraries is the targeted strategy, which involves cloning a particular genetic determinant. The cloned fragment, or a portion of it, is then used as a specific probe. In this case the function of the gene is usually known, and the function often determines the choice of clones for use as a probe. Probes developed using a targeted cloning strategy are often used as function-specific probes to detect similar genes, such as those encoding heavy metal resistance (Barkay, Fouts, and Olson 1985) or degradative phenotypes (Pettigrew and Sayler 1986; Sayler et al. 1985) in different bacteria. They can also be used as species-specific probes if the cloned-gene sequence is conserved within a species but is highly variable between species (Knight et al. 1991). Probes targeted toward the pathogenic determinants of enteropathogenic bacteria are an example of this (Echeverria et al. 1982; Hill, Payne, and Aulisio 1983; Kaper and Levine 1981; Moseley et al. 1982; Nataro et al. 1985; Rubin et al. 1985; So, Dallas, and Falkow 1978).

Cloned-DNA determinants encoding ribosomal RNA have significant application as species-specific or phylogenetic probes. These probes can be used to detect either rRNA determinants or the rRNA molecules themselves. Determinants coding for 5 S, 16 S, and 23 S rRNA subunits contain regions in which the sequence is highly conserved between species and regions in which it is highly variable. The variable regions comprise sequences which can be used as probes specific for a particular species, while the conserved regions offer probes with broader specificity [e.g., family or kingdom (Delong, Wickham, and Pace 1989)]. rRNA-directed probes are potentially much more sensitive than DNA-directed probes, because a single cell can have thousands of copies of an rRNA target, depending upon the stage of growth of the cell and how many ribosomes it contains (Delong et al. 1989). The application of rRNA sequences in the field of microbial ecology has been reviewed by Pace et al. (1986) and by Ward (1989). Successful use of rRNA probes to detect and enumerate specific organisms in different environments has been reported by many investigators (Delong, Wickham, and Pace 1989; Devereux et al. 1989; Kohne et al. 1986; Krauss, Ludwig, and Schleifer 1986; Regensburger, Ludwig, and Schleifer 1988; Schleifer et al. 1985; Stahl et al. 1988). Of special note is the technique described by Delong, Wickham, and Pace (1989), in which fluorescently labeled rRNA probes are used as stains for visualization of specific groups of organisms with an epifluorescent microscope.

Synthetic oligonucleotide probes comprise the third major group of probe types. Automated machines are now available commercially which allow easy and rapid synthesis of single-stranded oligonucleotides of defined sequence and specificity (Caruthers 1985; Galt 1984). Synthetic oligonucleotide probes usually range from 15 to 50 bases in length. They generally are more specific than restriction fragment probes, because the instability introduced into a hybrid with a mismatched base pair is greater for short hybrids than for long ones. Under stringent conditions, an oligonucleotide probes can distinguish a single base pair change in a DNA or RNA sequence (Wallace et al. 1979). A probe with broader specificity can be constructed by synthesizing a mixture of oligomers with different nucleotides at one or more positions (Giovannoni et al. 1988). Other advantages of oligonucleotide probes are their stability over time and their rapid rate of hybridization. The major disadvantage is that they usually are less sensitive than restriction fragment probes, because less label can be incorporated per molecule.

The sensitivity of a gene probe assay is determined largely by the labeling and detection technique used. Several reviews of DNA labeling strategies are available which discuss details of methods and their relative advantages and disadvantages (Matthews and Kricka 1988; McInnes and Symons 1989*a*, *b*). Methods for detecting the binding of a probe to the target DNA fall into four basic categories: radioactive labels, biotin–avidin systems, enzyme-conjugated probes, and serological detection.

Until recently, the most commonly used detection system was the insertion of radioisotopically labeled nucleotides into the probe by a technique known as nick translation (Rigby et al. 1977). One of the strands of DNA to be labeled is broken (or "nicked") at frequent intervals by the addition of a DNAse. Then radiolabeled (usually with ^{32}P) nucleotide bases are added, and the damaged strands are repaired with the enzyme DNA polymerase I, which inserts the radioactive bases into the DNA. After hybridization, the binding of probe DNA to the target can be detected by autoradiography or scintillation counting. This labeling method is highly sensitive, detecting less than 1 pg of DNA (Tenover 1988). Radiolabeling, however, is not well suited for routine use in water analysis laboratories, due to the short half-life of ^{32}P (making storage impractical), expense, and safety concerns.

Fortunately, the desire to eliminate radioactive labels from laboratory procedures has led to the development of alternative detection systems for gene probes. The first of these to be described was the biotin–avidin procedure (Al-Hakim and Hull 1986; Chan, Fleming, and McGee 1985; Langer, Waldrop, and Ward 1981; Leary, Brigati, and Ward 1983). Biotin is attached via a linker arm to the pyrimidine ring of the dUDP or UTP. The biotinylated nucleotides can then be incorporated into the probe DNA using nick translation and the appropriate polymerases. Oligonucleotides can be biotinylated by terminal labeling (Murasugi and Wallace 1984; Riley, Marshall, and Coleman 1986). The binding of the biotinylated probe to its target can be monitored colorimetrically by using complexes consisting of avidin or streptavidin (proteins with high affinity for binding biotin), which are linked to enzymes such as horseradish peroxidase or alkaline phosphatase. Biotinylated probes are extremely stable, and improvements in technology have made them almost as sensitive as radiolabeled probes (Chan, Fleming, and McGee 1985; Chollet and Kawashima 1985; Goltz et al. 1990; Matthews et al. 1985). Several variations of this technique have been developed, including the use of a photoreactive substrate that can be measured with a fluorometer (Forster et al. 1985), binding of fluorescently labeled antibodies to the avidin (Pinkel, Straume, and Gray 1986), and amplification systems that use multiple enzyme molecules coupled to each biotin (Chollet and Kawashima 1985) or multiple layers of avidin, biotin, and fluorescent antibodies (Pinkel, Straume, and Gray 1986).

Other nonradioactive labeling techniques have also been developed and improved, including direct attachment of enzymes to the probe DNA (Jablonski et al. 1986; Ruth 1984; Seriwatana et al. 1987), use of monoclonal antibodies to recognize the structure of DNA–RNA hybrids (Boguslawski et al. 1986), sulfonation and mercuration (McInnes and Symons 1989*b*), and chemiluminescent detection systems (Goltz et al. 1990). Many already are being incorporated into commercial packages. Significant improvements will continue to be seen in the next few years. Nonradioactive detection systems for nucleic acid probes have only recently begun to be tested in environmental applications, and it should be recognized that a number of interfering factors currently restrict their use with environmental samples (Atlas and Sayler 1988).

Application to Environmental Samples

Two main approaches have been used for application of gene probes to environmental bacteria. The first is known as colony hybridization, or colony blot (Ford, Simpson, and Olson 1987; Sayler et al. 1985, 1987). This method, originally developed by Grunstein and Hogness (Grunstein and Hogness 1975) for screening bacterial colonies for cloned genes, has been widely used for rapid detection of target sequences in colonies growing on agar plates. Colonies are transferred from the plate onto a nitrocellulose or nylon membrane by placing the membrane directly on the plate and lifting off the cells. Alternatively, the membrane can be placed on the agar plate and inoculated directly. The cells are lysed with NaOH so that the DNA is released and denatured and can bind directly to the membrane. The membrane is then incubated with a labeled probe, which hybridizes with complementary target sequences, and the hybrids are detected by a method appropriate for the probe label. Although it is relatively sensitive and specific, the colony hybridization method still requires that the organisms be cultured before analysis and, therefore, is subject to the limitations of the culture techniques used.

An alternative strategy is hybridization of direct DNA extracts (Knight et al. 1990*a*). Direct isolation and purification of nucleic acids eliminates the need for cultivation of bacteria and, therefore, bypasses the major source of sampling error inherent in the colony hybridization technique. It allows detection and enumeration of stressed, nonculturable pathogens and indicator organisms in aquatic environments. The technique is rapid, because analyses are performed directly on the environmental samples, with no laboratory incubation time required.

One of the simplest methods of probing environmental nucleic acids directly is the dot blot or slot blot, in which denatured nucleic acid is applied to nitrocellulose or nylon membranes in discrete dot- or slot-shaped areas using a vacuum manifold apparatus. Results of hybridization of probes with dot blots are quantitative as well as qualitative. Quantitative dot blot hybridization is described by Kafatos, Jones, and Efstradiadis (1979). Attwood et al. (1988) describe its use for enumerating specific microorganisms in a mixed community. The disadvantages of probing direct DNA extracts are that sensitivity is limited by the efficiency of DNA recovery from the sample, and DNA from organisms present in the community in small numbers may fall below the detection limits for the technique (Atlas and Sayler 1988). Sensitivity has been improved in some cases by using a solution hybridization method (Steffan and Atlas 1987). Solution hybridization also proceeds faster than mixed-phase (filter) hybridization. Syvanen, Laaksonen, and Söderlund (1986) developed a system which combines solution hybridization with a novel sandwich hybridization system, which uses two different probes for capture and detection of target sequence (Ranki et al. 1983).

Another gene probe method worthy of mention is one that was developed for use with MPN procedures (Frederickson et al. 1988; Van Vorris, Frederickson, and Ligotke 1987). In this method, multiple serial dilutions of a sample are placed into microtiter plates, and enrichment medium is added to the wells. After growth, the cells are lysed, and the released nucleic acid is transferred to a membrane for hybridization. The pattern of positive and negative hybridizations at various dilutions is used to calculate a most probable number of target organisms in the sample. Again, this technique is not useful for nonculturable bacteria.

Many probes specific for enteric bacterial pathogens have been constructed. Table 9-2 lists those that have been reported in the literature. They include probes directed at both DNA and RNA targets in strains of pathogenic *E. coli*, *Salmonella*, *Shigella*, *Vibrio*, *Yersinia*, *Listeria*, *Bacteroides*, and *Campylobacter*. Many of these probes have played a valuable role in epidemiological studies of diarrheal pathogens, particularly of *Vibrio* species (Kaper et al. 1982; Lin et al. 1986; Nishibuchi et al. 1986) and have contributed significantly to our understanding of their distributions. For the most part, however, these probes have been developed for, and tested in, clinical specimens such as stool samples. Most, in fact, have been used in colony hybridization assays which rely on culturing of the test organisms (Knight et al. 1991; Tenover 1988). Echeverria et al. (1982) used probes for enterotoxigenic *E. coli* to detect these pathogens in contaminated well water, but their method still required culture of colonies on filters.

Recently, gene probes have begun to be used extensively in environmental microbiology and biodegradation research, as well as to track genetically engineered microorganisms in the environment (Atlas and Sayler 1988; Knight et al. 1991; Ogram and Sayler 1988). As techniques for direct detection and enumeration of both culturable and nonculturable bacteria in environmental (and, particularly, water) samples are improving and becoming more routine, investigators are now beginning to apply this technology to studies of indicator bacteria and enteric pathogens in both freshwater and marine environments. Recently, Knight et al. (1990*a*) described the use of a DNA probe to detect and enumerate *Salmonella* spp. in nucleic acids extracted from polluted estuarine water. The method was shown to detect culturable and nonculturable *Salmonella* with equal efficiency.

A relatively new technique which vastly increases the sensitivity of gene probe assays, thereby allowing detection of organisms that are present in a sample only in very small numbers, is the polymerase chain reaction (PCR). PCR uses a thermostable DNA polymerase to amplify selectively the target sequence in a mixed DNA sample (Saiki et al. 1988). Two oligonucleotide primers specific for the 5′ flanking regions of a double-stranded DNA target sequence are added to the sample of DNA. The mixture is then heated to denature the DNA, i.e., to separate the strands. Upon cooling, the primers anneal, and the heat-stable DNA polymerase is added and the sample heated to 72°C, the optimal temperature for enzyme activity. New strands of target DNA are synthesized from the primers by the polymerase, and the mixture is again heated to denature the DNA. Repeated cycles of heating and cooling result in a geometric increase in the number of target sequences. Several millionfold increases in single-copy target sequences can be achieved (Saiki et al. 1988). Heat blocks which automatically cycle the temperature for PCR are readily available commercially. Using PCR it is possible to replicate specific DNA sequences to desired concentrations in short periods of time.

PCR techniques have been used successfully to amplify target sequences from environmental samples. Steffan and Atlas (1988) used PCR to amplify DNA from soils. The technique also is now being used to develop sensitive assays for bacteria in water. Somerville et al. (1988) developed a method for increasing sensitivity of hybridization assays based on the use of specific probes as primers for chain elongation. They showed that sensitivity could be enhanced by at least three orders of magnitude over conventional probing methods. Bej et al. (1990) used a portion of the *lacZ* gene of *E. coli* as a PCR target for the detection of coliform bacteria in water, and a portion of the *lamB* gene for selective detection of *E. coli*, *Salmonella*, and *Shigella*. They reported that incorporation of PCR into the assay resulted in a sensitivity of 1–5 *E. coli* CFU per 100 ml. Knight et al. (1990*b*) developed a function-specific PCR strategy which targets the heat-labile enterotoxin genes of *Vibrio cholerae* and *E. coli*. Several workers have reported inhibition of PCR by background flora and various dissolved and suspended substances in environmental samples (Alexander and Morris 1990; Knight and Colwell 1991; Paul, Cazares, and Thurmond 1990).

In summary, gene probes that are specific for the complementary nucleic acid base sequences in enteric microorganisms provide a reliable, powerful, and specific rapid detection and monitoring tool. Studies have demonstrated them to be particularly useful in identifying the source of disease outbreaks. And PCR methods are now available which allow detection of coliform bacteria and *E. coli* at sensitivities equivalent to, or greater than, standard culture techniques. Routine use of nucleic acid probes for water quality assessment currently is limited by the relative complexity of assay methods and the requirement in some cases for radioactive labels. The technology is advancing rapidly, however, and commercial development of prepackaged, sensitive, nonradioactively labeled gene probe kits for environmental monitoring can be expected in the near future.

9

Detection of Nonculturable Indicators and Pathogens

Marianne Walch

Center for Marine Biotechnology, University of Maryland System, 600 East Lombard Street, Baltimore, MD 21202

Rita R. Colwell

Department of Microbiology, University of Maryland, College Part, MD 20742

INTRODUCTION

For many years, conventional wisdom in microbiology has held that the sole reservoir of most human gastrointestinal pathogens was the human intestinal tract. Until the last decade, such pathogens generally were thought to be highly host-adapted and incapable of surviving longer than a few hours or days in the environment, outside of the intestine (Felsenfeld 1974). Recent work, however, has confirmed the autochthonous nature of pathogens such as *Vibrio cholerae* in estuarine environments (Colwell 1985) and has demonstrated that this and many other pathogenic bacteria may enter a dormant, viable, but nonculturable, state when exposed to stressful, low-nutrient environments (Roszak 1987). This chapter will review the current state of research on the viable, nonculturable state in pathogenic bacteria and indicators will summarize new methods which are useful in their detection and enumeration in marine and estuarine environments.

THE VIABLE BUT NONCULTURABLE STATE

The general inability to culture enteric bacteria soon after their introduction into an estuarine or marine environment is well known, and a considerable body of research has been conducted to understand the factors that control this "die-off" phenomenon. Colwell and coworkers (Colwell et al. 1985; Grimes and Colwell 1986; Palmer et al. 1984; Roszak, Grimes, and Colwell 1984; Xu et al. 1982), however, have shown clearly in the past decade that enteric bacteria introduced to seawater remain viable and metabolically active for days to weeks but rapidly lose the ability to be cultured. Microcosm experiments comparing acridine orange direct counts (AODC), direct viable counts (DVC) and plate counts of organisms exhibiting this phenomenon typically show that while AODC and DVC counts remain constant over time after introduction to the microcosm, plate counts decline rapidly to undetectable levels (Byrd et al. 1991). The nonculturable stage in the life cycle of these bacteria appears to be a strategy for survival when the cells are exposed to less than optimal conditions for growth and division.

Entry into a nonculturable state upon exposure to the stresses of low-nutrient aquatic environments has been demonstrated for a number of bacterial species (Table 9-1). These include representatives of both autochthonous and allochthonous microorganisms. Bacteria demonstrating the viable but nonculturable state to date include primarily gram-negative species, because the direct viable count—a method commonly used to enumerate viable cells in aquatic populations—depends on the sensitivity to naladixic acid (Goss et al. 1964). Recently, however, Byrd et al. (1991) demonstrated that at least some gram-positive bacteria enter a viable but nonculturable state after incubation in drinking water.

Much of the recent research in this area has focussed on bacteria of the genus *Vibrio*, many of which are indigenous in estuarine environments. Ecological studies have revealed a general inability to culture estuarine vibrios from cold water (Kaneko and Colwell 1973; Kelly 1982; Kaysner et al. 1987; Brayton et al. 1987; Kelley and Stroh 1988; Oliver 1989; Chowdhury et al. 1989, 1990; O'Neill et al. 1990). Studies have shown that in many cases the populations are not absent but merely in a dormant state (Stevenson 1978; Roszak and Colwell 1987). Natural bodies of water are known to be important both as reservoirs for and modes of transmission of *Vibrio cholerae* to humans. Isolation of *V. cholerae* from water in endemic areas, however, is often not successful, particularly during interepidemic periods (Colwell et al. 1981; Brayton et al. 1987; Huq et al. 1990). For example, epidemiologic surveillance in the intensively monitored Matlab area in Bangladesh failed to isolate *V. cholerae* O1 (CT^+) from water sources between cholera seasons or from water that was not subject to contamination by active cholera cases (McCormack et al. 1969). The finding by Colwell et al. (1985) that viable *V. cholerae* O1 present in the environment may not grow on laboratory media routinely used for its isolation was a pivotal point in the debate concerning *V. cholerae* ecology. Direct viable counts have

TABLE 9-1. Bacteria for Which a Viable But Nonculturable State Has Been Demonstrated

Organism	Reference
Aeromonas salmonicida	Allen-Austin et al. (1984)
Agrobacterium tumefaciens	Byrd et al. (1991)
Campylobacter jejuni	Rollins and Colwell (1986)
Enterobacter aerogenes	Byrd et al. (1991)
Escherichia coli	Xu et al. (1992) Colwell et al. (1985) Byrd and Colwell (1990)
Klebsiella pneumoniae	Byrd et al. (1991)
Micrococcus flavus	Byrd et al. (1991)
Salmonella enteritidis	Roszak, Grimes, and Colwell (1984)
Salmonella spp.	Knight et al. (1990)
Shigella spp.	Colwell et al. (1985)
Streptococcus faecalis	Byrd et al. (1991)
Vibrio cholerae	Singleton et al. (1982) Xu et al. (1982, 1984) Colwell et al. (1985) Brayton et al. (1987) Huq et al. (1990)
Vibrio vulnificus	Linder and Oliver (1989) Oliver and Wanucha (1989) Oliver, Nilsson, and Kjelleberg (1991)

clearly demonstrated the existence of nonculturable, viable cells (Brayton et al. 1987; Huq et al. 1990; Tamplin et al. 1990) in the environment which are capable of initiating disease (Colwell et al. 1990).

Despite the importance of the viable but nonculturable state in aquatic bacteria, little is known about the parameters which influence entry into that state. The time for entry into nonculturability varies for different strains of the same species (Xu et al. 1982; Byrd and Colwell 1990) and even within the same strain (Hoff 1989; Linder and Oliver 1989; Oliver and Wanucha 1989). Recent studies have shown that development of the nonculturable state in enteric bacteria depends on the conditions under which they were grown or maintained prior to their introduction into seawater (Gauthier et al. 1989). Thus, the history of their growth in the enteric environment and wastewaters could be important.

High salinity appears to be one factor which can initiate the viable but nonculturable state in *E. coli* (Xu et al. 1982; Hoff 1989). In oligotrophic seawater, enteric bacteria lose the ability to grow on culture media with low osmolarity (Xu et al. 1982; Gauthier, Munro, and Mohadjer 1987; Gauthier et al. 1991). Cells which are preadapted to high salinity tend to be more resistant to the detrimental effects of seawater, suggesting that osmoregulation may play a role in their survival in the marine environment (Munro et al. 1989). Loss of culturability can be temporarily blocked by using saline media (Gauthier, Munro, and Mohadjer 1987) or through the addition of glycine betaine to water samples before counting (Roth et al. 1989). The levels of potassium and glutamate, which are involved in osmoregulation, in the media in which *E. coli* cells are grown and transported to the sea can also affect subsequent survival (Gauthier et al. 1991).

Studies by Feacham et al. (1981) have shown an opposite effect of salinity for *Vibrio cholerae*. They found that the decline in cell number, as determined by culture methods, was slower in seawater than in fresh surface water. The conclusion from these studies that survival (i.e., culturability) is enhanced in saline waters is consistent with the observation by Singleton and Colwell (1982) that *V. cholerae* has an absolute requirement for sodium for growth.

In the case of *Vibrio vulnificus*, the viable but nonculturable state has been shown to be highly temperature-dependent. Cells of this species enter into nonculturability at 5°C but not at 10°C or higher (Oliver, Nilsson, and Kjelleberg 1991). Nonculturability occurs at 5°C even in the presence of high levels of nutrients (Oliver and Wanucha 1989). At 25°C, *V. vulnificus* exhibits a typical starvation response rather than a viable but nonculturable response (Oliver et al. 1991). A similar dependence of the viable but nonculturable state upon temperature has been suggested for *Vibrio cholerae* (Colwell and Spira, in press; Baker, Singleton, and Hood 1983; Xu et al. 1982).

Oliver et al. (1991) have shown that stationary-phase cells of *V. vulnificus* require a longer time to enter nonculturability than logarithmic-phase cells, suggesting that nutrient starvation, or another stationary-phase-induced stress response, may play a role in the nonculturable process. Of the many parameters examined in this study, the physiological age of cells inoculated into estuarine microcosms was the only factor shown to have a major effect on the rate at which *V. vulnificus* entered the nonculturable state. Various gram-negative bacteria have been shown to become viable but nonculturable when exposed to environmental or nutrient stresses (Grimes et al. 1986).

The response of bacteria to nutrient deprivation appears to be one of many examples of a common strategy for survival among bacteria in nutrient-poor environments (Roszak, Grimes, and Colwell 1984). Novitsky and Morita (1976, 1977, 1978) showed that cultures of the marine psycrophilic *Vibrio* species ANT-300 responded to starvation by increasing cell numbers and producing cells which were significantly decreased in volume and morphologically altered from rods to coccoid cells. The endogenous respiration of the small coccoid cells was less than 1% of that of the nonstarved cells, but they remained culturable.

These and other studies (Jones and Rhodes-Roberts 1981; Torrella and Morita 1981; Kjelleberg, Humphrey, and Marshall 1983; Amy and Morita 1983*a*, *b*) have suggested that certain gram-negative bacteria respond to conditions of nutrient deprivation by (1) continuing cell division and increasing the cell numbers while the total biomass remains constant (known as "reductive division"), (2) forming small coccoid cells, and (3) reducing the overall metabolic rate. These changes are associated with long-term survival and maintenance of culturability under conditions of extreme nutrient deprivation. The small cells produced by starvation under laboratory conditions resemble closely the "ultramicrobacteria" that occur naturally in marine and estuarine water and exhibit low metabolic rates (Torrella and Morita 1981; Oliver and Stringer 1984; Schaechter 1961; MacDonell and Hood 1984; Anderson and Heffernan 1965). Such microorganisms may be typical of the "true" autochthonous microflora of the marine environment.

The relationship between the starvation response and the viable but nonculturable response is not clear. Colwell and coworkers (Colwell et al. 1985; Huq et al. 1990; Roszak, Grimes, and Colwell 1984; Xu et al. 1982; Brayton et al. 1987) have demonstrated both in the laboratory and in the field that *Vibrio cholerae* undergoes conversion to a viable but nonculturable state, in which the cells are reduced in size and become coccoid. But, in contrast to starved cells, they do not grow at all on standard culture media.

Similarly, studies by Oliver, Nilsson, and Kjelleberg (1991) have shown that *Vibrio vulnificus* cells exposed to a temperature downshift to 5°C decreased in culturability and also underwent a morphological change from rods to small cocci. They did not appear, however, to undergo the reductive division seen in the starvation response that occurs in room temperature microcosms. Starved populations typically maintained cell densities of 10^4–10^5 CFU/ml even after 90 days at room temperature. In contrast, the entry of *V. vulnificus* into a viable but nonculturable state occurs at 5°C even at elevated nutrient levels (Oliver and Wanucha 1989). These data suggest that the viable but nonculturable process may be distinct from the starvation response reported for other marine vibrios (Kjelleberg et al. 1987; Morita 1985, 1988; Nystrom, Flärdh, and Kjelleberg 1990).

The viable but nonculturable state may be a second survival mechanism along with the starvation response. The two responses appear to be linked in *Vibrio vulnificus*, with starvation preventing the nonculturable response (Oliver, Nilsson, and Kjelleberg 1991). Such an inhibition may be due to the synthesis of starvation-induced stress proteins (Nystrom et al. 1989), which are known to be protective for a variety of other stresses (Matin 1990; Matin et al. 1989). Similarly, *V. cholerae* also undergoes a starvation response in which cells remain culturable for extended periods of time. But if the cells are subjected, in addition to nutrient depletion, to a reduction in temperature and/or an elevation in salinity, they rapidly become nonculturable, yet remain viable and potentially pathogenic (Colwell and Spira, in press; Colwell et al. 1985).

The interactions of individual stresses (nutrient deprivation and temperature downshift) are likely to be complex. As with the starvation response, nonculturable cells of *V. vulnificus* exhibit marked membrane fatty acid and ultrastructural modifications compared with their culturable counterparts (Linder and Oliver 1989). Such cells also demonstrate greatly reduced amino acid transport (Oliver and Wanucha 1989).

Oliver, Nilsson, and Kjelleberg (1991) demonstrated that the stationary phase allows only a partial repression of the viable but nonculturable response in *V. vulnificus*, whereas increased starvation for more than 24 h completely inhibits the response. This raises interesting questions concerning the role of the stationary phase or starvation-specific gene expression. The induction of genes specific for these stages in the bacterial life cycle is well documented (Lange and Hengge-Aronis 1991; Matin 1990; Matin et al. 1989; Nystrom, Flärdh, and Kjelleberg 1990; Tormo, Admiron, and Kolter et al. 1990). The coordinated response to the stationary phase or to starvation for one or more compounds has been reported to involve the induction of stimulons and regulons, some of which overlap with other known stress responses such as the heat shock regulon (Matin 1990; Nystrom et al. 1990; Nystrom, Flärdh, and Kjelleberg 1990). Starvation conditions induce these responses in *V. vulnificus*, and these may repress the viable but nonculturable program displayed in lower-temperature incubation (Oliver, Nilsson, and Kjelleberg 1991).

The existence of the viable but nonculturable state in bacteria in aquatic environments has implications for several areas of applied and environmental microbiology, including the release of genetically engineered microorganisms (Byrd and Colwell 1990), detection of pathogens and indicator bacteria in the environment, and general marine ecology. The phenomenon presents a new perspective and imparts a dynamic meaning to the term "survival" (Colwell and Spira, in press). More meaningful are the questions of adaptation and response to environmental conditions and stresses (Roszak and Colwell 1987). Cells may lose the ability to form colonies on culture media and yet remain functional as individual cells in the environment. Hoppe (1976) has suggested that nonculturable cells are primarily responsible for organic turnover in the open ocean.

The demonstration that nonculturable cells can be resuscitated and revert to a state of increased metabolism and growth when environmental conditions become more favorable (Nilsson, Oliver, and Kjelleberg 1991) suggests that the public health hazard of nonculturable pathogens in the environment may be significant. Colwell et al. (1990) reported that nonculturable cells of *Vibrio cholerae*, when ingested by human volunteers, caused clinical symptoms of cholera infection, and *V. cholerae* could be recovered from stools. *V. cholerae* can be present throughout the year in the aquatic environments of regions where cholera is endemic, yet culture methods routinely fail to isolate and enumerate the organism (Huq et al. 1990). Much of the historical information in the literature on the survival and ecology of pathogenic bacteria in aquatic environments has been based on culturing of cells and, therefore, must be interpreted with caution.

It is clear that a reassessment of the models currently used for predicting the bacterial quality of water subjected to fecal pollution is needed, in the light of the increasingly recognized importance of the viable but nonculturable state. The actual rates of disappearance of enteric bacteria released into the environment must be determined (Garcia-Lara et al. 1991) and the processes responsible for their death or survival identified. Methodologies for the enumeration and detection of indicators and pathogens in aquatic environments must rely less on the requirement for culture of the organisms. Fortunately, new approaches to the enumeration and detection of nonculturable cells are emerging and gaining acceptance by microbiologists.

METHODS FOR THE ENUMERATION AND DETECTION OF VIABLE BUT NONCULTURABLE BACTERIA

Direct Count Methods

The major drawback of most of the methods currently available for detecting and enumerating indicator microorganisms and pathogens in the environment is that they require culturing of the organisms to be monitored. Given the importance of viable but nonculturable cells in aquatic environments, many populations will be underestimated, or not detected at all, using these methods. Direct methods for enumeration, on the other hand, provide a greater efficiency of detection. Modifications of these techniques have become available which do allow an estimation of the proportion of the total cells which are growing, metabolizing, or respiring. These have been reviewed by Roszak and Colwell (1987*b*) and Byrd and Colwell (1991) and include the direct viable count (DVC) method to detect nutrient-responsive cells (Brayton et al. 1987; Brayton and Colwell 1987; Kogure, Simidu, and Taga 1979; Peele and Colwell 1981; Roszak and Colwell 1987*a*) microautoradiography to detect the uptake of isotopically labeled substrates (Fliermans and

Schmidt 1975; Hoppe 1976; Meyer-Reil 1978; Peroni and Lavarello 1975; Ramsay 1974; Roszak and Colwell 1987*a*; Tabor and Neihof 1982*b*), and reduction of 2-(*p*-iodophenyl)-3-(*p*-nitrophenyl)-5-phenyl tetrazolium chloride (INT) to detect respiring organisms (Dutton, Bitton, and Koopman 1983; King and Parker 1988; Maki and Remsen 1981; Tabor and Neihof 1982*a*; Zimmerman et al. 1978).

The coupling of direct methods with highly specific labeled antibodies allows one to target specific microorganisms very precisely without the need to culture them first (Wright 1991). Modifications of the procedures permit the assessment of both viability and metabolic activity. A simple and particularly useful method has been developed which combines fluorescent antibody (FA) staining with the direct viable count (DVC) method of Kogure et al. (Kogure, Simidu, and Taga 1979; Peele and Colwell 1981). The DVC technique allows a direct estimation of viable bacteria in water by preincubating samples with low concentrations of yeast extract and nalidixic acid. Nalidixic acid prevents cell division in gram-negative bacteria by inhibiting DNA replication and, subsequently, cross-wall formation (Goss, Deitz, and Cook 1965). This results in the formation of elongated, metabolically active cells in the presence of the nutrient, which can easily be detected by fluorescent staining.

The FA–DVC method has been used successfully to enumerate viable but nonculturable cells of *Campylobacter pylori* in water (Shahamat et al. 1989). The method has also been used extensively in the field to detect and enumerate viable *Vibrio cholerae* O1 (Brayton et al. 1987; Huq et al. 1990). Fluorescent antibody procedures have been found to be very useful in detecting *V. cholerae* in plankton, with which it is commonly associated and occurs in large numbers in the nonculturable stage (Tamplin et al. 1990).

Nucleic Acid Hybridization Methods

The tools of molecular genetics offer new and very powerful ways to detect and study organisms in their natural environment without the need for culturing organisms of interest. The extreme specificity of nucleic acid hybridization reactions has been exploited to produce a new set of powerful, rapid, and reliable techniques for the detection and identification of microorganisms. Gene probe techniques have been used widely in the clinical setting and, more recently, for environmental applications. The previous chapter describes gene probe strategies and methodologies in detail. This section will focus more specifically on the use of nucleic acid hybridization techniques for the detection and enumeration of viable but nonculturable indicators and enteric pathogens in aquatic environments.

Extraction of Nucleic Acids from Water and Environmental Samples

Direct isolation of bacterial DNA from environmental samples is becoming an important tool for ecological studies, permitting the detection of genes which encode important metabolic pathways (Walia, Khan, and Rosenthal 1990), tracking of genetically engineered organisms (Holben et al. 1988; Jansson, Holben, and Tiedje 1989; Atlas and Sayler 1988; Morgan et al. 1989), and *in situ* studies of microbial diversity (Torsvik, Goksogr, and Daae 1990*a*; Torsvik et al. 1990*b*). Application of nucleic acid extraction methods to environmental samples eliminates the need for cell cultivation. Methods of detection and enumeration based on this technology, therefore, are well suited for viable but nonculturable bacteria.

Methods for extracting and purifying DNA and RNA from laboratory cultures of microorganisms are highly developed (Sambrook, Fritsch, and Maniatis 1989), but techniques for obtaining pure, intact nucleic acids directly from environmental samples are still in their infancy. Several techniques for the direct extraction of DNA from aquatic environments have been described (DeFlaun, Paul, and Davis 1986; Fuhrman et al. 1988; Paul and Carlson 1984; Paul, Jeffrey, and DeFlaun 1985; Paul and Myers 1982; Somerville et al. 1989). Methods have also been developed for soils and sediments (Tsai, Park, and Olson 1991; Holben et al. 1988; Ogram, Sayler, and Barkay 1987; Torsvik 1980). These are described in the previous chapter.

The success of gene probe methods for the detection and identification of bacteria in a given environment, thus, depends largely on the ability to recover representative samples of nucleic acid from that environment. Alternative approaches may be needed for the optimal extraction of nucleic acids from starved or nonculturable bacteria (Brauns, Hudson, and Oliver 1991). Bacteria in aquatic systems often are very small (< 0.6 μm) and very dilute (ca. 10^6 per ml), making nonselective collection of a sufficient number of cells and quantitative DNA extraction difficult (Fuhrman et al. 1988). Nonculturable cells may exhibit an increased resistance to lysis (Brauns, Hudson, and Oliver 1991). Nystrom and Kjelleberg (1989) demonstrated significant alterations in the cell wall peptidoglycan of a starved marine vibrio, which resulted in thickened cell walls and an increased resistance to sonication. Loss or rearrangement of DNA within targeted gene sequences may also occur as cells enter the nonculturable state, resulting in a decreased probing efficiency (Brauns, Hudson, and Oliver 1991; Sadouk et al. 1990).

Special difficulties are also encountered in the extraction of bacterial DNA from shellfish meats (Hill et al. 1991). Commercially available column systems for rapid purification of bacterial DNA are rapidly clogged by viscous shellfish homogenates. Hill et al. (1991) successfully developed a method for the extraction of DNA from oysters which is based on lysis of bacteria using guanidine isothiocyanate, followed by chloroform extraction.

Gene Probes for Bacteria of Environmental Significance

Nucleic acid probes (or gene probes) are molecules of single-stranded DNA or RNA which have been labeled, either radioactively or chemically, so that they can be detected in a mixture of renatured nucleic acids. The probes are incubated under appropriate conditions with denatured DNA or RNA (e.g., the total cellular DNA extracted from a water sample) and hybridized specifically with regions of the target nucleic acid which are complementary. Excess, unhybridized probe is selectively removed, and probe–target hybrids are detected using methods appropriate to the particular labeling strategy.

Two main approaches have been used for the application of gene probes to environmental bacteria. The first, colony hybridization (Ford, Simpson, and Olson 1987; Sayler et al. 1985, 1987), is inappropriate for environments containing many nonculturable bacteria, because it requires cultivation of the organisms prior to probing. Hybridization of direct DNA extracts (Knight et al. 1990*a*), on the other hand, obviates the need for culturing. It allows the detection and enumeration of stressed, nonculturable pathogens and indicator organisms in aquatic environments. The technique is rapid, because analyses are performed directly on the environmental samples, with no laboratory incubation time required.

Strategies and techniques of gene probe construction are reviewed in the preceding chapter of this volume. It is important in selecting an appropriate probe that a unique base sequence be targeted that is specific to the organism or function of interest. Probes that have been developed for enteric pathogens and indicators typically are directed toward genes encoding a toxin or virulence factor (e.g., Moseley et al. 1980; Nataro et al. 1985; Kaper and Levine 1981) or to functional genes (Bej et al. 1990) or chromosomal or ribosomal sequences (e.g., Fitts et al. 1983; Attwood et al. 1988) that are unique to the organism(s) of interest. Table 9-2 summarizes nucleic acid probes for enteric pathogens and indicator bacteria that have been reported in the literature to date. They include probes directed at both DNA and RNA targets in strains of pathogenic *E. coli*, *Salmonella*, *Shigella*, *Vibrio*, *Yersinia*, *Listeria*, *Bacteroides*, and *Campylobacter*. Many of these probes have played a valuable role in epidemiological studies of diarrheal pathogens, particularly of *Vibrio* species (Kaper et al. 1982; Lin et al. 1986; Nishibuchi et al. 1986), and have contributed significantly to our understanding of their distributions.

For the most part, however, these probes have been developed for, and tested in, clinical specimens such as stool samples. Most, in fact, have been

TABLE 9-2. Summary of Gene Probes Available for Indicators and Enteric Pathogens

Organism	Target Sequence	Reference
Escherichia coli	*lacZ* and *lamB* genes	Bej et al. (1990)
Enterotoxigenic	Heat-labile and heat-stable enterotoxin genes	Moseley et al. (1980, 1982)
		Echeverria et al. (1982, 1984*a*, *b*, 1985, 1986)
		Lanata et al. (1985)
		Thompson et al. (1986)
		Hill et al. (1986)
		Romick, Lindsay, and Busta (1987)
		Seriwatana et al. (1987)
		Knight et al. (1990*b*)
Enteroinvasive	*Shigella* invasiveness plasmid genes	Boileau, d'Hauteville, and Sansonetti (1984)
		Sethabutr et al. (1985)
		Sethabutr, Echeverria, and Taylor (1985)
		Sethabutr (1985 or 1986)
		Wood et al. (1986)
	E. coli invasiveness plasmid genes	Small and Falkow (1986)
		Wood et al. (1986)
Enterohemorrhagic	Plasmid DNA encoding fimbrial antigen	Levine et al. (1987)
Enteropathogenic	Adherence factor determinants	Nataro et al. (1985)
Uropathogenic	Gal-Gal-pilus-associated chromosomal genes	
Salmonella		
S. typhi	Chromosomal Vi antigen gene	Rubin et al. (1985)
S. typhimurium	Plasmid pIMVS1 genes	Whiley et al. (1988)
Salmonella spp.	Chromosomal sequences	Fitts et al. (1983)
		Fitts (1985)
		Tompkins et al. (1986)
		Gopo et al. (1988)
		Knight et al. (1990*a*)
	Ribosomal DNA	Wilson et al. (1989)
	E. coli lamB gene	Bej et al. (1990)
Vibrio		
V. parahaemolyticus	Thermostable direct hemolysin gene	Nishibuchi et al. (1985, 1986?)
V. cholerae	*E. coli* heat-labile enterotoxin gene	
	CT toxin gene	Kaper and Levine (1981)
		Kaper et al. (1982)
		Lin et al. (1986)
		Knight et al. (1990*b*)

TABLE 9-2. (*continued*)

Organism	Target Sequence	Reference
V. vulnificus	Hemolysin gene	Morris et al. (1987)
Shigella spp.	Plasmid with entero-invasiveness genes	Boileau, d'Hauteville, and Sansonetti (1984) Sethabutr, Echeverria, and Taylor (1985) Small and Falkow (1986) Wood et al. (1986)
	E. coli lacZ and *lamB*	Bej et al. (1990)
Yersinia		
Y. enterocolitica	Virulence plasmid	Hill, Payne, and Aulisio (1983) Jagow and Hill (1986) Miliotis et al. (1989) Robbins-Brown et al. (1989) Kapperud et al. (1990)
	Ribosomal DNA	Chan et al. (1988)
Yersinia spp.	Vwa plasmid DNA	Gemski et al. (1987)
	inv and *ail* chromosomal loci	Robbins-Brown et al. (1989)
Enterobacteriaceae	OmpA gene	
Coliforms	*lacZ* gene of *E. coli*	Bej et al. (1990)
Listeria monocytogenes	*β*-hemolysin gene	Datta et al. (1987) Colburn et al. (1990)
Bacteroides spp.	Chromosomal sequences	Attwood et al. (1988) Kuritza et al. (1986*a*, *b*) Salyers, Lynn, and Gardner (1983)
Campylobacter spp.		
C. jejuni	?	Korolik, Coloe, and Krishnapillai (1988)
	Chromosomal DNA	Picken, Wang, and Yang (1987)
	Ribosomal DNA	Rashtichian et al. (1987)
C. cinaedi	Whole-cell DNA	Totten et al. (1985)
C. fennelliae		
Campylobacter spp.	—	Tompkins and Krajdin (1986*a*) Claus, Moseley, and Falkow (1983)

TABLE 9-3. Summary of Reported PCR Methods for Indicators and Enteric Pathogens

Organism	Sample Type	Reference
Escherichia coli	Environment	Bej et al. (1990*a*, *b*, 1991) Knight et al. (1990*b*)
Escherichia coli	Clinical	Karch and Meyer (1989) Pollard, Candrian, and Luethy (1990) Furrer et al. (1990) Olive (1989) Olive, Atta, and Sethi (1988) Candrian et al. (1991) Frankel et al. (1989)
Shigella flexneri	Lettuce	Lampel et al. (1990)
Shigella spp.	Environment	Bej et al. (1991)
Vibrio cholerae	Environment	Knight et al. (1990*b*)
Vibrio vulnificus	Environment	Brauns, Hudson, and Oliver (1991)

used in colony hybridization assays which rely on culturing of the test organisms (Tenover 1988; Knight et al. 1991). As techniques for the direct detection and enumeration of both culturable and nonculturable bacteria in environmental (and, particularly, water) samples are improving and becoming more routine, investigators are now beginning to apply this technology to studies of indicator bacteria and enteric pathogens in both freshwater and marine environments. Recently, Knight et al. (1990*a*) described the use of a DNA probe to detect and enumerate *Salmonella* spp. in nucleic acids extracted from polluted estuarine water. The method was shown to detect culturable and nonculturable *Salmonella* with equal efficiency.

Detection by Polymerase Chain Reaction

A relatively new technique which vastly increases the sensitivity of gene probe assays, thereby allowing the detection of organisms that are present in a sample only in very small numbers, is the polymerase chain reaction (PCR). PCR uses a thermostable DNA polymerase to amplify selectively the target sequence in a mixed DNA sample (Saiki et al. 1988). Several millionfold increases in single copy target sequences can be achieved in short periods of time.

PCR techniques have been used successfully to amplify target sequences from environmental samples. Steffan and Atlas (1988) used PCR to amplify DNA from soils. The technique is now being used also to develop sensitive assays for bacteria in water. Somerville et al. (1988) developed a method for increasing the sensitivity of hybridization assays, based on the use of specific probes as primers for chain elongation. They showed that sensitivity could be enhanced by at least three orders of magnitude over conventional probing methods.

Bej et al. (1990*a*, *b*) have used a portion of the *lacZ* gene of *E. coli* as a PCR target for the detection of coliform bacteria in water, and a portion of the *lamB* gene for the selective detection of *E. coli*, *Salmonella*, and *Shigella*. In a more recent report (Bej et al. 1991*a*), they describe a PCR method for amplifying regions of the *uid* gene of *E. coli* that codes for β-glucuronidase, expression of which forms the basis for detection of fecal coliforms by the commercially available Colilert kit (see the previous chapter). Amplified sequences from as little as 10 fg of genomic *E. coli* DNA was consistently detected when primers and probes for the *uid* gene were used, which is equivalent to the detection of 1–2 bacterial cells (Bej et al. 1991*a*). Several PCR assays for the detection and identification of enterotoxigenic *E. coli*, based on the amplification of enterotoxin gene sequences, have been described (Furrer, Candrian and Luethy 1990; Olive 1989; Candrian et al. 1991). Knight et al. (1990*b*) developed a function-specific PCR strategy which targets the heat-labile enterotoxin genes of *Vibrio cholerae* and *E. coli*.

PCR techniques are uniquely suited for the detection of nonculturable pathogens in the environment. PCR allows a specific DNA sequence to be amplified more than a millionfold within hours (Saiki et al. 1988), thereby potentially allowing the detection of cells present in very low numbers. The success of the method depends only on the presence of target DNA, not culturable cells. The method also eliminates problems of separating bacterial DNA from a large background of eukaryotic DNA in shellfish meats (Hill et al. 1991).

Bej, Mahbubani, and Atlas (1991*b*) have used PCR to detect dying cells of *Legionella pneumophila* which could not be cultured following injury by hypochlorite. Brauns, Hudson, and Oliver (1991) recently reported the use of PCR amplification of a segment of the hemolysin gene to detect *Vibrio vulnificus* and compared its efficacy for both culturable and nonculturable cells. As little as 72 pg of DNA from culturable *V. vulnificus* could be detected. Sensitivity for nonculturable cells was reduced, with 31 ng of DNA being the lower detection limit. The reasons for the decreased amplification of nonculturable cell DNA are not understood. Nonculturable cells may contain less DNA than their culturable counterparts, or extraction efficiency may be reduced due to increased resistance to cell lysis (Brauns, Hudson, and Oliver 1991). Unidentified factors present in lysates from nonculturable cells may also interfere with amplification. Several workers have reported the inhibition of PCR by background flora and various dissolved and suspended substances in environmental samples (Alexander and Morris 1990; Knight et al. 1991; Paul, Cazares, and Thurmond 1990). Clearly, further study and optimization of the technique are needed.

FUTURE OUTLOOK

Gene probes for enteric microorganisms provide a reliable, powerful, and specific rapid detection and monitoring tool that does not depend on cultivation of the organisms of interest. PCR methods are now available, which allow the detection of coliform bacteria and enteric pathogens at sensitivities equivalent to, or greater than, standard culture techniques. These techniques are well suited to the detection of viable but nonculturable pathogens and indicator species in the environment. They are unlikely to replace all conventional biochemical and physiological tests, but they may be used in conjunction with these tests to provide a more complete description of aquatic microbial communities. The technology of nucleic acid hybridization is advancing rapidly, and commercial development of prepackaged, sensitive, nonradioactively labeled gene probe kits for environmental monitoring of many of these organisms can be expected in the near future. Ultimately, a combination of detection methods, such as gene probes coupled with fluorescent antibodies, to confirm the presence of viable but nonculturable bacteria will be the most accurate and useful.

The future may also see the development of entirely new approaches to the detection and monitoring of intestinal pathogens and indicator organisms. Direct detection of mammalian DNA in water samples, combined with probes specific for bacterial pathogens, is a potential direction for future development. Gene probe detection of human intestinal epithelial cells coupled with a *Salmonella* probe, for example, would give definitive evidence of human fecal contamination.

Protocols for the extraction of messenger RNA (mRNA) from soil and water are only recently appearing in the literature (Tsai, Park, and Olson 1991; Pichard and Paul 1991). Detection of specific mRNA directly isolated from seawater samples permits a determination of whether genes are actually functioning in the environment by targeting their expression products. Detection of mRNA in the environment has been difficult due to its instability and the presence of humic materials and other interfering compounds (Tsai, Park, and Olson 1991). PCR amplification of specific mRNA molecules (Kawasaki and Wang 1989), however, provides great potential for ecological studies of *in situ* gene expression, particularly as an indication of the function of viable but nonculturable bacteria.

Clearly, the existence of viable but nonculturable bacteria in marine and estuarine environments has many implications for the detection and enumeration of bacteria associated with fecal pollution. Monitoring techniques for indicator species as well as pathogens must be sensitive to cells in the nonculturable state, lest an incomplete picture of the potential public health risks of pollution inputs be obtained. Rapidly developing molecular technologies now provide powerful tools for detecting viable microorganisms in the environment, regardless of their culturability. The near future undoubtedly will find many of these techniques being incorporated into routine water quality analyses.

ACKNOWLEDGMENTS

The authors express their thanks to Dr. Russell Hill and Ishrat Husein for assistance in the preparation of this manuscript.

10

Factors Affecting Indicator and Pathogen Populations During Handling of Shellstock

Angela D. Ruple and David W. Cook*

Gulf Coast Research Laboratory, 703 East Beach Drive, Ocean Springs, MS 39564-7000

INTRODUCTION

The microorganisms found in shellfish at harvest represent both those that are filtered from the water during feeding and the natural commensals of the shellfish. With the possible exception of some marine vibrios, the bacteria that compose the commensal microflora are not pathogenic to man nor do they represent any of the bacterial species used as indicators of fecal pollution. The National Shellfish Sanitation Program (NSSP) (Food and Drug Administration 1989*a*) allows a measurable number of indicators to be present in shellfish-harvesting waters; thus, shellfish may contain indicator bacteria. Pathogenic bacteria and viruses have also been isolated from shellfish harvested in approved areas (Ellender et al. 1980; Hackney, Ray, and Speck 1980; Gerba and Goyal 1978; Goyal, Gerba, and Helnick 1979; Hood et al. 1981, 1983; Peixotto et al. 1979; Rutala et al. 1982; Sobsey et al. 1980; Tamplin et al. 1982; Thompson, Vanderzant, and Ray 1976).

Pathogenic microorganisms of concern with shellfish represent those (1) from human and animal body wastes and (2) naturally occurring pathogens from the environment. Indicator bacteria are used to predict the possible

* Present address: U.S. Food and Drug Administration, Gulf Coast Seafood Laboratory, Dauphin Island, AL 36528-0158

presence of pathogens in the first group. For the indicators to be valid, they must arise from the same source as the pathogens.

In shellfish from approved areas, the numbers of enteric bacteria are expected to be low. Enteric bacterial pathogens usually have large infective doses. Therefore, precautions must be taken to prevent these bacteria, if present, from multiplying to an infectious level. Similarly, autochthonous estuarine pathogens that may be present in the shellfish must be prevented from multiplying to reduce the danger that they pose.

When shellfish are disturbed during harvest, they close their shells to prevent the loss of shell liquids. As long as the shells are tightly closed, microorganisms cannot gain entry. However, if the shellfish are immersed in water, inadvertently or intentionally during wet storage, or if the shellfish are maintained in a warm moist environment, they may open their shells allowing the entry of indicators or pathogens that may be present.

This chapter contains information on shellfish-handling practices as they are employed in the industry and how these practices effect the microflora of the shellstock, with special reference to the microbial pathogens and indicators.

SHELLSTOCK-HANDLING PRACTICES

Prior to the advent of modern refrigeration and rapid transportation, shellfish were harvested only during the cool months of the year because they spoil rapidly in warm temperatures. Over the past 25 years, shellfish-harvesting seasons have been extended to year round in some states and this has presented both unique economic opportunities and public health problems.

Shellfish-harvesting areas may be remote from landing docks, where shellstock are unloaded from vessels. Few harvesting vessels have refrigeration capability and, thus, the shellstock remain on the deck of the vessel at ambient temperatures for several hours to several days before unloading into refrigerated areas. Most shellfish-harvesting states now require that shellfish harvested during warm months of the year be landed daily to reduce the exposure of the shellstock to warm temperatures.

At harvest, shellfish are given a brief wash with water from the harvest area to remove mud. Shellstock are packed in burlap sacks or open mesh containers and transported dry to prevent their shells from opening. The NSSP guidelines recommend that shellstock be transported in adequately refrigerated trucks when ambient air temperature and travel time are such that bacterial growth can occur (Food and Drug Administration 1989*b*).

MICROFLORA IN SHELLSTOCK AT HARVEST AND CHANGES DURING STORAGE

Indicator Bacteria—Coliforms, Fecal Coliforms, and *E. coli*

The group of bacteria referred to as coliforms are aerobic and facultative anaerobic gram-negative, rod-shaped bacteria that ferment lactose with gas formation within 48 h at 35°C (95°F).

The significance of the coliform group as an indicator organism has been questioned because some of the genera within the group are frequently associated with surface runoff rather than fecal pollution (Greenberg and Hunt 1985).

Fecal coliforms are more directly associated with fecal contamination from warm-blooded animals. The National Shellfish Sanitation Program (NSSP) has established guidelines using fecal coliforms as an indicator of fecal pollution in market-level shellfish (Food and Drug Administration 1989*b*).

The inability to distinguish between fecal coliform organisms found in animal or industrial effluents free from fecal contamination and those from fecal pollution sources has brought about a discussion of using *E. coli* as a better fecal indicator. *E. coli* inhabits the lower intestinal tract of vertebrates and is more abundant in feces than other coliforms or pathogenic bacteria. Therefore, its presence is a more direct indication of fecal pollution.

The usefulness of any group as an indicator is questionable if multiplication occurs within the shellfish after harvest. Reports of fecal coliforms multiplying in post-harvest shellstock oysters has posed some question as to the validity of the NSSP fecal coliform guideline for market-level shellfish as a true indicator of the health risk associated with consuming those shellfish. At the time of harvest, fecal coliform levels in oysters in approved areas are generally < 50/100 g and rarely exceed the NSSP guideline of 230/100 g. However, when these oysters are temperature-abused during shipment, the fecal coliform levels of the oysters arriving at the processing plants often exceed an MPN of 230/100 g, indicating multiplication of fecal coliforms after harvest and during transport (Cook and Ruple 1989; Food and Drug Administration 1983; Paille et al. 1987).

Controlled-temperature storage studies showed no multiplication of fecal coliforms or *E. coli* in post-harvest oysters held below 10°C (50°F) (Cook and Ruple 1989). At higher storage temperatures of 22°C (71.6°F) and 30°C (86°F), multiplication of fecal coliforms including *E. coli* was observed in some, but not all lots of oysters. Since fecal coliforms may multiply in temperature-abused post-harvest oysters, their levels no longer relate pollution levels in the harvest area. Therefore, the Food and Drug Administration has cautioned against the use of fecal coliform levels in post-harvest oysters to indicate the water quality of the harvest areas (Food and Drug Administration 1983).

Coliform levels were also shown to increase in Manila clams and Pacific oysters held at temperatures of 20°C (68°F) and 27.5°C (81.5°F) (Hoff et al. 1967). Fecal coliform MPNs remained stable at 20°C (68°F), while at 27.5°C (81.5°F) they increased in Pacific oysters. At 10°C (50°F), coliform and fecal coliform MPNs in Pacific oysters, Olympia oysters, Native Littleneck clams, and Manila clams remained relatively stable throughout a 15 day storage period.

In soft-shell clams from Chesapeake Bay, Chai et al. (1990) found that fecal coliform and *E. coli* levels were relatively constant throughout the summer harvest season with slight increases in fecal coliform levels during July and August. Clams sampled at dock side had slightly higher levels of total coliforms and fecal coliforms than those at harvest indicating possible multiplication. However, no significant difference in levels of *E. coli* between harvest and landing was observed.

Results from trace lot studies of soft-shell clams (*Mya arenaria*) harvested in Maryland and shipped to New York city indicated that the clams generally met bacteriological standards for market shellstock upon arrival in New York city if they were (a) harvested from growing waters meeting NSSP standards; (b) transported, handled, and stored under proper refrigeration; and (c) moved expeditiously through the transport system (Anonymous 1979). No relationship was found between fecal coliform levels and salinity or temperatures in the harvest area.

The post-harvest multiplication of *E. coli* may depend on factors other than temperature abuse. Differences in *E. coli* multiplication have been attributed to the fecal pollution events and/or the physiological state of the oysters (Food and Drug Administration 1983). Cook and Ruple (1989) found that increases in *E. coli* levels did not always accompany increases in fecal coliform levels during commercial handling. *E. coli* predominated only when fecal coliform counts were low, suggesting that other fecal coliforms are more competitive and reproduce faster than *E. coli* in the oysters. However, *E. coli* counts on two occasions increased from < 36 to $> 230/100$ g during 12 h on the boat with additional increases during land transport. Even though *E. coli* does not always multiply in post-harvest oysters under conditions that cause multiplication of other fecal coliforms, *E. coli* multiplication has been demonstrated in post-harvest oysters and care should be taken in interpreting the meaning of high-*E. coli* levels in post-harvest oysters.

E. coli is the principal fecal coliform present in post-harvest oysters only when water temperatures are below 22°C (38.2°F) (Cook and Ruple 1989; Paille et al. 1987). During the warm summer months, *Klebsiella* sp. are the predominate fecal coliforms. Boutin, Spaulding, and Twedt (1986) and Paille et al. (1987) have shown that the strains of *Klebsiella* isolated from oysters are environmental strains and are not a public health risk.

Enteric Pathogens and Viruses

Human enteric viruses may be isolated from oysters from approved harvesting areas (Ellender et al. 1980*a*, *b*; Fugate et al. 1975; Gerba and Goyal 1978; Goyal et al. 1979; Sobsey et al. 1980). Viruses do not multiply outside their host organism and may be concentrated from the water by shellfish during filter feeding. Once they are taken up by the oyster, polioviruses remain viable for the entire storage life of the oyster (DiGirolamo, Liston, and Matches 1970). In shellstock oysters artificially contaminated with 10^4 PFU of poliovirus, 13% of the virus remained viable after 30 days of storage at 5°C (41°F) even after the oysters were badly decomposed (Tierney et al. 1982).

Enteric bacteria including *Salmonella* (Son and Fleet 1980), *Yersinia* (Peixotto et al. 1979), and *Plesiomonas* (Koburger and Miller 1984) have been isolated from shellfish harvested from approved areas. Available evidence indicates that these bacteria generally do not multiply in the shellfish as long as the shellfish are not temperature-abused after harvest.

Salmonella typhimurium and *Salmonella senftenberg* when inoculated into shellstock oysters (*Crassostrea commercialis*) decreases by about 100-fold after 12 days of storage at 20–25°C (68–77°F) (Son and Fleet 1980). Naturally occurring *Salmonellae* may survive, but do not multiply for up to 14 days in shellstock oysters (*Crassostrea virginica*) stored at 8°C (46.4°F) (Hood et al. 1983).

Other enteric pathogens such as *Yersinia enterocolitica* and *Plesiomonas shigelloides* have been isolated from shellfish (Peixotto et al. 1979; Koburger and Miller 1984). However, information on the fate of these pathogens in post-harvest shellstock is unavailable. Studies with pure cultures (Miller and Koburger 1986) and inoculated shucked meats (Peixotto et al. 1979) have indicated that growth at refrigeration temperatures is very slow.

Vibrionaceae

Autochthonous bacteria from the family *Vibrionaceae* are frequently isolated from shellfish and have been shown to be pathogenic to man. Members of this family of public health interest in association with shellfish include *Vibrio cholerae*, *V. parahaemolyticus*, *V. vulnificus*, and *Aeromonas hydrophila*. The occurrence of *V. parahaemolyticus* in shellfish shows a seasonal variation, with higher numbers being present during the summer months (Hackney 1980). During certain times of the year, the organism has been isolated from all shellfish sampled (Cook and Ruple 1989; Earle 1975; Kaysner 1990).

Multiplication of *V. parahaemolyticus* in shellfish occurs at temperatures above 10°C (50°F) (Cook and Ruple 1989; Hood et al. 1983; Johnson, Salinger, and King 1973). When stored at temperatures < 10°C (50°F), *V. parahaemolyticus* may survive in shellstock, but does not multiply (Cook and Ruple 1989; Johnson et al. 1973; Thompson, Venderzant, and Ray 1976).

Like *V. parahaemolyticus*, the presence of *V. vulnificus* is seasonal (Kelly 1982; Cook and Ruple 1989). Increases in levels of *V. vulnificus* in shellfish of 1–2 orders of magnitude were observed after 1 day of storage at 22°C (71.6°F) and 30°C (86°F). At storage temperatures of 4°C (39.2°F), 0°C (32°F) and − 1.9°C (28.6°F), *V. vulnificus* levels decreased, but were still present after 14 days of storage (Cook and Ruple 1990).

Vibrio cholerae is isolated from shellfish less frequently than the other vibrios (Cook and Ruple 1989; Hood et al. 1983; Kaysner 1990). Because of the sporadic isolation of *V. cholerae* it is difficult to determine if the organism can multiply in shellfish. In artificially contaminated oyster homogenates, *V. cholerae* persisted at the level of inoculation for more than 14 days at 7°C (44.6°F) (Reily and Hackney 1985). This suggests that the organism survives, but does not multiply at refrigeration temperatures.

Aeromonas hydrophila is frequently isolated from both Gulf Coast and West Coast oysters (Cook and Ruple 1989; Abeyta 1990; Hood et al. 1983). Levels of *A. hydrophila* in shellstock oysters have been shown to increase with storage at temperatures ranging from 2°C (35.6°F) to 35°C (95°F) (Cook and Ruple 1989; Hood et al. 1983).

With the exception of *A. hydrophila*, multiplication of all of the *Vibrionaceae* mentioned above can be prevented during commercial transport by following the Food and Drug Administration recommendation of storing shellstock at 7.2°C (45°F).

Standard Plate Count

The standard plate count (SPC) measures the numbers of bacteria that grow on plate count agar medium when incubated at 35°C (95°F) for 48 h (Greenberg and Hunt 1985). The SPC is used by regulatory agencies as an indicator of the quality of shellfish but is not an indicator of public health consideration.

The SPC does not provide a total count of all the heterotrophic bacteria present in the shellfish because it eliminates halophilic bacteria, psychrotrophic bacteria and bacteria that may require special nutrients to grow. However, it has been established that as shellfish are stored, the SPC of the meats increase with time and that the quality of the meats is poor after the SPC has exceeded 500,000 CFU/g.

The SPC of Gulf Coast oysters at the time of harvest generally ranges from 10^2 to 10^5 CFU/g with a mean of 10^3 CFU/g. Rarely do oysters harvested from approved areas exceed a SPC of 500,000/g at harvest (Cook and Ruple 1989; Presnell and Kelly 1961). During the warm summer months the SPC of oysters may increase, while the oysters are held on the deck of the harvest boats and during land transport. However, if the oysters are shipped under proper refrigeration (< 10°C) (50°F), significant increases in SPC do not occur during transport (Food and Drug Administration 1983).

The SPC of soft-shell clams also generally meet the recommended guideline of < 500,000/g at the time of harvest with no significant differences in levels between harvest and landing (Chai 1990).

Trace lot studies of soft-shell clams (*Mya arenaria*) in Maryland (Anonymous 1979) revealed an inverse relationship between SPC and salinity in the growing areas, but no clear relationship between elevated growing water temperatures and elevated SPC. No significant difference in SPC was noted after harvest or during transport if proper refrigeration was employed.

Hoff et al. (1967) studied time–temperature effects on four species of live shellfish: Pacific Oysters (*Crassostrea gigas*), Olympia Oysters (*Ostrea lurida*), Native Littleneck Clams (*Prototheca staminea*), and Manila Clams (*Venerupis japonica*). At harvest all SPCs were between 10^3 and 10^4 CFU/g and after 15 days of storage at 10°C (50°F) all species had SPC $< 10^5$ CFU/g. At temperatures above 10°C (50°F) [20°C (68°F) and 27.5°C (81.5°F)] higher SPCs were observed.

Wet Storage of Shellstock

Live shellfish harvested from approved waters may be stored for short periods of time in seawater prior to sale. This practice is termed wet storage. Shellfish may be stored in baskets on the sea bottom or suspended off-bottom on floats in waters meeting approved area classifications. Small quantities of shellfish are sometimes stored in tanks of aerated sea water. Tanks may be equipped with a flow through seawater system if approved area water is available or fitted with filter and disinfection systems to maintain water quality. Wet storage is not intended to increase the safety of the shellfish.

Reasons for practicing wet storage include: (1) desanding of bivalve species that tend to accumulate sand to increase their palatability, (2) ease of retrieval, (3) increasing salt content of shellfish harvested from low-salinity areas, and (4) for display purposes in retail markets (Food and Drug Administration 1989*b*; Colburn et al. 1989).

Since shellfish are alive when placed in wet storage, they become active and filter water through their system. If the water in the storage area or tank contains indicator bacteria or pathogenic microorganisms, they will be filtered from the water, thus contaminating the shellfish (Vasconcelos et al. 1969).

The wet storage of shellfish in tanks for display purposes in retail markets presents potential microbiological problems. In such systems, lobster and crabs, which may be harvested from contaminated waters and may have ingested pathogens or accumulated them on their body surfaces, are frequently commingled with molluscan shellfish. Colburn et al. (1989) examined wet storage tanks in five retail markets and found that both indicator and pathogenic bacteria could be isolated from the shellfish and water. Pathogenic species recovered included *Salmonella typhimurium*, *Aeromonas hydrophila*, *Vibrio cholerae* non O1, and *Yersinia enterocolitica*. While this study did not attempt to demonstrate that shellfish became contaminated after being placed in the tanks, it clearly showed that the potential exists for contamination.

If waters in wet storage tanks are allowed to increase above 10°C (50°F), there is the potential for fecal coliforms and pathogenic marine vibrios and aeromonads to increase in number. The inclusion of effective UV-disinfection systems in holding tanks for the wet storage of warm-water shellfish may be necessary to control the growth of bacteria.

SUMMARY

Oysters, clams and mussels harvested from waters meeting the approved shellfish-harvesting area classification generally meet the National Shellfish Sanitation Program (NSSP) microbiological guidelines for market-level shellfish. This guideline addresses fecal coliform and standard plate count levels in both shellstock and shucked product.

Bacteria present in shellfish at harvest are capable of multiplying in the shellfish after harvest. The storage temperature is important in controlling this multiplication. Storage temperatures below 10°C (50°F) are documented to prevent or slow the growth of most indicator and pathogenic bacteria that might be present in the shellfish in low numbers at the time of harvest. Since it is not possible to store shellfish on harvest boats at temperatures which control bacteria multiplication, the length of time the shellfish remain on the boats should be kept as short as possible. Thus, harvest and transport of shellstock in a timely manner, and under conditions recommended by the NSSP will minimize changes in the microbial flora of the shellstock.

11

Factors Affecting Indicator and Pathogen Populations During Processing and Handling of Shucked Shellfish

David W. Cook* and Angela D. Ruple

Gulf Coast Research Laboratory, 703 East Beach Drive, Oceans Springs, MS 39564-7000

OVERVIEW OF RAW PROCESSING

Raw processing involves these steps: (1) separating the halves of the shell (valves) to expose the meat, (2) cutting the adductor muscle, which holds the valves together, to free the meat from the shell, (3) washing the meat to remove mud, sand, and shell debris, and (4) placing the washed and drained meats in containers prior to chilling.

Oysters are eco-morphic resulting in shells that are of various shapes which do not lend themselves to being opened by automated mechanical means. Some automated equipment is available for opening clams and muscles which are more uniform in shape. However, the shucking of raw shellfish is principally a hand operation which uses a knife to separate the valves and cut the adductor muscle.

As shellfish meats are separated from the shell, they are collected in buckets, which, when filled, are transported to the packing room. The time the shellfish remain in the shucking buckets at room temperature depends on

* Present address: U.S. Food and Drug Administration, Gulf Coast Seafood Laboratory, Dauphin Island, AL 36528-0158

the speed of the shucker in filling the bucket. During the summer, ice is sometimes added to the buckets to cool the shucked meats.

In the packing room, shucked meats are washed and packed into containers for sale. Washing may be accomplished by placing the meats on a stand-supported perforated tray called a skimmer and spraying them with potable water to remove sand and mud. Shell fragments are removed by hand. "Blowing" is an alternative method of washing which involves placing the shucked meats in a tank of potable water and agitating them by introducing air into the bottom of the tank. After being drained thoroughly on a skimmer, the meats are placed into glass, plastic or metal containers before packing in crushed ice to promote rapid chilling.

Good sanitation in the shucking and packing areas, rapid movement of the shellfish through the process, and rapid chilling of the packed product are keys to preventing contamination of and minimizing bacterial multiplication in the shellfish during processing. The procedures described in the National Shellfish Sanitation Program (NSSP), Manual of Operations, Part II (Food and Drug Administration 1989) are designed to achieve this goal.

MICROFLORA CHANGES DURING RAW PROCESSING

The microorganisms in shellfish occur in the shell liquor, in the intestinal tract of the shellfish and are associated with the mucous on the exterior surfaces of the animal. Shucking and washing of the meats will remove the microbial flora associated with the shell liquor and the mucous.

When shellfish are processed under sanitary conditions, the numbers of bacteria in the meat, as measured by the standard plate count (SPC) (Greenburg and Hunt 1985) and the 25°C (77°F) aerobic plate count (APC), are reduced by about 90%. However, the number of indicator bacteria, vibrios, and *A. hydrophila* which are concentrated in the digestive system of the shellfish may not be significantly changed by processing (Hood, Baker, and Singleton 1984; Thompson et al. 1976; Vanderzant et al. 1973). The type of washing has little effect on bacterial reductions. In studies with the clam *Mya arenaris*, blowing was not found to be superior to spray washing in reducing the bacterial counts (Ward et al. 1969).

Chlorine is sometimes added to the water in the blowing tank in an attempt to reduce the bacterial load on shucked meats. Chai, Pace, and Cossaboom (1984) and Pace, Wu, and Chai (1988) demonstrated that blowing oysters in 150 ppm chlorine water for 20–30 min caused a reduction in the APC and altered the species composition of the oyster microflora. *Vibrio* sp. were apparently least affected by the chlorine treatment and increased from 23.0% to 63.7% of the population. This corresponded to the finding of

Motes (1982) who demonstrated that blowing in chlorinated water did not significantly reduce the concentration of *V. cholerae* in oyster meats.

MICROFLORA CHANGES DURING HEAT-AIDED PROCESSING

Heat Shock

When the internal temperature of live shellfish is increased above its maximum growing temperatures, the shellfish relaxes its adductor muscle. If the temperature is further increased, the muscle is relaxed to the point where the shell will actually gape. Seafood processors occasionally use heat-aided processing or "heat shock" to facilitate opening shellstock that are otherwise difficult to shuck.

Heat shock is accomplished by dipping the shellstock in hot water at 62.8–65.6°C (145–150°F) for 1–3 min or by exposing the shellfish to steam for short periods of time. Exposure times and temperatures are dependent on the species of shellfish, their size and shell thickness, and should be held to the minimum required to facilitate opening to prevent cooking the shellfish.

In addition to facilitating the shucking of the shellfish, heat shock brings about a reduction in the numbers of bacteria in the shellfish meats. Under controlled experimental conditions, immersing of oysters in 65.6°C (150°F) water for 3 min reduced the coliform counts by 82.8–97.8%, fecal coliform counts by 79.6–83.0% and plate counts by 75.0–99.4%. However, in studies of commercial "hot-dip" heat shock methods, reductions of coliforms and fecal coliforms were 50% and 25%, respectively (Pringle 1964).

Heat shocking oysters with steam to an internal temperature of approximately 40°C (104°F) facilitated opening the oysters and brought about a reduction in the SPC of 61–85% in four separate trials. Following shucking, these oysters generally retained a lower SPC over a 14 day storage period than did oysters that were not exposed to heat (Cook et al. 1980).

Care must be taken to shuck the oysters immediately after heat shock and to chill the shucked meats to a temperature below 7.2°C (45°F) immediately after shucking, or any advantages gained in bacterial reduction will be quickly lost. The heat shock process is not intended to improve the safety of the shellfish and in some instances may reduce the indicator bacteria level without destroying the pathogens.

Pasteurization

Pasteurization as it refers to shellfish is a heat treatment process which provides conditions sufficient to destroy organisms of public health significance. Pasteurization times and temperatures that have been used in experimental studies with oysters differ from the classical heating times and temperatures used for milk.

Goldmintz, Ernst, and Rasekh (1978) determined that steaming of shell oysters to an internal temperature of 60°C (140°F) reduced the total plate count sharply and eliminated coliforms. No aerobic spoilage or public health significant organisms could be detected and the remaining aerobic bacteria were *Bacillus*. While pasteurization of oysters in the shell shows promise for reducing the levels of bacteria, it is difficult to handle the oysters after treatment in a manner which would prevent contamination by spoilage organisms.

Pasteurization of shucked oyster meats which have been sealed in plastic bags to protect them from recontamination has been investigated. Chai et al. (1984) established that a pasteurization time of 8 min after the meats had reached an internal temperature of 72°C (161.6°F) was adequate to destroy most bacteria and extend the shelf life of the oysters up to 3 months. Immediately after pasteurization, only *Bacillus* could be isolated from the oysters. However, after storage at 0.5°C (32.9°F) other gram-positive bacteria, which can grow under low oxygen tension, were found. These included species of *Clostridium*, *Corynebacterium*, *Listeria*, *Peptostreptococcus*, and *Staphylococcus* (Pace, Wu, and Chai 1988).

While pasteurization of oysters in sealed containers under the conditions described above will improve the safety of the shellfish with respect to gram-negative bacteria usually associated with pollution, the finding of *Clostridium*, *Listeria*, and *Staphylococcus* in the oysters after storage raises questions about the safety of the stored product. Additional research will be necessary to ensure that the long-term storage of pasteurized shellfish is a safe practice (Pace et al. 1988).

Heat treatment may be used to selectively eliminate pathogens from shellfish. *V. vulnificus* can be eliminated from raw oysters by heating them for 10 min in a water bath at 50°C (122°F) (Cook and Ruple 1990). This treatment is sufficient to bring about a reduction in the SPC and fecal coliform count, but is not sufficient to eliminate fecal coliforms and bacteria that are thermal resistant (Cook and Ruple unpublished results).

EFFECT OF TIME–TEMPERATURE ON MICROFLORA OF RAW SHUCKED PRODUCT DURING STORAGE

Indicator Bacteria

Coliforms as a group have a wide range of growth temperatures. Their numbers have been shown to increase in raw shucked shellfish meats at all storage temperatures above freezing. However, the rate of increase is correlated directly with storage temperature. Fecal coliforms including *E. coli* require higher temperatures for growth than do coliforms and fail to multiply in shucked meats at temperatures less than 5°C (41°F) (Cox 1965; Hoff et al. 1967; Hood, Baker, and Singelton 1984; Kelly 1964; Ward et al. 1969; Wilson and McCleskey 1951). At storage temperatures approaching 10°C (50°F), fecal coliform increases have been noted (Hoff et al. 1967). Higher storage temperatures cause shellfish to spoil rapidly. Because fecal coliforms can multiply in temperature-abused shellfish meats, fecal coliform levels alone are not useful as an indicator of fecal contamination in processed meats.

Enteric Bacterial Pathogens

Information on the fate of enteric bacterial pathogens that may be in shellfish when processed is limited. However, it is unlikely that enteric pathogens will multiply in shellfish if they are stored below 7.2°C (45°F), as recommended by Food and Drug Administration (1989). Studies with pure cultures of *Salmonella* inoculated in tryptic soy broth show that they grow poorly or not at all at 6°C (42.8°F) (Matches and Liston 1968*b*). In seafood, low-temperature storage permits the psychrotrophic bacteria to grow rapidly, thus suppressing the growth of salmonellae and preventing them from reaching high numbers (Matches and Liston 1968*a*). Presnell and Kelly (1961) observed that *Salmonella* inoculated into shucked oysters held at 4–7°C (39.2–44.6°F) failed to grow within 6 days. However, when the oysters were held at 20°C (68°F), the *Salmonellae* grew rapidly.

None of the 40 strains of *Plesiomonas shigelloides* studied by Miller and Koburger (1986) were capable of growing at 5°C (41°F) in pure culture, and strains that could grow at 8°C (46.4°F) required 10 days to produce detectable growth. *Yersinia enterocolitica* developed very slowly in inoculated shucked oysters held at 0–2°C (32–35.6°F), increasing approximately 2 log units during 14 days of storage (Peixotto et al. 1979).

Vibrionaceae

The fate of *Vibrio parahaemolyticus* in processed oysters has been investigated intensively. Studies have included meats inoculated with laboratory strains via injection (Johnson and Liston 1973), surface contamination (Goatcher et al. 1974) and natural uptake (Johnson and Liston 1973). Meats naturally contaminated before harvest were also studied (Hood, Baker, and Singelton 1984; Thompson et al. 1976). Storage temperatures in these studies ranged from 1°C to 11°C (33.8–51.8°F). *V. parahaemolyticus* levels did not increase in any of the studies and typically showed a rapid decline. A reduction of the salt level in the oyster meats caused by washing in combination with cold temperature (Covert and Woodburn 1972) is probably responsible for the decrease in *V. parahaemolyticus.* However, *Pseudomonas* species which produce *V. parahaemolyticus*-inhibiting substances have been isolated from oysters and may be partly responsible for the repression of *V. parahaemolyticus* (Goatcher and Westhoff 1975).

Oliver (1981) studied the survival of *Vibrio vulnificus* in sterile oyster homogenate held at 4°C (39.2°F). He found that oyster tissues contained components that brought about a dramatic decrease in *V. vulnificus* not attributable to cold shock. In whole oysters inoculated with *V. vulnificus* and held at 0.5°C (32.9°F), a reduction in numbers of *V. vulnificus* occurred, but it was no greater than the reduction caused by cold shock alone. Kaysner et al. (1989) also observed reductions in *V. vulnificus* in artificially contaminated shucked oysters held at 4°C (39.2°F), with counts decreasing by 1–4 log units in 6 days.

Ruple et al. (1989) found that the numbers of naturally occurring *V. vulnificus* in shucked oysters were reduced by greater than 90% after 3 days storage at 0°C (32°F). Further reductions were time-dependent and attempts to isolate *V. vulnificus* from shucked oysters held in ice storage for greater than 7 days usually failed.

V. cholerae unlike *V. parahaemolyticus* and *V. vulnificus* is not affected by cold shock. Reily and Hackney (1985) demonstrated that laboratory-grown stains of *V. cholerae* persisted at high levels in oyster homogenate for 21 days at 7°C (44.6°F).

Aeromonas hydrophila, unlike the other species of *Vibrionaceae* tested, increased in oyster meats in cold storage. Significant increases were found in processed oysters after seven days storage at 4°C (39.2°F) (Hood, Baker, and Singelton 1984).

Standard Plate Counts

During the first few days of cold storage, bacterial populations undergo a change as indicated by a reduction in the SPC. Bacteria unable to tolerate the cold die, leaving a population of spoilage bacteria adapted for the storage temperature. Bacterial populations then increase in a logarithmic fashion. When the SPC exceeds 10^6/g, the meats are of substandard quality (Liebman et al. 1957). Kelly (1964) found the time required for the SPC in Pacific and Eastern oysters to reach 10^6/g at 0°C (32°F), 3°C (37.4°F), and 10°C (50°F) to be approximately 16, 10, and 3 days, respectively. Similar findings were reported for the Olympia oyster (Hoff et al. 1967).

Shucked soft-shell clam meats have a shorter shelf life than oysters. Clams take on a rancid, fishy odor after 5–7 days of storage at 1°C (33.8°F) and exceed a SPC of 10^6/g at 9–14 days (Cox 1965).

EFFECT OF FREEZING ON MICROFLORA OF RAW SHUCKED PRODUCT

Freezing of shellfish meats eliminates the concern of pathogen and indicator bacteria multiplication. However, freezing does not eliminate all pathogens that are present and in some cases, pathogens survive longer in frozen shellfish meats than in meats stored above freezing.

DiGirolamo et al. (1970*a*) studied the survival of polio viruses in oyster (*Crassostrea gigas*) meats which had been inoculated by injecting the virus into the body mass prior to freezing. After 4 weeks of storage at − 17.5°C (− 0.5°F), the number of viruses were reduced to less than 10%. Virus levels in live oysters (*Ostrea lurida*) held in dry refrigeration at 5°C (41°F) were reduced by 6% in 15 days. Although the difference in the two oyster species may have affected the results, it is clear that viruses survive well in frozen oysters and it appears that they survive better in frozen than live oysters.

Vibrio cholerae survived better in sterile oyster homogenate held at − 20°C (− 4°F) than in homogenate held at 7°C (44.6°F). At both temperatures, the organism was recovered after 21 days of storage (Reily and Hackney 1985).

During freezing, the numbers of naturally occurring *V. vulnificus* in oyster meats is reduced by > 99.9%. Further reductions are seen with increases in storage time, but *V. vulnificus* can be recovered from frozen oysters after 12 weeks of storage at − 20°C (− 4°F) (Cook and Ruple 1990).

DiGirolamo et al. (1970*b*) concluded that *Salmonellae* in oysters was highly sensitive to freezing. The numbers of *Salmonellae* were reduced by 99% from the original level present in the unfrozen oysters. *E. coli* was found to be less sensitive to freezing and about 5% of the cells remained viable after 14 days of frozen storage.

SUMMARY

The shucking, washing, and packing of shellfish under NSSP guidelines (Food and Drug Administration 1989) prevents the entry of and multiplication of pathogens and fecal coliform indicator bacteria in the shellfish. Further, the total microflora in the shellfish as measured by the SPC or APC does not increase and may decrease during processing.

The fate of bacteria in shellfish meats after processing is temperature-dependent. If shucked shellfish are stored at temperatures of 7.2°C (45°F) or below as recommended by the NSSP guidelines, pathogens and fecal coliform indicator bacteria will not multiply and usually decrease with time. Lower storage temperatures retard the multiplication of psychrotrophic bacteria in the shellfish meats and extend the shelf life.

12

Sanitary Surveys of Growing Waters

Mary Jo Garreis

Water Management Administration, Standards and Certification Division 2500 Broening Highway, Baltimore, MD 21224

INTRODUCTION

Bivalve shellfish are filter feeders which strain detritus, algae and microorganisms from the water. In waters contaminated with human sewage, bivalves can concentrate pathogenic microorganisms. When the shellfish containing pathogenic organisms are consumed raw or slightly cooked, diseases may be induced in humans. In order to assure that shellfish are safe for human consumption, public health officials need to protect shellfish harvest areas from sewage contamination. However, because of the bivalves' ability to bioconcentrate other pollutants such as heavy metals and chlorinated hydrocarbons, shellfish-harvest waters also need to be protected from unacceptable concentrations of bioaccumulatable substances. This protection is accomplished through the sanitary survey.

The sanitary survey is a thorough examination of the watershed draining to a shellfish-harvesting area to identify and, where possible, eliminate or minimize the actual or potential sources of serious human contamination. Its objective is to identify and evaluate those factors influencing the sanitary quality of shellfish harvest areas. The information generated in the survey is used to ascertain what waters are safe for the harvest of shellfish for direct human consumption; what waters require that additional shellfish treatment, such as depuration, be applied before human consumption of the shellfish; and what waters are unsafe for shellfish consumption under any circumstances. It is a powerful weapon in the public health arsenal and it pays water quality and public safety dividends which extend beyond protecting shellfish for human consumption.

The sanitary survey originated in the late 19th and early 20th century as an effort to protect drinking water supplies. Originally, it was a field survey where observations and examinations of the water were made to identify the conditions that affect or could affect the sanitary quality for drinking water use (Fair and Geyer 1963). Observations included location of sources such as raw or treated sewage discharges and chemical discharges. Visual effects of the discharges on the receiving water such as water discoloration and floating materials were noted. Other factors such as the effects of drainage patterns or climatic events on the water body were reported. Where possible, discharges of public health significance to the water supply were eliminated to assure that the water supply was safe.

Because of the success of the sanitary survey in protecting water supplies, its use was extended in the early 20th century to gauge the health of shellfish waters. Watersheds draining to shellfish waters were subjected to an intense review. Shorelines were examined for waste discharges from human activities. The banks of streams and rivers draining to harvest areas were walked to ascertain if the streams and rivers carried contaminants from inland sources. Correction orders were issued to remove untreated or poorly treated sewage discharges from harvest areas (Maryland Board of Health 1925–1952). As early as 1936, the city of Baltimore, Maryland, built the first sewage treatment plant in the United States, constructed expressly for the protection of shellfish waters as the result of a sanitary survey which identified raw sewage discharges.

The importance of field observations and field measurements to shellfish harvest area protection cannot be overemphasized. Field observations by technically trained persons who can reliably evaluate the effect of point and nonpoint source discharges on harvest areas is the most critical aspect of the survey. It is these observations that provide the best interpretation of the probable public health impact of sources on shellfish waters. Only with a thorough, careful sanitary survey, supplemented by periodic bacteriological sampling, can a proper assessment be made to determine whether areas are safe for the harvest of shellfish for direct human consumption. Areas subject to periodic human sewage contamination, which can cause the shellfish to become possible vectors in disease transmission, must be identified and restricted to harvesting. Areas where shellfish have concentrated organic and inorganic pollutants to unacceptable levels must be evaluated properly and the public warned of the unacceptable health risk. Periodic bacteriological or chemical examination of water samples alone cannot be substituted for the sanitary survey, although it is an integral part (Federal Security Agency 1946).

The frequency with which a sanitary survey should be performed is a matter of debate. The Interstate Shellfish Sanitation Conference guidelines (1990) recommend the completion of the full survey at least every 12 years. Periodic review of the critical contaminant sources is recommended at a minimum of every three years. States with aggressive sanitary survey programs generally believe that this frequency is too low, particularly in areas experiencing rapid development. These states frequently use a five- to six-year cycle for the full survey, and perform a critical source review annually or biennially.

The sanitary survey must be performed by trained technical personnel capable of evaluating the probable impact of sewage, industrial and nonpoint source discharges on a shellfish harvest area. Technical expertise is needed from many disciplines. A thorough survey requires the cooperation of many agency units with widely varying missions. The unit responsible for the public health safety of the harvest area acts as a synthesizer, which pulls together its own investigations and the information from these diverse sources into a interpretative report. This report provides the basis for the assessment of the probable public health risk associated with the consumption of shellfish and results in the area being identified as producing shellfish acceptable or unacceptable for human consumption.

PRELIMINARY ASSESSMENT

The sanitary survey begins with the preliminary assessment of the existing information for each of the survey's three major components. These components are: the shoreline survey (identification, and where possible, the minimization or elimination of sources of contaminants); the periodic bacteriological examination of water samples; and the bacteriological and chemical examination of shellfish. The process of identifying sources of contamination has three components: evaluation of actual and potential sources, evaluation of climatic factors and an assessment of the hydrodynamic factors that may affect distribution of pollutants throughout a harvest area. In this assessment, the following factors must be considered:

(1) The quantity of pollution from each source.
(2) The degree of treatment given to an effluent before it is discharged, with particular emphasis on the destruction of pathogenic organisms where human sewage is involved.
(3) The dilution available with clean water and the way in which the effluent disperses.
(4) The time lapse and distance between the discharge and the shellfish harvest area.
(5) The availability of natural self-purification.
(6) The quality of the harvest area water at predetermined stations on different days under varying weather and tide conditions.

The first step in the sanitary survey is the identification of the watershed which drains to the harvest area being scrutinized. A good first cut may be made from topographical maps of the area. The map delineation should be confirmed by field visits to the area, particularly in coastal plains with extensive marsh systems. Drainage patterns in these areas are difficult to detect and verify. Depending on the size of the watershed, it may be necessary to break the survey area into smaller, more manageable segments. Segment delineation using the subbasins of the various tributary creeks and streams is generally the most useful in compiling the final assessment.

Agency resources for sanitary surveys are very limited and most watersheds contain a large number of properties. Consequently, every effort should be made during the initial field visits to identify only those portions of the watershed with the potential to affect water quality in the shellfish harvest area. This practice sharply reduces the actual number of properties to be surveyed within the watershed and focuses agency resources on areas of greatest concern. Once the final boundaries of the survey are defined and mapped, several copies of the map should be reproduced. These maps will provide guidance to field personnel making the survey as well as providing useful tools for recording information.

The complexity of waste sources in many developed and rural watersheds requires the acquisition and evaluation of information from many disciplines. Information developed for other purposes exists in many local, state and federal agencies. Survey field work can be substantially reduced by ferreting out this information, particularly for point source discharges.

Point source information may be obtained from the agency responsible for National Pollution Discharge Elimination System (NPDES) permits, generally the state department for environmental protection or the water control board. The available information includes the location of the discharge pipes, discharge volume, treatment type, contaminant concentrations in the discharge, inspection reports, and the history of treatment failures. Some additional field work may be required for updating the information or for verification, but most preliminary information is readily available.

In some cases, hydrodynamic data, based on dye studies, drogue studies or other similar studies, are available for point sources. These studies are invaluable as they can provide detailed information as to how contaminants are dispersed into a shellfish harvest area. Dye studies, in particular, can provide information concerning contaminant concentrations over time and distance, and can assist with estimates of microbial dilution and die-off. Hydrodynamic studies for major dischargers are becoming more common because of their use in determining NPDES permit limits. Where these studies are not available, efforts should be made to obtain them. Their provision should be mandatory for the construction of new facilities discharging in or near shellfish waters.

Information concerning nonpoint sources (NPS) is more difficult to obtain. Frequently, the biennial 305(b) report prepared by each state will provide information concerning NPS discharges. In some states, watershed plans have been developed and will contain NPS information. In most cases, however, the agency responsible for protecting shellfish harvest areas will be forced to develop the necessary NPS information.

One frequently overlooked source of information concerning shellfish and shellfish harvest water quality is various studies performed by academic institutions. These institutions often investigate the uptake by aquatic organisms of the various pollutants in the environment, and may provide some background information. Comprehensive surveys such as the National Oceanic and Atmospheric Administration's annual mussel watch report (NOAA 1989) provide some information concerning contaminants in shellfish from various harvest areas.

The most important aspect of information gathering in the preliminary assessment phase is the establishment of an interagency, interdisciplinary information network. Open lines of communication need to be established among key staff so that information flows freely from one agency to another. Ordinarily, it is a matter of the shellfish regulatory unit making its needs known and persuading other agency units to routinely provide the information. In exchange, the shellfish unit can offer to provide information compiled in the survey process to support the planning, regulatory or enforcement action of other units. This information exchange can provide an invaluable tool for agencies to work together to correct pollution problems and recover valuable resources for public use.

RECORD KEEPING

Maintenance of good records is an essential part of the sanitary survey. Records serve three purposes. First, they document the total number of properties visited, including those where no discharges were found. Second, the records provide a clear statement of findings at those properties where undesirable discharges occur for subsequent evaluation of the total impact to the shellfish harvest area. Third, the records serve as the basis for initiating corrective action to minimize or eliminate pollution sources.

Because of the numerous individual properties involved in the sanitary survey, an effective record keeping system is essential for tracking properties visited and the findings associated with each property. One of the best methods is the use of a unique identification number for each basin or subbasin surveyed and the addition of a unique number for the property. Since surveys are repeated at specified frequencies, inclusion of the year in the identifier number is also useful. An example of this numerical system is as follows: 91-104-0010. The year identifier is 91; the 100 series designates the Cherokee River; 104 designates Bluebird Creek, a subbasin of the Cherokee River; 0010 designates the Harris Ford property, the tenth property visited in the survey. Proper property identification is essential for evaluation of impacts and corrective enforcement action.

Prior to the initiation of the sanitary survey, field record forms need to be developed. Generally, the most efficient method is to use a log of properties. The log provides entries for 5–10 properties (depending on format) on each page and summarizes the findings at each property. Each page of the log is identified by the unique basin identifier (91-104 in our example). Each property is identified by its unique number (0010 in our example). Each property entry includes the name and address of the property owner, the name and address of the tenant if different from the owner, the exact location of the property, and the nature of the discharge if one is found. This approach eliminates the need for numerous individual forms for many properties where no discharges are found and makes final report compilation easier.

A detailed form is needed for those properties where a discharge is discovered. The completion of this form is critical as it provides the basic information for evaluating the impact of the discharge. It may also serve as the basis for subsequent enforcement action. This form should include the name and address of the property owner, the name and address of the tenant, where appropriate, and the exact location of the property. For residences or apartments, additional required information is whether it is a seasonal or year-round residence, number of water-generating appliances (dishwashers, washing machine, etc.) and the type of sewage system if appropriate. The exact nature of the discharge should be reported. Space should be provided for the surveyor to sketch the location of the discharge with respect to the buildings and other property features. The distance the discharge must travel to the harvest area should also be reported.

Specialized survey forms may need to be developed for certain types of property use. One example is marina facilities where, in addition to sewage facilities, it is important to note the number and type of boats served, vessel pumpout facilities and other features unique to marinas. Another example is farms, where the numbers and types of animals maintained and the best management practices installed are critical to the effect of agricultural runoff on adjacent shellfish waters.

Surveyors should be instructed to inspect their assigned survey area in a systematic manner. The surveyor should indicate, on his field map, the property at which he begins the survey as property 0001. From that point, the surveyor should work systematically either to his right or left for the duration of the entire survey. This practice enables anyone revisiting any of the properties for a reevaluation or an enforcement action to locate the property.

In addition to the property records, a systematic approach to tracking the data generated through water column and shellfish meat testing is also needed. Identifying sample locations by the unique basin/subbasin identifier makes subsequent information integration into the final assessment much easier. For purposes of our example, sample station one in Bluefield Creek would be identified as 104-001; station two as 104-002 and so forth.

One useful tool in record keeping is to perform preliminary data assessments. This technique consists of reviewing the data being generated through the three components of the sanitary survey at periodic intervals. Transference of survey findings to the master survey map often assists in identifying needs for additional or expanded investigation. This is particularly true for direct point source discharges, where dye studies or similar tools may be used to estimate contamination dispersion in a harvest area. Disagreements among the data generated among the three survey components may be identified early and corrected through changes in survey practices. This time investment is well worthwhile. It can eliminate the need to repeat parts of the survey at a later date and prevent delays in completing the final survey assessment because of inadequate data gathering.

Computers are invaluable tools for establishing and maintaining systematic records for the sanitary survey. Computerized data bases facilitate the development of preliminary survey findings and tracking of pollution abatement. Software packages exist which provide good mapping capabilities, and the programming to organize data into tables, charts, and graphs for detailed report presentation. Wherever possible, shellfish agencies should make use of the many excellent software packages available to store, retrieve, and compile survey information.

POINT SOURCE ASSESSMENT

Point sources are the easiest contaminant sources to locate and assess. Point sources are discharges from fixed locations carried to a waterbody by pipes, conduits or similar devices. Typical point sources are sewage treatment plants, combined sewer overflows, factories, processing plants and other industries. In this discussion, sewage treatment plants (STPs) will be used as representative point sources since they are nationally the most important sources affecting shellfish waters.

Critical to the sanitary survey is an assessment of the treatment provided to the effluent from a point source. The higher the degree of treatment and the more reliable the treatment, the less is the impact on the shellfish harvest area. Treatment assessment consists of the knowledge of the concentrations to be achieved, monitoring information which demonstrates what concentrations are achieved consistently, the safeguards within the treatment system to assure its reliability, and the measures available to provide early warning in the event of system failure.

In 1972, the Clean Water Act established secondary treatment as the minimum treatment level acceptable for STPs. Most STPs nationwide discharging to shellfish waters now meet this minimum requirement. Addition of good disinfection practices, and additional treatment such as sand filtration prior to disinfection, can provide high-quality effluents with severely reduced pathogen loadings. In some states, such as Maryland, the NPDES discharge permit limitation for fecal coliform is established at the water quality standard adopted for shellfish waters (Luckman 1990). This practice assures that the STP maintains good treatment and disinfection practices. Achievement of high quality can easily be verified with review of the NPDES required daily monitoring reports and with periodic inspection and monitoring by the regulatory agencies.

The ability to achieve high-quality effluent and the reliability with which an STP achieves high-quality effluent are not the same thing. High reliability can be defined as a major failure of major STP components no more than once in every three to five years. High reliability is critical to shellfish harvest area protection since microbial agents may be taken up rapidly by shellfish and may induce disease in humans within hours of shellfish consumption. High reliability is achieved through elimination of overflow devices and infiltration in collection systems, flow equalization, equipment redundancy, increased disinfection contact times, sand filtration and alternate power sources. These technologies are recommended by the United States Environmental Protection Agency (USEPA 1974). With the exception of their extensive use in Maryland since 1968, these technologies have not been widely applied to shellfish waters nationally. Their presence and good operation enhance the ability of the STP operator to address a myriad of normal operating problems, without compromising the quality of the effluent. For example, with equipment redundancy, in the event of the failure of one primary clarifier, the other primary clarifiers could accept the additional loading and provide treatment while the first clarifier is repaired. There would be no change in the quality of the final effluent. Presence and operating history of all or any of these high-reliability components should be factored into the sanitary survey evaluation. The shellfish regulatory agency should work with its sister agencies to assure that new facilities incorporate as many of these components as necessary. In existing plants, these components should be included in future STP modifications.

The measures available to provide early warning in the event of STP system failure play an important role in the sanitary survey assessment. The purpose of the early warning measures is to assist a plant operator in the detection of a plant malfunction, to provide time for the operator to notify the shellfish regulatory agency, and to provide time for the shellfish agency to stop shellfish harvesting, if necessary. These measures include 24 h alarm systems on important STP components, additional wet well storage at pumping stations, and holding ponds with the capacity to store the STP's effluent for 24 h. Twenty-four hour holding ponds are required and in place at all major STPs discharging to shellfish waters in Maryland and are now being considered at two facilities in Massachusetts (Hickey 1990). When components of the STP fail, wastes are diverted to the 24 h holding pond and stored until repairs are completed. This eliminates the need to bypass the wastes to the receiving water to accomplish repairs or system maintenance. Shellfish waters remain protected and available for harvesting since most temporary corrections and repairs can be achieved in 24 h. These early warning measures, augmented by the information generated in the hydrodynamic studies, form the basis for the establishment of the Closed Safety Zone (CSZ).

The CSZ is that part of a shellfish harvest area which lies adjacent to an outfall or other area of contamination and where shellfish harvesting is prohibited. It has two time-driven objectives. The first objective is to assure public health protection by preventing pathogen contamination of shellfish-harvesting waters from normally operating treatment works. As the pathogens which survive the treatment process travel through the CSZ, pathogen concentrations are decreased by mixing with receiving waters, dilution and the occurrence of natural die off. The CSZ is designed to achieve acceptable pathogen concentrations at the edge of the zone under normal operating conditions. The second CSZ objective is to provide sufficient time for public officials to notify harvesters to stop shellfish harvesting in the event of a STP malfunction or failure. The degree of wastewater treatment at the STP, the STP's reliability, and the early warning systems, particularly the 24 h holding pond, are factors which can drastically increase or reduce the size of the CSZ.

In the sanitary survey, there are two distinct zones established with respect to a STP which discharges directly to shellfish harvest waters. The first zone is the CSZ which will remain restricted to harvesting under normal STP operating conditions. The second zone is the total area affected in the event of a major failure at the STP, for example, loss of treatment capability associated with a power failure following a storm event. This second zone is closed to harvesting in the event of a major STP failure. It would reopen to harvesting only when STP repairs were completed, the STP was operating normally, and sufficient time had elapsed for natural cleansing of the shellfish.

Where STP discharges are well removed from the immediate shellfish harvest area, treatment, time of travel and the dilution available are the major factors in evaluating STP discharge potential impact. Pathogen survival in sewage declines rapidly with time. The fresher the sewage, the greater is the associated public health risk. Sewage discharges, which are separated from shellfish harvest areas by several days travel time, may have little or no impact to the harvest area. This is particularly true where the STP effluent has been subjected to good sewage treatment. The potential for impact is greatly influenced by the volume of the discharge, the volume of the receiving streams, and stream flow. Under these circumstances, the greatest impact may occur when climatic events, such as flooding, reduce the travel time and hasten the delivery of relatively fresh sewage to the harvest area.

Assessment of the impact of industrial point source discharges is more difficult than that of STPs. Wastes from food processing plants, particularly meat and poultry processing operations, and from animal waste treatment facilities have characteristics similar to human sewage and should be approached in the same manner as STPs. Among the remaining industrial discharges, those of greatest importance are those associated with toxic substances which can be bioaccumulated by shellfish. Examples of bioaccumulatable substances are methyl mercury and chlorinated hydrocarbons such as chlordane, dieldrin and DDT. Bioaccumulation is a long-term process with concentrations building in the shellfish tissue over time. Ability to concentrate varies with shellfish species (Garreis and Pittman 1981; Unger, Hall, and Sullivan 1990) and environmental factors in the harvest area. Effluent testing can provide estimates of the total loading of the bioaccumulatable substance discharged to the harvest area. Hydrodynamic studies are useful in predicting the total area affected and relative distribution of contaminant concentrations within the affected area. Reliable scientific approaches for estimating the actual concentrations in shellfish tissue are, however lacking. Therefore, the best indicators are the shellfish. Measurements of concentrations in shellfish tissue near these waste sources at predetermined locations and time intervals provide the most accurate information concerning shellfish quality.

Unlike the immediate health risk associated with bacterial contamination, the risk associated with the concentration of toxic substances by shellfish generally requires a long-term exposure. For example, consumption of carcinogen-contaminated shellfish at a fixed rate over a 70-year period of life is normally used in estimating the increased cancer incidence in an exposed population. Because the impact is not as dramatic as that associated with microbial infections, this aspect of shellfish-harvest water protection is frequently overlooked or given low priority. Once the discharge of bioaccumulatable substances is suspected or established, the best survey assessment technique is the creation of a long-term monitoring network based on shellstock tissue. Trends indicating the increase in shellfish tissue of bioaccumulatable toxic substances should be immediately addressed by source control. Where the shellstock testing establishes that tissue concentrations exceed acceptable levels, appropriate harvesting restrictions are required.

NONPOINT SOURCE ASSESSMENT

Nonpoint sources (NPS) may be simply defined as all sources that are not point sources. This definition, however, ignores the magnitude and complexity of the problem. NPS is a broad category which includes agricultural runoff from feed lots and farmed fields, urban stormwater discharges, freshwater intrusion into estuaries, nitrate contamination in water bodies, atmospheric deposition of contaminants, failing septic systems, recreational boats and a host of other sources. It is characterized by sources which are diffuse and dispersed. NPS pollution is not easily identified and is not easily addressed and corrected. The problem is further complicated by our inability to quantify the public health risk associated with NPS pollution.

Identifying and locating NPS discharges is manpower-intensive and time-consuming. This process is usually characterized as the "shoreline survey" by most agencies responsible for shellfish harvest area safety. It requires that watersheds be delineated using topographical maps and field visits to determine the geographical area constituting the watershed that could potentially transport contaminants to shellfish-harvest areas. Each property within the watershed is visited and examined for potential waste discharges. Each potential waste discharge is then evaluated. Because of the interaction required with the public in gaining access to the property and the need to examine and evaluate potential NPS pollution discharges, a surveyor can complete only ten to 15 residential properties a day. Where large agricultural operations are involved, a single farm may require a full day or more. In urban areas, identifying the contributors to large storm drains may take several days.

The ability of the surveyor to communicate with the public and to correctly identify and evaluate the impact of NPS pollution is critical. Since private property rights are zealously guarded in the United States, the surveyor must elicit the cooperation of the property owner before the property may be surveyed. Often, this involves a one-to-one public education effort. The surveyor must clearly convey to the property owner the purpose of the survey, what types of discharges he will be looking for on the property, and the consequences should any unacceptable discharges be identified. Generally, most property owners are cooperative, but securing that cooperation is time-consuming.

Some shellfish agencies advertise the advent of their sanitary surveys in the local newspapers. The advertisement provides public notice as to the purpose of the survey, the type of information sought by the surveyors. The theory supporting this public notice is that, if the public understands why the survey is being made and what is expected of the property owner, much time can be saved during the survey. Differing opinions exist as to the validity of this approach. Some officials maintain that advertisements make accomplishing the survey easier, while others maintain that advertisements increase public resistance. Other officials believe that advance notice enables individuals to conceal important discharges. The value of advertising may vary from community to community within a region. Each agency has to weigh the pros and cons and apply its best judgement concerning advertising of the survey.

Discovering NPS pollution often requires ingenuity and persistence in the surveyor. Often, sources are not obvious. Septic systems frequently require dye testing to detect failure. Washing machines and bathing facilities may discharge to waterways, while the toilet facilities are connected to properly functioning septic systems. Dye testing only the septic systems does not capture these discharges. Some septic systems fail only under high water table conditions, i.e., the wet winter months. The surveyor needs to be sensitive to these anomalies to ferret out impacts.

Failing septic systems are frequently cited as major sources of contamination in shellfish waters. Septic systems are the most common on-site individual sewage disposal systems. They are built where public sewerage is not available to serve residences, schools, camps, and other small establishments. The septic system is simple in construction as well as operation. Sewage is detained and treated anaerobically in the septic tank for 24 h. Suspended solids removal of 50–70% is mainly the result of sedimentation. The effluent is discharged to tile fields, dry wells or seepage pits, where the effluent is absorbed into the soil. Ehlers and Steele (1965) provide an excellent discussion of septic system mechanics.

Among some public health officials, the perception exists that the majority of septic systems fail most of the time. The evidence does not necessarily support this conclusion. In Maryland in 1988, a total of 4461 properties served by septic systems or other on-site waste disposal systems were surveyed. Thirteen properties (0.3%) were documented with failing septic systems. The 1989 figures provide a similar story. Out of 4692 properties surveyed, 68 properties (1.4%) with failing septic systems were reported (DiStefano 1990).

This situation is not peculiar to Maryland which has a long history of repeated sanitary surveys combined with an aggressive enforcement program to correct failing systems. Communication with the Massachusetts Division of Marine Fisheries provides similar septic system failure rates. Properly installed septic systems in good soils average less than 2% failure rate. Failure rates are higher where rocky or impervious soils exist or where old improperly designed systems were installed. In these areas, installation of public sewerage is required to correct the problem (Hickey 1990).

The time of the year plays an important role in detecting failing septic systems. Many systems, which malfunction in wet weather months when the ground water table is high, perform well during the hot, dry summer months. On the other hand, the off season survey of resort communities primarily inhabited during the summer months does little to document the septic system failure rate. Local habitation patterns need to be determined prior to survey. Time of year may also be important in evaluating the impact of NPS discharges on the shellfish harvest area. Discharges from failing septic systems associated with a summer resort community may not be important in a harvest area where harvesting occurs only in the winter months. Shellfish resource in close proximity to these areas can be safely harvested during winter months in northern states after summer activity ceases. Adequate time must be allowed for natural shellstock cleansing prior to harvest.

As a general rule, properties located along the immediate shoreline should have their on-site septic system dye tested. Confirmation of a dye test may require several revisits to the property unless the discharge is immediately evident. Visual inspection is usually sufficient for septic systems located on inland properties. Discretionary dye testing can be used on any property where waste disposal practices are questionable.

The impact on shellfish harvest waters from individual or aggregations of failing septic systems can be estimated. Estimates range from 75 to 150 gallons per day per person in the household, with the average for a private residence seldom exceeding 500 gallons per day (Teutsch 1990). The usual estimate for fecal coliform discharge is 2×10^9 per person per day in raw sewerage (USEPA 1985; ISSC 1990). In cases of direct discharge, simple dilution calculations can provide an estimate of the area affected. Where the effluent is indirectly discharged to drainage ditches or inland streams, estimates should provide for the percentage of the effluent reaching the waterway, bacterial die-off, dilution and travel time to the harvest area. Unless the discharge is directly to an active harvest area, the impact of a single septic system is generally minimal. An aggregation of failing septic systems draining to a single stream or harvest area, however, duplicates the impact of the failure of a small sewage treatment plant. In the final survey assessment, the location and distribution of the failing septic systems is frequently more important than the number of failing systems.

Failing septic systems can be corrected in most cases with minimum effort. In older, developed areas where there are poor soils and minimum existing house lot sizes, public sewerage may be the only solution. The responsible shellfish agency should develop a close working relationship with local health authorities and other agencies responsible for correcting problems of this type. Sanitary survey information can be invaluable in providing the justification for public sewerage which will reduce the risk to the affected community and recover a valuable shellfish resource for harvest by the industry and use by the shellfish consumer.

Impacts from animals are more difficult to assess. There have been no cases of shellfish-borne illness directly traced to animal pollution. Most scientists agree that waters contaminated by domestic or wild-animal wastes offer a public health risk that is orders of magnitude less than that associated with human sewage. The precise relationship between human health risk and animal pollution, however, remains undefined (Stelma and McCabe 1990). The greatest potential risk for disease transmission in shellfish-harvest waters appears to be associated with large herds of domestic animals. Tillamook Bay in Northern Oregon, for example, is an important shellfish harvest area. The Bay watershed also contains 120 dairies with an estimated population of 19,000 cows, producing in excess of 256,000 tons of manure annually. By contrast, the human population is 11,000 residents (Jackson 1985). Both septic system failure and dairy runoff have been identified as important pollution sources. No known incidence of shellfish-associated disease, however, has been reported from this Bay (Jackson 1990). Effective evaluation and elimination of agricultural pollution problems associated with animal wastes requires development of a cooperative working relationship between the shellfish regulatory unit and the Soil Conservation Service, the Soil Conservation Districts, and the state's agriculture department. Working together, these entities can educate the farmer concerning the need to control animal waste discharges, foster good animal waste management practices, and elicit cooperation to protect shellfish water quality.

To evaluate properly the impact from animal herds, the surveyor needs to be familiar with a wide range of best management practices (BMPs) which are used to control animal waste. One of the most popular and most successful is to provide animal watering facilities on high ground away from streams. Fencing is provided to protect the stream bank and a vegetative buffer is used to impede erosion and filter pollutants. This BMP reduces animal waste discharges to negligible impact, particularly where herds of 100 animals or less are involved. In areas with moderate slopes, a grass-covered earth berm is frequently constructed outside the fence and prior to the vegetated buffer to further impede runoff. Farmettes where a few large animals such as horses, cows, or pigs are maintained have negligible impact if the animals are fenced away from the water, vegetative buffers are used, and manure piles are maintained so that pile drainage will not reach a waterway.

Manure disposal may be a major NPS if good BMPs are not used. Covered storage facilities are necessary for large operations such as feed lots and poultry houses. Provision of a cover eliminates the undesirable runoff associated with manure during precipitation events. For smaller operations, such as farmettes, location of the manure pile well away from the watercourse, and preferably on flat ground or ground sloping away from the water, is generally sufficient.

Another BMP is to plow the manure into fields. Since this practice is highly dependent on the time of the year to provide the needed crop fertilization, it is most effective when combined with the storage practices described above. Manured, plowed fields should be separated from the waterway by vegetated buffers to filter any pollutants which may escape from the fields. Vegetated berms may be useful as boundaries on fields with moderate slopes. Pastures with good grass cover are another effective BMP. Grass cover retains pollutants in the field and provides retention for absorption into the soil.

The importance of wildlife waste contributions to the sanitary quality of shellfish waters has been a source of controversy for years. The arguments are basically the same as those associated with domestic farm animal wastes, although wild animal sources are believed to be less important than domestic animal sources. As is the case with agricultural runoff, no shellfish-borne illnesses have been directly traced to wildlife contamination. Because some pathogens have been isolated from various wildlife species, it has been hypothesized that the potential for disease transmission exists. Estuaries with large populations of aquatic mammals and birds are believed to have the greatest potential for shellfish-borne illness (Stelma and McCabe 1990).

The NPS impact of recreational boats is particularly difficult to estimate. Despite a widespread perception among some public health officials and some segments of the public that recreational boats discharge large quantities of raw sewage and, therefore, constitute a public health risk to shellfish waters, the available epidemiological evidence does not support this perception. No outbreaks of shellfish-borne disease have been traced to the discharge of sewage from recreational boats. Given the public perception of the large number of recreational boats which discharge sewage while moored or plying shellfish-harvesting waters, this observation is surprising. Some public health officials are quick to point out that the reported outbreaks account for only a small fraction of all shellfish-borne illness (as is true with all foodborne illnesses), and that in a large percentage of the reported incidents, the origin of the implicated shellfish is unknown. Others believe that the risk is grossly overestimated and use the absence of solid epidemiological evidence to substantiate this point.

Recreational boat discharges that may have some potential impact to shellfish harvest areas are generally associated with marinas. A marina is a facility that provides structures for docking and mooring boats. As a general rule of thumb, ten or more boats constitutes a marina although the definitions in use among the states vary widely. The full range of impacts associated with marinas and their assessment is beyond the scope of this paper. Highlights are summarized below. For full discussion, the reader is referred to the *Coastal Marinas Assessment Handbook* (USEPA 1990).

Discharges from boats moored at a marina are intermittent in nature making their impact difficult to quantify. In estimating the potential volume of sewage that could be discharged, three factors are important. These factors are the number of boats equipped with marine toilets, the number of persons occupying each boat equipped with a toilet, and the number of boats occupied. Boats possessing marine toilets may be equipped to discharge wastewater or may store the wastes in a holding tank. The capability to discharge, however, does not assure discharge. Boaters may elect to use dockside facilities or portable toilets which are emptied into onshore waste disposal facilities. These are the same options available to boaters using vessels which do not have marine toilet facilities.

Little documentation concerning the number of boats equipped with marine toilets is available. The Chesapeake Bay Recreational Boat Pollution Work Group (1990) estimates that only 20% or 40,000 of the more than 200,000 boats using the Chesapeake Bay are large enough to be equipped with marine toilets. Actual surveys have shown a small percentage of boats at individual marinas to be equipped with installed marine toilet facilities designed for overboard discharge. Ragged Point Marina in Virginia reports that of 89 boats surveyed, only 28 (31%) boats had installed marine toilets. Only seven (0.08%) of the 28 used approved marine sanitation devices which discharged overboard. The remaining 21 boats were equipped with holding tanks to collect the waste for later discharge into land-based vessel pumpout facilities (Barker 1990).

Information generated at Mystic Cove Marina in Connecticut shows a similar distribution (Nielsen 1990). Among the 71 boats surveyed at Mystic Cove Marina in Connecticut, only 46 boats (67%) were equipped with marine toilets. One boat was equipped with a marine toilet that discharged directly overboard. The other 45 boats (65%) with installed toilets used holding tanks. Fifteen boats (21%) used portable toilets transported dockside for sewage disposal, while ten boats (14%) had no toilet facilities of any type. At Cedar Island Marina, Connecticut, the distribution was similar. Among 167 boats surveyed, 24 boats (14%) were equipped with marine toilets that discharged overboard; 137 boats (82%) used holding tanks and 11 boats (7%) had no toilet facilities (Nielsen 1990). Again, these figures support the contention that few boats are equipped to discharge overboard.

Estimates of the number of persons occupying a boat vary. The Interstate Shellfish Sanitation Conference (ISSC 1990) recommends using two persons per boat. Most shellfish agencies nationwide use two persons per boat as their occupancy figure. The suggested EPA estimates vary from two to three persons per boat (USEPA 1984). Actual boater surveys in Maryland documented two persons per boat (Gibson and Arnold 1987), while surveys in Delaware support a general occupancy figure of 3.3 persons (Pingree 1990).

Estimating the number of occupied boats with toilets which discharge overboard is extremely difficult. In the absence of actual survey figures, the Interstate Shellfish Sanitation Conference recommends using the estimate that 50% of the boats are occupied (ISSC 1990). Based on the information generated in Delaware and Maryland, this figure appears to be highly conservative. Surveys in Delaware have established 6.5% of the slips in the marina as the potential sewage discharge rate for the average marina (Pingree 1990). In Maryland, Gibson and Arnold (1987) reported that an average of 9% of the slips in the marina had persons on board and, therefore, a potential for discharge. Another Maryland survey made in Anne Arundel County, an area with one of the highest Maryland concentrations of marinas, reported a 13% occupancy rate, i.e., persons present on the boats in the slips (Maryland Department of the Environment 1987).

Because of the intermittent nature of the discharge from each boat, the potential volume of waste discharged at any one time is difficult to quantify. Generally, the assumption is made that all occupied boats will discharge at one time. This assumption provides an estimate of a worst case situation that overestimates what actually happens. Some reduction in the estimated volume of discharge should be provided where restrooms are provided dockside, and vessel sewage pumpout facilities are available. Dilution of the estimated volume of waste is usually calculated using the volume of water available within the marina boundaries. This approach allows the agency to estimate whether the potential area of impact is confined to the boundaries of the marina or extends beyond the marina boundaries. Impact beyond the marina boundaries may require additional dilution calculations to establish the total area of impact. Dye studies can also be used to determine the extent to which water from a marina is dispersed over adjacent shellfish waters (Cubet Engineering Limited 1984).

Nationwide, waters within marinas are usually restricted to shellfish harvesting during the boating season as a precautionary measure, even though sewage estimates and water quality data may not support the need. Additional area beyond the marina boundary may be restricted at some locations depending upon the waste discharge estimates and the available dilution water. Shellfish resource within or in close proximity to these areas can be safely harvested during winter months in northern states after summer activity ceases. Adequate time must be allowed for natural shellstock cleansing prior to harvest.

Boats which are used as year round living accommodations provide another nonpoint source. These units should be evaluated as residences. In most states, house boats and large vessels with live-aboard accommodations are unregulated. Discharges at marinas and mooring site occupied by live-aboard boats can be controlled by requiring installation of holding tanks on the vessels or by requiring their connection to public sewer or on-site septic systems. Vessels specifically designed for use as houseboats are required to be connected to sewage systems in Maryland and in some New Jersey townships (Tudor 1985).

Information concerning the public health risk associated with bioaccumulation of toxic substances in shellfish in and around marinas is equivocal. Reported differences in shellfish tissue concentrations may be influenced by species differences (Garreis and Pittman 1981; Unger, Hall, and Sullivan 1990). Investigations by Hall et al. (1988) report significant increases in tributyltin levels in marina waters, suggesting that the potential for significant bioaccumulation by shellfish exists. Uptake, however, may also be influenced by factors such as marina location and configuration. A 1985 study in New Jersey marinas found that heavy-metal concentrations in hard-clam tissue was influenced by the marina flushing rate. No significant increases in heavy-metal tissue levels was found at the well-flushed marina site. At the moderately flushed and poorly flushed marina sites, significant uptakes by hard clams of lead, copper, and zinc were reported (Osborne 1990).

One of the most difficult nonpoint sources to evaluate is urban storm water runoff. In many older sewer systems, sewage and storm water runoff are collected and transported in the same sewers. These are known as combined sewerage systems. This practice results in the oversizing of sewers. Oversizing causes the settling of solids during dry-weather flow conditions. During wet weather, the flow is greatly enhanced by runoff and is too large to treat at the sewage treatment plant. Combined sewerage systems are usually designed so that the combined sewage normally bypasses the treatment plant when the flow reaches some selected value greater than the dry-weather flow. Initial combined sewer flushing following a rainfall event can have an unusually high solids content as compared to that found in separate sewers. The only corrective measure available is to rebuild the system with separate storm sewers and sanitary sewers. During sanitary surveys, combined sewer overflows should be assessed as significant intermittent sewage discharges.

Storm drains carry runoff, and yard and street drainage in urban areas. In rural areas, they may carry runoff from wooded areas, plowed fields, and roadways. The Northern Virginia Planning District Commission (1978) demonstrated fecal coliform densities, comparable to those associated with raw sewage discharges, in stormwater runoff from disturbed soils, shopping center parking lots, single-family dwellings and townhouse complexes. Olivieri, Kruse, and Kawata (1977) report that there are many literature citations describing high, total, and fecal coliform indicator organism densities in runoff from forested, agricultural, rural, suburban, and urban areas. These investigators also point out that many factors affect the levels of microorganisms in runoff and that the relationship between the factors is complex. The magnitude, duration, and intensity of the storm event are important. The flow and temporal discharge characteristics of the storm drain will vary with each storm event. The discharge from each drain will be affected by land use, topography, extent of impervious surface, soil characteristics and existing vegetation. All of these variables, together with other factors, too numerous to cite, provide wide variabilities in the densities of observed indicator organisms.

Relatively, little work has been done on the presence of human pathogens in runoff. Olivieri, Kruse, and Kawata (1977) attempted to provide information on the presence and concentrations of selected pathogenic bacteria and viruses in urban runoff from areas served by combined and separate sewers. The densities of enteric pathogenic bacteria and viruses were relatively low, although they were consistently recovered in urban stormwater. They concluded that "there is little justification based on benefit–risk considerations to warrant a program of disinfection of billions of gallons of urban storm runoff."

If the surveyor discovers a nonpoint source discharge at a property, he must make a decision as to its relative importance to the water quality in the shellfish harvest area. Three possible options exist. The discharge may have a direct impact, an indirect impact or no impact on the harvest area. A direct impact is any waste discharge which would immediately affect the quality of a shellfish harvest area (e.g., a failing septic system discharging within ten feet of an active harvest area). The potential for disease transmission is obvious. A small area restricted to shellfish harvesting should be established until the necessary repairs are complete and the septic system discharge is eliminated. The location of the property and its discharge must be recorded on the survey area map for inclusion in the final assessment.

An indirect impact is any discharge that has a secondary impact on the harvest area or might eventually reach the harvest area in a roundabout way. Some potential for impact exists but its occurrence is unlikely. An example is a failing septic system which drains to a roadside ditch. The ditch runs 500 yards and intersects with a fast-flowing stream. The stream enters the shellfish harvest area one mile downstream. The effluent generally lays in the drainage ditch but a minor rainfall may transport it to the stream. Based on the estimated volume discharged daily, the aging of the sewage in the ditch, the stream travel time and dilution in the stream under normal rainfall conditions, it is possible to determine that the sewage is aged and diluted to acceptable levels. The potential for the shellfish in the harvest area to concentrate pathogens, if any are present, from this discharge is negligible.

These properties should also be recorded on the survey area map for review in the final assessment.

A discharge is identified as having no impact if there is no way the discharge could reach the harvest area. An example is a failing septic system which drains onto a flat lawn. During a precipitation event, there may be some minor runoff to a drainage ditch. The drainage ditch, however, runs several blocks and stops, providing no drainage to another ditch or to a waterway.

BACTERIOLOGICAL EXAMINATION OF WATER QUALITY

The second major component of the sanitary survey is the bacteriological examination of water samples. The purpose of this component is to confirm the tentative conclusions derived from the shoreline survey that a harvest area is uncontaminated, moderately contaminated or grossly contaminated. The degree of contamination is equated to potential public health risk of disease transmission.

Diseases of concern associated with shellfish ingestion follow the anal–oral transmission route. Disease transmission requires that a susceptible host be exposed to a sufficient number of pathogenic organisms. The key elements in the transmission of microbial disease by a water route are exposure and number. The mere presence of a pathogenic microorganism is not sufficient to cause disease. The pathogen must reach the host, enter the host's defenses and proliferate to the degree necessary to cause disease. Historically, the coliform indicator group (total and fecal) have been used to characterize the potential risk for disease transmission by measuring the amount of fecal material present in the water column.

Any bacterial indicator used as part of the sanitary survey for the classification of a shellfish harvest area should have several features. The indicator should detect the presence of fresh, fecal contamination at relatively high dilutions; differentiate between microbial flora of public health significance and flora with little or no significance; be more resistant to water and wastewater treatment processes than the pathogens of concern; respond to natural conditions in the environment in the same or similar way as the pathogens of concern; be easy to isolate, identify, and enumerate; and correlate with the sources of pollution as determined by the shoreline survey (Dutka 1973; Hunt 1977; Cabelli 1978). To date, no single indicator group meets all of these requirements.

In order to understand the proper use of the coliform indicator group, it is important to understand the philosophy behind its use. The use of the coliform indicator is based on a series of assumed relationships between pathogenic organisms and other organisms (such as coliforms) associated with human sewage. The underlying premise is the assumption that since fecal material sometimes contains pathogens, the more fecal material in the water, the greater the risk of persons contracting disease while using the water or shellfish. The underlying quantitative relationships in this premise have not been validated, but have been widely implemented in public health programs (Pipes 1978).

Implementation of this premise has been to use an indicator organism (coliform group) to measure the degree of fecal contamination in water. A postulated chain of relationships is assumed to occur. The first relationship is that the more fecal material in the water, the higher the densities of fecal indicator bacteria in the water. The second is that a more or less stable ratio exists between the density of a particular fecal indicator and the density of a particular pathogen in sewage from a large human population. This ratio is a function of the incidence of disease in the contributing population. The third relationship is that the higher the density of the pathogens in the water being used, the greater the incidence of disease in the user population. Although accepted in practice and implemented, these implied quantitative relationships have not been verified (Pipes 1978).

Current research is demonstrating that these relationships do not hold true for the coliform, particularly in the open environment, in the absence of a *large* discharging human population such as that associated with the raw or poorly treated wastes from a large population center. There are many reasons. Both the total and fecal coliform groups contain organisms that are not intestinal in origin (Dutka 1973; Pipes 1978). In addition, coliform bacteria may multiply in the marine environment (Lear 1962; Dutka 1973; Pipes 1978). Furthermore, the coliform group does not distinguish between human wastes, the primary health hazard, and the substantially lesser risk associated with animal wastes (Presnell and Miescier 1971). Lastly, the coliform water quality criteria used to determine harvest water acceptability are not based on sound scientific studies that relate the level of coliform in the water column in the open environment to levels of pathogens or to potential for disease transmission. This is particularly true for viruses (Metcalf 1968; Berg et al. 1978; Melnick, Gerba, and Wallis 1978; Gerba et al. 1979; Metcalf 1979). Recent studies in recreational swimming waters have shown little correlation between ambient total or fecal coliform levels and the incidence of disease associated with swimming (Cabelli 1976, 1978; Dufour 1984; Dufour, Calderon, and Mood 1985).

Uncertainty concerning the validity of the coliform indicator group raises the question as to why indicators are used. The solution appears to be to test in the water column for the presence of the pathogens of concern and demonstrate clearly the presence or absence of disease risk. As early as the turn of the century, investigators acknowledged that efforts to isolate pathogenic microorganisms from water were not cost-effective. Prescott and Winslow (1904) noted that, with respect to isolation of specific pathogens, positive results are always open to doubt and negative results prove nothing. The wide variety of pathogens which may be present in contaminated waters and the low numbers of pathogenic microbes compared to the natural microbial populations severely limit the utility of pathogen enumeration. Although dramatic strides have been made in the science of microbiology since 1904, the isolation of specific pathogens is still not cost-effective.

A test for the presence of any pathogen in a specified volume of water (a sample) cannot prove the absence of the pathogen in the larger body of water from which the sample was taken. Therefore, the practice arose to use a surrogate microbial determination to indicate the presence of fecal contamination. This is an important point. The primary function of the surrogate microbial indicator is to indicate the *amount* of fresh fecal material in the water. Since fecal material sometimes contains pathogens, the qualitative assumption is made that the more fecal material in the water, the greater the risk of contracting disease to the people using the water. This qualitative assumption leads to the quantitative assumptions (chain of relationships) discussed earlier. Early investigators who established the foundation principles of both water supply and shellfish harvest area protection recognized the qualitative nature of data derived using the coliform bacteria. These investigators promoted the requirement for knowledge of the surroundings and local geographic area to properly interpret bacteriological data. Sanitary engineers, following the early guidance, have traditionally viewed the presence of the indicator group in water only as an indication of the need for a full survey to determine the origin of the coliform group and to establish the means for its elimination. Unfortunately, in the past two decades, these basic tenets have been forgotten and the presence of the indicator bacteria is taken as unrefutable evidence of a serious health risk.

Under the existing Interstate Shellfish Sanitation Program guidance (ISSC 1990), a state may elect to use either the recommended total or fecal coliform standard or both. How accurately the ISSC recommended total or fecal coliform standard measures the amount of fecal pollution is open to debate. The first official harvest area standard was the total coliform standard which appeared in the 1946 Manual of Recommended Practices for the Sanitary Control of the Shellfish Industry (Federal Security Agency 1950). There is no formal documentation of how it was developed but it is popularly believed that it is an extension of the work of Kehr et al. (1941) in Raritan Bay. The investigation of a typhoid fever outbreak associated with hard-clam consumption indicated that consumption of raw clams harvested from waters contaminated with diluted raw human sewage where the total coliform density did not exceed 70 MPN per 100 ml of water would probably not induce disease. The investigators were careful to point out that the recommended 70 MPN was essentially a compromise

> . . . between the potential dangers of intestinal infections as developed by the theoretical conditions presented and the indicated comparative rarity, and possibly complete absence, of actual infections due to shellfish from approved sources under present conditions. The recommendations are based upon meager data, and the recommended limiting standards should be more accurately evaluated by extensive studies at the earliest opportunity.

Extensive studies have not been performed to this day.

The extension of the suggested limit of 70 MPN to harvest areas nationwide has several weaknesses. First, it assumes that the typhoid organism can serve as a "model organism" for the transmission of disease through a shellfish vector. Second, it assumes that all species of bivalves experience the same uptake of pathogens. Third, it ignores the difference in bacterial die-off caused by salinity, temperature and other environmental factors in various harvest areas nationwide. Fourth, it assumes that the risk associated with diluted raw human sewage is the same as the risk associated with treated human sewage or with animal wastes. Fifth, it assumes that the coliform group originates only with human sewage. Today, many of these assumptions are questioned by scientists and regulators in the light of new scientific evidence.

The ISSC recommended fecal coliform standard is a statistical derivation from the 70 MPN total coliform standard. It was derived using bacteriological data from the waters immediately adjacent to shellfish harvest area restriction lines established using the coliform standard (Hunt and Springer 1974; Hunt 1977). Because of the imperfections of the coliform standard from which it was derived, the validity of the fecal coliform standard is also questionable.

The ISSC total and fecal coliform standards are assumed to provide equivalent measures of fecal contamination, but actual practice has not supported this assumption. In 1974, Maryland was able to reopen extensive areas to shellfish harvesting by converting from the recommended total coliform to the recommended fecal coliform standard. Data collected in Louisiana suggested that the fecal coliform standard was less protective than the total coliform standard (Robichaux 1975). Furthermore, preliminary analyses of bacteriological data from New Jersey has indicated that as much as 10–15% of the waters which meet the total coliform standard would not meet the fecal coliform standard (Eisele 1990). Hunt (1977) maintains that the equivalency holds only when measuring levels of diluted sewage related to point sources.

The validity of the total and fecal coliform recommended standards are further open to question because of problems inherent in the multiple-tube procedure used to determine the most probable number (MPN) value as a measure of coliform density. The multiple-tube procedure is performed using replicate tubes of a selected test medium inoculated with decimal dilutions of a water sample. The greater the number of replicates of each sample volume in a dilution series, the greater the test precision. Five tube decimal dilution tests are more precise than three-tube decimal dilution tests; therefore, the number of decimal dilutions selected can have serious consequences in meeting the standard. Furthermore, the MPN probability tables were designed to include a positive bias. This can result in the true sample density in the five-tube test being over reported by 23%. The MPN value is not a precise measurement and a given test may range from 31% to 289% of the absolute value (Pipes 1978). Because of the imperfections in the methodology, the ISSC-recommended standards use a median or geometric mean value based on 15 samples collected over 3 years. In addition, 10% of the samples may exceed the 90th percentile value established for the number of replicates used for the decimal dilutions in the MPN test. This practice acknowledges that the MPN test at intervals reports higher values than the median although densities not more than the median value are present (Hunt 1977).

As indicated earlier in this discussion, the coliform group (total or fecal) does not satisfy the criteria for a good indicator. As early as the late 19th century, bacteriologists recognized its deficiencies. Its presence in large numbers in feces and sewage, however, provided a useful tool for estimating the extent of human sewage contamination in water when the presence of coliform densities was interpreted in the light of the information provided by the shoreline survey. At this point, it is important to note that the qualitative and quantitative determination of bacterial indicator density was never intended to be the only criteria for the evaluation of the health hazard associated with the quality of a particular water body. Lacking any sophisticated technology, early workers depended heavily on the shoreline survey and field observations to evaluate sewage contamination of water supplies and shellfish waters. As late as 1944, the U.S. Public Health Service continued to stress the importance of the survey in its *Manual of Recommended Water Sanitation Practice*. This emphasis on the survey was reiterated in 1946 recommendations of the U.S. Public Health Service for the sanitary control of shellfish harvest areas (Federal Security Agency 1950). In 1977, Hunt emphasized the use of the coliform indicators as adjuncts to the shoreline survey in the evaluation of point sources.

This discussion is not intended to argue that the use of the coliform indicator group be abandoned, nor that the recommended ISSC standards be discarded. The discussion attempts to sensitize the reader to the need to interpret the bacteriological data with discretion based on the findings of the shoreline survey and the investigation of contaminant sources, particularly point sources associated with human sewage discharges. Failure to meet the recommended ISSC total or fecal coliform standard is not *prima facie* evidence of sewage contamination or serious health risk. Unfortunately, because of the manpower and resource demands of the shoreline survey, and the need for its interpretation using best professional judgement, administrative expediency has been advanced as the compelling argument for the ascendancy of a bacteriological standard. Consequently, the practice in many states today is to rely on the current recommended ISSC bacteriological standard as a "magic number" for evaluating shellfish harvest water quality.

Because the coliform group does not satisfy the requirement of a good microbial indicator and because of the problems inherent in the MPN test methodology, the health significance of the presence of the total and fecal coliform indicator organisms continues to be questioned by scientists and regulators. This is particularly true where the primary coliform source is animal sources (urban, agricultural, and wildlife). Available evidence suggests that human enteric viruses are not passed from animals to humans through shellfish. Human enteric viruses are currently the most prominent etiologic agents in shellfish-borne disease (U.S. General Accounting Office 1984). High densities of coliform indicator organisms in water, particularly following runoff events, in watersheds with little or no human sewage discharges, has resulted in thousands of acres of shellfish harvest areas being restricted.

National concern has been expressed that the coliform indicator organism is unreliable. In some cases, its use may afford no reliable estimate of the potential for viral disease transmission. In other cases, its use may overestimate the potential health risk associated with nonhuman, nonpoint source runoff. Because of the national concern with reliable estimate of health risk and the economic consequences of erroneous assessments, research is needed to resolve the question of potential public health risk. A collaborative epidemiological study by the U.S. Environmental Protection Agency (EPA) and the U.S. National Oceanic and Atmospheric Administration is currently investigating the relationship between microbial indicators and incidence of disease. The initial phase uses sites affected by sewage treatment plants (human point sources) (Dufour 1990). In 1987, Congress appropriated the first year funding for a proposed National Collaborative Shellfish Pollution Indicator Study. The study will examine several sites nationwide affected by human point and nonpoint sources. One objective of the study is to validate the relationships between indicators and disease through epidemiological studies. This validation will provide much needed information to help bridge the gap in our understanding of shellfish-borne disease transmission. An underlying misunderstanding, particularly in the shellfish industry and the regulatory agencies, is the belief that these studies will lead to a single, universal indicator of acceptable shellfish-harvesting water quality. No single indicator by itself will disclose all the potential hazards associated with shellfish harvest waters, or provide all the information needed for harvest area classification, surveillance against potentially hazardous situations, or investigation following a disease outbreak associated with shellfish consumption. The best that can be hoped for is a suite of indicators that may be used concurrently for more definitive interpretation of shoreline survey findings and provide a dose–response relationship for some of the more common shellfish-associated diseases. Because of the variability in conditions from location to location, however, the final decision will continue to be based on best professional judgement.

Despite their many shortcomings, the total and fecal indicator groups can be used to provide useful information as part of the sanitary survey. Total, and particularly fecal coliform densities can be used to track the dispersion and dilution of certain sewage effluents in shellfish harvest areas. The utility of these organisms is directly related to the degree of waste treatment provided. The greater the degree of waste treatment and disinfection, the less useful the coliform group. Collection of bacteriological data from ditches, streams and similar systems under dry-weather and wet-weather conditions can provide estimates of coliform loading to harvest areas. These estimations can be related to sewage discharge through the shoreline survey and can be useful in determining relative source contribution to water quality. The results from point sources or nonpoint source effluent testing for the coliform group can be used with appropriate dilution calculations, bacterial die-off curves and models to approximate source impacts on harvest areas. Repeated testing for the presence and density of coliform organisms, particularly in the portions of harvest areas where approved and restricted waters converge, can be useful in establishing the boundary separating the two area classifications.

Repeated water column sampling at a series of fixed locations can also be used to substantiate the findings of the shoreline survey. This approach is most effective with point sources. In areas where nonpoint sources dominate, the coliform indicator group frequently provides mixed or false signals concerning the amount of serious sewage contamination reaching the harvest area. Periodic sampling at fixed locations can also be used as a quick check between surveys to ferret out the introduction of new sources which may offer potential health risks.

The selection of the location of the sample stations is critical to effective use of the coliform indicator. Sampling locations should be selected to monitor critical point sources, the effects of normally occurring adverse meteorologic and hydrodynamic conditions, and the quality of the water in active harvest areas. Information obtained during the shoreline survey should be used to site sampling stations. Periodic adjustments to sampling sites should be made as new or expanded point sources occur, hydrologic regimes change, or other contaminant sources are introduced.

MARINE BIOTOXINS AND NATURALLY OCCURRING INFECTIOUS AGENTS

In some states, certain shellfish harvest areas are subject to periodic contamination by marine biotoxins. Marine biotoxins are produced by certain naturally occurring dinoflagellates. These dinoflagellates are normally present as resting cysts on the marine sediments. When certain environmental conditions are right, the cysts suddenly vegetate or "bloom" and begin increasing in numbers. The exact conditions that cause the sudden blooms are not clearly understood. Several years or several months may elapse between bloom episodes. The blooms seem to be associated with certain salinity and temperature conditions as well as bottom upwellings. Upwellings bring the cysts to the surface and expose them to light which probably stimulates growth. The day length (photo period) may also play a role in fostering algae blooms. When the unusually dense concentration of dinoflagellates associated with a bloom occurs, the water often appears reddish brown, causing "red tides".

Over 60 species of dinoflagellates are known to cause red tides. Most red tides have no direct effects on humans. Only about 30 dinoflagellate species produce toxins. Of the 30 species, only a few contaminate shellfish (Texas Sea Grant Program 1986). The two major concerns in humans are neurotoxic shellfish poisoning (NSP) and paralytic shellfish poisoning (PSP). NSP causes nausea, dizziness, tingling sensations in the extremities, dilated pupils and hot–cold reversals. The symptoms generally disappear in a few days. PSP has a more severe onset of similar symptoms and can cause death within 2 or 3 h of shellfish ingestion.

As filter feeders, bivalve shellfish strain dinoflagellates as food from the waters and concentrate the biotoxins in their visceral organs. Because bivalves filter large quantities of water, the shellfish may become contaminated with concentrated toxin when the density of dinoflagellates is insufficient to form a visible red tide. Since the occurrence of the dinoflagellate blooms is unpredictable and because the toxins can be only detected by mouse bioassay or chemical analysis, states with historic marine biotoxin problems have elaborate monitoring programs to detect these algae blooms in their early stages.

State shellfish agencies, with affected harvest waters, develop a marine biotoxin control plan for each harvest area. This plan establishes the administrative procedures to be followed in identifying the public health risk, surveillance strategies and control procedures. An early warning monitoring program which uses representative locations is maintained to predict peak periods of algae bloom. Representative shellfish samples, other reliable biological specimens, and certain environmental indices may be used individually or in combination in this monitoring program. When the shellfish sample assays show the concentration of paralytic shellfish poison equals or exceeds 80 μg per 100 g of the edible portion of raw shellstock, the area is closed to shellfish harvesting. The findings of neurotoxic shellfish poison at any detectable level also results in the area being closed to harvesting. The area remains closed until the biotoxin levels have subsided to acceptable levels.

In determining whether an area may be reopened to harvesting following an episode of marine biotoxin contamination, the shellfish agency weighs many factors. Repeated testing of shellstock must demonstrate that the toxin levels in the shellfish have declined to levels acceptable for human consumption. A general rule of thumb is three acceptable samples over a 14-day period. Shellfish samples from adjacent areas are evaluated to determine if these levels are increasing or decreasing. Environmental factors such as water temperature, bottom sediment upwelling and the number of toxic cysts in the sediments are evaluated to determine if environmental conditions can be expected to remain stable.

Episodes of paralytic shellfish poisoning or neurotoxic shellfish poisoning have been documented in New England, Canada, the Gulf states, and the West Coast states. Because of the seriousness of this type of contamination, shellfish agencies react swiftly to close harvest areas when biotoxin levels increase in shellfish meats. The management of areas affected by marine biotoxins is an extensive and expensive responsibility for shellfish control agencies. Maine, for example, spends over $150,000 annually for biotoxin monitoring activities. This cost is in addition to the $216,500 spent on the more traditional shellfish sanitation activities (Leonard, Broutman, and Harkness 1989).

Pathogenic vibrios are marine bacteria which occur naturally in the environment and which are capable of inducing disease in man when transmitted through a shellfish vector. The frequency of occurrence is generally very low and usually affects persons with compromised immunological systems and medical disorders.

Non-O1 *Vibrio cholerae* are widely distributed in the environment. They are found most frequently in estuarine waters with salinities ranging from 0.4% to 1.7% (Colwell and Kaper 1978). Almost all cases on non-O1 Vibrio in the United States have been associated with raw oyster consumption. Although isolated frequently from many seafoods, the reported incidences of seafood-associated illness is very low (Hackney and Dicharry 1988).

Vibrio vulnificus is widely distributed in the environment and has been isolated from most estuarine waters of the coastal United States. The occurrence of *V. vulnificus* infections associated with shellfish most commonly occurs with the consumption of raw oysters originating in the warm weather months (April to October) from the Gulf of New Mexico states. Several cases of death have been reported in the United States in the past ten years (FDA 1988).

There are presently no known public health measures which can limit the harvesting of shellfish to those areas free from *V. vulnificus* and non-O1 *V. cholerae*. Good refrigeration practices following harvest and during the handling and transport of shellstock and rapid transport of the shellstock from the harvest area to the consumer, are believed to be the best available controls.

SHELLFISH EXAMINATION

The third component of the sanitary survey is the examination of shellfish meat to assure its wholesomeness. This component has two broad categories, microbial and chemical. No national recommended criteria exist for microbial contaminants at the shellfish-harvest level. The Interstate Shellfish Sanitation Conference (1990) only requires shellfish meats to be routinely tested when shellfish from restricted areas are subjected to natural or artificial purification processes. The purpose of the testing is to establish existing coliform concentrations in the meats as a comparison level to evaluate whether the purification process has achieved a reduction in indicator organism levels.

Few states routinely test their shellstock as an indicator of the appropriateness of harvest area classification. In Maryland, shellstock is both randomly and systematically collected from approved harvest areas and tested for fecal coliform concentrations. This testing is used to monitor harvest area conditions, as the bivalves' ability to bioconcentrate microorganisms provides a better indicator of long-term harvest area water quality than an instantaneous grab water sample. Shellfish meat testing can also be used to provide an effective means of verifying the appropriateness of a closed safety zone. Low concentrations of fecal coliform in shellstock collected around the perimeter of a closed safety zone can be used to confirm its size and adequacy. Shellfish meat tissue concentrations can be used in a similar manner to verify the boundary line separating approved and restricted harvest areas.

Evaluation of chemical contaminants in shellfish meats requires the establishment of a long-term monitoring network. In establishing the network, careful consideration should be given to what toxic substances will be tested.

Bioaccumulatable substances have the most important public health significance. Bioaccumulatable substances may be defined as those substances which accumulate in shellfish tissue over time. These toxic substances can have public health significance if, at certain concentrations, they induce systemic or carcinogenic effects in man. Examples of bioaccumulatable substances are chlordane, polychlorinated biphenyls (PCBs) and methyl mercury.

Shellfish agencies should use the information gathered in the point and nonpoint source evaluation to help select the important chemicals to be included in tissue monitoring. Substantial cost savings in the monitoring effort can be achieved if some forethought is given to the substances to be monitored. If certain toxic substances are not present in discharges, or are discharged at levels too low to be bioaccumulated, there is no need to include the substances in the monitoring network. If certain toxic substances are not bioaccumulated to any appreciable extent by shellfish, they also can be eliminated from the network. This point can be an important consideration with certain classics of substances, e.g., pesticides. Unlike historic pesticides such as DDT, chlordane and dieldrin, the new pesticides are generally chemically engineered to preclude bioaccumulation. In establishing the network, the shellfish agency should identify the most heavily used pesticide in the watershed and whether these pesticides have a tendency toward bioaccumulation. Federal agencies such as the Environmental Protection Agency, the Fish and Wildlife Service, and the National Oceanic and Atmospheric Administration, as well as the epidemiologists from state health departments can assist with this effort.

On the other hand, the shellfish agency may wish to include certain historic toxic substances in its first monitoring efforts. PCBs, chlordane, DDT, dieldrin, toxaphene, and other substances have been banned in use. Their residues, however, are extremely persistent in the environment. Although nationwide trends are showing a sharp decline in some of these substances in the environment (Schmitt, Zajicek, and Peterman 1990), the shellfish agency may wish to confirm that these substances are not a problem in a particular harvest area.

The species of bivalve inhabiting the harvest area is an important consideration in selecting the chemicals to be monitored. Garreis and Pittman (1981) showed a marked difference in bioaccumulation of heavy metals, between the eastern oyster and the soft-shell clam. Unger, Hall, and Sullivan (1990) have shown an order of magnitude difference in bioaccumulation of tributyltin between soft-shell clams and the American oyster.

Once the chemicals to be tested are identified, a determination must be made as to the availability of test procedures and the capability of laboratories to perform the analyses. What concentrations of each toxic substance are acceptable should also be determined in advance. These are not minor points. The availability of laboratory methodologies to extract toxic substances from shellfish meats is very limited. Standard methods do not exist for most chemicals. Where recognized methods exist, their performance may be beyond the capability of most laboratories. For example, only five commercial laboratories nationwide can reliably extract dioxin from fish tissue. The availability of methods and laboratories, and the expense of the analyses plays an important role in establishing the network.

Acceptable concentration levels in foods, particularly shellfish, has not been established for most toxic substances. The United States Food and Drug Administration (1985) has established food tolerance levels in shellfish for only 15 toxic substances. Estimates of acceptable concentrations for certain other toxic substances can be gleaned from the recommendations of the World Health Organization (1972, 1977, 1981). The epidemiologists at state health departments can also assist with this effort.

The location of sampling sites within the network should be carefully selected to reflect the chemical contaminant loadings associated with point and nonpoint source discharges. In areas affected by point sources, the hydrodynamic studies describing the dispersion of the discharge plume can be very useful in station selection. In areas where the sources are primarily nonpoint, stations should be associated with key drainage areas.

When the locations of the sample sites are selected, the sites should be tested in the shortest time frame possible. Results of the first tissue contaminant analyses form the baseline for evaluation of area acceptability. Sites with chemical concentrations in excess of acceptable levels should be immediately resampled in an intensive pattern to determine the extent of the contamination. Information from sites with acceptable concentrations serve as a benchmark for comparison with future monitoring results to document trends in the increase or decrease of toxic substance accumulation in shellfish tissues.

Once the baseline is established, the frequency of future monitoring should be decided. This decision is a judgement call influenced by the baseline information, resources and manpower. A good rule of thumb for areas where no tissue data exists is to monitor annually or semiannually for three years. If no discernible increases in tissue concentration are identified in the three-year period, then monitoring may be reduced to biennially or triennially. In areas with major municipal or industrial point sources, quarterly sampling may be appropriate until the shellfish agency is satisfied that bioaccumulation is not occurring.

Historically, chemical contamination of shellfish meats has not been a serious problem in shellfish harvest areas. Whether this is a reflection of the national program emphasis on microbial contaminants, or the lack of resources in the states for this type of sophisticated monitoring, or actual national environmental conditions, is not known. Generally, shellfish agencies have responded to site-specific chemical contamination problems. An example is the 1970–1971 mercury contamination in Lavaca Bay, Texas. A nationwide survey of possible mercury contamination in seafood products was conducted in 1970 by the Food and Drug Administration (FDA). Oysters from Lavaca Bay in Texas showed high mercury levels in excess of the FDA food tolerance level for mercury of 1 ppm. Subsequent intensive sampling in 1970 by the Texas Department of Health showed mercury levels in oysters which ranged from < 0.1 to 8.0 mg/kg.

Over 11,000 acres of Lavaca Bay were restricted to shellfish harvesting because of mercury contamination. The mercury source was industrial wastes discharged by the Aluminum Company of America (ALCOA). ALCOA changed their discharge practices to retain their mercury wastes in settling ponds built on company property. The oysters showed a marked decrease in mercury levels once ALCOA stopped its discharge in July 1970. Oyster samples collected in 1971 ranged from < 0.1 to 0.66 mg/kg. In the fall of 1971 an area of the Bay reopened to shellfish harvesting (Garreis and Pittman 1981).

Very few states have long-term monitoring programs for chemical contaminants in shellfish meats. In 1986, the Florida Department of Natural Resources initiated an environmental monitoring program to detect heavy-metal loadings in shellfish meats in its harvest areas (Heil 1986). The findings were inconclusive as to potential health risk. The program has not been continued because of the lack of financial resources (Thompson 1990). The current Maryland program, which has been in place since 1974, substantiates the contention that overall chemical contamination is not a serious problem in shellfish waters. This network monitored 85 to 120 stations annually until 1987. Shellfish meats were analyzed for nine heavy metals and 17 chlorinated hydrocarbons. The continuous 13 year record showed no widespread levels of concern and a minor increase in cadmium contamination at one location. Monitoring efforts have subsequently been reduced to once in every three years (Murphy 1990).

The California State Water Resources Control Board (Stevens 1988) uses a somewhat different approach. Since 1977, the quality of California coastal waters, bays and estuaries, have been monitored, using clams and mussels as indicators. These long-term monitoring activities comprise the California State Mussel Watch Program. Mussels (*Mytilus* spp.) are used in saltwater and estuarine environments and clams (*Corbicula fluminea*) are used in freshwater and brackish waters. The data base is used to pinpoint areas of higher than normal levels of certain toxic pollutants, and to provide a basis for follow-up studies and regulatory actions. In addition, the data are used to track changes in levels of selected pollutants over time and distance. Presumably, the data would also be used to assess public health risk in areas of commercial or recreational shellfish harvesting.

DATA ANALYSIS

When the field work and the information gathering for each of the three components is complete or nearly complete, the final steps in data integration can be taken. The process is best accomplished in phases which test tentative conclusions. The initial step is to locate all sources of pollution which impact the shellfish harvest area on a single-area map of the watershed. If the shellfish agency has been performing preliminary data assessments as recommended in the record keeping portion of this paper, this exercise can be completed fairly rapidly. During this exercise, the distribution of wastes from each point source is plotted on the harvest area map using any results from effluent testing, dye studies or similar tests, and models. The impact from any intermittent discharges should be included in this exercise as well as any changes in the magnitude of impact caused by climatic events.

Nonpoint sources (NPS) are also located on the master map and the impact of their discharges plotted. Location of NPS which discharge directly to the harvest area are always mapped. Information concerning indirect discharges is reviewed, and may or may not be plotted, depending upon whether or not there is an impact on the harvest area. Once the information from the shoreline survey is plotted, a picture of the dispersion of contamination in the harvest area should emerge. Based on this information, a tentative determination of the area affected and the degree of impact can be made. Based on this determination, tentative conclusions concerning the microbial or chemical quality of the harvest area water can be reached and preliminary restriction lines may be established, if necessary. The validity of these preliminary conclusions may be tested using the median or geometric mean values from the bacteriological analysis of coliform densities in the water column or in shellfish meats. Where wastes which are not sewage-related are involved, the results of chemical water column or shellfish meat analysis should be used.

If the data from the microbial and chemical analyses support the conclusion derived from the evaluation of the shoreline data, the shellfish agency can be reasonably confident that the water quality assessment is correct. When the shoreline survey data and the microbial or chemical data do not agree, it is frequently useful to review and reassess the information gathered in the shoreline survey. In some cases, field visits may be helpful. If this reassessment continues to suggest negligible amounts of serious contamination, particularly where nonpoint source runoff is involved, the circumstances under which the microbial and chemical data were collected should be reviewed and the test protocols assessed. If that review shows no discrepancies, then the agency must rely on best professional judgement in weighing the relative importance assigned to the information generated in each of the sanitary survey's components.

Some typical situations where best professional judgement is required are provided as examples. If the shoreline survey indicates that negligible amounts of sewage contamination reach the harvest area, particularly where nonpoint source runoff is involved, but coliform densities marginally exceed recommended standards, the shellfish agency may determine that the bacteriological indicator data is overestimating the amount of serious sewage contamination reaching the harvest area. The area would remain open to harvesting. On the other hand, the shoreline survey may indicate that serious sewage contamination reaches the harvest area but is not reflected in the coliform densities because of its intermittent nature or other factors. In this case, the shellfish agency would restrict the area to harvesting. Best professional judgement in these situations often involves the combined factors of training and experience as well as a historical perspective on the harvest area involved.

The exercise of best professional judgement often requires the weighing of the relative potential health risk associated with various impacts in the harvest area. The rule of thumb which can be used in determining relative health risk (ranked from most important to least important) is described below.

Environmental Sources

a. Point Source Dischargers	
Marine biotoxins	Highly significant risk as toxins may cause severe illness and death
Vibrio vulnificus	Highly significant risk to immunologically compromised individuals
Private/municipal sewage plant	Most significant risk because of diverse contributing human population and volume; type of treatment important
Industrial waste sources (meat processing plants, etc.)	Significant risk if wastes involve pathogens capable of causing human disease, or chemicals which can be bioaccumulated; important primarily because of volume of wastes

Environmental Sources	
Combined sewer overflows	Significant risk because of untreated human waste contribution and volume
Animal feedlots/poultry houses	Potential human risk because of *large* aggregation of animals and ability of some domestic animals (pigs, fowl, cattle) to transmit human diseases
b. Nonpoint Source Dischargers	
Failing septic systems	Potential risk due to presence of human pathogens; considerably less risk than point sources because of lesser volumes and diffusion of source; highly site-specific, localized effect
Waste discharges from boats	Potential risk due to possible intermittent discharge of small quantities of raw sewage
Storm drains, street runoff	Potential risk because of some potential for human sewage discharge; risk significantly less than combined sewers
Rural land with domestic animals (farms, pastures, etc.)	Significantly less risk than direct human sources
Forest, marsh (dominated by wild animals, wild birds, and other "environmental sources")	Significantly less risk than human sources until proven otherwise

The application of these rules of thumb vary in some circumstances because of certain site-specific constraints.

SHELLFISH HARVEST AREA CLASSIFICATION

There are three primary classifications of shellfish waters: approved, restricted, and prohibited. Shellfish harvested from waters classified as approved are sold directly for human consumption (direct marketing). In restricted areas, shellfish waters do not meet the approved area criteria consistently, but shellfish may be harvested if subjected to an effective natural or artificial purification process. Harvesting of shellfish from waters classified as prohibited is not allowed at anytime.

The Interstate Shellfish Sanitation Conference recognizes three types of restricted areas: conditionally approved, restricted, and conditionally restricted. Conditionally approved areas are closed when coliform concentrations exceed the approved area criteria, but are available to harvesters when the coliform concentrations are within acceptable limits. Conditionally approved areas are usually associated with predictable events. One example is the conditionally approved area in Narragansett Bay. Shellfish harvesting is restricted for five days following any rainfall greater than or equal to one inch because of the combined sewer overflows into the shellfish waters. Seven days is the time frame established for effective natural purification. The area automatically reopens to harvesting on the eighth day (Speaker 1990). Another example is the conditionally approved areas in Maryland. These areas are restricted for three days following any one inch rainfall because the runoff from forests, fields, and plowed agricultural lands cause the coliform levels to marginally exceed the ISSC recommended standard.

A restricted area is an area where shellfish waters do not meet approved area standards. Shellfish may be harvested if subjected to a suitable, effective purification process before being sold for consumption. The purification process may be natural or artificial. Natural purification, generally known as relaying, consists of transporting bivalves from restricted waters to approved harvest areas. The shellfish are placed overboard either on the bottom or in specially designed containers. The shellfish cleanse themselves of microbial contaminants by pumping clean water. Fourteen days is the usual amount of time allotted for this process.

Artificial cleansing or depuration involves the use of tanks, closed water systems and treated water. Shellfish are held under conditions approximating natural for 48 h to cleanse the animals of microbial contaminants. The animals are tested for coliform levels to determine the effectiveness of the process before being released for human consumption.

A conditionally restricted area is a harvest area which is prohibited when the coliform concentrations exceed the criteria acceptable for restricted areas. When the waters meet the restricted area criteria, bivalves may be harvested for use in relaying or depuration operations.

Based on the public health risk assessment generated through the sanitary survey, each harvest area is assigned a classification. The classification dictates the limits placed on the shellstock use and the regulatory controls established to assure that the limits are met. For example, prohibited harvest areas are patrolled to assure that no illegal harvesting occurs. Restricted areas are patrolled and controlled to assure that shellstock is harvested and transported to appropriate purification sites and is not sold directly to the public. The objective of the classification system is to allow safe, public use of the resource.

PUBLIC HEALTH RISK ASSESSMENT

In determining the proper classification for a shellfish harvest area, there is only one practicable process on which to base judgement concerning the probable freedom of shellfish harvest areas from the potential presence of pathogenic organisms. This process, the sanitary survey, involves the building of a careful inventory of all actual and potential sources of pollution which may affect the area, and the careful weighing of the probability that natural purification will protect shellfish harvest areas. The conclusions reached in this manner can be tested by the evaluation of a series of bacteriological samples of water and shellfish or the chemical analyses of shellfish meats.

Unfortunately, the tools available for this assessment are no better today than they were in the 1920s and 1930s when the foundations of the drinking water and shellfish protection programs were established. Early workers, lacking sophisticated technology, depended heavily on the sanitary survey to evaluate the amount of fecal contamination in drinking water supplies and shellfish-harvest waters. Because of the poor sanitation practices and the endemic diseases in the American public at that time, these workers emphasized the control of human sewage wastes. Efforts were directed toward the identification and control of serious contamination of human origin. Judgements were made based primarily on shoreline survey findings, and harvest areas were approved or restricted for use. Significant declines in shellfish disease transmission occurred. Today, we can detect and enumerate more microorganisms with finer resolution and can measure more precisely certain characteristics of wastewater and its dispersion. Increasingly sensitive analytical methods have been developed to detect smaller and smaller concentrations of chemicals in a variety of media. We still lack, however, the bacteriological, chemical, or physical indicator system that provides a documented reasonable association with disease transmission through shellfish consumption. Implicit in this indicator system assumption is the concept that the indicator will allow us to determine "acceptable risk."

Acceptable risk is a difficult concept to define. We are frequently at risk in whatever we do. Sometimes risk is obvious and sometimes it is unsuspected. At other times, we refuse to consciously acknowledge the consequences of the risk, such as is the case with the confirmed smoker or the individual who refuses to use seat belts. No protective mechanism derived by man can protect him against all contingencies. The society collectively, at a conscious or unconscious level, makes a decision as to what level of protection is acceptable based on a wide range of factors including scientific information, availability of technology, economic welfare, and political feasibility.

Whether we choose to acknowledge it or not, we intuitively know that there are very small, finite risks associated with all food and water consumption. Despite the numerous controls in place, and the many chemical and microbial tests used to determine quality, a small chance exists for transmission of disease even when all indicators and systems are performing within acceptable limits. This chance is greater for certain high-risk groups within any population, and in the case of microbial-induced diseases, it increases during enteric disease epidemics where potential exists for raw sewage wastes to come into contact with food or water.

Risk assessment is the characterization of the probability of potentially adverse health effect from human exposures to environmental hazards. It involves the collection, organization and evaluation of data characterizing the risk. It strives to formulate formal quantification of probability and uncertainty. Uncertainties arise at each stage of the process, and often an assumption is required to fill the information gap. Specific rules are followed in characterizing the assumptions in calculating exposure, in identifying sensitive populations and in assessing the impact of multiple exposures. Risk assessment studies seek to provide quantitative measures of risk levels, where risk refers to the possibility of uncertain adverse consequences.

The sanitary survey is a primitive form of the more formal risk assessment process used in estimating the consequences of exposure to toxic substances. Generally, the sanitary survey exhibits crude and simplistic estimates of the magnitude, likelihood, and time distribution of contaminants, especially fecal material, in the environment. It is a structured attempt to deal with masses of data, which are often unstructured and contain substantial uncertainties. The use of the more formal assessment process in characterizing the potential for disease transmission and in assisting shellfish agencies in establishing levels of acceptable risk is severely hampered by the existing available indicators of sanitary quality. No quantifiable relationship has been established between the density or concentration of the current indicators in shellfish waters or shellfish meats and the potential human health risks involved. Cabelli (1978) provides an excellent perspective on this issue in his arguments for new standards for risk assessment associated with enteric bacteria.

Although no quantifiable relationship has been established between the coliform indicator group and the risk of sewage-related disease in shellfish waters, there has been a marked trend among shellfish agencies since the late 1960s and 1970s to use the coliform index organism as the sole arbitrator of the quality of shellfish waters. Many state shellfish agencies have no full sanitary surveys to support the classification of their shellfish waters (Broutman and Leonard 1988; Leonard, Broutman, and Harkness 1989).

Extensive reliance is placed on meeting the current ISSC coliform standards (1990) as the only measure for shellfish harvest area quality. This practice, still encouraged and endorsed by the Food and Drug Administration, is beginning to change.

Both scientists and regulators have voiced concern that the coliform indicator group is unreliable. In some cases, its use may afford no reliable estimate of the amount of serious human fecal contamination reaching harvest areas (Richards 1985). Underestimation of risk can have serious consequences. For example, 278 persons, in Texas, were found to have oyster-associated hepatitis in 1973. The oysters came from areas identified as meeting the criteria for approved harvesting areas as measured by the coliform indicator group (Portney et al. 1975; Mackowiak, Caraway, and Portney 1976).

On the other hand, many regulators and scientists believe that where animal sources (urban and agricultural runoff and wildlife) of fecal coliform exist, the use of the coliform group substantially overestimates risk, resulting in unnecessary restrictions. Coliform levels from wildlife have caused 29% of the restrictions in the Southeastern United States (Broutman and Leonard 1988). Agricultural runoff was a factor in 36% of the restrictions in the Southeast and 17% in the Mid-Atlantic. Overestimation of risk can have important direct and indirect social and economic consequences. First, it unnecessarily deprives the public of the use of an important food resource. Second, it may cause economic hardship, particularly in small towns and villages where shellfish harvesting is an important industry. Third, it causes the public to perceive extensive tracts of water where no contamination may exist as "polluted".

The primary reasons given by most shellfish agencies for their reliance on the coliform indicator is lack of manpower and resources to do the full survey. This situation is changing because of recent intense scrutiny of the reasons for harvest water classifications and their social, economic, and political consequences. Some shellfish agencies are receiving more resources to do full surveys (Broutman and Leonard 1988). Political pressure brought to bear by citizens concerned over water quality and the affected industry is resulting in the upgrading of shellfish agency programs. Some funding at the national level is being made available to perform the necessary epidemiological studies. Taken together, these efforts will improve the climate for sound risk assessment in shellfish harvest waters.

Since the development of the correct type of indicator system will take many years, what do we do in the meantime? The best solution is the proper performance of the sanitary survey and the use of the coliform indicator as a part of that survey. Regulators must recognize and accept the fact that existing coliform criteria for shellfish harvest waters is best used in situations involving the discharge of treated or untreated wastes associated with large human populations. These situations most closely approximate the conditions under which the existing criterion were derived.

Regulators should apply the existing recommended coliform criteria as they were intended to be used, i.e., as criteria, not standards. A criterion designates a means or rule by which a condition is assessed to form a correct judgement concerning the situation. Unlike a standard, it carries no legal connotation of immutability, but requires the use of judgement in its application. Criteria are quantitative, but not inflexible, and depend on considerable exercise of best professional judgement.

Shellfish agencies need to recognize that the current coliform indicators are inadequate for assessing most sources of nonpoint source pollution. In the evaluation of nonpoint sources, best professional judgement is the final arbitrator. This concept is difficult for many regulators to accept because it requires them to relinquish the use of "magic numbers". This approach also requires a much more substantive manpower effort to arrive at a sound judgement.

Regulators need to assure that sound sanitary surveys are made in all shellfish-harvest areas. Only after a careful, thorough inventory of all the actual sources of pollution which affect the harvest area or contribute indicator organisms can the shellfish agency weigh the potential health risk associated with the use of the area. Until a quantifiable relationship between some microbial, chemical, or physical indicator and the probability of disease is established, shellfish agencies must continue to use the evidence generated through the sanitary survey and best professional judgement to assess the probability of health risk associated with shellfish consumption from a particular water body.

SANITARY SURVEY DIVIDENDS

In many states, sanitary surveys are not performed or are done to a limited degree. Broutman and Leonard (1988) reported that no Gulf states had completed sanitary surveys of all their waters by 1985. Leonard, Broutman, and Harkness (1989) documented the lack of sanitary surveys for shellfish waters in several East Coast states. The primary reasons given were financial and manpower limitation.

Under the Interstate Shellfish Sanitation Conference guidelines (1990), any shellfish harvest area which does not have a current sanitary survey must be classified as prohibited to harvesting. This requirement has led to thousands of acres of shellfish harvest waters being restricted nationwide. Broutman and Leonard (1986) report that approximately one-fourth of the prohibited classification acreage shown in the 1985 National Shellfish Register of

Classified Estuarine Waters was classified as prohibited because of a lack of sanitary surveys. This lack can have important ramifications for public use of the resource. In Connecticut, where the state has not done the required shoreline surveys, the newly developed oyster industry is unable to harvest oysters from waters believed to be available if the proper surveys were done (Leonard, Broutman, and Harkness 1989).

The importance of the sanitary survey in shellfish water protection cannot be overemphasized. In many states, however, the data are not used to its fullest advantage. In addition to its use in providing the necessary information for shellfish water classification, the sanitary survey data can be a powerful tool for recovering valuable shellfish resources and improving water quality. Information generated in the survey identifies the contaminant sources responsible for restrictions on shellfish harvesting and certain types of water pollution. Mitigation or elimination of these sources can result in substantial recovery of shellfish resources for public use. Too often, the shellfish agencies consider their function complete when the area classification is established.

States, which extend the use of the sanitary survey to include the minimization or elimination of contaminant sources, have made impressive improvements in shellfish waters. By aggressively using the survey data as demonstration of the need to correct failing septic systems and upgrade sewage treatment plants, Maryland was able to reduce its restricted waters from 28% of its harvest area in 1972 to 6% in 1977. In 1973, Maine began a project together with the U.S. Environmental Protection Agency to identify and correct sewage violations affecting shellfish waters. Improvements made in 1980–1981 under this program resulted in the harvest of soft-shell clams worth more than $600,000 (Leonard, Broutman, and Harkness 1989). At Dauphin Island, Alabama, 774 acres of shellfish harvest waters were upgraded from the prohibited classification after a new sewage treatment plant replaced septic systems (Broutman and Leonard 1988).

Sanitary survey data can be used to engender citizen support for correcting contaminant sources. In Maryland, sanitary survey data has been used to obtain high-priority funding for areas with failing septic systems and wastewater treatment plant upgrades. In Massachusetts, the Shellfish Bureau is using sanitary survey data to educate citizens as to why harvest waters are restricted for harvesting. This approach has resulted in the funding of one, and possibly a second, wastewater treatment plant with a high degree of treatment and reliability to replace failing septic system problems (Hickey 1990). The same information can also be used to persuade water quality agencies to seek corrective action against dischargers, or to implement remedial actions to correct failing septic system and agricultural waste problems.

The dividends from actively seeking remedies to minimize or eliminate the discharge identified in the sanitary survey extend far beyond shellfish water protection. Maintenance of the high quality necessary for shellfish harvesting assures the public of clean, safe recreational waters. Removal of contaminants and impacts associated with poor sewage treatment, such as high BOD, suspended solids, and dissolved oxygen sags can have a marked influence on aesthetics, property values, fish and shellfish spawning and habitat, algae bloom control, and general environmental quality. Minimization or elimination of heavy metals and organic chemical discharges can increase fish and shellfish spawning success, protect fish and shellfish from accumulating undesirable substances in their tissues, protect wildlife such as osprey, eagles, and ducks, and assure better general water quality.

The American public is vitally interested in protecting its waters. Opinion polls nationwide have indicated that the public is willing to pay additional revenues to protect our natural resources, particularly in coastal states. The sanitary survey is a powerful tool to help with the identification of water quality problems which need correction, and the targeting of public monies toward corrective actions.

13

Depuration and Relaying of Molluscan Shellfish

Gary E. Roderick and Keith R. Schneider

Food Science and Human Nutrition Department
359 Food Science Building
University of Florida, Gainesville, FL 32611

INTRODUCTION

The technical information contained in this chapter is derived from research, pilot plant studies and existing commercial plants. Most of the information presented is directed mainly towards: (1) defining the optimal environment for shellfish filter-feeding processes, (2) National Shellfish Sanitation Program guidelines, criteria (NSSP 1990*a*, *b*) and regulations for depuration and relaying.

Successful depuration and relaying is a consequence of complete purging of the shellfish tissue in an allotted time period using carefully controlled conditions. Both relaying and depuration can be optimized by harvesting the cleanest shellfish available and lowering the microbial content of the meat in the shortest amount of time. Therefore, it is necessary in depuration to present the shellfish with a seawater environment that optimizes their filter-feeding activity.

HISTORY

Historically, commercial depuration of molluscan shellfish has been practiced for at least 75 years (Canzonier 1991). The association between typhoid fever and raw mussel consumption led to the development of the first depuration facility located at Conwy, North Wales through the pioneering work of R. W. Dodgson (1928). Specifically, seawater was disinfected by the use of hypochlorite and then neutralized by adding sodium thiosulfate to eliminate the residual chlorine. Other disinfecting methods for seawater were investigated several years later, and included ozone and ultraviolet light. Ozone was first introduced by Violle in 1929 to disinfect seawater and reported that it required 5–6 days to "sterilize" oyster.

A major advocate of molluscan depuration in the U.S. early this century was an outspoken, and sometimes controversial, individual by the name of William F. Wells (Canzonier 1991). He was influenced by the work of Phelps (1911) of Rhode Island and Johnstone (1914) of England. He performed extensive work in New York that paralleled Dodgson's work in Wales (Canzonier 1991). He conducted demonstration projects for the U.S. Public Health Department and was involved in some of earliest experimental depuration research. The first attempt of commercial depuration was in 1921 at Inwood, New York, and remained operational until 1925 (Canzonier 1991).

During the late 1920s there was much discussion concerning the improvement of shellfish sanitation. This was brought about because of a series of typhoid fever outbreaks associated with the consumption of raw oysters. These outbreaks were a major catalysis for preventive action that lead to the establishment of the NSSP (Canzonier 1991).

In the 1930s, Major Richard E. Messer conducted a collaborative study between the state health authorities of Virginia and the U.S. Public Health Service on Willoughby Split near Norfolk Virginia (Canzonier 1991). Other shellfish purification activities during the 1930s and 1940s included those of Tarbett in 1934 and Fisher in 1941.

During the 1950s, P. L. Goggins and J. W. Hurst (1955) performed experimental laboratory depuration research on the soft clam near Boothbay Harbor, Maine. Later a soft-shell clam depuration plant was established in Maine and was operational for many years for both depuration and for the determination of the physiological and microbiological requirements for depuration. At about the same time, a depuration plant in Newburyport, Massachusetts was constructed and operated to depurate soft-shelled clams (Canzonier 1991).

The use of ultraviolet light to disinfect seawater and depurate shellfish was introduced by Wood (1961*a*, *b*) in the United Kingdom and by Arcisz and Kelly (1955) in the United States approximately 30 years ago. The exposure of seawater to ultraviolet light allowed a continuous dosage of seawater to the irradiation. During the past 20 years, the ultraviolet light assisted depuration has been refined. For example, Ayres in 1978 introduced the recirculating constant volume design of depuration systems and the technology of the UV light delivery systems are quite advanced and efficient. It should be pointed out that there was a 15-year gap (1946–1961) when very little was done in the area of depuration. Some investigators attribute this to Galtsoff's work showing that chlorination inhibited shellfish activity. The use of UV treatment renewed the interest in the depuration of shellfish. During the past 10 years, a renewed interest in depuration research and development has occurred (Otwell et al. 1991). The major reasons for this occurrence seem to revolve around the diminishing acreage of approved shellfish harvesting waters and the increased sources and levels of pollution and the increasing number of reported shellfish associated illnesses that has focused consumer attention on the potential danger of raw shellfish consumption. Political, economic, legal, and social forces are currently changing the U.S. shellfish industry, and depuration is one method that is seriously being reevaluated (Richards 1990).

TYPES OF DEPURATION PLANTS

Two general types of molluscan shellfish depuration plants exist and are either (1) flow-through or (2) recirculating or closed systems. Molluscan shellfish can be effectively depurated of contaminants by either type of plant design. Flow-through systems are of choice when the depuration facility is close to a reliable and consistently clean source of seawater to pass over the shellfish. In addition, filtration and ultraviolet treatment of the seawater pumped into the depuration plant can enhance the seawater quality. However, in some instances the quality of seawater may change (red-tide bloom) before being filtered and treated with UV light. Therefore, long-term reliability of the source of seawater is important.

Recirculating or closed depuration systems require the same consistent high-quality source of seawater. However, the seawater is recirculated for at least 24 h and then released into the environment. One advantage that the closed systems have over the flow-through is that the inadvertent introduction of toxic or any adverse chemicals that may suddenly appear in the seawater can be eliminated. Closed systems do have potential limitations. Specifically, there is always the potential of buildup of toxic metabolic end products such as ammonia. In addition, one should be more aware of the biological oxygen demand of the shellfish, pH changes and the buildup of both feces and pseudofeces, which is a primary source of bacteria.

Recently, there has been some concern over the disposal of the seawater used for the depuration of shellfish. For the most part, the city, county, state and federal agencies have the regulatory authority to issue permits approving the dumping of the seawater into the environment. In most cases information concerning the quality of the seawater being disposed is of importance. In some instances, the seawater is treated with ultraviolet light before being released into the environment. All washings, feces and pseudofeces and any other waste products are usually disposed of through public water and waste treatment plants.

IMPORTANCE OF SEAWATER QUALITY

Although depuration is a natural process to molluscan shellfish, man can enhance the process if he can modify the seawater environment that the shellfish are exposed to in the depurating tanks. The physiological activity, pumping rate, and behavioral responses of molluscan shellfish will vary in response to changes in their seawater environment. Successful depuration depends upon knowing how the environmental parameters affect the depuration process. Those factors associated with the seawater that have been shown to significantly alter the depuration process are: (1) temperature, (2) salinity, (3) dissolved oxygen content, (4) turbidity, and (5) phytoplankton concentration.

Water Temperature

The temperature of the treated seawater entering the depuration should be within the range of 14–29°C (Neilson et al. 1978). However, different shellfish species will pump at different ranges of temperature. The genetic makeup and geographical location of the molluscan shellfish will determine the range of temperature in which pumping will occur.

The temperature range is important in the location of a depuration plant where seasonal water temperatures may range from 0°C to 32°C. Depuration is not recommended when water temperature is below 14°C (Neilson et al. 1978). Studies have shown that oysters can be depurated at 10–14°C if their initial fecal coliform level in the meats is 200 or less per 100 g. Moreover, if depuration is to be practiced when seawater temperatures are below 14°C, it is recommended that the seawater be heated if the fecal coliform counts are greater than 200 per 100 g of meat (Neilson et al. 1978). It should be pointed out that as the temperature of seawater falls to below 14°C, the fecal coliform levels often decline to very low levels. Therefore, depuration may be feasible in spite of a depression of pumping and associated physiological activities of molluscan shellfish. As a rule, the seawater temperature for depuration should not vary more than 20°C from that of the seawater harvest area.

Numerous depuration studies and reviews concerned with the effects of seawater temperature upon bacterial reduction in various shellfish meats have been conducted (Arcisz and Kelly 1955; Hefferman and Cabelli 1970; Cabelli and Hefferman 1971; Souness and Fleet 1979; Rowse and Fleet 1984; Richards 1990; Rodrick and Schneider 1991). Table 13-1 lists the various microorganisms and the optimal temperature for their elimination from the hard clam, eastern oyster, soft clam, and the Sydney rock oyster. Direct comparisons between various studies are difficult because of temperature gaps and overlaps.

TABLE 13-1

	Optimal Temperature (°C) for Depuration
Hard Clam	
Fecal coliforms	22–25
E. coli	20
V. parahaemolyticus	15
V. cholerae	22–24
V. vulnificus	22–24
Polio type 1	20
Polio type 3	18–20
Coxsackie B4	20
Eastern Oyster	
Fecal coliforms	24–29
Total coliforms	24–29
V. cholerae	22–24
V. vulnificus	22–24
Polio type 1	12–23
Polio type 3	18–20
Coxsackie B4	20
Hepatitis A	12–23
Sydney Rock Oyster	
Total coliforms	18–20
E. coli	13–20
Salmonella spp.	18–22
Soft-shelled Clam	
Total coliforms	8–16
E. coli	8–16
Salmonella spp.	13

Salinity

The salinity of the seawater used for depuration of molluscan shellfish is of critical importance. Specifically, the salinity of the depuration processing water that the molluscan shellfish are placed in can affect the pumping rate and other physiological processes (see Table 13-2). Since the salinities of the depuration processing water may differ from the shellfish harvesting area, shellfish should be acclimated to their new seawater environment. The period of acclimation is variable but should be long enough to allow for adequate pumping activity to be restored. For reasons that still remain unclear, higher salinities seem to enhance the depuration process while lower salinities inhibit the process. It is recommended that the salinity of the depuration water does not vary more than 20% of that of the water where shellfish were harvested (NSSP 1990*a*).

TABLE 13-2

	Optimal Salinities (ppt) for Depuration
Hard Clam	
E. coli	20–30
Polio type 1	31
Eastern Oyster	
Total coliforms	16–18
Fecal coliforms	16–18
Hepatitis A	8–28
Sydney Rock Oyster	
E. coli	23–36 & 43–47
Salmonella charity	23–36 & 43–47

Dissolved Oxygen

Molluscan shellfish require oxygen to maintain normal physiological activities. The source of oxygen is from their seawater environment. Various factors influence the dissolved oxygen levels of the seawater in the depuration tank. These factors include: (1) the surface area of the seawater exposed to air; (2) the flow rate of seawater during depuration; (3) the number of shellfish to be depurated; (4) the physiological activity of the shellfish during depuration; (5) aeration; and (6) the salinity and temperature of seawater used for depuration. Because of the demand for oxygen by the depurating shellfish, oxygen must be replaced in the seawater during depuration at a constant and adequate rate. This is usually accomplished by either aeration or spraying the seawater on the surface during its delivery into the depuration tank or by the use of in-line venturi tubes. Both methods are commonly used in molluscan shellfish depuration plants. Problems can arise through aeration if the process disturbs the settled feces and/or pseudofeces excreted by the depurating shellfish. In addition, the formation of gas bubbles in the seawater can inhibit the gas exchange in the gill tissue and the respiration of the depurating shellfish.

TABLE 13-3. Dissolved Oxygen Content in Seawater

Salinity (ppt)	Temperature (°C)		
	10	20	30
10	10.7	10.1	9.4
20	8.6	8.1	7.6
30	7.2	6.7	6.4

Source: Neilson et al. (1978).

While the oxygen concentration range for depuration is wide, it is recommended that the oxygen level should not fall below 2 mg/l of seawater during depuration. There is also an upper limit of oxygen concentration that is determined by the solubility of oxygen in seawater. The solubility of oxygen decreases with the rise of temperature and with an increase in the salinity of the seawater. Table 13-3 shows the dissolved oxygen content in seawater and its relationship to salinity and temperature.

Turbidity and Total Suspended Solids

Molluscan shellfish are filter feeders and most of the suspended solids in the seawater are filtered from the seawater and excreted on the bottom of the depuration tank as feces or pseudofeces. These feces are commonly referred to as "bio-deposits" and are a reservoir of bacteria that can recontaminate the depurating shellfish. These deposits constitute a potential bacterial and toxic hazard, since the oxygen content is high. If the flow of seawater in the depuration tank is stopped, the dissolved oxygen levels can drop quickly and hydrogen sulfide created by the decomposition of these deposits could kill the shellfish in the depuration tank.

Excessive turbidity will reduce the penetrating power of the ultraviolet light used to disinfect the seawater. The source of seawater used for depuration can be obtained from salt water wells. Salt water wells are advantageous because of their constant salinity, temperature, and clarity of the water. If a well is not possible the source of seawater should be taken from an approved water area where waves, currents, or other factors do not cause excessive turbidity.

TYPES OF SEAWATER TREATMENT

Depuration is a process of purification in which filter-feeding shellfish are placed in a clean environment; and allowed to pump, thus ridding themselves of pathogens and other bacteria (Blogoslawski and Stewart 1983). The depuration of contaminated shellfish was addressed as early of 1911 (Phelps 1911). Four methods exist for the disinfection of seawater used in depurating molluscan shellfish. These methods include the use of (1) chlorine, (2) ultraviolet light, (3) ozone and activated oxygen, and (4) iodophors.

Chlorine/Hypochlorite

The earliest method used to cleanse shellfish in an onshore controlled environment was chlorine. The first use of chlorine in the depuration of shellfish was in 1914 by Johnstone (1914). Bacterial levels in mussels were reduced after keeping them in water disinfected by chlorine. It was chosen because of its known disinfecting ability and ease of use and measurement (Blogoslawski 1980). Disinfection using hypochlorite method was first used by the U.S. Public Health Service in Jamaica Bay, NY, in 1916. Carmelia (1921) reported the purification of oysters using hypochlorite at a concentration of 4–6 mg/l for 12–18 h. The use of free chlorine species has several drawbacks that limit its effectiveness as a method of shellfish purification. Dodgson (1928) observed results differing from the previous study, due to the inhibitory effect of chlorine on the mussel's ability to continue pumping. He found that low levels of free chlorine, approximately 0.2 ppm, severely limited or completely shut down the filter-feeding activity of oysters. This effect was alleviated by quenching free chlorine with sodium thiosulfate. A final bath containing 3 ppm of free chlorine was used to cleanse the mussel shells (Dodgson 1930). These recommendations were later supported by Galtsoff (1946) and Allen et al. (1950). The use of sodium thiosulfate to dechlorinate the treated seawater was noted by Blogoslawski (1980) as a method of restoring the natural level of pumping activity. However, some evidence indicates that sodium thiosulfate may itself reduce or inhibit the level of pumping as reported by Kelly (1961). In addition, the chlorine depurated shellfish imparted a chemical taste and a chewy texture (Voille 1929). Depuration using chlorine, while still in use in certain areas in Europe (West et al. 1985) has for the most part given way to other methods of disinfection such as ozonation and ultraviolet irradiation.

Ultraviolet Irradiation

The use of ultraviolet (UV) light irradiation in the depuration of shellfish is still the most often used method in the United States and in the United Kingdom. Numerous studies have shown UV light to be effective in inactivating bacteria from infected shellfish (Arcisz and Kelly 1955; Huntley and Hammerstrom 1971; Cabelli and Heffernan 1970; Cortelyou 1954; Kawabata and Harada 1959; Kelly 1961; Chang et al. 1985) and virus (Fogh 1955; Hill et al. 1969, 1970).

Ultraviolet light irradiation has been known for many years to possess bactericidal effects (Herrington 1991) by disrupting unsaturated bonds, in particular, the purine and pyrimidines found in DNA. These changes in the nucleic acid components can cause progressive injury and lethal biochemical changes in the bacterial cell (Huff et al. 1965). For most species of bacteria, the bactericidal effect is greatest between 250 and 260 nm (Harm 1980). In most systems the primary source of UV energy is from low-pressure mercury arc lamps. These bulbs emit approximately 85% of their energy monochromatically at 253.7 nm (Herrington 1991). In most UV depurating systems, seawater is exposed to UV light in a flow-through or recirculating system for disinfection before being passed over the shellfish.

Kelly (1961) examined the limitations of using UV light systems, and found that turbidity, color, and dissolved iron salts can affect the transmission of UV light by the aqueous medium, thereby reducing the bactericidal action in the water column and severely limit a UV system's effectiveness. It is well documented the UV bactericidal activity is directly related to the depth of penetration (Kelly 1961). Particulate matter in the water can shield organisms from the UV light source and reduce penetration. Water depth in a UV contactor should be limited to a few millimeters and must be prefiltered to remove particulate matter greater than 20 μm for optimum effectiveness. In addition, the UV bulb or protective quartz sleeve of the contactor must be continually cleaned of organic slime and inorganic materials which accumulate and block the penetration of UV radiation.

Son and Fleet (1980) tested a number of pathogens to determine which species could be depurated as effectively as fecal coliforms from the Sydney rock oyster. Their results indicated that all artificially introduced pathogens readily depurated from the oyster under conditions that were conducive to the removal of *E. coli.* However, the total plate counts did not decrease below 10^6 cells per 100 g which indicated a retained indigenous microbial flora. Further research of the microbial flora revealed that several were predominant. These included *Pseudomonas, Vibrio, Aeromonas, Moraxella, Acinetobacter, Flavobacterium,* and *Cytophaga* species. Research by Vasconcelos and Lee (1972) on *Crassostrea gigas* displayed similar results with commensal flora remaining at levels comparable to control oysters. These findings corroborate other studies on *C. gigas* (Colwell and Liston 1960) and *C. virginica* (Tripp 1960). Work by Souness and Fleet (1979) have also shown the buildup of UV-resistant bacteria in the treated water. This might pose a health risk if significant numbers of resistant bacteria survive and replicate, though not much is known on the life cycle of these bacteria. Despite these limitations of UV light assisted depuration systems still remains cost effective and popular in the United States.

Ozone Depuration

Ozone, the triatomic form of oxygen, is a colorless gas, approximately 1.5 times as dense as air at sea level and is a strong oxidizing agent. The first commercial plant to use ozone was constructed in France in 1963 (Fauvel 1963). The system was so successful that all of the major depuration plants in France now use ozone (Fauvel, Pons, and Legeron 1982). Ozone depuration is also currently being used in Australia. Voille (1929) reported that ozone could be used to disinfect seawater and depurate oysters. The major advantage of ozone assisted depuration over chlorination is that it does not change the taste or appearance of the shellfish.

When ozone comes in contact with bacteria it can react directly to inactivate the organism. Ozone may also react in seawater with bromide that can be oxidized and become a disinfecting agent. It is also important to consider oxidant contact time, concentration, temperature, and pH of the disinfection system (Pichet and Hurtubise 1976).

The literature contains numerous reports on ozone's ability to inactivate bacteria (Blogoslawski and Monasterio 1982), viruses (Englebrecht and Chian 1985; Sproul et al. 1985). Combs and Blogoslawski (1975) reported that pathogenic yeasts were inactivated in ozonated seawater. Suspensions of 3.0×10^6 cells/ml were subjected to ozonation for periods up to 240 s. The flow rate was set at 100 ml/min, though no information was given on oxidant concentration. Their results show that both pathogenic and nonpathogenic yeasts were effectively removed from seawater. Some reactivation was observed.

Blogoslawski et al. (1975) designed a pilot-scale ozonation facility. Seawater at 20°C with a flow rate of 7 gallons/min was subjected to ozonation in a chamber that allowed for 30 s of contact time. The pH of the system remained at 7.8 with no change after ozonation. After two hours at an oxidant concentration of 0.56 mg/l, most marine microorganisms were inactivated. Only *Pseudomonas* spp. were isolated from the ozonated water. Some organisms thought to be killed, were in fact recovered after several days.

Farooq and Akhlaqu (1983) examined the effects of ozonation on mixed cultures of bacteria and viruses in a phosphate-buffered solution. The organisms used and their titers were *Mycobacterium fortuitum* (8.1×10^5 ml^{-1}), poliovirus type 1 (2.32×10^5 ml^{-1}), *Candida parapsilosis* (1.55×10^6 ml^{-1}), *E. coli* (2.09×10^5 ml^{-1}), and *Salmonella typhimurium* (6.56×10^5 ml^{-1}), respectively. The organisms were placed in deionized phosphate buffer solution with an ozone residual of 0.23–0.26 mg/l. The maximum detention time was 120 s. The order of resistance was *M. fortuitum*, *C. parapsilosos*, *E. coli*, and *S. typhimurium*.

Englebrecht and Chian (1985) studied the inactivation of viruses by ozone. All experiments were at 20°C, and pH 7.2. Six enteroviruses were inactivated in the continuous flow mixed reactor. The relative resistance of the viruses were as follows: poliovirus 2, echovirus 1, poliovirus 1, coxsackievirus B5, echovirus 5, and coxsackievirus A9. The inactivations rates were influenced by changes in ozone residual concentrations, temperature, and pH. As ozone and temperature were increased, an increase in the biocidal effect was noted. Experiments with varying pH levels demonstrated an increased biocidal activity at pH 7.2 as opposed to 5.9 and 4.3, respectively.

In addition to ozone's ability to inactivate bacteria and viruses, it is capable of oxidizing marine dinoflagellate toxins (Thurberg 1975). Blogoslawski and Stewart (1977) reported ozone's ability to degrade tetrodotoxin (TTX). Blogoslawski et al. (1979) studied the effects of ozone on potent neurotoxins produced by *Gonyaulax tamarensis*. These toxins are collectively known as paralytic shellfish poison (PSP). Clams exposed to a *G. tamarensis* bloom were collected and transported to a pilot-scale depuration facility. The oxidant concentration was measured at 1.2 or 0.5 mg/l in two experimental sections. The clams were kept in the system for 72 h, at which time the clam meat was homogenized and toxin extracted. Detoxification was determined using the biomouse assay. Results showed that detoxification occurred in both the control and experimental chambers, with most detoxification occurring in the experimental section.

A variant of ozone, photozone, or activated oxygen has recently been promoted commercially of Photozone (Water Management Inc., Englewood, CO). The photozone gas mixture consists of oxygen (O_2), ozone (O_3), hydrogen peroxide (H_2O_2) and hydrogen dioxide (HO_2) produced by the photooxidation of oxygen via a U.K. lamp emitting a peak maxima of 180 nm (Product literature, Water Management Inc., Englewood, CO). Several claims have been made by Water Management for the effectiveness of photozone, though no reviewed literature is currently available.

Iodophors

Little information is available on the use of iodophors for the depuration of molluscan shellfish. Iodophore disinfection of seawater in depurating molluscan shellfish has been practiced commercially in Italy. Fleet in 1978 reported rapid bacterial reductions occurred in depurated tank water containing

TABLE 13-4. Comparative Aspects of the Four Methods of Seawater Disinfection

Operation or Condition	Ultraviolet Light	Chlorine/ Chlorine cpds.	Ozone	Iodophores
Capital costs	Low	Medium	High	Medium
Operating costs	Lowest	Low	High	Low
Installation	Simple	Complex	Complex	Complex
Maintenance	Easy and low cost	Moderate and low cost	Difficult and costly	Moderate and low cost
By-products	None	THM	Toxic by-products	THM
Contact time	1–5 s	30–60 min	10–20 min	30–60 min
Shellfish toxicity	None	Yes	Yes	Yes

THM = Trihalomethanes.

0.1–0.4 mg iodophore/l. No adverse effects on shellfish physiology or organoleptic characteristics for shellfish depurated with iodophore-disinfected water was reported.

Table 13-4 compares the four different methods discussed for seawater disinfection.

RULES AND GUIDELINES FOR CONTROLLED PURIFICATION

Controlled purification or depuration is a process in which shellfish are introduced into a controlled seawater environment, and allowed to pump or filter feed. The intent is to reduce the concentration of bioaccumulated pathogens that may have been present in the shellfish harvesting areas that could be described as moderately polluted (restricted) waters. This method offers several advantages over natural purification processes such as relaying, in that depuration does not suffer from possible recontamination from changing environmental conditions. Depuration can be used for shellfish that have been harvested from clean waters that would, normally, not need further processing in order to reduce the amount of sand, yielding a less gritty mouthful. Controlled purification should not be employed on shellfish harvested from heavily polluted (prohibited) waters. This process is also not intended to reduce the concentration of poisonous or harmful substances that may be bioaccumulated from the surrounding waters. The continuing acceptance of controlled purification is contingent of the State Shellfish Control Agency (SSCA) overseeing all phases of the purification process.

It is the SSCA responsibility to supervise the administrative duties concerning the controlled purification, or depuration of shellfish harvested from restricted waters. The state controlling agency should provide a detailed control plan which outlines the following procedures:

a. Monitoring the harvest of shellfish to assure they come from approved or restricted water and not from prohibited waters. Also, if the restricted water are classified as conditionally restricted, then some means of establishing when areas are harvestable must be outlined and a means of notifying or posting an area as restricted must be implemented.
b. The transportation to an approved depuration plant. The definition of a depuration is a facility that contains one or more depuration units. The depuration unit is further subdefined as a tank or group of tanks that is supplied by a single source of clean disinfected water. These systems can include the use of ultraviolet light, chlorine, ozone, and several other types of water sterilization. Each category of disinfection process has several sub-groups of guidelines governing their use, and will be discussed later.

c. The approval of plant design and their operation. These guidelines should include any changes or modification to existing plants and the routine inspection and subsequent certification of these facilities. These guidelines should be in accordance with the requirements of NSSP Manual of Operations and any local laws or restrictions regarding the handling and processing of shellfish.
d. The ability to monitor and prohibit interstate shipments of shellfish in the event that the above guidelines have not been met and that the shellfish product may possess some public health risk. The NSSP manual defines the nonconformance of the aforementioned rule as anything that is found that compromises the validity of the process. The manual also calls for a Memorandum of Understanding (MOU) to be developed between appropriate agencies when more than one state agency is involved in the control plan (NSSP 1990*b*).

Compliance to the above items (a)–(d) occurs when the following have been put into operation:

a. The controlling agency must establish criteria and methods for determining classification of harvesting waters. Laws must implement and establish guidelines for the proper use and restriction of the shellfish harvesting area. The state agency must have the resources and equipment to enforce the laws governing the harvest of shellfish from restricted areas. The SSCA must also be able to monitor the transportation of shellfish to a depuration facility. This is to prevent product from purposely being diverted from a purification plant and channeled directly into the marketplace, increasing the associated health risks of possible contaminated shellfish to the consuming public.
b. The SSCA should be advised and approve all plans for construction or updating of all depuration facilities. Approval should be obtained before the construction of any new depuration facility.
c. A written control purification plan should be prepared by the state controlling agency. This document should outline in detail the depuration procedures required under the guidelines established in the NSSP Manual for the depuration of shellfish. The operating plan should be updated as needed, especially if certain techniques are found to be inadequate or if new technology is found to be more effective.
d. Each shellfish harvester should possess a separate permit issued by the SSCA for the delivery of shellfish to a depuration facility. This permit should specify the terms that govern the license to harvest shellfish. A statement should be included that any violation of the terms outlined may result in loss of the permit. Laborers working under the supervision of a person who has been certified and has a valid permit may be exempt from

applying and obtaining an individual permit. There are additional requirements and can be found in Part I, Section E, Paragraphs 2a and 2b of the NSSP Manual.

e. The design of the depuration plant, its construction and subsequent operation of the purification process have to be approved by the SSCA. The evaluation of the controlled purification facilities should be in accordance with the requirements of the NSSP Manual prior to certification. This procedure must also be followed if one wants to change the design of the plant or in the operation of the purification process.
f. Depuration plants are to be inspected and certified in accordance with the procedures outlined in the NSSP Manual. These are to ish harvesting areas are issued two certification numbers. These will include a DP certificate for depurated product and an SS, RS, SP, or RP certificate for shipped, reshipped, shucked, or repacked, respectively.
g. Once certified, the depuration facilities are to be inspected to ensure operating compliance with the NSSP regulations. The time between inspections should be sufficient so to assure that each site is within the recommended guidelines outlined in the NSSP Manual. Each facility will receive an inspection rating score. The minimum inspection frequency for a depuration plant should be no less than once a month. If any noncompliance to the guidelines is found, the plant operator's must make immediate corrections. If the violations are not corrected immediately, the plant's operating permit can be suspended, or the certification can be cancelled until the corrections are made.
h. Analysis of the depuration process data and other relevant data should be conducted by the SSCA once a month to verify that the plant and methodology are effective. The evaluation should be able to establish if the final-product criteria for the depuration plant or for specific harvesting and growing areas are being met. If warranted, the frequency of inspection may be increased to adequately determine the operational efficiency of the depuration facility.
i. The state controlling agency should maintain adequate records for each depuration plant. Files that are under the above mentioned items a, b, and c should be maintained indefinitely and items d, e, and f should be maintained for a minimum of two years. Items a–f include the following:

1. a control plan that has been approved by the state controlling agency;
2. sanitary survey reports and other data for the shellfish harvest areas;
3. data for the pre-certification process verification and any reports that are used to establish the design and operational specifications;
4. analysis of the process data taken at specific time intervals;
5. sanitary inspection reports for the purification plant; and

6. a final report generated on verification that the operator's records have been reviewed and the process has been evaluated.

HAZARD ANALYSIS OF CRITICAL POINTS

Moreover, if a state shellfish control agency is to be effective in managing controlled purification several key areas need to be addressed. The use of HACCP (hazard analysis of critical control points) should be employed as to identify the key areas that pose the largest problems in providing clean shellfish to the consumer. There exist a number of such control points, any one of which if not properly managed, could result in the marketing of contaminated shellfish. One such area to be controlled is harvesting location. In order to assure that the shellfish come from restricted areas and not prohibited shellfish lots is of major importance. Prohibited area can be so polluted that any purification scheme put in place will not be adequate to cleanse the product. Next the design and construction must be regulated so that only a facility that is properly planned and designed will go into operation. The operation and management of a depuration plant must meet operation guidelines established by the state regulatory agency. These guidelines consist of operational criteria based upon scientific studies conducted on the process of depuration as to assure a safe product to the consumer. The last item that needs to be addressed is the need for an inspection program. All inspection programs should be designed in such a manner as to avoid or detect any deviations that may lead to the marketing of suspect shellfish.

The NSSP Manual calls for a scheduled controlled purification process (SCPP) to be initiated. The state agency responsible for the shellfish program is required to establish design and operational guidelines. These guidelines may be established by other qualified persons outside the state agency and/or in cooperation with persons who have experience in the designing and operating of a controlled depuration facility. Approval for any process shall be based upon an extensive scientific research in the area of shellfish depuration. The effectiveness of the plant operations will be evaluated before certification is given. The state agency involved should take into consideration that several types and design might be put into operation.

The operator will be in compliance with the above items when the following have been put into operation: a SCPP has been agreed upon by the state controlling agency and/or in cooperation with a group or individual(s) that has experience in the area of controlled purification. The processing plan should be based on the scientific experimentation that support methods that adequately reduce the number of bacteria that are present in shellfish harvested from restricted waters. The same guideline for the purification process may be used at other sites. However, the subsequent plants must first be approved by the state controlling agency. Once evaluated, the plant may begin commercial operations if it meets the criteria set for a purification facility and demonstrates its effectiveness in reducing the number of microorganisms found in shellfish. Critical control points need to be identified and a monitoring program put in place. All parameters governing the depuration process must be clearly outlined. These operating guidelines should include all tests indicating the bacteriological assay that are to be used in determining a clean product.

When developing a scheduled controlled purification process, the following principles and procedures should be met: The development should be overseen by a knowledgeable individual(s) or in cooperation with persons knowledgeable in the field of controlled purification. The design, construction, and operation of the depuration plants should also be under their direct supervision.

The operating parameters should be able to handle any variation in the process that might occur such as environmental changes. These environmental process variables may be encountered during the course of depuration and should be studied carefully and in depth to assure a safe shellfish product. There exist many different processing variables that can effect the outcome of depuration. The list includes factors such as (1) the shellfish species being depurated, (2) seasonal effects on both the shellfish and the surrounding environment, (3) the water temperature, (4) salinity of the water, (5) dissolved oxygen (DO), and (6) turbidity. The area in which the shellfish are harvested and where the processing water originates is very important. How the depuration processing water is to be treated (which is directly related to the variable of where it comes from) is of critical importance in the purification facility design and construction. In addition, design aspects should pay strict attention to flow characteristics, the clearance between shellfish-holding trays and tank walls, processing time, and flow rate. Bacteriological monitoring of both untreated and treated shellfish is needed to determine if the depuration process is meeting the endpoint criteria set in the SCPP plan. Finally, a schedule for plant sanitation and the monitoring and recording of these efforts is needed.

Each depuration facility should establish a maximum zero-hour level of fecal coliforms. This is determined by comprehensive research studies that verify that the process will in fact work. These studies should be conducted over a time period that will take into account all environment changes and stresses that might occur in a harvesting area. The ultimate objective is to establish a maximum zero-hour level of fecal coliforms in shellfish that when subjected to the purification treatment will depurate to end-point criterion set for the marketing of a safe shellfish product and do so consistently. Studies on the initial bacterial (pathogen) concentration should be continued during normal commercial operations to assure a consistent product that meets or exceeds the standard set, which is not effected by the changing environmental conditions. These changes should be monitored for both harvesting area and the depuration facility. If continuing studies point to a deficiency in the operating procedures, then a change in zero-hour tolerance can be made along with modification in plant design or operating parameters. If any changes in plant design or operating parameters are made, then new studies must be conducted along with continuous monitoring to assure a safe product.

The depuration facility should provide a copy of the operating procedures or SCPP to the controlling state agency, in addition to copies of all scientific studies conducted that were used in establishing the effectiveness of these guidelines. These documents should be retained indefinitely in the central file by the SSCA. Operational manuals for the processing plan should be maintained at the purification site. Plant managers and/or responsible individuals trained in the application of the process should have ready access to the purification scheme and should be familiar with scientific work in this area.

As stated earlier, the depuration of molluscan shellfish is a complex biological process. Each species of shellfish may require different operational parameters to effectively cleanse harmful pathogens. If several species are to be depurated at the same facility, operation guidelines including turbidity, salinity, temperature, stacking height, and tank design must be addressed for each. Many different designs for depuration facilities have been proposed along with parameters for the effective purification of shellfish. Without an established standard for plant design and construction, it will be necessary for the plant operator and/or the SSCA to establish the effectiveness of each plant, in relationship with each harvesting area (including environmental changes that occur) and for each species. From these studies a written SCPP for each purification facility will be drafted. This processing plan must specify the maximum zero-hour criterion for fecal coliforms in the shellfish to be depurated.

In the developing of a SCPP, persons involved should be knowledgeable in the design, construction and operation and maintenance of a depuration facility. Any process agreed upon by the state controlling agency must take into account HACCP analysis. The process must be able to function properly under any condition that may arise. All operating parameters involved should be supported by scientific studies that show that the SCPP will consistently depurate shellfish which meet the end-point criterion set for a particular plant and process. Variables to be examined should include the following: shellfish species; water temperature, salinity, and turbidity;

harvesting area and source of processing water; treatment of aforementioned processing water; tank design, construction, and hydraulics; processing time, raw-product quality, end-point criteria, process monitoring, and general plant sanitation (NSSP 1990*b*). This list is by no means all inclusive. HACCP analysis may uncover other variables that may require as much or more attention than the ones listed above. Each plant must go through the process of analysis and evaluation before going into operation.

DEPURATION PLANT LOCATION, DESIGN AND CONSTRUCTION

The location of a depuration facility should be in an area that will not be affected by conditions that may result in the contamination or recontamination of shellfish while in the depuration plant or in storage. The building construction should be of an adequate size, construction and design as to prevent an adulteration of the shellfish product (either treated or processed) by animals and other pests. The facility should be designed to isolate the untreated from the treated shellfish (cross-contamination). The construction should be so as to facilitate adequate sanitizing, operation, and maintenance of the depuration facility. All areas of the facility, including the processing tanks, processing trays, plumbing, and conveyances, are housed within a protective structure.

Satisfactory compliance with the above items occurs when the facility grounds on which the depuration plant is constructed and operated meet the following requirements. The stored equipment, litter, waste, and other garbage do not pose a health risk nor do they act as an attractant or breeding place for rodents, insects, or other pests. In addition, any weeds and/or uncut grass around or in the vicinity of the process plant that might cause the same problems should be removed.

The depuration plant should not be placed near roads, yard or parking facilities that may pose the possibility of dust or other contaminates reaching the treated or untreated product. Finally, construction and design should allow adequate drainage, that will prevent contamination by seepage or footborne filth that could provide a conducive environment for microbial growth and insect propagation.

The construction of the depuration plant should include effective barriers such as roofs, screens, walls, and doors to prevent any access to the product by animals and other pests. The type of barriers will depend on the facility constructed and the environment stresses it may encounter.

Each plant should provide separate holding and storage areas for depurated and undepurated shellfish. This is to prevent any cross-contamination.

Separate washing and culling facilities should also be provided for purified and unpurified product and, finally, each plant should have ready access to potable water supply.

The NSSP Manual (1990), Section D, outlines the requirement for construction and operations of depuration plant. These operational guidelines cover the following items: floors, wall, ceilings, lighting, heating and ventilation, water supply, insect control, plumbing, and sewage disposal.

All floors which are in an area where shellfish are shucked or packed should be made from a impenetrable, corrosion-resistant material. This material should be easily cleanable with conventional disinfectants. The surface should be graded to facilitate drainage and free cracks or areas which would hamper proper disinfection and/or drainage. This is also true for areas in which utensils and equipment are to be cleaned.

The walls and ceilings should be made from a impenetrable, corrosion-resistant material. This material should be easily cleanable with conventional disinfectants. They should be constructed with light-colored materials as not to mask filth and be maintained as to prevent contamination of the product during processing and packing. This is also true for areas in which utensils and equipment are to be cleaned.

The depuration facility should have adequate lighting for the normal operations of the plant. Any glass objects such as light bulbs, fixtures or skylights should made from a break-resistant material to protect against adulteration of shellfish due to breakage.

The ventilation system should able to remove or minimize any odors, noxious fumes, condensation in areas where they can come in contact with the product. The ventilation system (including heating and cooling) should not produce an environment which is conducive to the growth or propagation of any airborne pests.

Water used in the processing of shellfish should originate from a safe, potable source. This source or area should be free from contamination. Any system used to move or transport water should be constructed, maintained, and operated according to the state's regulations for a potable water usage. Water pressure and temperature must be maintained at a suitable level throughout the depuration plant where needed. The state regulatory agency is required to collect samples periodically for bacterial examination. Sinks must be equipped with hot and cold water. Hand washing sinks are only required to have cold and warm water. If steam is to be used in contact with food and food-contact surface, it must be free of any materials or additives. The only exceptions are outlined in the Code of Federal Regulations (CFR 173.310).

The plumbing system is required to supply sufficient quantities of water to all locations throughout the depuration facility. It must be able to remove all sewage and waste from the plant, preferably to a public sewage or treatment facility whenever possible. If transport to a treatment facility is not possible, a private on-site disposal plant may be constructed. Any such plant must meet the regulation set by the SSCA. All plumbing must be constructed as not to contaminate food or food-contact surfaces. Finally, drainage must be adequate to avoid flooding or standing water during normal operations and cleaning.

It is also very important to avoid any cross-contamination between potable water and nonpotable sources. The plumbing system must be designed to avoid any backwash or siphonage due to pressure changes in the systems. Hand-washing facilities must be adequate in size and number and be readily available to workers. These facilities should be provided with warm water outputs of 100°F (with a maximum of 115°F) or equipped with hot and cold outlets. They should also have hand-washing detergent or soap and a supply of disposable towels or hand-drying device. Bathroom facilities should be provided for each sex (except for family shucking operation or plants with ten employees or less). It is the supervisor's obligation to observe and make sure that all employees wash their hands before beginning work and after each interruption.

Finally, no drainpipes of wastepipes should be located above any processing or storage areas. This is also true for any areas containing packing material that will be used in the packing or transporting of the product.

The primary rule in the designing and operation of depuration plants is that the shellfish are consistently purified and protected from recontamination and spoilage. Several scientific studies dealing with controlled purification process and from experience gained from commercial facilities now in use, have led to several generally accepted standards. These standards are generally considered to be good operating practices for the depuration of shellfish (Furfari 1966; Hefferman and Cabelli 1971; Ayres 1978). There does not exist any clearly defined criteria for the design and construction of depuration plants. The guidelines set in these papers should be considered as general guidance in the development and construction of a purification facility. Finally, any facility should be designed and constructed in such a manner as to facilitate operation; so, adequate cleaning and sanitizing can be performed (CDF Title 21; Parts 170–199; 1986).

SOURCE OF SHELLFISH TO BE DEPURATED

Depuration may only be used on shellfish harvested from waters classified as approved, conditional approved or restricted. Shellfish may not be harvested for any reason from prohibited waters, regardless of whether depuration is to be used or not. The water quality requirements for approved, conditionally approved and restricted areas are noted in NSSP Manual (Part 1) and presented below.

Approved. The total coliform median or geometric mean MPN of the seawater does not exceed 70 per 100 ml. Also, no more than 10% of the samples exceed an MPN (5-tube) value of 230 per 100 ml or an MPN (3-tube) value of 330 per 100 ml. Fecal coliforms in the water should not exceed 14 per 100 ml and 10% of the samples do not exceed a 5-tube MPN of 43 per 100 ml or a 3-tube MPN value of 49 per 100 ml. A minimum of 15 samples must be taken before an approved classification can be given.

Conditionally Approved. The classification of these areas are subjected to bacterial pollution events. In these areas levels of bacterial contamination may rise above the allowable limits set by the states. These events include the following: (1) heavy rainfall; (2) effluent discharge for a waste or industrial facility; or (3) seasonal population fluctuations. If any of these events occur the SSCA or other agency can perform a sanitary survey. These are conducted to determine if the harvesting area is safe for shellfish collection. If the areas do not meet the recommended sanitary standards, the harvesting area will remain closed until further notice. Studies must be conducted for each of the pollution events in each unique area to determine the effects and the time needed to return to approved levels.

Restricted. These areas have been classified by a sanitary survey as having a limited degree of bacterial pollution. Shellfish may be harvested and sold from these areas only if they undergo some form of controlled purification process such as depuration or relaying. Studies must be conducted by the processor to show the effectiveness to reduce bacterial levels to an acceptable limit. The maximum value allowable MPN value as stated by the NSSP Manual is 700 per 100 ml, with no more than 10% of the samples exceeding 2300 per 100 ml for a 5-tube MPN or 3300 per 100 ml for a 3-tube MPN. Fecal coliform levels in the seawater must not exceed 88 per 100 ml and no more than 10% of the samples exceeding 260 per 100 ml for a 5-tube MPN or 300 per 100 ml for a 3-tube MPN (NSSP Manual 1990*b*).

All shellfish from the aforementioned areas should meet the zero-hour criteria for bacterial loading established in the process verification study. Shellstock should be harvested and transported in such a manner so as not to induce stress. Undue stress may adversely affect the shellfish and reduce their ability to depurate effectively. Care should be taken when handling and labeling shellstock as to avoid the mixing of depurated product and undepurated product. All shellstock delivered to a depuration plant should be properly tagged. Tags should identify the: (1) harvest area; (2) date of harvest; and (3) harvester or group of harvesters. A file should be maintained for all tags and should be readily accessible for traceability for at least 90 days.

Compliance to the above items will be satisfied when the following have been put into operation:

1. the close supervision of both the harvesting and transportation of shellfish from the restricted area,
2. the shellfish intended for depuration meet the zero-hour product specifications established in the process verification study,
3. the protection of shellfish from contamination during harvesting and storage according to the requirements of Sections B and E of the NSSP Manual.
4. the possession of complete and accurate harvest transaction records specifying, at the minimum, harvest area (designated by the SSCA), harvester's permit numbers, date of harvest and the amount shipped are required. (Other information may also be required.)

Several studies (Hefferman and Cabelli 1970, 1971) have shown that consistent depuration can be achieved when the levels of fecal coliforms did not exceed 1000 and 2000 per 100 g for the eastern hard clam, *Mercenaria mercenaria* and the soft clam, *Mya arenaria*, respectively. Fecal coliform limits for oysters and other species of clams have not been established. Until such limits are determined, field trials for each species and for various climatic conditions should be conducted as part of the process verification study. A maximum zero-hour criterion for fecal coliforms in shellfish to be depurated must be established.

Both thermal and physical abuse are known to affect adversely the pumping action of shellfish and, hence, reduce the rate of elimination of microorganisms (Hefferman and Cabelli 1971; Son and Fleet 1980). Contamination of the shellstock during harvest could raise the level of contamination to such a level that adequate purification will not occur. Temperature abuse of the harvested shellfish may cause increased bacterial levels at which depuration may not be effective in 48 h (Rowse and Fleet 1984). Temperature abuse of the shellstock can be prevented by (1) shading harvested shellstock during warm weather, (2) providing refrigeration during transit, (3) transporting as quickly as possible to the depurating plant, (4) cold weather freezing, (5) not breaking of the shells, and (6) optimizing both holding or storage times before depuration.

Shellfish harvested from restricted areas for depuration must be so controlled that they cannot be illegally diverted and sold. Therefore, special methods for monitoring harvest operations and the tagging of shellfish are required and may include the use of labels, colored containers, and/or various colored and shaped tags. If bulk transport of shellfish from restricted areas is needed, other identification may be needed and would include: (1) continuous surveillance of the boat or truck; (2) transporting shellstock in trucks sealed with a serially numbered tamper-evident seal; and (3) a count by the SSCA of the quantity shipped and quantity received at the depuration plant.

EQUIPMENT CONSTRUCTION AND DEPURATION FACILITY DESIGN

Excellent papers discussing depuration plant design exist (Ayres 1978; Furfari 1966, 1976, 1982; Nielson et al. 1978). The basic requirements of a facility include: (1) an adequate loading/unloading area; (2) washing and culling area; (3) depuration tanks; (4) seawater system including source, pumps, piping, and treatment; and (5) storage areas. In addition, depending on the states requirement, an office, laboratory, and toilet space may be necessary. The importance of seawater and types of seawater treatment have been discussed earlier.

Although the seawater quality is one of the most important elements of successful shellfish depuration, without proper processing tanks and plumbing, satisfactory depuration may not be achieved. Depuration tank design, dimensions, and construction should allow a uniform hydraulic flow of the treated seawater with minimum turbulence. Nonuniform flow of water in the depuration processing tank can create "dead spots" and oxygen deletion, all of which result in inefficient depuration at certain parts of the processing tank. The processing tanks and plumbing lines must be fabricated from nontoxic corrosion-resistant materials that are easily accessible for cleaning and inspection. Moreover, they should be self-draining and installed for effective physical or chemical cleaning and sanitation. Clearance between shellfish containers and tank walls (especially the tank bottom) must be specified for each particular tank design and depuration water input systems. A mesh-type shellfish container that allows water flow to all shellfish in the containers is required. Flowmeters are advised in order to measure the flow rate of the process water. The NSSP defines process water as the water in depuration tanks during the time that shellfish are being depurated.

Process water must meet physical, chemical, and microbiological parameters required for safety and normal physiological activity of the shellfish species. These parameters have been discussed earlier and include (1) temperature, (2) salinity, (3) dissolved oxygen, (4) pH, and (5) turbidity.

Moreover, factors such as water chemistry, tank dimensions, rate of flow, tank loading, and clearance between shellfish containers and tank walls must be well thought-out and controlled if the shellfish are to pump effectively and eliminate microorganisms that may be present.

It is important to point out that processing tanks and containers that have cracked, rough or inaccessible surfaces, or that are of improper composition may harbor accumulations of organic material in which bacteria, including pathogens, may reside and grow. This situation can allow for regular introduction of bacteria into the depuration system from contaminated shellfish. This can be prevented by remembering that surfaces must be smooth and easily cleanable if bacteria are to be flushed out in the cleaning and sanitizing process.

PLANT OPERATIONS

Operating procedures for depuration plants are required to follow an established SCPP that has been developed in accordance with Section I.2 (NSSP; 1990*b*). In addition, the procedures must also comply with applicable sanitation requirements of Sections B–F of the NSSP Manual and with the Current Good Manufacturing Practices in 21 CFR 110.

Specifically, shellfish must be washed and culled prior to depuration using a safe seawater meeting the requirements of Section B.2.b of the NSSP Manual of Operations. Basically, the shellfish are culled (removes dead, broken, and cracked shellfish). The culled shellfish are disposed of in such a manner to prevent their consumption. Throughout the whole process, the shellfish are stored and handled so that both the physiological status and bacteriological quality are not adversely affected.

Shellfish harvested from approved areas cannot be stored or processed in the same plant where shellfish harvested from restricted areas for depuration are held, unless all shellfish are to be depurated.

In addition, different lots of shellfish cannot commingle during washing, culling, processing, or packing. In other words, the identity of each harvest lot must be maintained throughout the depuration process and final packing. If shellfish in different tanks are at different stages of depuration, each tank should be labeled to show when depuration began.

Different shellfish species cannot be processed in the same depuration unit. A depuration unit is defined as a tank or series of tanks supplied by a single process water system.

There are specific requirements for the flow rate during the depuration process. The minimum flow rate of process water in each depuration tank must be one gallon per minute per standard U.S. bushel (bushel) of shellfish. The minimum volume of process water in depuration tanks shall be 8 cubic feet of water per bushel for hard clams and oysters, and 5 cubic feet per bushel for soft clams, based on the effective empty-tank volume (NSSP 1990*b*). Effective empty-tank volume is defined as the wetted volume of the unloaded tank of the overflow level (NSSP 1990*b*).

Shellfish placed in containers shall not exceed 3 in (7.6 cm) for hard clams and oysters, 8 in (20.3 cm) for soft clams. The maximum depths specified may be increased if justified by data from process verification studies.

Shellfish must be depurated for at least 48 h. Longer depuration times may be necessary under certain conditions (NSSP 1990*b*). Modifications of these criteria can only be allowed if studies show that efficient depuration occurs and results in a depurated product that contains a consistent and acceptable amount of bacteria.

Simultaneous depuration of a shellfish species from restricted areas and the processing and/or holding of the shellfish species harvested from approved areas in the same physical facility creates the potential of contaminated shellfish not being processed. Because of this problem, the ISSC in 1984 recommended against this practice. However, this practice does not preclude simultaneously processing or holding of other species in the same facility under a separate certification number or the processing or holding the same species from approved areas where there is an adequate time interval to assure that all restricted area and depurated shellfish have been removed from the premises, and thereby preventing the combining of purified and unpurified shellfish (NSSP 1990*b*).

MAINTENANCE AND CLEANING OF DEPURATION FACILITIES

Only chemical agents necessary for plant operations should be present in the plant. The need for proper and thorough daily cleaning and sanitization is well known (Hefferman and Cabelli 1971; NSSP 1990*b*). Lack of cleaning and/or improper cleaning of walls, floors, and ceilings can lead to major bacterial contamination problems. Only FDA-approved sanitizing agents should be used. In addition to the walls, floors, and ceilings, the processing tanks and containers should be cleaned and sanitized before another depuration run is started. Disinfection of the UV unit is important and should be rinsed free of any residual disinfectant. The required frequency for cleaning and bulb replacement will depend upon factors such as water temperature, turbidity, and growth of fouling organisms.

QUALITY ASSURANCE

An adequate routine sampling program must be developed and maintained to assure that the depuration plant continues to operate in accordance with the SCPP, and (2) that each group of depurated shellfish produced meets the current bacterial end-point criteria established by NSSP for the shellfish species being processed. Bacterial analyses of the samples taken must be conducted by a laboratory approved by the SSCA as described in the NSSP Manual, Part I, Section B. This SSCA approved laboratory must use APHA. AOAC, or ISSC approved methods for their analysis. The laboratory is inspected as part of the plants annual certification.

Specific sampling programs must be developed and applied to each batch of shellfish to document adequate depuration has been achieved and that NSSP end product fecal coliform criteria have been met. Shellfish samples must be collected from various locations within the tank and analyzed before and after depuration to determine whether the depuration process is working. The minimum sampling is shown in Table 13-5.

With the approval of the SSCA, both zero-hour and 24 h samples may be taken and analyzed bacteriologically. Results from these analysis can be used to predict if the shellfish are acceptable for sale. This approach can only be used when there is both the appropriate and statistical analysis and history to ensure an acceptable product.

For the most part though, evaluation of the plant's depuration efficiency must be continuous on the basis of end-point bacteriological assays of the depurated product using a geometric mean and upper 10% level (i.e., no more than 10% of the samples analyzed can exceed this value) for the species listed in the following Table 13-6 (NSSP 1990). All final depurated product data from every group of 10 consecutive process batches must be used. For depuration plants processing less than 10 batches in a three-month period, if

TABLE 13-5. Minimum Sampling Schedule

		Minimum Number of Samples	
No. of Harvest Areas	Pollution Variability[2]	Incoming Shellfish	Final Product[1]
Single area	Low	Periodic[3] single samples	Single samples of each harvest lot[4] in each process batch[5]
Single area	High	One sample from each harvest lot	Duplicate samples of each harvest lot in each process batch
Multiple area	Variable	Periodic single sample from each area	At least one duplicate sample from each highly variable area each week in each processing batch. Single samples of each process batch at other times during the same week

Source: NSSP Manual of Operation, Part 2, Section 1.

[1] If final-product criteria are met prior to 48 h, the product may be released after 48 h of depuration without a finished product sample.

[2] *Pollution variability* is determined by sanitary survey and water quality data evaluation as part of the SCPP.

[3] The frequency of sampling is determined by the SSCA based on the information from the SCPP.

[4] A *harvest lot* is defined as all shellfish harvested from a particular area at a particular time and delivered to one depuration plant. The designation of areas is left to the SSCA.

[5] A *process batch* is a quantity of shellfish used to fill each separate depuration unit. A depuration unit is a tank or series of tanks supplied by a single process water system.

TABLE 13-6. End-product Standards for Overall Depuration Plant Performance Evaluation

	Fecal coliforms per 100 g	
Species	Geometric Mean[1]	Upper 10%
Soft Clam (*Mya arenaria*)	50	130
Hard Clam (*Mercenaria mercenaria*)	20	70
Oyster (*Crassostrea virginica*)	20	70

Source: NSSP Manual of Operations, Part 2, Section 1.

[1] Each sample having an indeterminate low value will be assigned a value of 10 for computational purposes.

TABLE 13-7. End-product Standards for Each Process Batch of Shellfish

		Fecal Coliforms per 100 g		
No. of Samples	Shellfish Species	Geometric Mean not to Exceed	One Sample May Exceed	No Sample To Exceed
1	S.C., O.,	—	—	170
	H.C.	—	—	100
2	S.C., O.,	125	—	170
	H.C.	75	—	100
3	S.C., O.,	110	—	170
	H.C.	45	—	100
5	S.C., O.,	50	100	170
	H.C.	20	45	100
10	S.C., O.,	50	130	170
	H.C.	20	70	100

Source: NSSP Manual of Operations, Part 2, Section 1.

S.C. = Soft Clam, O = Oyster, H.C. = Hard Clam, — denotes no value.

any *one* sample exceeds the upper 10% level or if the geometric mean is in excess of the value specified, the plant is not in compliance with NSSP end-product standards.

When shellfish are harvested from new areas or are from a harvest area that historically has highly variable water and/or shellfish quality, more intensive sampling is warranted.

The end-product standards for each process batch of shellfish are shown in Table 13-7. The depurated shellfish should not be released for market unless the number of fecal coliforms per 100 g of shellfish meet the criteria established by the SSCA are met at 24 or 48 h for each process batch.

All shellfish quality assurance records for both the depuration plant and the laboratory must be maintained for at least 2 years.

In summary, the development of an adequate sampling program is essential to determine if critical environmental conditions are suitable for purification and if the depurated shellfish being released for market meet accepted NSSP criteria. Because of variations in plant size and design, a sampling program must be developed for each plant. The sampling program should include at least the following: (1) total coliform level of both the raw water source and treated water in the depuration tank; (2) fecal coliform levels of incoming (zero hour) and final-product (end-product) shellfish as earlier; (3) temperature, turbidity, and salinity of the processing water; and (4) dissolved oxygen levels at the depuration tank outlet.

PACKING AND LABELING OF DEPURATED SHELLFISH

Depurated shellfish shall be washed and culled after depuration and packaged in clean containers fabricated from safe materials. Different harvest lots of shellfish shall not be commingled during packing. Each container of depurated shellstock shall have securely affixed to it a durable, waterproof tag or label bearing information that will identify the shellfish as being depurated and will allow the shellfish to be traced back to a specific batch and specific harvest lot.

The satisfactory compliance to the above items will be satisfied when the following have been put into operation:

After depuration, the shellfish are packaged in clean shipping containers fabricated from safe materials and cannot be repackaged after leaving the depuration plant. Use of tamper-evident seals is highly recommended to prevent repackaging.

Moreover, shellfish from different harvest lots are packaged separately to prevent commingly and are not permitted to remain unrefrigerated for long periods of time, especially during warm weather. The shellfish should be held at a temperature no higher than 50°F (10°C).

Durable, waterproof tags or labels are securely affixed to each package of depurated shellstock. These tags or labels must be approved by the SSCA and include the following information in legible and indelible form: (1) the scientific name, or the common name of the shellfish; (2) quantity of shellfish; (3) name, address, and depuration plant certification number of the processor; (4) date of processing; (5) depuration cycle number, if any; and (6) the identity of the harvest area.

SUPERVISION AND RECORDKEEPING

Management shall clearly designate a supervisor who is knowledgeable and competent. This individual should be present at the plant to assure that both operating procedures and personal hygiene practices are followed. The supervisor should also maintain a complete and accurate record that will permit each package of depurated shellfish to be traced back to its original. A current copy of the depuration plant operating procedure should also be kept at the plant site.

A record should be established for each depuration cycle. This is usually accomplished by assigning each cycle a sequential number so that the monitoring and sampling data for that cycle can be identified with certain cycle numbers. A cycle may include more than one harvest lot of shellfish provided all lots begin the depuration cycle at the same time and the conditions of depuration are the, and harvest lot separation is maintained.

Complete and accurate records must be maintained for at least 90 days. This enables a package of depurated shellstock to be traced back to the process batch and harvest lot. Purchases should be recorded in a permanently bound ledger book or stored in a computer file for easy access.

It is essential that detailed identification information be maintained on all harvest lots of shellfish and shipping containers of depurated shellfish. This information will allow the tracing of implicated shellfish in the event that a disease outbreak occurs back to a specific depuration cycle, and to the harvest area. Moreover, maintaining complete and accurate records of all transactions enhances business integrity of the harvesters, processors, and dealers. Records of product samples and critical parameters within the should be kept for at least 2 years.

INTERNATIONAL DEPURATION

The controlled purification of shellfish is practiced in many countries around the world. Countries such as Australia, New Zealand, Southeast Asia, United Kingdom, Spain, Italy, France, Denmark, Canada, and Turkey use depuration on both the mandatory and voluntary levels. The current state of the technology used is in some cases more advanced than what is practiced in the United States.

Australia, New Zealand, and Southeast Asia

The Sydney rock oyster (*Saccostrea commercialis*) is the principal shellfish species harvested in Australia. In New South Wales alone the estimated revenue is in excess of 30 million Australian dollars (Department of Health, N.S.W.) and a total industrial values has been estimated as high as 200 million Australian dollars (Ayres 1991). Depuration is mandatory for all such shellfish before being sent to market. The Pacific oysters (*Crassostrea gigas*) harvested in the states of Tasmania and Victoria do not have to be depurated if they meet the bacteriological standards set by the National Health and Medical Research Council (1987).

Australia uses ultraviolet light in the depuration of its shellfish. The use of ozone was also used but recently the plant was shut down (C. Phillips, personal communication). The depuration facilities range in size from full size plants to several trays located in a corner of a boat house. New Zealand harvests and exports several types of molluscan shellfish. These include mussels (*Perna canaliculus*), oysters (*Saccostrea glomerata, C. gigas*), scallops (*Pecten novaezelandiea*) and to a lesser extend clams (*Paphies subtriangulata* and *P. donacina*) and the cockle (*Chione stutchburyi*). The total export tonnage reached 7775 tons of shellfish worth in excess of 36 million New Zealand dollars (Fletcher and Hayden 1991).

Both Australia and New Zealand monitor temperature, salinity, turbidity, flow rates, and fecal coliform levels in the shellfish. The minimum depuration time for the Sydney rock oyster in Australia is 36 h. In New Zealand, unless a plant has proved its effectiveness in reaching require bacteriological levels, all shellfish are required to be depurated 48 h for all species. The lone exception is the New Zealand cockle (*Chione stutchburyi*). This shellfish actively pumps only for 4–5 h every 12 h tidal cycle, thus requiring 120 h of processing to ensure 48 h of pumping.

The ASEAN community include Malaysia, Singapore, Thailand, The Philippines, and Indonesia. Ayres (1991) reported that shellfish landings in 1985 were over 2.8 million metric tons which represented some 65.5% of the total world landings. The main species of bivalves include members of the families *Ostreidea, Mytilidae*, and *Arcidea* and oyster species belonging to the genera *Crassostrea* and *Saccostrea*. In certain areas, *Ostrea* sp. are of local importance. Mussels of the genera *Mytilius* and *Perna* and cockle species *Anandara* and *Arca* are also cultured. Ultraviolet light assisted depuration is the system of choice.

Europe

The United Kingdom, Spain, Italy, France, and Denmark all require mandatory depuration of all molluscan shellfish (Richards 1991). The U.K. probably has the smallest shellfish industry in the European community, centered mainly in England and Wales. The commercially important species include oysters (*O. edulis* and *C. gigas*), mussels (*M. edulis*), cockles (*C. edulis*), clams (*M. mercenaria*) and scallops (*P. maximus* and *C. opercularis*) (West 1991). The 1988 production value for shellfish harvested in England and Wales was approximately 1.1 million English Pounds. Scotland's shellfish industry netted for the same period roughly half of that, with 570,000 English Pounds (West 1991).

Historically, depuration has been practiced in the U.K. for 75 years. The first plant, which used chlorine to sterilize seawater, went into operation in 1914 in North Wales. Today ultraviolet-light-assisted depuration is the system of choice.

Spain has the highest shellfish consumption per capita in the European community. Approximately 225,000 to 275,000 metric tons are produced annually. The commercially important species are flat oysters (*O. edulis*) and mussels (*M. galloprovincialis*) that are cultured on suspended "rafts". Harvested species include grooved, pullet and banded carpet shells (*Venerupis decussata, V. pullastra,* and *V. rhomboideus*), striped venus (*Venus gallina*), and the cockle (*C. edule*) (Monroy et al. 1991). Most of the 63 authorized depuration plants in Spain (as of 1988) utilize gaseous chlorine, and a few smaller plants use ultraviolet light or ozone (Monroy et al. 1991).

Like Spain, Italy also used chlorine as the major disinfectant in its shellfish purification facilities. Currently, over 50 depuration plants are authorized, though not all are active at all times (Canzonier 1991). These plants are of flow-through design. Chlorinated seawater is passed through activated carbon filters to remove residual oxidants.

France, in contrast to Spain, utilizes ozone as the preferred method of depurated shellfish. To a lesser degree, chlorine, bromine, and ultraviolet light are also used and regulated. Some 200,000 tons of oysters of cultured, harvested, and sold annually with a revenue approximately 2300 million Francs, which ranks France as the largest oyster producer in Europe. The major types of shellfish cultured include oysters (*C. gigas* and to a lesser extent *O. edulis*), mussels (*M. edulis* and *M. galloprovincialis*), and some *Veneridea* or clams (*Ruditapes philippinarum*).

Denmark utilizes UV light to sterilize the processing water used in depuration. Originally, *O. edulis* was the endigenous species but a series of oyster diseases in the 1980s nearly wiped the entire population. The main shellfish species today is the maricultured and more resistant *C. gigas*. The depuration treatment time for oysters is 7 days.

Turkey

Turkey produces approximately 625,000 tons of seafood annually, of which only about 30,000 tons are shellfish. Significant species include mussels (*M. galloprovincialis*), clam (*T. decussatus*), oysters (*O. edulis*), horse mussel (*Modiolus barbatus*), Kidonya (*V. verucosa*), and Lupina (*V. gallina*). Only one modern depuration unit is currently in operation. More plants will be needed to satisfy the demand being created by the European market (Alphaz and Balaban 1991).

Canada

The depuration of shellfish is fairly new in Canada. Container relaying was the method of choice for the purification of shellfish during the 1970s and early 1980s but was abandoned due to high operating costs and PSP contamination (Doncaster 1991). Commercial depuration of the soft shell clam (*M. arenaria*) began in 1984. Presently, only two registered depuration plants are currently in operation. Both are located on the Atlantic coast in New Brunswick. Only 10,000 bushels are processed annually.

SHELLFISH RELAYING

This section will review the basic concepts and the U.S. NSSP guidelines and rules for relaying. Shellfish have the ability to purge themselves of certain abiotic and biotic substances when placed in clean seawater. The rate of purging depends on the type of contaminant, species of shellfish, water quality, length of relay, method of relay, and various environmental factors. The basic concept of relaying involves the harvesting of shellfish and their transferal from restricted areas to approved areas for natural biological purification. Relays can be operated by state agencies or the shellfish industry. The overall relay operation must assure that the relayed shellfish are all actually relayed to approved or conditionally approved waters and sufficiently purified. Specific relay control procedures include: (1) issuing special permits for harvesters; (2) sampling of shellfish; (3) marking of approved areas; (4) monitoring environmental parameters; (5) keeping records; (6) establishing interagency agreements; (7) imposing quarantine measures; and (8) increasing patrols (NSSP 1990).

The state shellfish agency and the State Shellfish Control Agency (SSCA) have developed and written specific relay operating procedures. These procedures will include: (1) the source of species of shellfish harvested; (2) information on the quality of shellfish and/or water quality before harvest; (3) information on the water quality and/or shellfish meats indigenous to the relay area; (4) information of the quality of relayed shellfish after relaying; (5) specific periods of the year during which relaying may or may not be conducted; (6) acceptable methods of marking the approved or conditionally approved areas used for relaying; (7) the acquisition of a special harvesting permit to licensed harvesters as described in Section F of the NSSP manual of operations, Part I (1990); (8) the method of transportation to the relay site and the method of deposition for the to be relayed shellfish; (9) the method for adequate separation of different lots of relayed shellfish; and (10) that records will be maintained and filed by permitted relayer/harvester.

Relaying for Natural Depuration

This technique is effective for bacterial depuration, and is practiced in the United States. Erdman and Tennant (1956) showed an average of 85% reduction in the number of coliform bacteria in the soft-shell clam (*Mya arenaria*). The major problem associated with relaying is that it is a labor-intensive process, and up to 50% of the original harvest can be lost (Furfari 1976). In addition, the relocation areas are subjected to varying pollution and physical conditions that might interfere with depuration (Blogoslawski 1983; Furfari 1979).

14

Using Indicator Information for Managing Risks

Dale Hattis

Environment, Technology and Society Programs
Clark University, Worcester, MA 01610-1477

INTRODUCTION

A recent National Research Council report has estimated that current monitoring programs for marine pollution cost $133 million/year (NRC 1990). Overall, this expenditure was considered to be reasonable in relation to the $5–9 billion/year that the report estimated to be spent in the control of discharges to the marine environment. However, the authors of the report were critical of the institutional arrangements tying the design of various monitoring systems to both basic research on causal processes producing potential effects of social significance, and the needs of the array of people who make real choices related to the protection of both public health and the marine environment:

> Monitoring is generally not well coupled with research programs designed to improve the appropriateness of routine measurements and allow interpretation of the implications of monitoring results. Most marine environmental monitoring programs are technically sound; it is the overall design and institutional context that limits the usefulness of the information that results (NRC 1990, pp. 3–4).

Previous chapters in this book document a range of opportunities to do different kinds of monitoring than are now routinely performed. The NRC report, although hailing the traditional use of measurements of coliform bacteria as having successfully led to appreciable reductions in risks from fecal-derived pathogens in shellfish since the 1920s, also recognized the potential desirability of developing monitoring approaches that are more attuned to the presence/concentrations of pathogenic microorganisms themselves, rather than *E. coli* and related proxy "indicator" organisms.

This chapter addresses the issue of how exactly "indicator" information is to be tied to the practical needs for decision making by public agencies. To do this, it is first necessary to define some different categories of decisions that public agencies make, which might be informed by imaginable kinds of indicator information. Then, for each decision category, distinctive properties of helpful information will be identified.

TYPES OF DECISIONS WITH DIFFERENT REQUIREMENTS FOR INFORMATION

There are, of course, a wide range of "decisions" that are made by public agencies. However, for our purposes it will be convenient to define only three broad categories into which a wide array of decision subtypes can be grouped:

- *Generic "standard-setting" decisions*—in which a general rule is set-defining, for example, water quality criteria, standards for permissible levels of contaminants in products for public sale.
- *Individual adjudicatory decisions*—in which, for example, specific beaches are closed to swimming, specific batches of product are seized as unhealthful, or requirements for depuration are defined for shellfish harvested in a particular area.
- *Priority-setting decisions*—in which the limited time of inspectors is allocated to sampling specific types of growing waters, monitoring the performance of specific types of processing plants, or where limited research budgets are allocated to researching specific problems.

The discussion below begins with the second of these. This is because the first decision type, standard setting, largely contributes to the management of risk by setting up systems for making individual adjudicatory decisions. Similarly, the third, priority setting, largely contributes to risk management by fostering efficient implementation of standards.

INDIVIDUAL ADJUDICATORY DECISIONS

One basic requirement for measurements that are to be used in enforcing legal standards is that they are capable of supporting decisions that will withstand possible legal challenge. A starting point for this is that the measurements satisfy the familiar scientific criteria of "validity" and reliability" as measurements.[1] Classically, a measurement is "valid" if it in fact represents what it purports to measure, without known sources of systematic error. For example, if a grocer is weighing meat, the measurement would be "invalidated" if he were to put a thumb on the scale (in that case, the scale would be recording the pressure exerted by the thumb, in addition to the actual weight of the meat). On the other hand, a measurement is said to be relatively "reliable" if the random error in the measurement is within reasonable bounds in comparison to the existing technical state of the art. Thus, we would say that if the grocer were to use a bathroom scale to weigh the meat (which only gives readings that can be read within half a pound or so), that would be an unacceptably "unreliable" measure, by the standards of accuracy we have come to expect in food stores.

More formally, the reliability of a measurement is sometimes expressed as the ratio of the actual variation in a group of things being measured (in the form of its "true" variance, σ_{T}^2) to the variation that is observed (the sum of the variances attributable to random measurement error, σ_{E}^2, and the "true" variation):

$$R = \frac{\sigma_{\mathrm{T}}^2}{\sigma_{\mathrm{E}}^2 + \sigma_{\mathrm{T}}^2}$$

As discussed by Fleiss (1986), this measure of reliability is useful in that the reciprocal of R directly affects the sample size required to have a given probability of detecting an effect at a given level of statistical significance.

Beyond basic acceptability of a particular type of measurement as a measurement, of course, we should want some rational relationship to actual risk. Other things being equal, the indicator should be as closely coupled with the causal process involved in producing harm as can be managed, and there should be as good a correlation as possible between the level of the indicator and the presence of risk or the probability and/or severity of harm. The better this correlation is, the greater will be the opportunities to minimize the bad consequences of "incorrect" decisions—both "false positive" findings in which the indicator signals a need for protective action when no such need is actually present, and "false negative" findings in which the indicator fails to give a signal when preventable harm is in fact likely. By analogy with the terminology frequently used in connection with medical diagnostic tests, we can define the "sensitivity" of an indicator test as how accurately the test

1. In addition, there are some technical requirements for law enforcement use related to documentation of the chain of custody of samples, documentation of any calibration procedures on the days the samples were processed in the laboratory, etc. These needs will not be discussed further here.

TABLE 14-1. Results of 1000 Tests, Applied in a Clinical Setting

	Test Outcome		
True Disease Status	Positive	Negative	Total
Positive	285	15	300
Negative	35	665	700
Total	320	680	1000

identifies cases where a true hazard (or an unacceptable probability of harm) is actually present—thus avoiding "false negatives". By contrast, "specificity" can be defined as how accurately the indicator test identifies cases where no hazard (or only an "acceptable" probability of harm) is present—thus avoiding "false positives."

One point that is not always appreciated in setting the threshold levels on a test that will be used to initiate protective action is that the relative numbers of false positives and false negatives—and the resulting costs in the form of mistaken decisions of the two types—will differ depending upon the baseline incidence of "true positive" cases in the population of things being sampled. This will be clarified by carrying the analogy with medical diagnostic tests a little further.

First, let us imagine that we have a test for a rare bacterial disease that is fairly good at distinguishing people with the disease from those without the disease. Consider that, using a particular cutoff level to define "positive,"[2] the test gives a positive result for 95% of those with the disease and a negative result for 95% of those without the disease. This test is initially developed for application in a clinical setting; there is already good reason to believe that patients to be tested might have the disease—a true incidence of cases of the disease in the tested population of 30%. For every 1000 people who are given the test, we would then expect an outcome of the type shown in Table 14-1. In other words, of those who tested positive, most but somewhat short of 90% (285/320) actually have the disease. Of those who test negative, nearly all (665/860) are actually free from the disease. In a clinical setting, this might prove to provide reasonable guidance on the issue of who might benefit from the use of a particular antibiotic that treats the bacterium in question effectively, if the potential health cost of the relatively few "false negatives"

2. In this case, such a cutoff level can be either explicitly defined as the number of colonies of a particular type, after growing on a specific selective medium, or implicitly defined, as the presence or absence of *any* colonies of a distinctive type, after a specific volume of a test fluid has been sampled and grown in a selective medium.

was not very great. (If, for example, these could be detected in a followup round of testing for those who test "negative" but still manifest symptoms a day or two later.)

Let us now imagine that this disease is found to be transmitted via consumption of raw shellfish, and that a public agency wishes to adapt the test for screening samples of shellfish meats grown in different areas—with the "positive" samples required to be either discarded or put through a depuration process prior to marketing. The public agency may well be tempted to use the same culture procedures and the same cutoffs for defining "positive" samples as were initially defined for the clinical setting. For simplicity, suppose that the intrinsic sensitivity and specificity are found to be the same as in the clinical setting (95% of truly positive samples test positive and 95% of truly negative samples test negative), but that both the incidence and the public health significance of positive samples is greatly changed from the clinical setting. Consider that only one sample in 1/1000 now is a "true positive". In this case, for each 100,000 samples screened we could expect an outcome of the type shown in Table 14-2. One can immediately see that the relative numbers of false positive and false negative findings have changed dramatically. Seeing these data, a production manager's response might well be: "You mean to say that you want to throw away or further process 5% of my total production (5190 "positive" testing lots out of 100,000), even though 98% of these (4995/5190) are perfectly harmless "false positives?" On the other hand, a public health advocate, or an attorney for the affected liability insurance company of the shellfish processor might well react, "You mean to say that even after passing the test, you expect that over one lot in 20,000 (5/94,910) could be expected to be dangerous?"

In practice, implicitly or explicitly, those who would make this adaptation of a clinical test to the lot-screening context, in setting the cutoff level to be used in defining a "positive," would need to decide the relative value they wished to assign to false positives and false negatives. Defining a lesser value on the test as "positive" would be expected to reduce the number of

TABLE 14-2. Results of 100,000 Tests, Applied in a Shellfish Lot-Screening Setting

	Test Outcome		
True Disease Status	Positive	Negative	Total
Positive	95	5	100
Negative	4995	94,905	99,900
Total	5190	94,910	100,000

contaminated lots that escaped detection, while increasing the percentage of output that is classified by the screening test as suspect and requiring further processing or disposal. Conversely, defining a higher value on the test as required to classify a lot as "positive" would have the effect of increasing the number of contaminated lots that escaped detection, while lowering the percentage of output that is classified by the screening test as suspect. A third option, of course, would be to improve the test (perhaps by coupling a screening threshold change with a confirmatory supplemental test to be applied to either the positive or the negative samples)—lowering both the percentages of false positives and false negatives, albeit somewhat increasing the costs of the testing program in the process. A formal decision analysis would seek to determine the combination of testing procedures, cutoffs for defining "positives," and modes for treatment of "positive" lots which minimized the aggregate of anticipated costs (including testing costs, both health harm and monetary costs from false negatives, and increased processing costs/value reductions for the false positive lots of product).

Posing this example illustrates the importance of (a) the expected frequency of positive findings (b) the assigned "value" of false positive findings, and (c) the assigned "value" of false negative findings as key factors that should affect a rational decision maker in selecting the cutoff value to define positive findings on a test. A less obvious implication is that where decision makers have reason to believe that one or more of these factors differs across easily distinguishable subcases, minimizing the sum of all anticipated costs may well require different standard levels for the subcases. For example, different temperature conditions in winter vs. summer, or different types of sites for collection (fresh water, vs. estuarine, vs. open coast) may well give rise to greatly different expected frequencies of "true positive" cases of contamination of shellfish from a particular type of growing site; different types of shellfish may have vastly different incidences of contamination; and different contaminating organisms may well be expected to pose vastly different public health risks per unit of concentration. In the light of such differences, it would be a surprise if the same standard level for defining contamination were capable of minimizing relevant costs across different seasons, different organisms used for seafood, or different contaminating species of microbes.

One final point needs to be made about the connectedness of test procedures to putative causal pathways along the road to harm. The use of more easily measurable quantities that are "proxies" for the true conditions that might cause harm inherently carries the danger that the connection between the proxies and the harm-producing conditions might become further weakened in practice as commercial procedures are changed to reduce the frequency of positive findings. It is easy for those in the field who are implementing a specific test procedure to make the mistake that the things that are counted by that procedure are the things that count. The implementors need to be alert to situations in which a change of harvest or processing conditions is likely to change the performance of a test used for screening in ways that might diverge from changes in the risks the test is intended to help prevent. And in selecting the things that are to be counted in screening tests, decision makers must pay attention to the incentives for change in commercial practices that are thereby encouraged, and whether or not these changes are likely to change risks and costs in the most helpful and desirable ways.

PRIORITY-SETTING DECISIONS

A second common type of decision is how a public agency should allocate its scarce resources among a number of competing potential targets for attention. How should an inspectorate divide its time among different types of vessels or processing plants? How much sampling of beaches of different types should be done, and which kinds of organisms should receive relatively more or relatively less attention?

Frequently, such discussions begin and end with an appraisal of where the worst problems are—often expressed in the most simple kind of statistical units, such as the incidence of reported disease per 1000 pounds of meat, etc. What this type of analysis leaves out, unfortunately, is any appraisal of the relative amount of abatement of the hazard that might result from the allocation of specific types of preventive effort across different candidates for action. The appropriate measure to guide targeting is not how much evil is there in different parts of the world, but how much *reduction* in evil might I expect *per unit of agency resources expended* in different parts of the world. More formally, a good "figure of merit" for priority setting should have two elements:

- A numerator which represents a value-weighted change in the world (e.g., reduced restricted-activity days due to sickness, or cases of gastroenteritis prevented). Often it is easier to infer this from the internal reports of previous agency activity than from general health statistics. For example, how often does an inspector of a particular type of establishment find a condition that is deemed worth citing and correcting? How frequently does a water sample from beach of a specific type come back positive, and what is the estimated benefit of the preventive actions that are taken following such a finding?
- A denominator which represents the expenditure of whatever limiting type of agency resources (inspector time, dollars) is to be allocated among candidates for attention.

In setting up such a system, the agency should, of course, carefully consider what is actually limiting to its effective prevention activity. There may well be a need to make an initial sort of different kinds of candidates for resource expenditure that are limited by fundamentally different, noninterconvertible resources. For example, agency efforts to set standards to control different kinds of organisms may well be importantly limited by the available funds to contract out key pieces of research, test development, etc. On the other hand, the time of skilled inspectors may well limit many kinds of enforcement activity.

A final caveat that is needed here is that it will generally be unwise to follow any specific priority-setting scheme too religiously. The Internal Revenue Service, for example, while devoting the bulk of its resources to auditing tax returns that its discriminant system has rated as having a high likelihood of yielding increased tax revenue, saves at least a modest quantity of auditor time to monitor returns that do not carry the highest priority rating. This is because it helps encourage "voluntary compliance" if those who know or suspect they are not in the highest-priority category know that they, nonetheless, face at least some modest risk of an audit.

STANDARD-SETTING DECISIONS

Of our three types of decisions, standard-setting carries the highest social profile, involves the greatest inputs of resources per individual decision, and clearly provides the context that can benefit from the most extensive inputs of basic scientific information. The aim is to develop as clear as possible an understanding of the likely consequences of different options available to the decision maker—where the effective "options" can differ in terms of the nature of the parameters that are chosen for standard setting (e.g., fecal coliforms vs. total coliforms vs. direct enumeration of specific pathogens; growing water monitoring vs. monitoring of bacterial contamination in meats from newly harvested shellfish), the levels of the selected parameters that are considered to warrant different actions, and the range of applicability of standards (e.g., freshwater vs. estuarine vs. unprotected ocean beaches).

To evaluate the competing policy options, the analyst needs predictive models that represent the actual causal structure of the world as it would be influenced by the standards being evaluated. Arriving at such models, of course, involves the whole panoply of the scientific method, from the generation of theories reflecting reasonably possible alternative concepts of the state of relevant causal relationships, to the testing of these alternative theories/models with observations in the natural world and laboratory experiments. The NRC report cited earlier considered the validation of predictive models to be one of the key uses of monitoring information—citing as examples the original work of Streeter and Phelps (1925) on dissolved oxygen as a key parameter determining needs for wastewater treatment for Ohio River dischargers, and the validation of more recent water quality models for Boston Harbor, New York Harbor, the Sacramento-San Juaquin Delta (NRC 1990, pp. 23–24).

As every good graduate student finds out, part of becoming a skilled scientist is learning to recognize when causal assumptions are being made uncritically without full examination and testing of alternative mechanistic possibilities. For example, one causal presumption that seems to have been made in the early days of standard setting for microbiological standards for shellfish growing waters, and in defining requirements for the performance of depuration plants, is that there are thresholds in the dose–response relationships for the risk of infection from pathogenic organisms. Earlier in this volume, Kator and Rhodes report that those who developed the original growing water standard for shellfish derived their target density for total coliforms using information on the ratio of the coliforms to pathogens of concern in a large point source of domestic sewage, and based on "coliform data showing that bivalves (i.e., hard clams) grown in water at or below that standard indicator level would not bioconcentrate pathogens to densities exceeding the presumed infective dosage." In two recent articles on depuration, Richards (1988, present vol. 1993) has included the statement, "The presence of trace amounts of bacterial pathogens in depurated shellfish does not generally confer illness to consumers because threshold levels required to cause illness are seldom reached." The alleged threshold nature of dose–response relationships for bacterial pathogens is contrasted by Richards with enteric viruses, which are reported to be "infectious even in very low numbers (Plotkin and Katz 1967; Schiff et al. 1984)."

Working in a very different context, I have recently had the opportunity to compare expectations under an alternative mechanistic theory with infection and illness dose–response data for an appreciable number of experiments in which different doses of a variety of pathogenic microorganisms were administered to people in clinical settings (Hattis 1990).[3] I examined dose–response data for 14 human experiments in which 18 infection or disease endpoints were measured for a total of nine different organisms (five viruses, three bacterial genera, and one protozoan).[4] In this diverse group of studies, there is not a single case where the data indicate threshold-like dose–response behavior (e.g., a finding that a large proportion of individuals are infected at

3. My analysis extends and corrects an earlier analysis of some of these data using three different mathematical models by Haas (1983).

4. This is not the entire world literature of such experiments. In particular, I have recently obtained a review by Blaser and Newman (1982), which cites a number of other experiments with various species of *Salmonella* administered in various vehicles.

TABLE 14-3. Dose Response Relationship for the Production of Typhoid Fever by Oral Administration of *Salmonella typhosa*[a]

No. of Bacteria Administered	No. of Volunteers Challenged	No. with Disease	Incubation Periods[b]	
			Median	Range
10^9	42	40 (95%)	5	3–32
10^8	9	8 (89%)		
10^7	32	16 (50%)	7.5	4–56
10^5	116	32 (28%)	9	6–33
10^3	14	0		

a. Source: Hornick et al. (1970).
b. Days to disease manifestation (fever).

one dose, and that many fewer are affected at the next lower dose than would be expected on the basis of a simple one-hit dose–response model[5]). Quite the contrary, very often the data have quite shallow dose–response slopes, such as the Hornick et al. (1970) data for *Salmonella typhosa*, shown in Table 14-3. In that data set, it can be seen that a 100-fold reduction in dose from the apparent ED_{50} (10^7 organisms) suffices to reduce the incidence of disease only by about 2-fold. This reduction in incidence is very much less than would be expected if one had a simple one-hit model of disease incidence, let alone a model with a high requirement for cooperation among organisms (and, hence, a steep dose–response slope).

Figures 14-1–14-9 show simple log probit plots of some of these data sets. The assumption underlying such plots is that susceptibilities for infection or illness are lognormally distributed in the population. A lognormal distribution of susceptibilities would be produced if (a) the causes of individual differences are numerous, and (b) the action of specific factors contributing to individual differences tends to be multiplicative. In such plots, the probit slope (the coefficient of x in the regression line) is a measure of the overall interindividual variability in susceptibility suggested by the data set. (The reciprocal of the probit slope is the standard deviation of the distribution of the logarithms of individual infection susceptibilities.) Relatively low values of the probit slope correspond to greater interindividual variability. A probit slope of 1 indicates a situation in which the susceptibilities of 95% of the population are spread out over a 10,000-fold range—from 100-fold less than the median to 100-fold greater. In general, the amounts of interindividual variability indicated in these data sets are enormous in comparison with those observed for toxic effects of chemicals (see Ballew and Hattis 1989).

5. The data of Minor et al. (1981) on infection of infants with poliovirus 1 (Sabin) come closest to providing evidence of such a pattern, but this experiment simply has too few subjects to give a clear indication of threshold-like behavior.

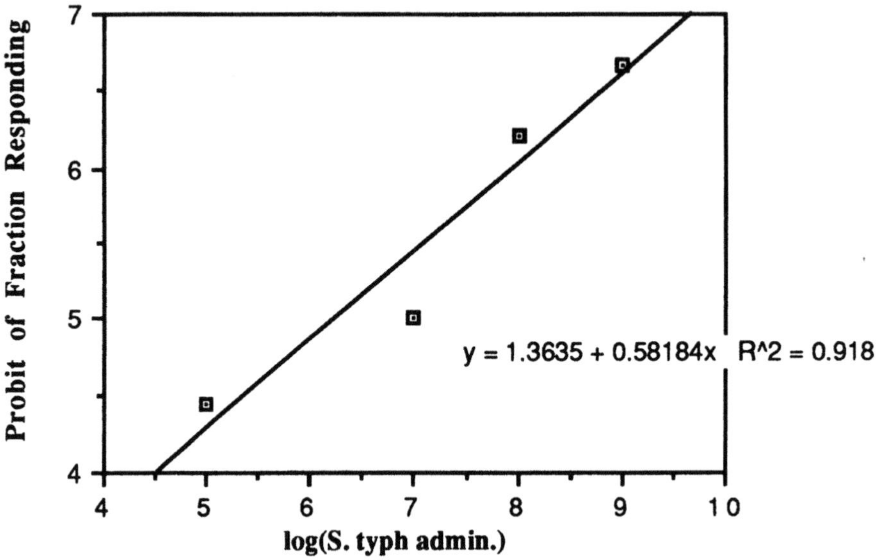

FIGURE 14-1. Log Probit Dose Response for Typhoid After *Salmonella Typhosa* Administration (Data of Hornick et al.)

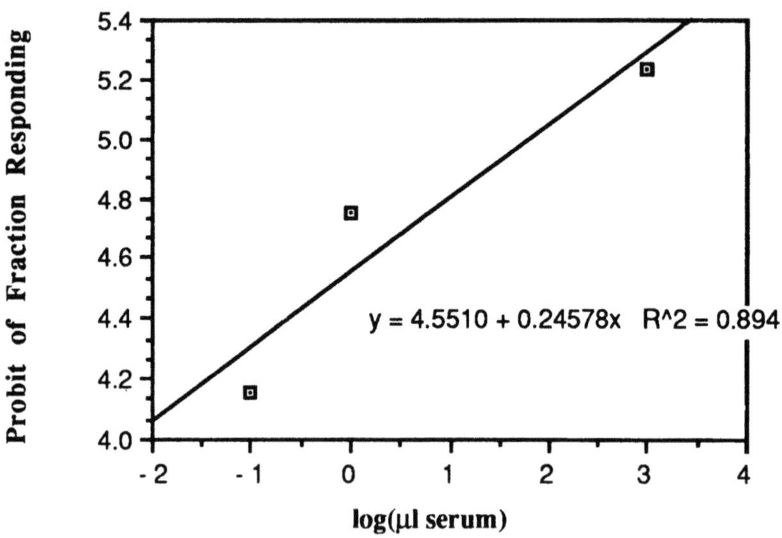

FIGURE 14-2. Log Probit Dose Response for Barker and Murray Clinical Hepatitis Cases

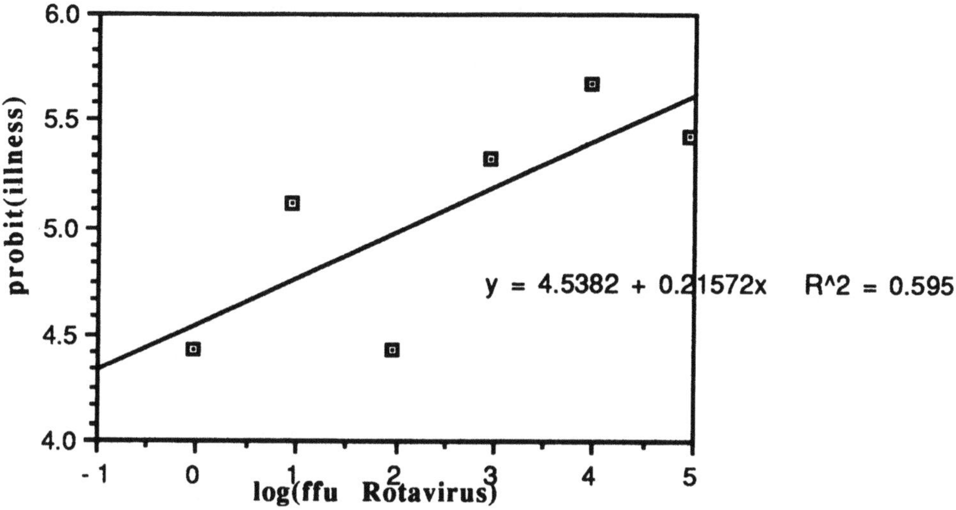

FIGURE 14-3. Log Probit Plot of Illnesses Resulting from Rotavirus Administration (Data of Ward et al., 1986)

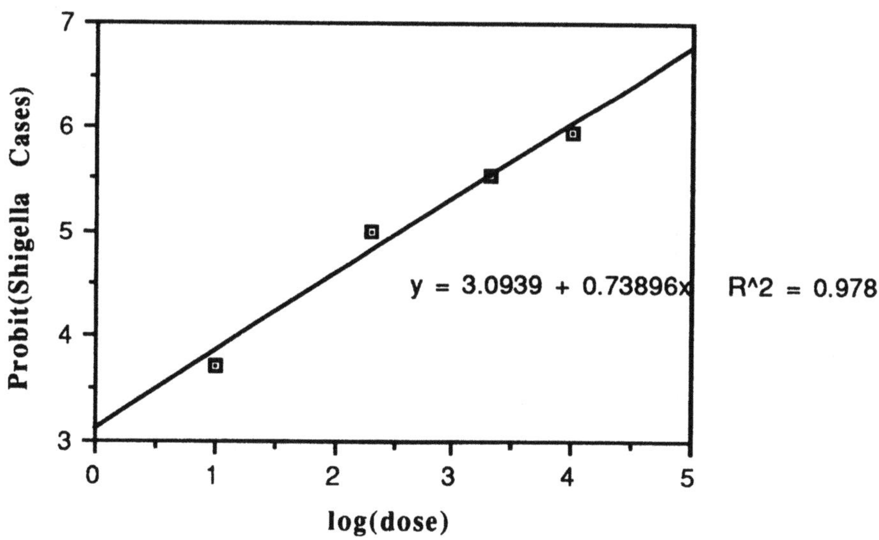

FIGURE 14-4. Log Probit Plot of *Shigella Dysentariae* Illnesses (Levine et al., 1973)

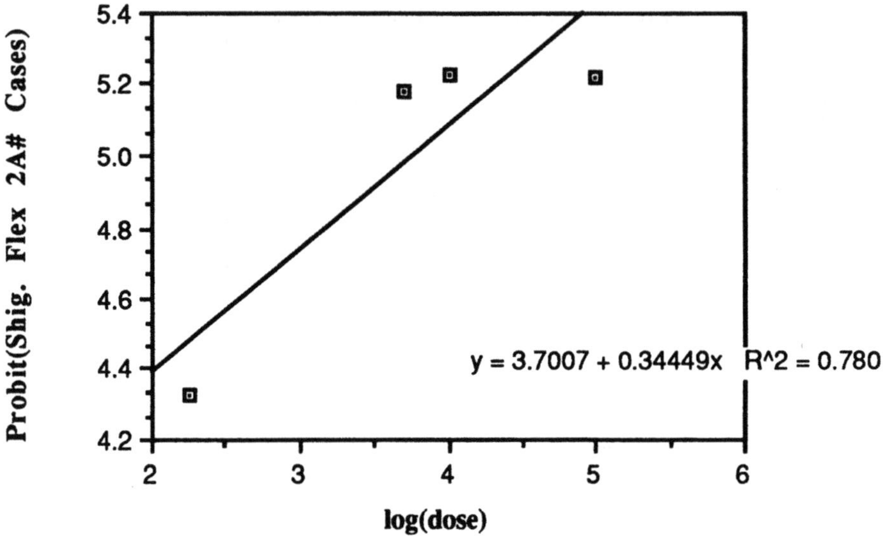

FIGURE 14-5. Log Probit Plot of *Shigella Flexneri* 2A#. Disease Data from Dupont et al.

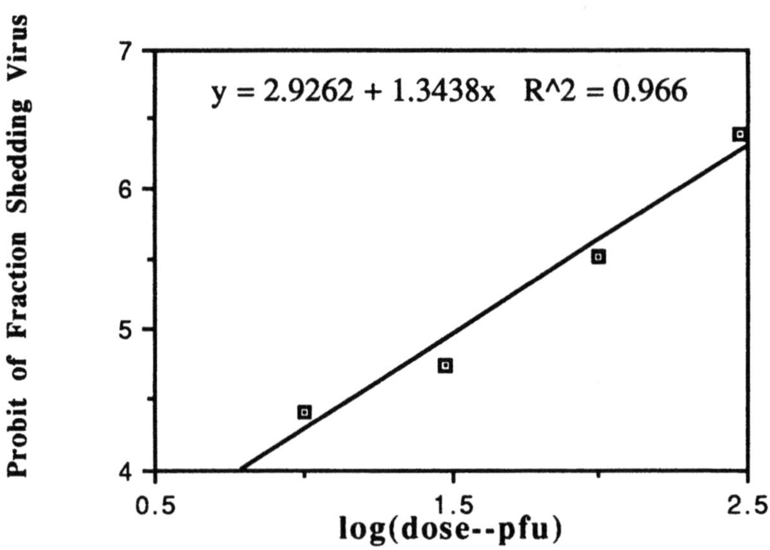

FIGURE 14-6. Log Probit Dose Response for Ward and Akin (1984) Echovirus 12 Infections

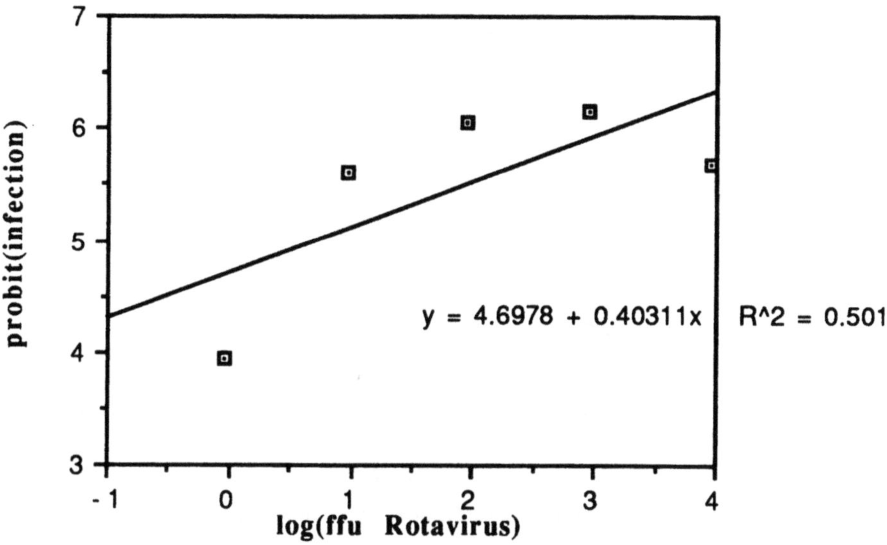

FIGURE 14-7. Log Probit Plot of Infections Resulting from Rotavirus Administration (Data of Ward et al., 1986)

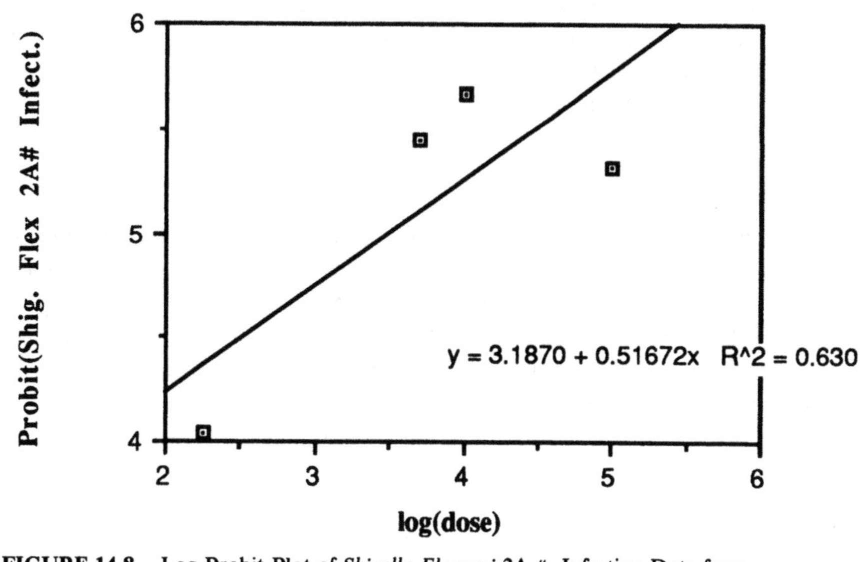

FIGURE 14-8. Log Probit Plot of *Shigella Flexneri* 2A#. Infection Data from DuPont et al.

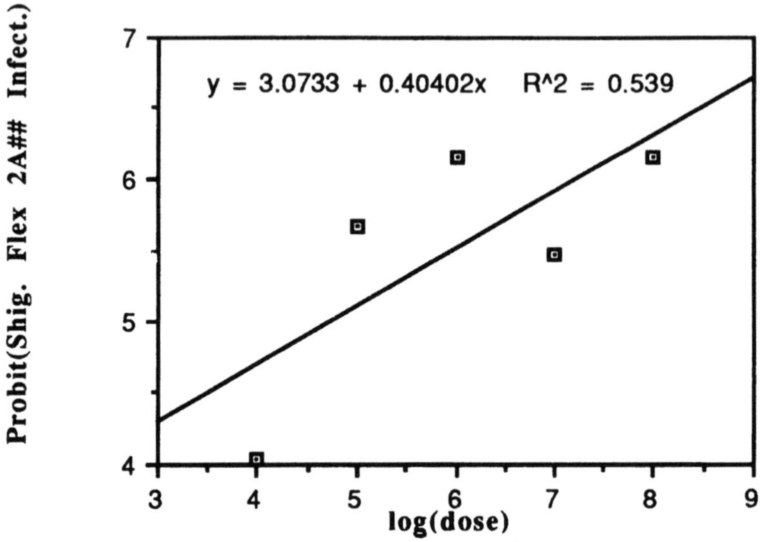

FIGURE 14-9. Log Probit Plot of *Shigella Flexneri* 2A#. Infection Data of DuPont et al. as Quoted by NAS (1977)

The results of the simple probit plots shown in these figures can be criticized in that they represent least-squares fits in which the points are weighted equally. To make a more sophisticated determination of the probit slopes, the data were fit to the lognormal risk model in Clement Associates' ToxRisk2 package. The model fit is given by

$$P(d) = 1 - e^{[A_0 + (1 - A_0)F(A_1 + A_2 \log 10 d)]}$$

where $P(d)$ is the risk at a particular dose (expressed in terms of the number of organisms administered, or related indices such as focus forming units or Tissue Culture Infectious Dose$_{50}$ units), A_0 is the "background" frequency of response (set to zero for purposes of this analysis), $F(x)$ is the cumulative distribution function of the standard normal distribution, A_1 is a fitted term that essentially adjusts the equation to give the best central tendency (ED_{50}) value, and A_2 is analogous to $\log_{10}$ probit slope values—the number of standard deviations of the distribution of individual susceptibilities per tenfold change in dose. This is essentially a one-hit model (individual organisms act independently with some probability to cause infections) with a lognormal distribution of susceptibilities in exposed subjects.

Table 14-4 shows the resulting coefficients for all the data sets studied, and parameters assessing the goodness of fit of the data sets to the lognormal model, both individually, and (through summation of the X^2 values and the

TABLE 14-4. Parameters and Fit Statistics for Lognormal Infectious Dose Response Models[a]

Organism	Endpoint	Data Source	A1	A2	χ^2	Degrees freedom[b]	p
Hepatitis B	Disease	Barker and Murray (1972)	− 0.774	0.351	1.83	4	0.77
Rotavirus	Disease	Ward et al. (1986)	− 0.805	0.362	5.70	6	0.46
Salmonella Typhosa	Disease	Hornick et al. (1970)	− 3.330	0.530	4.89	3	0.18
Shigella Dysenteriae	Disease	Levine et al. (1973)	− 1.942	0.749	0.17	2	0.92
Shigella Flexneri 2A #	Disease	DuPont et al. (1972)	− 1.351	0.376	3.09	2	0.21
Cholera	Severe Disease (IV support)	Cash et al. (1974)	− 4.538	0.645	0.97	1	0.32
Cholera	Moderate Disease	Cash et al. (1974)	− 4.301	0.221	0.24	1	0.62
Cholera	Infection	Cash et al. (1974)	0.099	0.224	0.07	1	0.79
Shigella Flexneri 2A #	Infection	DuPont et al. (1972)	− 1.942	0.60	15.98	2	0.0003
Shigella Flexneri 2A ##	Infection	DuPont et al. (1969), cited by NAS (1977)	− 1.134	0.267	3.50	3	0.32
Rotavirus	Infection	Ward et al. (1986)	− 0.624	0.640	11.59	6	0.07
Poliovirus 1 (Sabin)	Infection	Minor et al. (1981)	− 9.117	4.922	8.10	10	0.62
Poliovirus 1 (Sabin)	Infection	Lepow et al. (1962)	− 1.819	0.369	0.32	1	0.57
Poliovirus 3 (Fox)	Infection	Katz and Plotkin (1967)	− 0.615	0.858	0.24	1	0.62
Polio (SM strain)	Infection	Koprowsky et al. (1956)	− 1.675	6.995	0.00	2	1.00
Echovirus 12	Infection	Ward and Akin (1984)	− 1.830	1.187	0.78	2	0.68
Resp. Syncyt. Virus	Infection	Hall et al. (1981)	− 3.620	0.842	0.64	1	0.42
Endamoeba Coli	Infection	Rendtorff (1954)	− 1.050	0.411	3.69	3	0.30
Sum, all organisms					61.8	51	0.14

degrees of freedom) as a group. Although the model is rejected with some vigor ($p = 0.0003$) for one of the 18 data sets, the overall degree of departure of the data from the model is quite compatible with the operation of statistical sampling error-type fluctuations. This indicates that on the basis of the clinical experimental infectious dose–response data that have been accumulated in recent years, a one-hit mechanism for infection and the production of infectious disease cannot be rejected. As might be expected, quite large amounts of variability in susceptibility are indicated in many cases (relatively low values of A_2), and this implies that there may well be finite amounts of risk of infection to appreciable numbers of people even at doses of pathogenic microorganisms that are orders of magnitude below those that would be expected to cause infection or disease in the majority of people. The basic point, however, is that a model that assumes a basic one-hit mode of action of infectious organisms, and a great deal of human interindividual variability in response, is entirely consistent with the available data. Above one organism per host, there is no justification in the data to assume that there are true "threshold" doses below which there is zero risk of infection or illness.

In conclusion, it seems very likely that there is room for improvement in both the fundamental information used for all three of our decision types with respect to food safety, and in the paradigms that are used to analyze such information and relate it to different options for decision making on protective actions. This chapter has suggested a number of directions both for the creative expansion of the basic scientific and technical information relevant to decision making in this area, and for the use of more formal policy analysis techniques to help inform practical choices by decision makers.

15

Epidemiologic Studies

J. Glenn Morris, Jr.

Department of Medicine and Department of Epidemiology and Preventive Medicine, University of Maryland School of Medicine
10 South Pine Street, Baltimore, MD 20201

INTRODUCTION

The methodology for detection of microbiologic hazards in shellfish is becoming increasingly sophisticated, providing expanding lists of shellfish-associated microorganisms that are potentially pathogenic (NAS 1990). Increasingly sophisticated indicator systems for harvest waters are also being developed, improving our ability to predict when potential pathogens will be present in shellfish samples. At a more basic level, however, we still do not have a good understanding of how these microorganisms or indicators correlate with actual human disease occurrence. The simple presence of a microorganism does not guarantee that illness will occur: disease is dependent on a complex set of variables, including the presence or absence of virulence factors in the infecting microorganism, the infectious dose, the route of transmission, and the status of host defense mechanisms.

Epidemiology can be defined as the dynamic study of the occurrence, distribution, and control of health and disease in a population (Table 15-1). Epidemiologic studies provide a means of linking specific microorganisms or indicators with disease occurrence; when combined with appropriate microbiologic and pathophysiologic data, these studies can help us understand why and how people become ill after eating shellfish. Much of our current knowledge of shellfish-associated illness has come from community-based epidemiologic studies. However, such studies may be difficult to conduct, expensive, and, for outbreak investigations, are dependent on chance occurrences. Volunteer feeding studies provide an alternative means of looking at the same basic questions. They allow a more controlled assessment of potential virulence factors and infectious dose for specific microorganisms,

TABLE 15-1. Definitions

Epidemiology: The dynamic study of the occurrence, distribution, and control of health and disease within a population.

Types of Studies

a. Case-Control Study: A retrospective study using cases of disease as its basis. Persons with a disease or condition are identified, and their risk factors for development of the disease compared with control persons who do not have the disease.

b. Cohort Study: A prospective study using risk factors as its basis. Groups with varying degrees of risk are identified, and then followed to determine the rate at which disease occurs among persons in the group.

and permit a direct linkage of indicator and disease occurrence data. However, volunteer studies may also be expensive, and may raise important ethical questions; subjects in volunteer studies also tend to be young and healthy, and, consequently, the results may not reflect the overall disease occurrence or disease burden in the general population.

This chapter summarizes methodologies involved in community-based epidemiologic studies (outbreak investigations, case-control studies of sporadic cases, cohort studies) and volunteer feeding studies. Strengths and weaknesses of each type of study are considered, together with their possible role in defining the epidemiology of shellfish-associated illnesses.

COMMUNITY-BASED EPIDEMIOLOGIC STUDIES

Outbreak Investigations

Outbreak investigations have traditionally provided the basis for evaluating public health risks associated with eating shellfish. Investigations of shellfish-associated outbreaks of typhoid fever in 1924–25 served as the impetus for the establishment of the Shellfish Sanitation Program (FDA 1989; Bundesen 1925). These and earlier outbreak investigations also provided data linking coliform counts in water with disease occurrence: it was believed that typhoid fever occurred only in association with shellfish harvested from water in which "not more than 50% of the 1 cc (cubic centimeter) portions of water examined were positive for coliforms." This translates into 70 MPN total coliforms/100 mL, which served as the first standard for shellfish growing waters (NAS 1990).

Subsequent outbreak investigations by state and local health departments and the Centers for Disease Control (CDC) have implicated shellfish as vehicles of transmission of *Vibrio* species (Blake et al. 1980; Lowry et al. 1989*b*; Klontz et al. 1987; Pavia et al. 1987; Barker 1974; Wilson et al. 1981), hepatitis A (Mackowiak 1976; Bostock, Mepham, and Phillips 1979; Ruddy 1969; CDC 1979), and Norwalk, Snow Mountain, and other viral diarrheal agents (Morse 1986; Grohmann 1980; Truman et al. 1987). While many of these studies have been descriptive, others have used epidemiologic techniques such as case–control studies (see definitions, Table 15-1) to evaluate risk factors for illness and associate shellfish consumption with human disease.

The correlation with indicators of water quality in these studies has been more problematic. In an investigation of clusters of cases of non-O1 *V. cholerae* and hepatitis A associated with shellfish from Appalachacola Bay, Florida, there was a suggestion that Bay water had fecal coliform counts above standard at the time implicated shellfish were harvested (Wilson et al. 1981; CDC 1979). Similarly, in studies of viral gastroenteritis outbreaks in the New England states it appeared that clams and oysters associated with disease were frequently harvested from "closed" or contaminated areas (Truman et al. 1987; Morse 1986). However, in other reports it has not been possible to identify clearly the growing waters (Ruddy 1969; Bostock, Mepham, and Phillips 1979), or, where harvest areas have been tentatively identified, to correlate disease occurrence with increases in the level of indicators such as fecal coliforms (Mackowiak 1976).

Outbreak investigations remain a potentially important source of data for assessing the public health significance of specific microorganisms and indicator systems. However, as these investigations indicate, there are inherent problems in using outbreak investigations as a source of epidemiologic data. These include:

1. Uniqueness of Outbreaks. Outbreak investigations are dependent on the occurrence of an outbreak; while there may be some expectation that outbreaks will occur in a particular setting, outbreaks are, by definition, unusual events, and generally reflect the breakdown of standard control measures. As such, outbreaks may not be representative of the overall problems or disease burden in the community. While outbreaks can highlight specific problem areas, one cannot "arrange" for an outbreak so that a specific hypothesis can be tested or question can be answered.

2. Difficulties in Outbreak Recognition. The first step in an outbreak investigation—the recognition that an outbreak has occurred—requires that an appropriate surveillance system be in place. An outbreak involving thousands of cases may be detected with a minimal surveillance system. However, identification of smaller outbreaks or clusters of cases is usually dependent on aggressive case finding. As an example, Moore swab surveillance (sampling of sewage systems) has been important in identifying cases of cholera in Louisiana (Barrett et al. 1980; Lowry et al. 1989). Similarly, the detection of outbreaks of shellfish-associated viral gastroenteritis in New York state has been facilitated by the establishment of procedures for systematic collection of epidemiologic and clinical data in suspect cases (Morse 1986).

3. Difficulties in Locating Harvest Areas. Even after a shellfish-associated outbreak has been recognized, it is often impossible to identify the location at which the implicated shellfish were harvested. Shellstock are tagged, but tags are frequently lost or destroyed. In instances where tags are available, there is often a long chain which has to be followed through a series of distributors—often with minimal documentation. If, ultimately, it is possible to identify the boats that harvested the implicated shellfish (something which has proven to be virtually impossible in most outbreak investigations), there may be uncertainties about where the actual harvesting took place; for example, Mackowiak, Caraway, and Portnoy (1976) in their investigation of the source of outbreaks of hepatitis A listed among their hypotheses the possibility that "bootlegged" oysters from contaminated areas had been delivered clandestinely to implicated boats.

4. Lack of comprehensive water quality indicators in harvest areas. Even if harvest areas are identified, data on water quality in the harvest area are generally restricted to fecal coliform counts. While this can provide some index to the utility of fecal coliforms as predictors of disease occurrence, it is not possible to say anything about other indicator systems unless those systems happened to be in use in the appropriate harvest area at the time the outbreak occurred.

Case–Control Studies of Sporadic Cases

Case–control studies of sporadic cases of disease (i.e., single cases occurring outside of an outbreak setting) have been useful in evaluating the possible link between shellfish consumption and some bacterial pathogens such as *Vibrio* (Morris 1981; Shandera et al. 1983; Johnston, Becker, and McFarland 1985; Tacket 1984), *Aeromonas* (Holmberg 1986*a*), and *Plesiomonas* (Holmberg 1986*b*). In these studies clinical and epidemiologic data were obtained from persons known to have been infected with the organism under study. Comparable data were obtained from matched control persons; risk factors for infection were then identified by comparing data from cases and controls. These studies have generally been conducted by state health departments or the CDC, using isolates submitted to state or CDC reference laboratories as a way to identify cases.

By studying a group of sporadic cases, it is possible to avoid having to wait for an outbreak to occur (see point 1, above). However, these studies tend to magnify some of the other difficulties inherent in outbreak investigations. Case identification becomes more difficult, as one is looking for single, unrelated cases rather than a case cluster. For an outbreak [or an unusual single case (Morris 1984; Lin et al. 1986)] it may be worth the effort to try to trace back and identify the area in which the implicated shellfish were harvested. This becomes a virtual impossibility when evaluating multiple cases: in none of the studies cited were harvest areas identified or questions of water quality addressed.

Cohort Studies

Cohort studies are designed to look prospectively at the occurrence of disease within a population at risk. This contrasts with case–control studies, in which persons who already have the disease are identified (i.e., persons with oyster-associated gastroenteritis), and one looks backward (retrospectively) to try to define risk factors (Table 15-1).

A cohort design was used to evaluate illness among a group of convention attendees in New Orleans (Lowry et al. 1989*a*). Twelve percent of the respondents in this study reported diarrhea, with the risk of diarrhea significantly associated with a history of having eaten raw or cooked oysters (15% vs. 9% among those not eating oysters); no data were available on shellfish source or water quality indicators. Cohort-type designs have also been used in seroepidemiologic studies. There have been two studies in which the frequency of antibodies to shellfish-associated pathogens (*Vibrios*, Norwalk agent) in persons with high shellfish exposure/consumption (water-workers, oyster shuckers, etc.) was compared to those among persons with minimal shellfish exposure (Seventh-Day Adventists, who abstain from eating shellfish for religious reasons) (Hunt et al. 1988; Lefkowitz et al. 1992). Again, data on shellfish source or water quality were not included in these analyses.

Cohort studies are generally regarded as providing optimal data on occurrence of disease among persons with specific risk factors or behaviors (such as eating shellfish). At a technical level, they permit direct calculation of relative and attributable risks associated with a particular outcome (Fox, Hall, and Elveback 1970; Friedman 1974). However, they can be difficult to conduct: they require careful selection of an appropriate population and intensive surveillance within that population over time. Cohort studies also tend to require large numbers of participants, particularly when dealing with an outcome event (such as diarrhea) that may not be commonly reported.

There have been proposals that community-based cohort studies be set up for other foodborne pathogens and vehicles. A National Academy of Sciences expert committee (NAS 1987) raised the possibility of establishing a system of sentinel county health departments to assess the importance of various chicken-associated pathogens. It was suggested that counties where broiler chickens were supplied almost entirely by a limited number of processing plants be identified, and within these counties continuing efforts be made to find all possible cases of chicken-associated disease; these data could then be correlated with microbiologic and other production data from the processing plant(s) supplying the county in an effort to identify risk factors for illness and assess the utility of various interventions to control microbial contamination. Establishment of a comparable system for evaluating risks associated with shellfish would require identification of a population who eat shellfish from a few, carefully defined harvest areas; it would also require maintenance of intensive surveillance within this population, with simultaneous monitoring of water quality indicators in areas from which shellfish eaten by this population were harvested.

VOLUNTEER FEEDING STUDIES

Feeding of Specific Pathogens

Volunteer feeding studies provide an alternative means of looking at associations between microorganisms or indicator systems and disease occurrence. Studies to date have fallen into one of two general groups. There have been several small studies that have evaluated the role of shellfish-associated microorganisms such as *Vibrio parahaemolyticus* (Sanyal and Sen 1973), non-O1 *Vibrio cholerae* (Morris et al. 1990), *Aeromonas* (Morgan et al. 1985), and *Plesiomonas* (Herrington 1987) as etiologic agents of disease in normal, healthy persons. These studies have involved direct feeding of different doses of these microorganisms to volunteers. This type of study, with relatively small numbers of volunteers, careful laboratory support, and close clinical observation, has provided invaluable information on virulence factors, infectious dose, and pathophysiologic mechanisms responsible for disease. By defining which organisms are pathogenic and in what doses they also provide a basis for interpreting studies correlating harvest water indicators with the presence of specific microorganisms in shellfish.

Feeding of Oysters

In a cohort-type design, oysters from different harvest areas (with different risks and/or different patterns of indicators) can be fed to groups of volunteers, and the volunteers observed for occurrence of illness. As oysters are supplied by the investigators, it should be possible to characterize carefully the potential pathogens present in the shellfish and water quality at the point of harvest. The one study of this type that has been conducted used three groups: 521 persons were fed oysters from an area of "borderline" human contamination in the James River, Virginia, 545 persons were fed oysters harvested in a "pristine" area in the Rappahannock River, Virginia, and a third group of 327 did not eat shellfish. Statistically significant differences in occurrence of diarrhea were said to have been found when persons eating oysters from the area with borderline contamination (ca. 7% of whom had diarrhea) were compared with those eating oysters from "pristine" areas (ca. 4% with diarrhea) and those who did not eat oysters (ca. 2% with diarrhea) (Personal communication, Harris White, NOAA). Unfortunately, this study did not clearly identify the etiology of diarrhea among ill volunteers: serologic studies were not conducted, only a limited number of stool cultures were available for bacterial culture, and stool samples that were available were not screened for viral pathogens. Data were available on water quality indicators in the harvest areas and in shellfish samples. However, potential pathogens in water or shellfish were not directly identified.

There is no question that properly designed volunteer feeding studies provide a powerful tool for associating specific indicators or microorganisms with disease occurrence. Again, however, the limitations of these studies should be recognized.

1. Ethical Considerations. There has been a great deal of concern within the last 10 years about ethical issues in volunteer studies. The basic criteria for use of human subjects, as formulated by the National Institutes of Health (DHHS 1983), are summarized in Table 15-2. Several of these criteria have potential implications for studies involving shellfish and shellfish-associated pathogens. The need to minimize risks to the subjects (and limit risks to those that are reasonable in relation to anticipated benefits) precludes studies with

TABLE 15-2. Basic Criteria for Volunteer Studies

1. That the risks to the subjects are minimized.
2. That the risks to the subjects are reasonable in relation to the anticipated benefits.
3. That the selection of subjects is equitable.
4. That an informed consent will be sought from each prospective subject or the subject's legally authorized representative.
5. That the informed consent will be properly documented.
6. That, where appropriate, adequate provisions are made for the safety of the subjects.
7. That, where appropriate, adequate provisions are made for the privacy of the subject and the confidentiality of data.

Source: Protection of Human Subjects, Code of Federal Regulations 45 CFR 46.111, March 8, 1983.

microorganisms that cause more than mild illness or that are not readily treatable. In direct oyster feeding studies this raises questions about use of oysters that might carry hepatitis or other pathogens associated with potentially severe disease. If it is not possible to minimize the risk of hepatitis by direct screening of shellfish for hepatitis virus, serious questions could be raised about the ethics of feeding oysters that have been harvested in areas of known human fecal contamination. Other potentially invasive organisms, such as *Vibrio vulnificus*, are probably of minimal risk to persons with a normal immune system (Morris 1988). Again, however, ethical considerations would require that all volunteers be screened for evidence of immune competence before consuming oysters known to carry *V. vulnificus* or other potentially invasive bacterial species.

2. Sample Size. As is characteristic of cohort-type studies, direct oyster feeding studies may require relatively large numbers of volunteers to attain statistical significance. Representative sample size calculations are given in

TABLE 15-3. Representative Sample Size Calculations

Estimated rate (%)		
Control Group	To Detect Minimum Difference Between Groups	Total Number of Volunteers Needed
1	6	332
1	4	568
1	3	848
1	2	1536
4	6	566
4	4	1104
4	3	1812
4	2	3726
8	6	852
8	4	1764
8	3	2998
8	2	6428
12	6	1110
12	4	2360
12	3	4072
12	2	8878

Calculations assume approximately equal numbers in control and "high-risk" feeding groups; power is 80%, at 95% significance level. Calculations performed using EPISTAT statistical package.

Table 15-3. Depending upon the rate of disease in the control group and the difference that one wants to be able to detect when comparing groups fed control and "high-risk" oysters, the total number of persons needed may run well into the thousands. While this does not preclude studies of this type, it does emphasize their potential expense and the logistic difficulties that they may involve.

3. Generalizability—Harvest Area. Data from a single group of persons fed oysters harvested from a single site at a single time point may not necessarily be generalizable to all harvest areas—i.e., there may be unexpected conditions or an unusual pathogen present which render the usual correlation between indicators and disease occurrence invalid. This problem can be solved in part by taking multiple samples from multiple sites; however, this requires multiple groups—with the potential for increasing the number of volunteers needed to tens of thousands.

This issue can also be addressed by looking carefully at the pathogens causing disease in persons who become ill after eating shellfish, and correlating these data with data on pathogens in the shellfish fed to the volunteers. For example, it may be possible to show that human illness always occurs when a specific pathogen is present at a certain dosage level in shellfish. With these data it may not be necessary to feed shellfish to volunteers from every site being evaluated: instead, indicators can be correlated with the presence or frequency of isolation of this pathogen from shellfish harvested from the site under study. While this approach is attractive from the standpoint of limiting the number of volunteers needed, it does require sophisticated laboratory support. There are a large number of potential pathogens (both viral and bacterial) present in shellfish: it may require very complex procedures to identify clearly the cause of illness in sick volunteers, and, in turn, identify the implicated microorganisms in shellfish samples.

4. Generalizability—Population Sampled. Volunteer feeding studies, due both to ethical and logistic considerations, tend to be restricted to young, healthy volunteers. Data obtained in these studies provide a guide to the response of this population to eating oysters. However, this does not take into account the health impact of shellfish-associated pathogens on other groups: the very young, the very old, or persons with chronic underlying health problems. There are no guarantees that an indicator system that predicts illness accurately in young adults will also provide protection for persons in these other groups. As an example, *V. vulnificus* may cause no disease in young healthy adults (Morris 1988); however, it is probably the major bacterial cause of shellfish-associated morbidity and mortality (NAS 1991).

COMMENT/SUMMARY

The epidemiologic evaluation of shellfish-associated disease has involved both community-based studies and volunteer feeding studies. However, the number of studies that have been conducted is limited. The current system of indicators is based on epidemiologic data that are 65 years old; more recent studies, while documenting the occurrence of illness, have provided few data that can be used to correlate directly indicators with human disease.

Community-based studies are necessary to evaluate disease burden in the general population. While it may not be possible or practical to follow an entire county, as was suggested for chicken-associated illness (NAS 1987), it may be worthwhile identifying certain groups in which intensive surveillance can be maintained, and for which the source of shellfish can be fairly well controlled and characterized (i.e., oystermen who take oysters only from a single, well-defined harvest site, and who only eat their own catch; Seventh-Day Adventists living in the same town). This would permit a dynamic study of the relationship between indicators, microorganisms, and disease, with changes in indicator levels correlated with occurrence of specific diseases among shellfish-eating community members. This type of study would require creativity, patience, and money; the return would be invaluable data on shellfish-associated disease within the population as a whole.

In the absence of good community-based cohort studies, there should be continued emphasis on investigation of outbreaks and sporadic cases of shellfish-associated illness. As has been noted, one of the major problems in these studies is the accurate identification of harvest areas. Efforts need to be made by industry and government to improve "trace-back" capabilities so that areas in which suspect shellfish were harvested can be accurately identified. When harvest areas are identified, there also need to be good water quality data available so that efforts can be made to correlate disease occurrence with a variety of indicators.

Volunteer feeding studies address a somewhat different set of issues and provide a setting for much more controlled studies. However, these studies require careful planning from both an ethical and scientific standpoint. Feeding studies involving oysters from multiple sites are possible, but generally require large numbers of volunteers. It may be more useful to analyze carefully the frequency and cause of diarrhea among persons eating oysters from two or three well-characterized harvest areas. This would still permit hypothesis testing, and provide data on shellfish-associated pathogens that could be used as a basis for interpreting microbiologic data from water and shellfish harvested at other sites.

The importance of epidemiologic studies cannot be underestimated; it is these studies that provide the scientific and public health basis for interpretating microbiologic and marine biologic data. However, there is no perfect epidemiologic methodology for evaluating whether an indicator system predicts occurrence of human disease adequately. Our current knowledge is based on studies conducted using a variety of methodologies, each with its inherent strengths and weaknesses. Expansion of this data base will require a variety of types of epidemiologic studies, with varying complexities and costs.

16

Statistical Sampling of Growing Waters

Harrison M. Wadsworth

School of Industrial and Systems Engineering
Georgia Institute of Technology, Atlanta, GA 30332-0205

INTRODUCTION

Sampling refers to the observation of a portion of a population for the purpose of obtaining some information about it. In this case, a population refers to an entire area (a bay or other waters) or to an entire shipload of shellfish or other seafood. In either case, it is usually not feasible to inspect the entire population; so, a sample is inspected. It must be remembered that sampling inspection does little or nothing to improve the quality of a population or the safety of the product. Its purpose is to assist in making a decision regarding the acceptability of the shipment or growing area. Indirectly, sampling inspection, by eliminating poor shipments or identifying harvest areas which are contaminated may improve the overall quality and safety of seafood products.

Sampling will always incur certain risks. Since we are not inspecting all elements of a population, we cannot know everything about it. We only know about the sample. Thus, we incur risks of making two types of errors in our accept, not accept decision. We can reject a shipment or declare an area unsafe when it should have been accepted. The risk of doing this is commonly called a *producer's risk* and the error is a Type I error or error of the first type. The second error is that we can accept a shipment or approve an area when it should not be accepted. The risk of doing this is often called a *consumer's risk* and the error is called a Type II error. Although sampling always incurs both of these risks, they can be measured and, thus, controlled, if proper statistical procedures are used.

Generally, the producer's risk is stated in terms of a probability or risk level denoted by the Greek letter α. This is the probability that a shipment whose quality, in terms of levels of contaminants, is at a certain, specified acceptable level or better is rejected by the sampling inspection procedure. The consumer's risk, denoted by the Greek letter β, is the probability that a lot whose quality is at or below a certain unacceptable level is accepted. The first (acceptable) level is generally called the *acceptable quality level or AQL*. The second is the *limiting quality level or LQL*. Both *AQL* and *LQL* are often expressed as percent nonconforming or number of nonconformities per hundred units, where a unit may be an individual seafood item or a unit of area such as a square meter in the case of sampling of harvest areas. In designing sampling plans, one goal will be to design a procedure resulting in shipments whose quality is equal to or better than the *AQL* being accepted $100(1 - \alpha)\%$ of the time. For ease of notation, the symbol p_1 will be used for the lot quality which has probability of acceptance of $(1 - \alpha)$. That is, in the case of contaminants at or below a certain specified level, p_1 is the level which has probability of acceptance of $(1 - \alpha)$. Similarly, the symbol p_2 will be used to designate that quality (number of contaminants) which has a probability of acceptance of β.

A word of caution should be stated here regarding the interpretation of the *AQL*. In order to allow anything other than complete (i.e., 100%) inspection of a shipment, the *AQL* must be greater than zero. This does not mean that we are approving that a certain fraction of the shipment can be nonconforming. It means that we are recognizing that, since we have not completely inspected the shipment, it is possible for a small number of nonconforming items to be present. If a series of shipments are submitted with contaminant level equal to or better than the *AQL*, the sampling procedure will reject only a fraction α or less of them. The *AQL* may be made as small as necessary but in doing so a larger fraction of the shipment must be inspected.

If the contaminant is of a critical nature such that its presence may cause death or serious illness and inspection is not destructive, complete inspection may be necessary. Even such complete inspection may not be sufficient, however, to assure us completely that no contaminants are present. This is because of errors in inspection. If there are a large number of seafood items to be inspected, the chance for such inspection errors increases rapidly. For this reason, complete inspection may actually result in more contaminants passing the inspection than sampling where more care may be made in inspection and the risks are known.

TYPES OF SAMPLING PLANS

Sampling plans may be classified either variables or attributes plans. Variables plans require a knowledge of the underlying frequency distributions of organisms in the seafood. This knowledge is seldom, if ever, available. This limits the usefulness of such procedures. Even though attributes plans are much less powerful in terms of the risks involved, they do not require a knowledge of the underlying distribution. For that reason, they are usually used for inspection of food products.

Attributes plans commonly used for food inspections are of two types: two-class and three-class plans. Two-class plans are those for which the inspected item either has less than the stated limit of organisms or more. Three-class plans have two limits. None of the inspected items may exceed the upper limit, but a limited number may exceed the lower limit but not the upper limit. Details and examples of both of these types of plans in use for seafood inspection will be discussed later.

In addition to designation of attributes plans, either two-class or three-class, we may designate them as single, double, or multiple plans. Single-sampling plans are the simplest and easiest to use and understand. That is, in the two-class attributes case, a sample of size n is selected from a shipment and the shipment is accepted if no more than c nonconforming items are found in the sample. The decision-making process is defined by these two numbers. For example, for the two-class plan $n = 10$, $c = 2$, we select ten items and as long as two or fewer items have organism concentrations above a limit, m, the shipment is accepted.

Double-sampling plans may also be used. These require the taking of a second sample when inspection of the first sample indicates the shipment is of marginal quality. For very good or very bad results, a smaller first sample will give the same protection against errors as a single sample. Shipments of intermediate quality may require a second sample and, thus, more inspection, but there will usually be a saving in the average sample size over single-sampling plans which provide the same protection. This savings may be counteracted by the increased complexity of a double-sampling procedure.

The concept of double sampling may be extended to multiple sampling and item-by-item sequential sampling. In multiple sampling, decisions may be made on any number of smaller samples. Usually, multiple plans require an accept–reject decision after no more than five to seven samples. Sequential sampling requires one of three decisions to be made after each item is inspected: (a) accept the shipment, (b) do not accept the shipment, or (c) inspect another item. This process is continued until one of the first two decisions are made.

Multiple sampling results in a smaller average sample size than double sampling, and sequential sampling results in an average sample size savings over multiple plans. However, each sampling procedure is progressively harder to administer, so that administrative costs must be balanced against cost savings due to decreased average sample size.

Most sampling inspection procedures developed by the Food and Drug Administration for use in inspecting seafood are single-sampling two-class plans. In some instances the FDA uses three-class single-sampling plans and in a very few cases double plans (two-class) are used. Details of all of these plans will be presented later but first we will discuss methods for evaluating risks for the plans. We will first discuss in some detail methods for evaluating two-class single-sampling plans.

TWO-CLASS SINGLE-SAMPLING PLANS

A two-class single-sampling plan is one in which the inspector is forced to make a decision concerning the acceptability of a shipment of seafood on the basis of the inspection of the items in one sample taken from that shipment. In addition, a level of acceptable concentrations of an organism is also established. This level separates the conforming items from those which do not conform. For example, if m units of an organism are allowed, a single-sampling plan might call for the inspection of 50 shellfish from a large shipment. If two or fewer of the shellfish have more than m units of the organism, the shipment is accepted; otherwise, it is not. Thus, we see that some decision must be made concerning the acceptability of the lot based on the results of a single sample.

This single-sampling situations may be expressed as a probability problem. Suppose we wish to take a random sample of size n from a shipment of size N, $n \leq N < \infty$. What is the probability of obtaining exactly x, $x \leq n$, nonconforming items in the sample if there are d, $x \leq d \leq n$, nonconforming items in the entire shipment? The answer to this question may be obtained using the hypergeometric distribution

$$P(x) = \frac{d!}{x!(d-x)!} \cdot \frac{(n-d)!}{(n-x)!(N-d-n+x)!} \div \frac{N!}{n!(N-n)!}$$

In virtually all cases about which we are concerned, the shipment size N is very much larger than the sample size n. In this situation the binomial distribution may be used. In particular, this approximation is quite satisfactory for situations where the lot or shipment size is at least 10 times the sample size. To use the binomial distribution, we introduce the parameter p, which is the true fraction of the shipment which is nonconforming. If we knew d and N, their ratio d/N would be p.

Using this distribution, the probability of obtaining x nonconforming items from a shipment with $100p$ percent nonconforming is

$$P(x) = \frac{n!}{x!(n-x)!} p^x (1-p)^{n-x}$$

The Operating Characteristic Curve

One of the most useful considerations of a sampling plan is its operating characteristic function. Whenever a statistical sampling plan is derived, its description is not complete until its operating characteristic function has been described, often by means of a curve called the operating characteristic or OC curve. This curve shows the probability of acceptance of lots with varying percents nonconforming that might be submitted for inspection under the specified procedure.

Figure 16-1 is the OC curve for the single two-class sampling plan $n = 50$, $c = 2$. The probabilities on this curve have been calculated using the binomial distribution. If the shipment size is large enough to be ignored, as discussed previously, this curve describes the ability of the sampling procedure to discriminate between acceptable and nonacceptable shipments without regard for the size of the shipment.

Three points are indicated on the OC curve of Fig. 16-1 as p_1, p_2, p_3. The first one, p_1, is a fraction nonconforming level that should be accepted with a relatively high probability. The probability that a shipment whose fraction nonconforming is equal to p_1 will be rejected is the producer's risk α. Recall that this value, α, is the probability of making a Type I error. The value of the OC function at $p = p_1$ is, therefore, $1 - \alpha$. The fraction nonconforming, p_1, is often called the *acceptable quality level or AQL.*

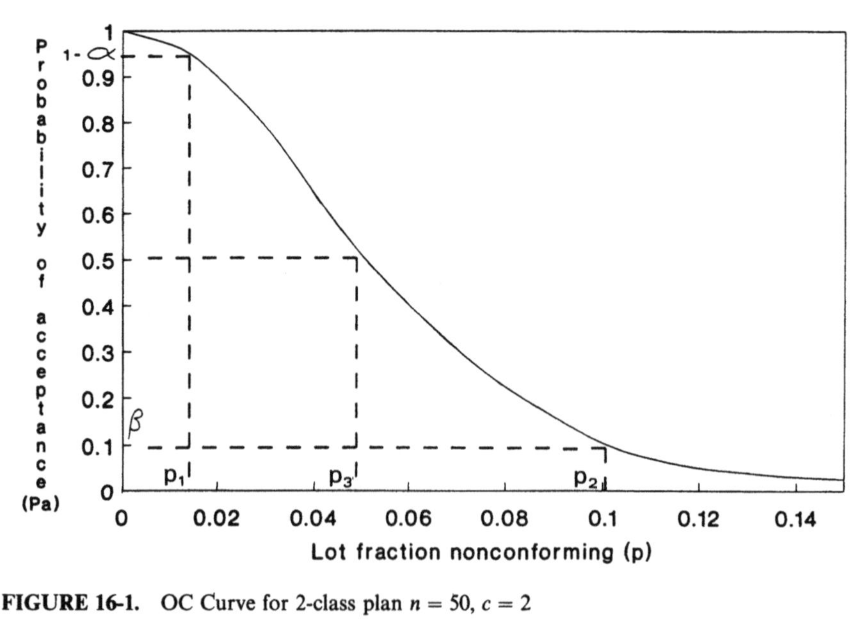

FIGURE 16-1. OC Curve for 2-class plan $n = 50$, $c = 2$

The second point, p_2, in Fig. 16-1 is a fraction nonconforming that should be accepted with a relatively low probability. The probability that a lot whose fraction nonconforming is equal to p_2 will be accepted is β. This corresponds to the probability of making a Type II error. The fraction nonconforming, p_2, is often called the *limiting quality level* (*LQL*).

The third point, p_3, in Fig. 16-1 is commonly called the *indifference quality level* (*IQL*). This is the lot quality for which there is a 0.50 probability of acceptance. In other words, there is an equal chance of accepting or rejecting a shipment whose fraction nonconforming is equal to p_3. Any two of these points with their associated probabilities may be used to derive a two-class single-sampling plan.

Effect on the OC Curve of Changes *n* and *c*

If the sample size is reduced, the OC curve tends to flatten out, as illustrated by Fig. 16-2. Actually, as can be seen from the figure, if n is reduced while keeping the ratio between n and c constant, both producer's and consumer's risks are increased. If the acceptance number, c, is changed without changing the sample size, the OC curve will move to the left or right without much change in curvature except for $c = 0$. Figure 16-3 illustrates this fact. From the figure we can see that reducing c alone increases α and decreases β. In particular, the curve for $c = 0$ indicates an extremely high value of α relative

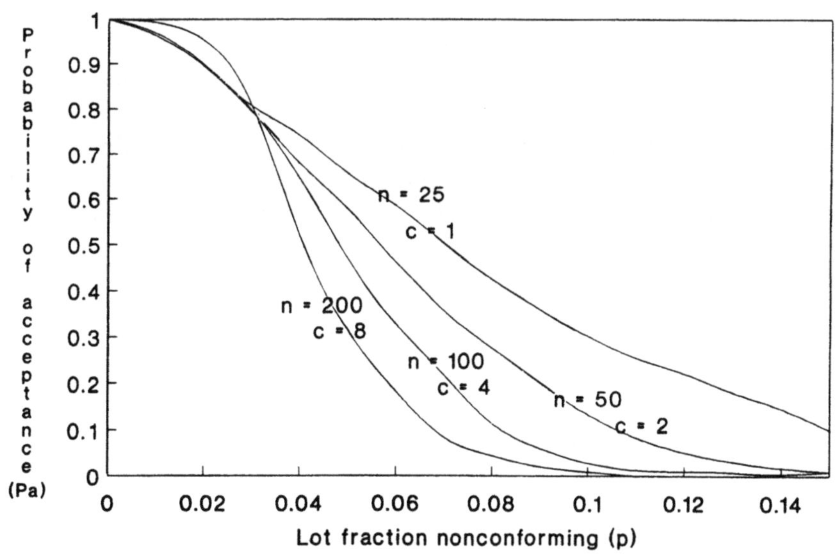

FIGURE 16-2. Effects on the OC Curve of Varying *n*

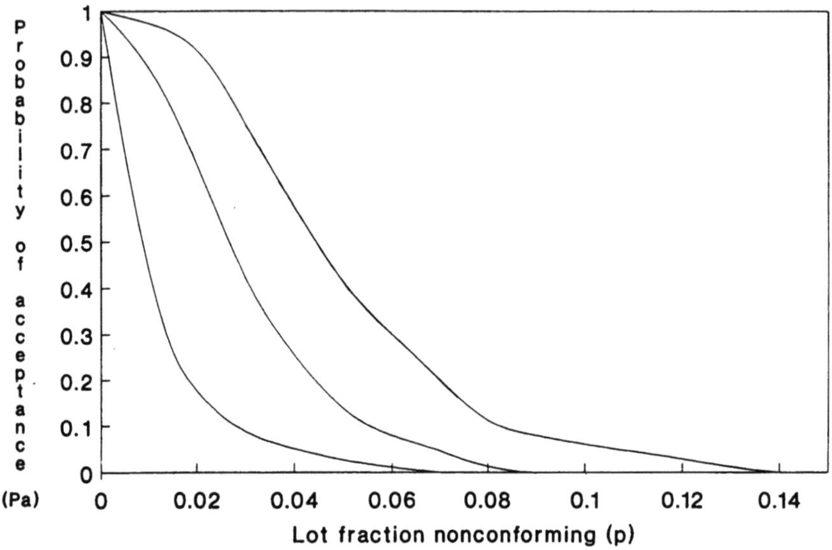

FIGURE 16-3. Effects on the OC Curve of Varying *c*

to the other curves. In other words, there is little protection for the producer when no nonconforming items are allowed in the sample. For this reason most sampling plans allow at least one nonconforming item in the sample.

To illustrate further these effects on the risks involved, Table 16-1 contains percentages of lots with varying percents nonconforming which would be expected to be accepted using samples of size 13 and acceptance numbers from 0 to 3. The table indicates that, using an acceptance number of 2, we would pass virtually all 2% nonconforming shipments and would also pass most (86%) of those shipments containing 10% nonconforming items. We could also pass over half (52%) of those shipments which contain 20% nonconforming items.

TABLE 16-1. Percentage of Lots Accepted for $n = 13$

Percent Nonconforming	Acceptance numbers 0	1	2	3
2	77	97	99.8	100
5	52	86	97	99.6
10	27	63	86	96
20	7.4	27	52	74

On the other hand, if we used an acceptance number of zero, we would reject virtually all 20% nonconforming lots and would pass only 77% of those lots with only 2% nonconforming items. If 2% is the value we wish to accept most of the time while we wish to reject 20% nonconforming lots most of the time, the plan with acceptance number 1 might be the best of the four in Table 16-1.

Derivation of Two-Class Plans

We might wish to derive a two-class single-sampling plan that comes as close as possible to some fixed values of the *AQL*, *LQL*, and their associated producer's and consumer's risks (α and β). This may be done quite easily using the values in Table 16-2. These values are derived using an approximation to the binomial known as the Poisson distribution. The approximation is quite good when the sample size is large and percent nonconforming is small. Since our choice of risk values is arbitrary and only approximate and since sample sizes and acceptance numbers must be integers, the error in using this approximation is of no real consequence.

To illustrate the use of Table 16-2, suppose we wish to find a two-class single-sampling plan for which the *AQL* is 2% and the *LQL* is 10%, with corresponding risks of 5% and 10%. That is, we wish to find a plan that will accept lots which are 2% nonconforming 95% of the time and lots which are 10% nonconforming only 10% of the time. In Table 16-2, $p_{0.95}$ represents the value (in decimals) of p that has a 95% chance of acceptance and $p_{0.10}$ is the value with a 10% chance of acceptance. Thus, for our example, $p_{0.95}$ is 0.02 and $p_{0.10}$ is 0.10.

If we calculate the ratio of these two values, we get 0.10/0.02 = 5.00. This corresponds to the values in the right-hand column of Table 16-2. The next

TABLE 16-2. Values for Two-Class Single Sampling Plan

c	$np_{0.95}$	$np_{0.10}/np_{0.95}$
0	0.0513	44.84
1	0.355	10.96
2	0.818	6.51
3	1.366	4.89
4	1.970	4.06
5	2.613	3.55
6	3.286	3.21
7	3.981	2.96

step is to find the number in the right-hand column closest to 5.00. In this case it is 4.89, corresponding to an acceptance number of 3.

We now need to determine the sample size n for our plan. Note that the middle column is headed $np_{0.95}$. Thus, if we divide the value in this column by $p_{0.95}$, we get n, the sample size. For our example we get, for $c = 3$,

$$n = np_{0.95}/p_{0.95} = 1.366/0.02 = 68$$

The sampling plan chosen would be $n = 68$, $c = 3$.

Now, for another example, we might choose $p_{0.10}$ as 0.20 with the same values of $p_{0.95}$, α, and β. For this case $np_{0.10}/np_{0.95}$ is 10.0 and, using Table 16-2, our acceptance number is $c = 1$. The corresponding sample size is

$$n = np_{0.95}/p_{0.95} = 0.355/0.02 = 18$$

The plan which has a 95% chance of accepting a 2% nonconforming lot and a 10% chance of accepting a 20% nonconforming lot is

$$n = 18$$

$$c = 1$$

With this simple procedure and the ability to choose these two quality levels, an appropriate plan may be easily derived.

THREE-CLASS ATTRIBUTES SINGLE-SAMPLING PLANS

This type of plan is used chiefly for compliance to microbiological standards. It is more complicated than a two-class plan and is not used as often. It requires the selection of two specification values (m and M). The lowercase m represents a target value or limit of microorganisms for product manufactured under Good Manufacturing Practices (GMPs). The uppercase M represents the limit of the same microorganisms which may be allowed. That is, more than M is unacceptable and a shipment should be rejected if any sample unit contains M or more of the microorganisms whereas some sample units with values between m and M might be allowed. The three classes are, therefore, less than m, greater than or equal to m but less than M, and greater than or equal to M.

As an example of such a plan, for *Staphylococcus aureus*, the FDA currently uses a sample of five. No sample unit may exceed 1000/g and one of the five items may be equal to or greater than 500/g but not as much as 1000/g. Thus, we have $m = 500$, $M = 1000$, $n = 5$, and $c = 1$.

TABLE 16-3. Operating Characteristics for Three-class Plan, $n = 5$, $m = 500$, $M = 1000$, $c = 1$

	Q (%)				
P (%)	0	10	20	30	40
0	100	59	33	17	8
10	92	53	29	14	6
20	74	41	21	9	4
30	53	27	12	5	1
40	34	16	6	2	< 1

Since this is really a combination of two sampling plans, a three-dimensional OC curve would be needed. As an alternative, the operating characteristic could be described by means of a two-way table. For this example, Table 16-3 contains some values of this operating characteristic. In Table 16-3, *P* represents the actual percentage of all items in the lot exceeding 500/g, i.e., *m*, but not 1000/g (*M*). *Q* represents the actual percentage of all items equal to or exceeding 1000/g. The percentage of lots accepted in a long series of such lots for each combination of *P* and *Q* are shown in the body of the table. The values in the table were calculated using the multinomial distribution (an extension of the binomial distribution).

Since we now have three classes instead of two, we will accept the lot if there are one or less in Class 1 or 2 (less than *M*) and none in Class 3 (greater than *M*). That is, if 10% of the shipment is in Class 3 and 20% in Class 2, 70% will be in Class 1 (less than *m*). We will accept the lot if all five items inspected are in Class 1 or four items in Class 1 and the fifth in Class 2. In either case, there must be none in Class 3. The (excepted) fraction of lots accepted will thus be, using the notation, $P(a, b, c)$, to mean the probability that there are *a* in Class 1, *b* in Class 2, and *c* in Class 3,

$$P_a = P(5, 0, 0) + P(4, 1, 0)$$
$$= (0.7)^5 + 5(0.7)^4(0.2) = 0.41 \text{ or } 41\%$$

In general, this is

$$P_a = P_1^n + \sum_{x=1}^{c} \frac{n!}{x!(n-x)!} p_1^{n-x} p_2^x$$

where p_1 is the fraction of items that are in category less than *m* (Class 1) and p_2 is the fraction of items that are in class greater than or equal to *m* but less than *M*.

From Table 16-3 we can observe that if a shipment of 40,000 lb of seafood is submitted to this inspection process, with $P = 20\%$ (8000 lb) and $Q = 10\%$ (4000 lb), the probability of lot acceptance will be 41%. Similar two-way tables can easily be prepared to describe the operating characteristic of any three-class plan.

SAMPLING PLANS OF THE FOOD AND DRUG ADMINISTRATION

Salmonella Sampling Plans

Sampling plans for *Salmonella* break food into three categories:

I. Foods that would normally be in Category II, except that they are intended for consumption by the aged, infirm or infants.
II. Foods that would not normally be subjected to a process lethal to *Salmonella* between the time of sampling and consumption.
III. Foods that would normally be subjected to a process lethal to the *Salmonella* between the time of sampling and consumption. Most seafood will be in Category II.

For all categories, a sample unit consists of a minimum of 100 g selected at random from a shipment. A 25 g analytical unit is analyzed from each 100 g sample unit for *Salmonella*. The analytical units may be composited with the maximum composite size of 375 g or 15 analytical units.

Table 16-4 describes the sampling plans for each of the food categories. The plans are all two-class plans. In the table, n refers to the number of sample units to be selected from each shipment. If the samples are composited, 60 sample units would yield four composite units. It is assumed that if one or more analytical unit in the composite contain *Salmonella*, the composite will test positive. The c values refer to the acceptable number of positive tests. For these plans, no positive tests are allowed.

The percentages under the $p_{0.95}$ column represent the percent nonconforming in a shipment which has a 95% chance of acceptance. Similarly,

TABLE 16-4. ***Salmonella* Sampling Plans**

Food Category	n	c	$p_{0.95}$ (%)	$p_{0.50}$ (%)	$p_{0.10}$ (%)
I	60	0	0.086	1.16	3.84
II	30	0	0.171	2.31	7.68
III	15	0	0.342	4.62	15.35

$p_{0.50}$ is the percent nonconforming with a 50% chance of acceptance and $p_{0.10}$ is the percent nonconforming with a 10% chance of acceptance.

For Category I, 95% of the lots for which 0.086% of the seafood contains *Salmonella*, 50% of the lots with 1.16% and 10% of the lots with 3.84% would be expected to be accepted.

For Category II, the corresponding percentages are 0.17%, 2.31%, and 7.68% while for Category III they are 0.342%, 4.62%, and 15.35%, respectively. Thus, in a 40,000 lb shipment, for Category II, 95% of the lots with 68 lb containing *Salmonella* would be accepted. Similarly, 50% of the lots with 924 lb containing *Salmonella* and 10% of the lots with 3072 lb containing *Salmonella* would be expected to be accepted. Similar figures may be developed for the other categories.

The Food and Agriculture Organization/World Health Organization (FAO/WHO) has another sampling plan for *Salmonella*. This plan calls for a sample of five units with acceptance number 0. For this plan, $p_{0.95} = 1\%$, $p_{0.50} = 13\%$, and $p_{0.10} = 37\%$. Compared to the FDA plans, these plans give little protections to the consumer.

Staphylococcus aureus Plans

The FAO/WHO plan for *Staphylococcus aureus* is a three-class plan as is the FDA plan discussed earlier. For this plan the sample size is also five, but $c = 2$, $m = 500/\text{g}$, and $M = 5000/\text{g}$. That is, for five sample units analyzed, no unit should exceed 5000/g and no more than 2 units should exceed 500/g. Otherwise, the shipment should be rejected. Table 16-5 describes the performance of this plan. In Table 16-5, Q is the percentage of items in a lot exceeding 5000/g or M. P is the percentage of items exceeding m (500/g) but not M.

Tables 16-3 and 16-5 cannot be compared directly because of the different values for M in the FDA and FAO/WHO plans. However, since M is larger

TABLE 16-5. Percentage of Lots Accepted for Several P and Q Values for *Staphylococcus aureus*

	Q (%)				
P (%)	0	10	20	30	40
0	100	59	33	17	8
10	99	58	32	16	7
20	94	54	29	14	6
30	83	47	24	11	4
40	68	36	16	6	2

for the FAO/WHO plans, we would expect the percentage of lots accepted to be higher in Table 16-5. The FDA plan is much more stringent than the FAO/WHO plan because of its lower value for M and the lower value for c.

Crabmeat—Adulteration with Filth Containing *Escherichia coli*

The FDA in Section 7108.02 of the Compliance Policy Guide (CPG) sets sampling requirements using two two-class plans. One plan calls for a sample size (n) of six with instructions to seize the shipment if two or more items contain *Escherichia coli.* The second calls for legal action to be taken if any fish contains adulteration. Thus, we have two different plans using the same sampling units. The plans and their performance are indicated in Table 16-6.

Table 16-6 indicates that there is a 50% chance that a lot 11% of which contains adulteration or 20% of which contains *Escherichia coli* will be accepted. Even if 51% of the seafood in the lot contains *Escherichia coli,* there is a 10% chance it will be passed.

Shrimp—Adulteration Involving Decomposition

Shrimp are classified by the FDA in CPG 7108.11 as follows:

Class 1—Very fresh to fishy odor characteristic of the product.
Class 2—Slightly odor-persistent and perceptible to an experienced examiner as pertaining to decomposition.
Class 3—Strong odor of decomposition that is distinct and unmistakable.

A subsample of 100 shrimp is classified as decomposed if one or more of the following criteria is met:

1. 5% or more are Class 3,
2. 20% or more are Class 2,
3. Percentage Class 2 + 4 (Percentage Class 3) $\geq$ 20.

TABLE 16-6. Sampling Plan Characteristics for Crabmeat

n	Ac	Re	$p_{0.95}$ (%)	$p_{0.05}$ (%)	$p_{0.10}$ (%)
6	0	1	0.85	11	32
6	1	2	6.3	26	51

Ac = Acceptance number.
Re = Rejection number.

Criterion 3 says that if, for example, 10% are Class 2 and 3% of the shrimp are Class 3, the 100-shrimp subsample would be classed as decomposed since $10 + 4(3) = 22$. A sample unit is a subsample of 100 shrimp and each subsample is classified as decomposed or not according to the above three criteria. The two-class plans shown in Table 16-7 are developed using this concept. In Table 16-7, the sample size (n) and acceptance/rejection numbers refer to numbers of subsamples of 100 shrimp. Thus, the first one says we take six subsamples of 100 shrimp each and, if two or more fail the criteria given above, the shipment will be rejected.

The percentage of lots expected to be accepted will depend on a combination of Class 2 and Class 3 decomposition since only these classes lead to a subsample being classified as decomposed. This makes the operating characteristics for these plans similar to the three-class plans discussed earlier. Because of this, we use a two-way table to describe them. Tables 16-8, 16-9, and 16-10 describe respectively, the three plans mentioned in Table 16-7.

The three tables indicate that the probability of lot acceptance decreases as the lot size increases. This seems to be a reasonable result since it would be more serious to accept a large lot of poor quality than a small one. For example, if lots are submitted with 3% Class 3 and 5% Class 2 decomposition, the percentage of lots accepted goes from 33 to 9 as the lot size increases.

These numbers can be reduced by either increasing the sample size or decreasing the acceptance number. If we do the latter, we increase the risk of

TABLE 16-7. Sampling Plans for Shrimp Inspection

Number of Cases in Lot	n	Ac	Re
1–20	6	1	2
21–100	12	2	3
101 or more	18	3	4

Ac = Acceptance number.
Re = Rejection number.

TABLE 16-8. Percentage of Lots Accepted for Lot Size 1–20 ($N = 6$, $c = 1$)

	Percentage Class 3		
Percentage Class 2	1	3	5
1	100	63	4
5	99	33	1
10	76	9	0

TABLE 16-9. Percentage of Lots Accepted for Lot Size 21–100 ($n = 12$, $c = 2$)

	Percentage Class 3		
Percentage Class 2	1	3	5
1	100	52	0
5	100	16	0
10	71	1	0

TABLE 16-10. Percentage of Lots Accepted for Lots Size 101 or more ($n = 18$, $c = 3$)

	Percentage Class 3		
Percentage Class 2	1	3	5
1	100	45	0
5	100	9	0
10	69	0	0

rejecting lots which are good. For example, if we use the same sample sizes but decrease the acceptance numbers from 1, 2, and 3 to 0, 1, and 2, respectively, the percentage of lots accepted for 3% Class 3 and 5% Class 2 changes from 33, 16, and 9 to 8, 5, and 3, respectively. However, the percentage of lots accepted for lots containing 1% Class 3 and 5% Class 2 will change from 99, 100, and 100 for the present plans to 86, 97, and 99, respectively, for the plans with reduced acceptance numbers.

CONCLUSIONS AND RECOMMENDATIONS

In the previous section, we have presented a sample of the sampling inspection plans used by the United States Food and Drug Administration for seafood. Along with the sampling plans, the operating characteristics of each plan were given. This indicates the amount of protection to both the producer and the consumer provided by the plans. A quick review of these results indicates that very poor protection is provided by any of the plans.

The current thinking in the control of quality and safety of any type of product is that it is better to control the manufacturing process than to attempt to inspect its results. The purpose of an acceptance sampling plan such as those we have been discussing is to make a decision about a lot or shipment as to whether it meets certain criteria. It seems far better to provide some assurance that suppliers are providing good-quality product than to attempt, by means of these imperfect sampling plans, to guard against the acceptance of unsafe product.

However, it may be difficult to control the suppliers' processes since most suppliers are small operators and any seafood house may deal with many suppliers. An alternative to this is the use of a sampling system which allows the customer to change the sampling plans used, depending upon the history of a supplier. This procedure would be applicable only when we buy a series of many lots from each supplier. Such a sampling system is used by the U.S. Department of Defense as well as most manufacturing companies throughout the world.

The sampling system to which we refer is MIL-STD-105E. This sampling system has been adopted with some modification by the American National Standards Institute as an American National Standard under the designation ANSI/ASQC Z1.4-1981. It is available from the American Society for Quality Control. The Standard has also been adopted in modified form as an international standard as ISO 2859-1 (1989).

The concept behind the system is that schemes are offered which are indexed by Acceptable Quality Level or *AQL* and a sampling code letter which depends on lot size. The *AQL* is the quality, in terms of percent nonconforming, which should be accepted most of the time. The definition of "most of the time" also varies with lot size. For the smallest lots, it is 88% while for the largest lots it reaches 99%. By offering many different *AQL*s versatility can be obtained. That is, a smaller *AQL* can be used for critical defects while larger ones can be used for safety hazards that are less critical to the health of the user.

An additional aspect of this system is that a sampling scheme is selected depending upon the lot size and the *AQL* selected. A scheme consists of a set of three plans, called normal, tightened, and reduced plans. Inspection starts with normal inspection and changes to tightened if indication is given that the supplier's quality has worsened (indication means that two out of five or fewer consecutive lots have been rejected). If the supplier's quality has improved (10 consecutive lots accepted), reduced inspection may be instituted. While on tightened inspection, normal inspection is reinstituted if five consecutive lots are accepted. Normal inspection is also reinstituted if a lot is rejected while on reduced inspection. If a total of five lots are rejected while on tightened inspection, purchase from that supplier is to be discontinued until an indication of significant improvement is shown.

Tightened inspection, in general, means that the acceptance number is reduced while the sample size remains the same. Reduced inspection means that the sample size and acceptance number are reduced.

For example, suppose that a shipment of 1000 items is submitted. The standard calls for the sampling plans shown in Table 16-11 for $AQL = 4\%$.

TABLE 16-11

Normal	Tightened	Reduced
$n = 80$ Ac = 7 Re = 8	$n = 80$ Ac = 5 Re = 6	$n = 32$ Ac = 3 Re = 6

Ac = Acceptance number.
Re = Rejection number.

TABLE 16-12. Plans for Lots of 1000 Items (AQL 4%)

	Sampling Level						
	S1	S2	S3	S4	I	II	III
Normal Plans							
n	3	3	13	20	32	80	125
Ac	0	0	1	2	3	7	10
Re	1	1	2	3	4	8	11
Tightened Plans							
n	5	5	20	20	32	80	125
Ac	0	0	1	1	2	5	8
Re	1	1	2	2	3	6	9
Reduced Plan							
n	2	2	5	8	13	32	50
Ac	0	0	0	1	1	3	5
Re	1	1	2	3	4	6	8

Ac = Acceptance number.
Re = Rejection number.

The meaning of the gap between the acceptance number and rejection number for the reduced plan is that, if there are three or fewer nonconforming items, the lot is accepted and reduced inspection is continued for the next lot submitted. If there are six or more nonconforming items in the sample, the lot is rejected and the next lot submitted is inspected using the normal plan. If there are four or five nonconforming items, the current lot is accepted but the next lot is inspected using the normal plan.

An alternative to reduced inspection is the use of a skip-lot sampling plan. Such a plan is found in ISO 2859-3 or in ANSI/ASQC S-1 available from the American Society for Quality Control. This type of plan calls for skipping the inspection of some lots but using normal inspection on those lots inspected. This approach is favored by the *Quality Management Program* of the

Department of Fisheries and Oceans, Gulf Region of Canada. For details of this procedure, the reader is referred to the standards.

There are several advantages to using these standards for sampling rather than the fixed plans of the FDA and other organizations. These are:

1. There is an incentive to the supplier to provide acceptable product through the use of either the reduced or the skip-lot plans.
2. There is protection to the consumer through the use of tightened inspection and the discontinue procedures.
3. There are several inspection levels that are available depending upon the amount of protection needed.
4. The *AQL* can be changed for various types of the nonconforming product.
5. The Standard is a universally accepted standard in use by virtually all countries in the world.

In addition to the *AQL* and lot size considerations given in the standard, different inspection levels may be chosen depending upon the level of the severity desired. For example, for lots of 1000 items seven different levels are available. The *S* levels are for special instances when particularly small samples are needed because inspection is costly or destructive. These are not recommended for the general case. The three plans for these general levels are indicated in Table 16-12.

References

Abeyta, C. Jr., M. Wekell, and T. Peeler. 1985. Comparison of media for enumeration of *Clostridium perfringens* in foods. *J. Food Sci.* 50(6):1732–35.

Abeyta, C. Jr., C. Kaysner, M. Wekell, J. Sullivan, and G. Stelma. 1986. Recovery of *Aeromonas hydrophila* from oysters implicated in an outbreak of foodborne illness. *J. Food Prot.* 49(8):643–46.

Abeyta, C. Jr., S.D. Weagant, C.A. Kaysner, M.M. Wekell, R.F. Stott, M.H. Krane, and J.T. Peeler. 1989. *Aeromonas hydrophila* in shellfish growing waters: Incidence and media evaluation. *J. Food Prot.* 52:7–12.

Abshire, R.L. and R.K. Guthrie. 1971. The use of fluorescent antibody technique for detection of *Streptococcus faecalis* as an indicator of faecal pollution in water. *Water Res.* 5:1089–97.

Abshire, R.L. 1976. Detection of enteropathogenic *Escherichia coli* strains in wastewater by fluorescent antibody. *Can. J. Microbiol.* 22:364–78.

Abshire, R.L. 1977. Evaluation of a new presumptive medium for group D streptococci. *Appl. Environ. Microbiol.* 33:1149–55.

Adams, J.C. and D.S. Farrier. 1982. The effect of some oil shale process waters upon the viability of indicator bacteria. *J. Environ. Qual.* 11:171–74.

Adams, M.H. 1959. *Bacteriophages.* New York: Wiley-Interscience.

Adams, M.R., S.M. Grubb, A. Hamer, and M.N. Clifford. 1990. Colorimetric enumeration of *Escherichia coli* based on *b*-glucuronidase activity. *Appl. Environ. Microbiol.* 56:2021–24.

Agbalika, F., P. Hartemann, and J.M. Foliguet. 1984. Trypsin-treated MA-104: A sensitive cell line for isolating enteric viruses from environmental samples. *Appl. Environ. Microbiol.* 47:378–80.

Ahmed, F.E. ed. 1991. *Seafood Safety.* Washington, D.C.: National Academy Press.

Alter, M.J., R.J. Gerety, L.A. Smallwood, R.E. Sampliner, E. Tabor, F. Deinhardt, G. Frosner, and G.M. Matanoski. 1982. Sporadic non-A non-B hepatitis: Frequency and epidemiology in the urban U.S. population. *J. Infect. Dis.* 145:886–93.

Al-Hakim, H.H. and R. Hull. 1986. Studies towards the development of chemically synthesized nonradioactive biotinylated nucleic acid hybridization probes. *Nucleic Acids Res.* 14:9965–72.

Al-Jebouri, M.M. and D. Trollope. 1981. The *Escherichia coli* content of Mytilus edulis from analysis of whole tissue or digestive tract. *J. Appl. Bacteriol.* 51(1):135–42.

Aldridge, K.E., B.B. Gardner, S.J. Clark, and J.M. Matsen. 1978. Comparison of Micro-ID, API 20E, and conventional media systems in identification of *Enterobacteriaceae. J. Clin. Microbiol.* 7:507–13.

Aleksis, S. and J. Bockemuhl. 1984. Proposed revision of the Wauters et al. Antigenic scheme for serotyping of *Yersinia enterocolitica. J. Clin. Microbiol.* 20:99–102.

Aleksis, S., J. Bockemuhl, and F. Lange. 1986. Studies on the serology of flagellar

antigens of *Yersinia enterocolitica* and related *Yersinia* species. *Zentralbl. Bakteriol. Mikrobiol. Hyg.* [A] 261:299–310.
Alexander, L.M. and R. Morris. 1990. PCR and environmental monitoring—The way forward? *Water Sci. Technol.* 22: (in press).
Allen-Austin, D., B. Austin, and R.R. Colwell. 1984. Survival of *Aeromonas salmonicida* in river water. *FEMS Microbiol. Lett.* 21:143–46.
Allen, K. 1899. The transmission of typhoid by sewage polluted oysters. *Rep. Am. Public Health Assoc.* 25:154–65.
Allen, L.A., G. Thomas, M.C.C. Thomas, A.B. Wheatland, H.N. Thomas, E.E. Jones, and J. Hudson. 1950. Reported reuse of seawater as a medium for the functioning and self-cleansing of molluscan shellfish. *J. Hyg.* 48(4):431–57.
Allred, J.N., J.W. Walker, V.C. Beal Jr., and F.W. Germaine. 1967. A survey to determine the *Salmonella* contamination rate in livestock and poultry feeds. *J. Am. Vet. Med. Assoc.* 151:1857.
Allsop, K. and D.J. Stickler. 1984. The enumeration of *Bacteroides fragilis* group organisms from sewage and natural waters. *J. Appl. Bacteriol.* 56:1524.
Allsop, K. and D.J. Stickler. 1985. An assessment of *Bacteroides fragilis* group organisms as indicators of human faecal pollution. *J. Appl. Bacteriol.* 58:95–99.
Alphaz, A. and M.O. Balaban. 1991. An overview of molluscan shellfish industry in Turkey. In *Molluscan Shellfish* Depuration, eds. W.S. Otwell, G.E. Rodrick, and R. Martin. Boca Raton, Fla.: CRC Press (in press).
Alvarez, R.J. 1984. Use of fluorogenic assays for the enumeration of *Escherichia coli* from selected seafoods. *J. Food Sci.* 49:1186–232.
Ambinder, R.F., P. Charache, S. Staal, P. Wright, M. Forman, S.D. Hayward, and G.S. Hayward. 1986. The vector homology problem in diagnostic nucleic acid hybridization of clinical specimens. *J. Clin. Microbiol.* 24:16.
American Public Health Association (APHA). 1980. *Standard Methods for the Examination of Water and Wastewater*, 15th ed. Washington, D.C.: American Public Health Association.
American Public Health Association. 1984*a*. *Compendium of Methods for the Microbiological Examination of Foods*, 2nd ed. Washington, D.C.: American Public Health Association.
American Public Health Association. 1984*b*. *Laboratory Procedures for the Examination of Seawater and Shellfish*, 5th ed. Washington, D.C.: American Public Health Association.
American Public Health Association. 1985*a*. *Recommended Procedures for the Examination of Seawater and Shellfish*. 5th ed, eds. A.E. Greenberg and D.A. Hunt, p. 144. Washington, D.C.: American Public Health Association.
American Public Health Association. 1985*b*. *Standard Methods for the Examination of Water and Wastewater*, 16th ed. Washington, D.C.: American Public Health Association.
American Public Health Association. 1989. *Standard Methods for the Examination of Water and Wastewater*, 17th ed. Washington, D.C.: American Public Health Association.

American Society for Quality Control. 1981. ANSI/ASQC Z-1.4, Sampling Procedures and Tables for Inspection by Attributes. Milwaukee, Wis.

American Society for Quality Control. 1987. ANSI/ASQC S1–1987, An Attribute Skip—Lot Sampling Program. Milwaukee, Wis.

American Society for Testing and Materials (ASTM). 1987. Standard Test Method for Enumeration of *Candida albicans* in water. In *ASTM Standards on Materials and Environmental Microbiology*, p. 113–18. Philadelphia, Pa.: American Society for Testing and Materials.

Amy, P.S. and R.Y. Morita. 1983. Protein patterns of growing and starving cells of a marine *Vibrio* sp. *Appl. Environ. Microbiol.* 45:1748–52.

Amy, P.S. and R.Y. Morita. 1983. Starvation-survival patterns of sixteen freshly isolated open-ocean bacteria. *Appl. Environ. Microbiol.* 45:1109–15.

Amy, P.S. and H.D. Hiatt. 1989. Survival and detection of bacteria in an aquatic environment. *Appl. Environ. Microbiol.* 55:788–93.

Anderson, J.I.W. and W.P. Heffernan. 1965. Isolation and characterization of filterable marine bacteria. *J. Bacteriol.* 90:1713–18.

Anderson, J. and A. Baird-Parker. 1975. A rapid and direct plating method for enumeration of *Escherichia coli* type one in food. *J. Appl. Bacteriol.* 39:111–17.

Anderson, I.C., M. Rhodes, and H. Kator. 1979. Sublethal stress in *Escherichia coli*: A function of salinity. *Appl. Environ. Microbiol.* 38:1147–52.

Anderson, I.C., M. Rhodes, and H. Kator. 1983. Seasonal variation in survival of *Escherichia coli* exposed in situ in membrane diffusion chambers containing filtered and nonfiltered estuarine water. *Appl. Environ. Microbiol.* 45:1877–83.

Andersen, P. and T. Fenchel. 1985. Bacterivory by microheterotrophic flagellates in seawater samples. *Limnol. Oceanogr.* 30:198–202.

Andrews, W.H. and M.W. Presnell. 1972. Rapid recovery of *Escherichia coli* from estuarine water. *Appl. Microbiol.* 23:521–23.

Andrews, W.H., C. Diggs, M. Presnell, J. Miescier, C. Wilson, C. Goodwin, W. Adams, S. Furfari, and J. Musselman. 1974. Coliforms as indicators of Salmonella in shellfish. Washington, D.C.: FDA, Div. of Microbiol. *Abstr. Ann. Meet. Am. Soc. Microbiol.* 74:65.

Andrews, W.H., C.D. Diggs, and C.R. Wilson. 1975. Evaluation of a medium for the rapid recovery of *Escherichia coli* from shellfish. *Appl. Microbiol.* 29(1):130–31.

Andrews, W.H., C. Diggs, M. Presnell, J. Miescier, C. Wilson, C. Goodwin, W. Adams, S. Furfari, and J. Musselman. 1975. Comparative validity of members of the total coliform and fecal coliform groups for indicating the presence of Salmonella in the Eastern oyster, Crassostrea virginica. *J. Milk Food Technol.* 38(8):453–56.

Andrews, W.H., C.D. Diggs, J.J. Miescier, C.R. Wilson, W.N. Adams, S.A. Furfari, and J.F. Musselman. 1976. Validity of members of the total coliform and fecal coliform groups for indicating the presence of Salmonella in the quahaug, Mercenaria. *J. Milk Food Technol.* 39(5):322–24.

Andrews, W.H., C.R. Wilson, P.L. Poelma, A. Romero, R.A. Rude, A.P. Duran, D. McClure, and D.E. Gentile. 1978. Usefulness of the stomacher in a microbiological regulatory laboratory. *Appl. Environ. Microbiol.* 35:89–93.

Andrews, W.H., C.R. Wilson, P.L. Poelma, L.K. Bullock, F.D. McClure, and D.E. Gentile. 1981. Interlaboratory evaluation of the AOAC method and the A-1 procedure for recovery of fecal coliforms from foods. *J. Assoc. Off. Anal. Chem.* 64:1116–21.

Andrews, W.H. 1985. A review of culture methods and their relation to rapid methods for the detection of *Salmonella* in foods. *Food Technol.* 39:77–82.

Anonymous. 1975. National interim primary water regulations. *Fed. Regist.* 40:59566–88.

Anonymous. 1979. Trace lot study of soft shell clam (*Mya aremaria*) from harvesting area to New York market. Baltimore, Md.: Environmental Health Administration, Maryland Dept. of Health and Mental Hygiene.

Anonymous. 1985. Bacterial food poisoning: Annual review. *New Zealand Med. J.* 98(788):863–65.

Anonymous. 1986. Changes in methods. *Salmonella* in foods: Enzyme immunoassay screening method. *J. Assoc. Off. Anal. Chem.* 69:381–84.

Anonymous. 1987. Changes in methods. *Salmonella* in foods: DNA hybridization method. *J. Assoc. Off. Anal. Chem.* 70:394–96.

Anonymous. 1989*a*. Changes in official methods. *Salmonella* in foods: Colorimetric polyclonal EIA screening method. *J. Assoc. Off. Anal. Chem.* 72:201–03.

Anonymous. 1989*b*. Changes in official methods. *Salmonella* in foods: Fluorogenic monoclonal EIA screening method. *J. Assoc. Off. Anal. Chem.* 72:203–05.

Applebaum, P.C., R.R. Arthur, M.E. Parker, G.L. Shugar, L.C. Von Kuster, and P. Charache. 1982. Comparison of two methods for same-day identification of *Enterobacteriaceae. Am. J. Clin. Pathol.* 78:351–55.

Applebaum, P.C., P.S. Chaurushiya, M.R. Jacobs, and A. Duffett. 1984*a*. Evaluation of the rapid strep system for species identification of streptococci. *J. Clin. Microbiol.* 19:588–91.

Applebaum, P.C., M.R. Jacobs, J.I. Heald, W.A. Palko, A. Duffett, R. Crist, and P.A. Naugle. 1984*b*. Comparative evaluation of the API 20S system and the automicrobic system gram-positive identification card for species identification of streptococci. *J. Clin. Microbiol.* 19:164–68.

Applebaum, P.C., M.R. Jacobs, W.M. Palko, E.E. Frauenhoffer, and A. Duffett. 1986. Accuracy and reproducibility of the IDS RapID STR system for species identification of streptococci. *J. Clin. Microbiol.* 23:843–46.

Appleton, H. and M.S. Pereira. 1977. A possible virus aetiology in outbreaks of food poisoning from cockles. *Lancet* i:780–81.

Appleton, H., S.R. Palmer, and R.J. Gilbert. 1981. Food-borne gastroenteritis of unknown aetiology: A virus infection? *Br. Med. J.* 282:1801–02.

Appleton, H. 1987. Small round viruses: Classification and role in food-borne infections. In *Novel Diarrhoea Viruses*. Ciba Foundation Symp., pp. 108–25. Chichester, U.K.: Wiley.

Arai, T. and N. Ikejima. 1980. A survey of *Plesiomonas shigelloides* from aquatic environments, domestic animals, pets and humans. *J. Hyg.* Cambridge, Mass. 84:203–11.

Archer, D.L. and F. Young. 1988. Contemporary issues: Diseases with a food vector. *Clin. Microbiol. Rev.* 1:377–98.

Arcisz, W. and C.B. Kelly. 1955. Self-purification of the soft clam *Mya arenaria. Public Health Rep.* 70(6):605–14.

Armon, R. and V.J. Cabelli. 1988. Phage f2 desorption from clay in estuarine water using nonionic detergents, beef extract, and chaotropic agents. *Can. J. Microbiol.* 34:1022–24.

Armon, R. and P. Payment. 1988. A modified m-CP medium for enumerating *Clostridium perfringens* from water samples. *Can. J. Microbiol.* 34:78–79.

Arumugaswamy, R.K. and R.W. Proudford. 1987. The occurrence of *Campylobacter jejuni* and *Campylobacter coli* in Sydney Rock oyster (Crassostrea commercialis). *Int. J. Food Microbiol.* 4(2):101–04.

Arumugaswamy, R.K., R.W. Proudford, and M.J. Eyles. 1988. The response of *Campylobacter jejuni* and *Campylobacter coli* in the Sydney Rock oyster (Crassostrea commercialis), during depuration and storage. *Int. J. Food Microbiol.* 7(3):173–83.

Aserkoff, B., S.A. Schroeder, and P.S. Brachman. 1970. Salmonellosis in the United States—A five-year review. *Am. J. Epidemiol.* 92:13–24.

Association of Official Analytical Chemists. 1985. Total coliform, fecal coliform and *Escherichia coli* in foods, hydrophobic grid membrane filter method. *J. Assoc. Off. Anal. Chem.* 68:404.

Association of Official Analytical Chemists. 1987. Changes in Methods. *Staphylococcus aureus* in Foods. *J. Assoc. Off. Anal. Chem.* 70:393–94.

American Society for Testing and Materials (ASTM). 1987. Standard Test Method for enumeration of *Candida albicans* in water. In *ASTM Standards on Materials and Environmental Microbiology*, pp. 113–18. Philadelphia, Pa.: American Society for Testing and Materials.

Atlas, R.M. 1982. Enumeration and estimation of biomass of microbial components in the biosphere. In *Experimental Microbial Ecology*, eds. R.G. Burns and J.H. Slater, pp. 84–102. Oxford: Blackwell Scientific Publishers.

Atlas, R.M. and G.S. Sayler. 1988. *Tracking Microorganisms and Genes in the Environment.* New York: Plenum Press.

Atlas, R.M. and A.K. Bej. 1990. Detecting bacterial pathogens in environmental water samples by using PCR and gene probes. In *PCR Protocols: A Guide to Methods and Applications*, eds. M.A. Innis, D.H. Gelfand, J.J. Sninsky, and T.J. White, pp. 399–406. San Diego, Calif.: Academic Press.

Attaway, D. and P.L. Parker. 1970. Sterols in recent marine sediments. *Science* 169:674–75.

Attwell, R.W. and R.R. Colwell. 1984. Thermoactinomycetes as terrestrial indicators for estuarine and marine waters. In *Biological, Biochemical and Biomedical Aspects of Actinomycetes*, eds. L. Ortiz-Ortiz, L.F. Bojalil, and V. Yakoleff, pp. 441–52. New York: Academic Press.

Attwood, G.T., R.A. Lockington, G.P. Xue, and J.D. Brooker. 1988. Use of a unique gene sequence as a probe to enumerate a strain of *Bacteroides ruminicola* introduced into the rumen. *Appl. Environ. Microbiol.* 54:534–39.

Balachandran, K.K. and P. Surendran. 1985. Purification of live clams for processing. *Seafood Export J.* 17:5–9.

Ballew, M. and D. Hattis. 1989. *Reproductive Effects of Glycol Ethers in Females—A Quantitative Analysis.* CTPID 89-7. Cambridge, Mass: MIT Ctr. for Technology, Policy, and Industrial Development.

Bamber, M., H.C. Thomas, B. Bannister, and S. Sherlock. 1983. Acute type A, B, and non-A non-B hepatitis in a hospital population in London: Clinical and epidemiological features. *Gut.* 24:561–64.

Barbay, J.R., H.B. Bradford, and N.C. Roberts. 1984. The occurrence of halophilic vibrios in Louisiana coastal waters. In *Vibrios in the Environment*, ed. R.R. Colwell, pp. 511–20. New York: Wiley.

Barcina, I., I. Arana, J. Iriberri, and L. Egea. 1986. Influence of light and natural microbiota of the Butron river on *E. coli* survival. *Antonie van Leeuwenhoek J. Microbiol.* 52:555–66.

Barcina, I., J.M. Gonzalez, J. Iriberri, and L. Egea. 1989. Effect of visible light on progressive dormancy of *Escherichia coli* cells during the survival process in natural fresh water. *Appl. Environ. Microbiol.* 55:246–51.

Barkay, T., D.L. Fouts, and B.H. Olson. 1985. Preparation of a DNA gene probe for detection of mercury resistance in gram-negative bacterial communities. *Appl. Environ. Microbiol.* 49:686–92.

Barker, L.F. and R. Murray. 1972. Relationship of virus dose to incubation time of clinical hepatitis and time of appearance of hepatitis-associated antigen. *Am. J. Med. Sci.* 263:27–33.

Barker, N.G. 1990. Vanasse, Hangen, Brustlin, Inc., Hayes, Va. (personal communication).

Barker, W.H. Jr. 1974. *Vibrio parahaemolyticus* outbreaks in the United States. *Lancet* i:551–54.

Barnard, R.J. and G.J. Jackson. 1984. *Giardia lamblia.* The transfer of human infections by foods. In *Giardia Giardiasis, Biology Pathogenesis and Epidemiology*, eds. S.L. Erlandsen and E.A. Meyer, pp. 365–76. New York: Plenum Press.

Barnes, E.L. 1986. Anaerobic bacteria of the normal intestinal microflora of animals. In *Anaerobic Bacteria in Habitats Other than Man*, eds. E.M. Barnes and G.C. Mead, pp. 225–38. Oxford: Blackwell.

Barnes, R., J.I. Curry, L.M. Elliott, C.R. Peter, B.R. Tamplin, and B.W. Wilcke Jr. 1989. Evaluation of the 7-hour membrane filter test for quantitation of fecal coliforms in water. *Appl. Environ. Microbiol.* 55:1504–06.

Baron, R.C., F.D. Murphy, H.B. Greenberg, C.E. Davis, D.J. Bregman, G.W. Gary, J.H. Hughes, and L.B. Schonberger. 1982. Norwalk gastrointestinal illness. An outbreak associated with swimming in a recreational lake and secondary person-to-person transmission. *Am. J. Epidemiol.* 115:163–72.

Baross, J. and J. Liston. 1970. Occurrence of *Vibrio parahaemolyticus* and related hemolytic vibrios in marine environments of Washington St. *Appl. Microbiol.* 20:179–86.

Barrett, T.J., P.A. Blake, G.K. Morris, N.D. Puhr, H.B. Bradford, and J.G. Wells. 1980. Use of Moore swabs for isolating *Vibrio cholerae* from sewage. *J. Clin.*

Microbiol. 11:385–88.
Barrett, T.J., C.M. Patton, and G.K. Morris. 1988. Differentiation of *Campylobacter* species using phenotypic characterization. *Lab. Med.* 19:96–102.
Barrett, T.J., M. Potter, and I. Wachsmuth. 1989. Bacterial endotoxin both enhances and inhibits the toxicity of Shiga-like toxin II in rabbits and mice. *Infect. Immun.* 57:3434–37.
Barry, A.L. and R.E. Badal. 1979. Rapid identification of *Enterobacteriaceae* with the Micro-ID system versus API 20E and conventional media. *J. Clin. Microbiol.* 10:293–98.
Barry, A.L., T.L. Gavan, P.B. Smith, J.M. Matsen, J.A. Morello, and B.H. Sielaff. 1982. Accuracy and precision of the Autobac system for rapid identification of gram-negative bacilli: A collaborative evaluation. *J. Clin. Microbiol.* 15:1111–19.
Barry, V.C., M.L. Conalty, J.M. Denneny, and F. Winder. 1956. Peroxide formation in bacteriological media. *Nature* 178:596–97.
Bartholomew, B., M. Stringer, G. Watson, and R. Gilbert. 1985. Development and application of an enzyme-linked immunosorbent assay for *Clostridium perfringens* type A enterotoxin. *J. Clin. Pathol.* 38:222–28.
Bartlett, P. 1987. Degradation of coprostanol in an experimental system. *Mar. Pollut. Bull.* 18:27–29.
Bates, S.S., C.J. Bird, A.S.W. deFreitas, R.A. Foxall, M. Gilgan, L.A. Hanic, G.R. Johnson, A.W. McCulloch, P. Odense, R. Pocklington, M.A. Quilliam, P.G. Sim, J.C. Smith, D.V. Subba Rao, E.C.D. Todd, J.A. Walter, and J.L.C. Wright. 1989. Pennate diatom *Nitzschia pungens* as the primary source of domoic acid, a toxin in shellfish from eastern Prince Edward Island, Canada. *Can. J. Fish. Aquat. Sci.* 46:1203–04.
Batish, V.K., D.R. Ghodeker, and B. Ranganathan. 1978. The thermostable deoxyribonuclease (Dnase) test as a rapid screening method for detection of staphylococcal enterotoxin in milk and milk products. *Microbiol. Immunol.* 22:437–41.
Baumann, P. and R.H.W. Schubert. 1984. *Vibrionaceae* in *Bergeys Manual of Systematic Bacteriology*, Vol. 1, eds. N.R. Krieg and J.G. Holt, pp. 516–17. Baltimore, Md.: Williams and Wilkins.
Bearns, R.E. and K.F. Girard. 1958. The effect of pasteurization on *Listeria monocytogenes*. *Can. J. Microbiol.* 4:55–61.
Beaudoin, E.C. and W. Litsky. 1981. Fecal streptococci. In *Membrane Filtration: Applications, Techniques, and Problems*, ed. B.J. Dutka, pp. 77–118. New York: Marcel Dekker.
Bedford, A.J., G. Williams, and A.R. Bellamy. 1978. Virus accumulation by the rock oyster Crassostrea glomerata. *Appl. Environ. Microbiol.* 35:1012–18.
Beerens, H. 1990. An elective and selective isolation medium for *Bifidobacterium* spp. *Lett. Appl. Microbiol.* 11:155–57.
Bej, A.K., R.J. Steffan, J. DiCesare, L. Haff, and R.M. Atlas. 1990*a*. Detection of coliform bacteria in water by polymerase chain reaction and gene probes. *Appl. Environ. Microbiol.* 56:307–14.

Black, R.E., M.M. Levine, M.L. Clements, T.P. Hughes, and M.J. Blaser. 1988. Experimental *Campylobacter jejuni* infection in humans. *J. Infect. Dis.* 157:472–79.

Blake, N.J., G.E. Rodrick, M. Tamplin, and T.R. Cuba. 1982. Validity of bacteriological standards for shellfish harvesting waters. In *Proceedings of 7th American Tropical and Subtropical Fisheries Technology Conference*, pp. 311–320. Sea Grant Pub. TAMU-SG-82-110. Austin, Tx.: Texas A & M.

Blake, P.A., D.T. Allegra, J.D. Snyder, T.J. Barrett, L. McFarland, C.T. Caraway, J.C. Feeley, J.P. Craig, J.V. Lee, N.D. Puhr, and R.A. Feldman. 1980. Cholera—A possible endemic focus in the United States. *New Engl. J. Med.* 302:305–09.

Blake, P.A. and R. Feldman. 1986. Shigellosis. In *Maxey–Rosenau Public Health and Preventive Medicine*, ed. J.M. Last, 12th ed., pp. 240–42. Norwalk, Conn.: Appleton–Century–Crofts.

Blankenship, L.C. 1981. Some characteristics of acid injury and recovery of *Salmonella barely* in a model system. *J. Food Prot.* 44:73–77.

Blaser, M. and L. Newman. 1982. A review of human salmonellosis: I. Infective dose. *Rev. Infect. Dis.* 4:1096–102.

Blaser, M.J., J. Cravens, B.W. Powers, F.M. LaForce, and W.L.L. Wang. 1979. *Campylobacter* enteritis associated with unpasteurized milk. *Am. J. Med.* 67:715–18.

Blaser, M.J. 1980. *Campylobacter fetus* subspecies *jejuni*: The need for surveillance. *J. Infect. Dis.* 141:670–76.

Blaser, M.J. and L.B. Reller. 1981. *Campylobacter enteritis. New Engl. J. Med.* 305:1444–52.

Blaser, M.J. and L.S. Newman. 1982. A review of human Salmonellosis: I. Infective dose. *Rev. Infect. Dis.* 4:1096–106.

Blogoslawski, W.J., C. Brown, E.W. Rhodes, and M. Broadhurst. 1975. Ozone disinfection of a seawater supply system. In *Proceedings of 1st International Symposium Ozone for Water and Wastewater*, pp. 674–87. Washington, D.C.

Blogoslawski, W.J. and M.E. Stewart. 1977. Detoxification of marine poisons by ozone gas. In *Proceedings of 3rd Congress of the International Ozone Institute, Paris, France.* Washington, D.C.: International Ozone Association.

Blogoslawski, W.J., M.E. Stewart, J.W. Hurst Jr., and F.G. Kern III. 1979. Ozone detoxification of paralytic shellfish poison in the softshell clam (*Mya arenaria*). *Toxicon.* 17:650–54.

Blogoslawski, W.J. 1980. Use of chlorination in the molluscan shellfish industry. In *Water Chlorination Environmental Impact and Health Effects*, Vol. 3, eds. R.L. Jolley, W.A. Brungs, R.B. Cumming, and V.A. Jacobs, pp. 487–500. Ann Arbor, Mich.: Ann Arbor Science Publishers.

Blogoslawski, W.J. and P.O. Monasterio. 1982. Bacterial depuration of the mexican scallop *Argopecten circularis.* In *Ozone: Science and Engineering*, Vol. 4, pp. 121–29. New York: Pergamon Press.

Blogoslawski, W.J. 1983. Influence of water quality on shellfish culture. In *Mariculture Committee, Ref. Shellfish Committee, International Council for the Exploration of the Sea, Statutory Meeting*, pp. 1–35. Gothenburg, Sweden.

Blogoslawski, W.J. and M.E. Stewart. 1983. Depuration and public health. In

Proceedings of 14th Annual Meeting World Mariculture Society, Vol. 14, pp. 535–45.

Boardman, G., T.R. McBrayer, and P. Kohlhepp. 1989. Detection and occurrence of waterborne bacterial and viral pathogens. *J. Water Pollut. Control Fed.* 61:1097–109.

Bobowick, A.R., J.A. Brody, M.R. Matthews, R. Roos, and D.C. Gajdusek. 1973. Creutzfeldt-Jakob disease: A case control study. *Am. J. Epidemiol.* 98:381–94.

Boehm, P.D. 1983. Coupling of organic pollutants between the estuary and continental shelf and the sediments and water column in the New York Bight Region. *Can. J. Fish Aquat. Sci.* 40:262–76.

Boguslawski, S.J., D.E. Smith, M.A. Michalak, K.E. Mickelson, C.O. Yehle, W.L. Patterson, and R.J. Carrico. 1986. Characterization of a monoclonal antibody to DNA–RNA and its application to immunodetection of hybrids. *J. Immunol. Methods* 89:123–30.

Boileau, C.R., H.M. d'Hauteville, and P.J. Sansonetti. 1984. DNA hybridization technique to detect *Shigella* species and enteroinvasive *Escherichia coli*. *J. Clin. Microbiol.* 20:959–61.

Bonde, G.J. 1977. Bacterial indication of water pollution. In *Advances in Aquatic Microbiology*, Vol. 1, eds. M.R. Droop and H.W. Jannasch, pp. 273–364. London: Academic Press.

Booth, S.J., R.L. van Tassell, J.L. Johnson, and T.D. Wilkins. 1979. Bacteriophages of *Bacteroides*. *Rev. Infect. Dis.* 1:325–36.

Boothby, J.C., Genigeorgis, and M.H. Fanelli. 1979. Tandem coagulase/thermonuclease agar method for the detection of *Staphylococcus aureus*. *Appl. Environ. Microbiol.* 37:298–302.

Borchardt, K.A. 1968. Scheme for screening and identifying enteric and other gram-negative bacteria using reagent-impregnated strips. *Am. J. Clin. Path.* 50:129–31.

Borrego, J. and P. Romero. 1985. Coliphage survival in seawater. *Water Res.* 19:557–62.

Borrego, J.J., M.A. Moriñigo, A. de Vicente, R. Cornax, and P. Romero. 1987. Coliphages as an indicator of fecal pollution in water. Its relationship with indicator and pathogenic microorganisms. *Water Res.* 21:1473–80.

Borrego, J.J., R. Cornax, M.A. Moriñigo, E. Martinez-Manzanares, and P. Romero. 1990. Coliphages as an indicator of fecal pollution in water. Their survival and productive infectivity in natural aquatic environments. *Water Res.* 24:111–16.

Borrego, J.J., R. Cornax, D.R. Preston, S.R. Farrah, B. McElhaney, and G. Bitton. 1991. Development and application of new positively charged filters for recovery of bacteriophages from water. *Appl. Environ. Microbiol.* 57:1218–22.

Bosley, G.S., R.R. Facklam, and D. Grossman. 1983. Rapid identification of *Enterococci*. *J. Clin. Microbiol.* 18:1275–77.

Bostock, A.D., P. Mepham, and S. Phillips. 1979. Hepatitis A infection associated with the consumption of mussels. *J. Infect.* 1:171–77.

Boutin, B.K., P.L. Spaulding, and R.M. Twedt. 1986. Evaluation of the Enteropathogenicity of *Klebsiella pneumoniae* Isolates from Summer-Harvested Louisiana Oysters. *J. Food Prot.* 49(6):442–44.

Buchanan, R.L. and S. Palumbo. 1985. *Aeromonas hydrophila* and *Aeromonas sobria* as potential food poisoning species. *Appl. Environ. Microbiol.* 47:1146–49.

Buck, J.D. 1977. *Candida albicans.* In *Bacterial Indicators/Health Hazards Associated With Water*, eds. A.W. Hoadley and B.J. Dutka, pp. 139–47. Philadelphia, Pa.: American Society for Testing and Materials.

Bucker, E.R. and S.E. Martin. 1981. Superoxide dimutase activity in thermally stressed *Staphylococcus aureus. Appl. Environ. Microbiol.* 41:449–54.

Bulstrode, H.T., Klein, and C.B. Thorne. 1896. Oyster culture in relation to disease. *The Br. Med. J.*, p. 1603.

Bulstrode, H.T. 1911. Report on shellfish other than oysters in relation to disease. Supplement to 39th Annual Report of the Local Government Board in Continuation of the Report of the Medical Officer for 1909–10.

Bundesen, H.N. 1925. Typhoid epidemic in Chicago apparently due to oysters. *J. AMA* 84:641–50.

Burkhardt III, W., S.R. Rippey, and W.D. Watkins. 1988. A comparison of classical and alternative indicators used for assessing the purification rates of depurating shellfish. *Abstr. Ann. Meet. Am. Soc. Microbiol.* 88:291.

Burkhardt III, W., S.R. Rippey, and W.D. Watkins. 1989. The behavior of coliphage and F+ Escherichia coli within whole clams and their homogenates. *Abstr. Ann. Meet. Am. Soc. Mibrobiol.* 89:348.

Burlingham, G.A., J. McElhaney, M. Bennett, and W.O. Pipes. 1984. Bacterial interference with coliform colony sheen production on membrane filters. *Appl. Environ. Microbiol.* 47:56–60.

Burow, H. 1974. *Clostridium perfringens* contamination of chilled sea fish and mussels in Turkey. *Archiv fuer Lebensmittelhygiene* 25:39–42.

Burrows, W. 1985. *Burrows Textbook of Microbiology*, 22nd ed. (revised by B.A. Freeman). Philadelphia, Pa.: W.B. Saunders.

Busta, F.F. 1976. Practical implications of injured microorganisms in food. *J. Milk Food Technol.* 39:138–45.

Busta, F.F. 1978. Introduction to injury and repair of microbial cells. *Adv. Appl. Microbiol.* 23:219–43.

Butnam, B.T., M.C. Plank, R.J. Durham, and J.A. Mattingly. 1988. Monoclonal antibodies which identify a genus-specific *Listeria* antigen. *Appl. Environ. Microbiol.* 54:1564–69.

Byrd, J.J. and R.R. Colwell. 1990. Maintenance of plasmids pBR322 and pUC8 in nonculturable *Escherichia coli* in the marine environment. *Appl. Environ. Microbiol.* 56:2104–07.

Byrd, J.J. and R.R. Colwell. 1991. Microscopy applications for analysis of environmental samples.

Byrd, J.J., H.-S. Xu, and R.R. Colwell. 1991. Viable but nonculturable bacteria in drinking water. *Appl. Environ. Microbiol.* 57:875–78.

Cabelli, V.J. and W. Heffernan. 1970. Accumulation of *Escherichia coli* by the Northern Quahaug. *Appl. Microbiol.* 19:239–44.

Cabelli, V.J. and W.P. Heffernan. 1971. Seasonal factors relevant to coliform levels in northern quahaugs. *Proc. Natl. Shellfish Assoc.* 61:95–101.

Cabelli, V.J. 1976. Indicators of recreational water quality. In *Bacterial Indicators Health Hazards Associated With Water*, eds. A.W. Hoadley and B. Dutka. Philadelphia, Pa.: American Society for Testing and Materials.

Cabelli, V.J. 1977*a*. *Clostridium perfringens* as a water quality indicator. In *Bacterial Indicators/Health Hazards Associated With Water*, eds. A.W. Hoadley and B.J. Dutka, pp. 65–79. Special Technical Publication 635, Philadelphia, Pa.: American Society for Testing and Materials.

Cabelli, V.J. 1977*b*. Indicators of recreational water quality. In *Bacterial Indicators/Health Hazards Associated With Water*, eds. A.W. Hoadley and B.J. Dutka, pp. 222–38. Special Technical Publication 635, Philadelphia, Pa.: American Society for Testing and Materials.

Cabelli, V.J. 1978*a*. New standards for enteric bacteria. In *Water Pollution Microbiology*, Vol. 2, ed. R. Mitchell, pp. 233–71. New York: Wiley.

Cabelli, V.J. 1978*b*. Obligate anaerobic bacterial indicators. In *Indicators of Viruses in Water and Food*, ed. G. Berg, pp. 171–200. Ann Arbor, Mich.: Ann Arbor Science Publishers.

Cabelli, V.J. 1978*c*. Swimming-associated disease outbreaks. *J. Water Pollut. Control Fed.* 50:1374–77.

Cabelli, V.J. 1979. Evaluation of recreational water quality, the EPA approach. In *Biological Indicators of Water Quality*, eds. A. James and L. Evison, pp. 14-1–14-23. Chichester, U.K.: Wiley.

Cabelli, V.J., A.P. Dufour, L.J. McCabe, and M.A. Levin. 1982. Swimming associated gastroenteritis and water quality. *Am. J. Epidemiol.* 115:606.

Cabelli, V.J., A.P. Dufour, L.J. McCabe, and M.A. Levin. 1983. A marine recreational water quality criterion consistent with indicator concepts and risk analysis. *J. Water Pollut. Control Fed.* 55:1306–14.

Calabrese, J.P. and G.K. Bissonnette. 1990*a*. Improved membrane filtration method incorporating catalase and sodium pyruvate for detection of chlorine-stressed coliform bacteria. *Appl. Environ. Microbiol.* 56:3558–64.

Calabrese, J.P. and G.K. Bissonnette. 1990*b*. Improved detection of acid mine water stressed coliform bacteria on media containing catalase and sodium pyruvate. *Can. J. Microbiol.* 21:1960–68.

Calambokidis, J., B.D. McLaughlin, and G.H. Steiger. 1989. Bacterial contamination related to harbor seals in Puget Sound, Washington. A final report to Jefferson County and Washington Department of Ecology, Cascadia Research, Olympia, Washington.

Caldwell, B.A., C. Ye, R.P. Griffiths, C.L. Moyer, and R.Y. Morita. 1989. Plasmid expression and maintenance during long-term starvation-survival of bacteria in well water. *Appl. Environ. Microbiol.* 55:1860–64.

Calkins, J. 1982. *The Role of Solar Ultraviolet Radiation in Marine Ecosystems*. New York: Plenum Press.

Candrian, U., B. Furrer, C. Höfelein, and J. Lüthy. 1991. Use of inosine-containing oligonucleotide primers for enzymatic amplification of different alleles of gene coding for heat-stable toxin type I of enterotoxigenic *Escherichia coli*. *Appl. Environ. Microbiol.* 57:955–61.

Centers for Disease Control. Sept. 1985. Foodborne Disease Outbreaks. Annual Summary 1982. USDHHS Publication No. (CDC) 85-8185.

Centers for Disease Control. Dec. 1989. Foodborne Surveillance Data for all Pathogens in fish/shellfish for years 1973–1987.

Centers for Disease Control. 1990*a*. Protection Against Viral Hepatitis. *Morbid. Mortal. Wkly. Rpt.* 39(RR-2):1–24.

Centers for Disease Control. 1990*b*. Viral agents of gastroenteritis. *Morbid. Mortal. Wkly. Rpt.* 39(RR-5):1–13.

Centers for Disease Control. 1990*c*. Waterborne Disease Outbreaks, 1986–1988 and food borne disease outbreaks, 5-year summary, March 1983–1987. 39(SS-1):1–59.

Chai, T.-J. 1983. Characteristics of *Escherichia coli* grown in bay water as compared with rich medium. *Appl. Environ. Microbiol.* 45:1316–23.

Chai, T.J., J. Pace, and T. Cossaboom. 1984. Extension of shelf-life of oysters by pasteurization in flexible pouches. *J. Food Sci.* 49:331–33.

Chai, T.J., T. Han, R. Cockey, and P. Henry. 1990. Microbiological studies of Chesapeake Bay soft-shell clams (Mya arenaria). *J. Food Prot.* 53(12):1052–57.

Chan, F.T., G. Stringel, and A.M.R. Mackenzie. 1983. Isolation of *Campylobacter jejuni* from an appendix. *J. Clin. Microbiol.* 18:422–24.

Chan, S., T. Pitman, J. Shah, W. King, D. Lane, and J. Lawrie. 1988. Nonradioactive DNA assay for detection of foodborne Yersiniae. In ed. Abstr. Annu. Meeting Amer. Soc. Microbiol.

Chan, V.T., K.A. Fleming, and J.O. McGee. 1985. Detection of subpicogram quantities of specific DNA sequences on blot hybridization with biotinylated probes. *Nucleic Acids Res.* 13:8083–88.

Chang, P.W., O.C. Liu, L.T. Miller, and S.M. Li. 1971. Multiplication of human enteric viruses in Northern quahogs. *Proc. Soc. Exp. Biol. Med.* 136:1380–84.

Chang, G.W., J. Brill, and R. Lum. 1989. Proportion of β-D-glucuronidase-negative *Escherichia coli* in human fecal samples. *Appl. Environ. Microbiol.* 55:335–39.

Chapman, N.M., S. Tracy, C.J. Gauntt, and U. Fortmueller. 1990. Molecular detection and identification of enteroviruses using enzymatic amplification and nucleic acid hybridization. *J. Clin. Microbiol.* 28:843–50.

Chesapeake Bay Recreational Boat Pollution Work Group. 1990. *Final Draft Recreational Boat Pollution and the Chesapeake Bay: Time for Action, A Report to the Chesapeake Executive Committee.* Annapolis, Md.: Implementation Committee of the Chesapeake Bay Program.

Chollet, A. and E.H. Kawashima. 1985. Biotin-labeled synthetic oligodeoxynucleotides: Chemical synthesis and uses as hybridization probes. *Nucleic Acids Res.* 13:1529–41.

Chowdhury, M.A.R., H. Yamanaka, S. Miyoshi, K.M.S. Aziz, and S. Shinoda. 1989. Ecology of *Vibrio mimicus* in aquatic environments. *Appl. Environ. Microbiol.* 55:2073–78.

Chowdhury, M.A.R., H. Yamanaka, S.-I. Miyoshi, and S. Shinoda. 1990. Ecology and seasonal distribution of *Vibrio parahaemolyticus* in aquatic environments of a temperate region. FEMS Microbiol. *Ecol.* 74:1–10.

Churchland, L.M. and G. Kan. 1982. Variation in fecal pollution indicators through tidal cycles in the Fraser River estuary. *Can. J. Microbiol.* 28:239–47.

Ciesielski, C.A., B. Swaminathan, and C.V. Broome. 1987. *Listeria monocytogenes*. A food borne pathogen. *Clin. Microbiol. Newslett.* 9:140–50.

Clark, C.S., G.G. Linnermann, G.L. Vanneer, G.M. Schiff, P.S. Gastside, A.B. Johnson, E.J. Phair, C.R. Buncher, D.L. Alexander, S.E. Trimble, and B.C. Bornett. 1981. Health risks of human exposure to wastewater. EPA-600/SI-81-002:1–4.

Clark, H.F., E.E. Geldreich, H.L. Jeter, and P.W. Kabler. 1951. The membrane filter in sanitary bacteriology. *Public Health Rep.* 66:951–77.

Claus, P., S.L. Moseley, and S. Falkow. 1983. The use of gene-specific probes for the identification of enteric pathogens. *Prog. Food Nutri. Sci.* 7:139–41.

Clausen, E.M., B.L. Green, and W. Litsky. 1977. Fecal streptococci: Indicators of pollution. In *Bacterial Indicators/Health Hazards Associated With Water*, eds. A.W. Hoadley and B.J. Dutka, pp. 247–64. Special Technical Publication 635, Philadelphia, Pa: American Society for Testing and Materials.

Cliver, D.O., R.D. Ellender, and M.D. Sobsey. 1983. Methods for detecting viruses in foods: Background and general principles. *J. Food Prot.* 46(3):248–49.

Cliver, D.O. 1988. Virus transmission via foods. *Food Technol.* 42(10):241–47.

Close, M.E., L.R. Hodgson, and G. Tod. 1989. Field evaluation of fluorescent whitening agents and sodium tripolyphosphate as indicators of septic tank contamination in domestic wells. *New Zealand J. Mar. Freshwater Res.* 23:563–68.

Closs, O. and A. Digranes. 1971. Rapid identification of prompt lactose-fermenting genera within the family *Enterobacteriaceae*. *Acta Path. Microbiol. Scanda.* Sect. B79:673–78.

Colburn, K., C. Kaysner, M. Wekell, J. Matches, C. Abeyta, and R. Stott. 1989. Microbiological quality of oysters (*Crassostrea gigas*) and water of live holding tanks in Seattle, Washington markets. *J. Food Prot.* 52:100–04.

Colburn, K.G., C.A. Kaysner, C. Abeyta Jr., and M.M. Wekell. 1990. *Listeria* species in a California coast estuarine environment. *Appl. Environ. Microbiol.* 56:2007–11.

Cole, M.T., M.B. Kilgen, L.A. Reily, and C.R. Hackney. 1986. Detection of enteroviruses, bacterial indicators and pathogens in Louisiana oysters and their overlying waters. *J. Food Prot.* 49:596–601.

Cole, M.T., M.B. Kilgen, and C.R. Hackney. 1986. Evaluation of methods for extraction of enteric viruses from Louisiana oysters. *J. Food Prot.* 49:592–95.

Collins, C.H. and P.M. Lyne. 1984. *Microbiological Methods*, 5th ed. London: Butterworths.

Colwell, R.R. and J. Liston. 1960. Microbiology of shellfish. Bacteriological study of the natural flora of Pacific oysters (*Crassostrea gigas*). *Appl. Environ. Microbiol.* 8:104–09.

Colwell, R.R., J. Kaper, and S.W. Joseph. 1977. *Vibrio cholerae, Vibrio parahaemolyticus* and other Vibrios: Occurrence and distribution in Chesapeake Bay. *Sci.* 198:394–96.

Colwell, R. and J. Kaper. 1978. *Vibrio cholerae, Vibrio parahemolyticus* and other vibrios: Occurrence and distribution in Chesapeake Bay. *Sci.* 198:394.

Colwell, R.R., R.J. Seidler, J. Kaper, S.W. Joseph, S. Garges, H. Lockman, D. Maneval, H. Bradford, N. Roberts, E. Remmers, I. Huq, and A. Huq. 1981.

Craven, S.E. 1980. Growth and sporulation of *Clostridium perfringens* in foods. *Food Technol.* 34(4):80.

Croci, L., G.D. Filip, S. Gizzarelli, L. Toti, S. Agrigento, S. Amato, F. Lombardi, F. Novello, and R. Santoro. 1984. Methodology for the determination of polio viruses in edible Lamellibranch molluscs. *Microbiol. Alimento Nutri.* 2:85–90.

Cromeans, T., M.D. Sobsey, and H.A. Fields. 1987. Development of a plaque assay for a cytopathic, rapidly replicating isolate of hepatitis A virus. *J. Med. Virol.* 22:45–56.

Crosa, J.H., J. Olarte, L.J. Mata, L.K. Luttrop, and M.E. Penaranda. 1977. Characterization of an R plasmid associated with ampicillin resistance in *Shigella dysenteriae* type 1 isolated from epidemics. *Antimicrob. Agents Chemother.* 11:553–58.

Crovari, P. 1958. Some observations on the depuration of mussels infected with poliomyelitis virus. *Ig. Mod.* 5:22–32.

Cubit Engineering Limited. 1984. *Datau Marina Study*. Charleston. S.C.

Cubitt, W.D. 1987. The candidate caliciviruses. In *Novel Diarrhoea Viruses*, Ciba Foundation Symposium, pp. 126–143. Chichester, U.K.: Wiley.

Cynar, F.J., K.W. Estep, and J. McN. Sieburth. 1985. The detection and characterization of bacteria-sized protists in "protist-free" filtrates and their potential impact on experimental marine ecology. *Microbiol. Ecol.* 11:281–88.

D'Amato, R., L. Eriguez, K. Tomfohrde, and E. Singerman. 1978. Rapid detection of *Neisseria gonorrhea* and *Neisseria meningitidis* by enzyme profiles. *J. Clin. Microbiol.* 7:77–81.

D'Aoust, J. 1989. *Salmonella*. In *Foodborne Bacterial Pathogens*, ed. M.P. Doyle, pp. 328–446. New York: Marcel Dekker.

D'Aoust, J.Y., R. Gelinas, and C. Maishment. 1980. Presence of indicator organisms and recovery of *Salmonella* in fish and shellfish. *J. Food Prot.* 43:679–82.

D'Aoust, J.Y. 1981. Update on pre-enrichment and selective enrichment conditions for detection of *Salmonella* in foods. *J. Food Prot.* 44:369–74.

D'Aoust, R.A. and W. Litsky. 1975. Pfizer selective enterococcus agar overlay method for the enumeration of fecal streptococci by membrane filtration. *Appl. Environ. Microbiol.* 29:584–89.

D'Aoust, Y., J. Giroux, L.R. Barran, H. Schneider, and W.G. Martin. 1974. Some effects of visible light on *Escherichia coli*. *J. Bacteriol.* 120:799–804.

Dfreth, S., R. Herrman, and K. Pecher. 1986. Tracing fecal pollution by coprostanol and intestinal bacteria in an ice-covered Finnish lake loaded with both industrial and domestic sewage. *Water Air Soil Pollut.* 28:131–49.

Daemer, R.J., S.M. Feinstone, I.D. Gust, and R.H. Purcell. 1981. Propagation of human hepatitis A virus in African green monkey kidney cell culture: Primary isolation and serial passage. *Infect. Immun.* 32:388–93.

Daggett, P.-M., T.K. Sawyer, and T.A. Nerad. 1982. Distribution and possible interrelationships of pathogenic and nonpathogenic *Acanthamoeba* from aquatic environments. *Microb. Ecol.* 8:371–86.

Dahlen, G. and A. Linde. 1973. Screening plate method for detection of bacterial b-glucuronidase. *Appl. Microbiol.* 26:863–66.

Dahling, D.R., G. Berg, and D. Berman. 1974. BGM, a continuous cell line more sensitive than primary rhesus and African green kidney cells for the recovery of viruses from water. *Health Lab. Sci.* 11:275–82.

Daily, O.P., S.W. Joseph, J.D. Gillmore, R.R. Colwell, and R.J. Seidler. 1981. Identification, distribution and toxigenicity of obligate anaerobes in polluted waters. *Appl. Environ. Microbiol.* 41:1074–77.

Daley, R.J. and J.E. Hobbie. 1975. Direct counts of aquatic bacteria by a modified epifluorescence technique. *Limnol. Oceanogr.* 20:875–82.

Daley, R.J. 1979. Direct epifluorescence enumeration of native aquatic bacteria: uses, limitations and comparative accuracy. In *Native Aquatic Bacteria: Enumeration, Activity and Ecology*, eds. J.W. Costerton and R.R. Colwell, pp. 29–45. Special Technical Publication 695. Philadelphia, Pa.: American Society for Testing and Materials.

Dalton, H.P., G.L. Archer, M. Slifkin, R.C. Harris, H.J. Welshimer, K.M. Sosnowski, H.C. Nottebart Jr., R.J. Duma, R.B. Clark, N.G. Warren, T.M. Kerkering, P.D. Swenson, A.L. Ager Jr., and R.G. May Jr. 1986. Blood specimens. In *Interpretive Medical Microbiology*, eds. H.P. Dalton and H.C. Nottebart Jr., pp. 28–29. New York: Churchill Livingstone.

Dalwara, S. and H.H. Ng. 1986. Evaluation of a rapid detection method for *Escherichia coli* in foods using fluorogenic assay. *Food Microbiol.* 3(4):373–77.

Datta, A.R., B.A. Wentz, D. Shook, and M.W. Truckess. 1988. Synthetic oligodeoxyribonucleotide probes for detection of *Listeria monocytogenes. Appl. Environ. Microbiol.* 54:2933–37.

Davanipour, Z., M. Alter, E. Sobel, D.M. Asher, and D.C. Gajdusek. 1983. A case control study of Creutzfeldt-Jokob disease. *Am. J. Epidemiol.* 122:443–51.

Davis, B.D., R. Dulbecco, H.N. Eisen, H.S. Ginsberg, and W.B. Wood Jr. 1989. *Principles of Microbiology and Immunology*, 3rd ed. New York: Harper and Row.

Davis, B.R., G.R. Fanning, J.M. Madden, A.G. Steigerwalt, H.B. Bradford Jr., H.L. Smith Jr., and D.J. Brenner. 1981. Characterisation of biochemically atypical *Vibrio cholerae* strains and designation of a new pathogenic species *Vibrio mimicus. J. Clin. Microbiol.* 14:631–34.

Davis, W.A. II, J.G. Kane, and V.F. Garagusi. 1978. Human *Aeromonas* infections. *Medicine* 57:267–77.

Dawe, L.L. and W.R. Penrose. 1978. "Bactericidal" property of seawater: Death or debilitation? *Appl. Environ. Microbiol.* 35:829–33.

De Moraes, E.C. and R.M. Tyrrell. 1983. Mutational interactions between near-uv radiation and DNA damaging agents in *Escherichia coli*: The role of near-uv induced modifications in growth and macromolecular synthesis. *Photochem. Photobiol.* 38:57–63.

Dean, A.G., Y. Ching, R.G. Williams, and L.B. Harden. 1972. Test for *Escherichia coli* enterotoxin using infant mice. Application in a study of diarrhea in children in Honolulu. *J. Infect. Dis.* 125:407–11.

Debartolomeis, J. 1988. Enumeration of F male-specific bacteriophages from sewage and fecally polluted waters. Ph.D. Dissertation. Kingston, R.I.: University of Rhode Island.

DeFlaun, M.F., J.H. Paul, and D. Davis. 1986. Simplified method for dissolved DNA determination in aquatic environments. *Appl. Environ. Microbiol.* 52:654–59.

Donnelly, C.W., G.J. Baigent, and E.H. Briggs. 1988. Flow cytometry for automated analysis of milk containing *Listeria monocytogenes*. *J. Assoc. Off. Anal. Chem.* 71:655.

Donnelly, C.W. 1988. Historical perspectives on methodology to detect *Listeria monocytogenes*. *J. Assoc. Off. Anal. Chem.* 71:644–46.

Dougherty, W.J. and R. Altman. 1962. Viral hepatitis in New Jersey 1960–1961. *Am. J. Med.* 32:704–16.

Dow, M.A., R. Fujioka, and P.C. Loh. 1984. Recovery of enteric viruses from Hawaiian marine waters using indigenous bivalve molluscs. *Abstr. Ann. Meet. Am. Soc. Microbiol.* 84:215.

Dow, M.A., P.C. Loh, and R.S. Fujioka. 1986. Nitrocellulose enzyme immunoassay (NC-EIA) for detection of enteropathogenic viruses in molluscs. *Abstr. Ann. Meet. Am. Soc. Microbiol.* 86:306.

Dowell, V.R. Jr. and T.M. Hawkins. 1968. *Laboratory Methods in Anaerobic Bacteriology*. U.S. Department of Health, Education, and Welfare. Atlanta, Ga.: National Communicable Disease Center.

Doyle, M. 1981. *Campylobacter fetus* subsp. *jejuni*—an old pathogen of new concern. *J. Food Prot.* 44.

Doyle, M.P. and D.J. Roman. 1981. Growth and survival of *Campylobacter fetus* subsp. *jejuni* as a function of temperature and Ph. *J. Food Prot.* 44:596–601.

Doyle, M.P., M.B. Hugdahl, and S.L. Taylor. 1981. Isolation of virulent *Yersinia enterocolitica* from porcine tongues. *Appl. Environ. Microbiol.* 42:661–66.

Doyle, M.P. and D.J. Roman. 1982*a*. Recovery of *Campylobacter jejuni* and *Campylobacter coli* from inoculated foods by selective enrichment. *Appl. Environ. Microbiol.* 43:1343–53.

Doyle, M.P. and D.J. Roman. 1982*b*. Response of *Campylobacter jejuni* to sodium chloride. *Appl. Environ. Microbiol.* 43:561–65.

Doyle, M.P. and J.L. Schoeni. 1987. Isolation of *Escherichia coli* 0157:J7 from retail fresh meats and poultry. *Appl. Environ. Microbiol.* 53:2394–96.

Doyle, M.P., K.A. Glass, J.T. Beery, G.A. Garcia, D.J. Pollard, and R.D. Schultz. 1987. Survival of *Listeria monocytogenes* in milk during high-temperature, short-time pasteurization. *Appl. Environ. Microbiol.* 53:1433–38.

Doyle, M.P. 1988. Effect of environmental and processing conditions on *Listeria monocytogenes*. *Food Technol.* 42:169–71.

Drahos, D., M.J. Brackin, and G. Barry. 1985. Bacterial strain identification by comparative analysis of chromosomal DNA restriction patterns. 75:1381.

Drake, J.F. and Tsuchiya, H.M. 1976. Predation on *Escherichia coli* by *Colpoda steinii*. *Appl. Environ. Microbiol.* 31:870–74.

Dressel, D.M. 1990. *Interstate Certified Shellfish Shippers List*. Oct. 1–Nov. 30. Washington, D.C.: FDA, PHS, U.S. Department of Health and Human Services, (ISSN 0364-7048).

Dubrou, S., H. Kopecka, J.M. Lopez-Pila, J. Marechal, and J. Prevot. 1990. Detection of hepatitis A virus and other enteroviruses in wastewater samples by gene probe assays. International Symposium on Health-Related Water Microbiology, Tubingen, Germany.

Duff, M.F. 1967. The uptake of enteroviruses by the New Zealand marine blue mussel, Mytilus edulis aoteanus. *Am. J. Epidemiol.* 85:486–92.

Dufour, A.P. 1980. A 24-hour membrane filter procedure for enumerating enterococci. *Abstracts of the 80th Annual Meeting of the American Society for Microbiology, Miami, Fla., May, 1980.*

Dufour, A.P., E.R. Strickland, and V.J. Cabelli. 1981. Membrane filter method for enumerating *Escherichia coli. Appl. Environ. Microbiol.* 41:1152–58.

Dufour, A.P. 1984. *Health Effects for Fresh Recreational Waters.* EPA 600/1-84-004 Cincinnati, Oh.: U.S. Environmental Protection Agency.

Dufour, A.P., R. Calderon, and E. Mood. 1985. Health effects and nonpoint sources of pollution. Paper presented at the Interstate Seafood Seminar, 29–31, October 1985, at the Fenwick Inn, Ocean City, Md.

Dupont, H.L., R.B. Hornick, A.T. Dawkins, M.J. Snyder, and S.B. Formal. 1969. The response of man to virulent *Shigella flexneri* 2a. *J. Infect. Dis.* 119:296–99.

Dupont, H.L., S. Formal, R. Hornick, M. Snyder, J. Libonati, D. Sheahan, E. LaBrec, and J. Kalas. 1971. Pathogenesis of *Escherichia coli* diarrhea. *New Engl. J. Med.* 285:1–9.

Dupont, H.L., R.B. Hornick, M.J. Snyder, J.P. Libonati, S.B. Formal, and E.J. Gangarosa. 1972. Immunity in Shigellosis. II. Protection induced by oral live vaccine or primary infection. *J. Infect. Dis.* 125:12–16.

Durairaj, S., G. Chinnaswamy, and S. Mohamed. 1983. A bacteriological study of the natural flora of edible oyster, Crassostrea madrasensis. *Fish. Technol.* 20:111–14.

Düreth, S., R. Herrman, and K. Pecher. 1986. Tracing fecal pollution by coprostanol and intestinal bacteria in an ice-covered Finnish lake loaded with both industrial and domestic sewage. *Water Air Soil Pollut.* 28:131–49.

Dutka, B.J. 1973. Coliforms are an inadequate index of water quality. *J. Environ. Health.* 36:39–46.

Dutka, B.J. 1979. Microbiological indicators, problems and potential of new microbial indicators of water quality. In *Biological Indicators of Water Quality*, eds. A. James and L. Evison, pp. 18-1–18-24. Chichester: Wiley.

Dutka, B.J., K. Jones, K.K. Kwan, H. Bailey, and R. McInnis. 1988. Use of microbial and toxicant screening tests for priority site selection of degraded areas in water bodies. *Water Res.* 22:503–10.

Dutton, R.J., G. Bitton, and B. Koopman. 1983. Malachite Green-INT (MINT) method for determining active bacteria in sewage. *Appl. Environ. Microbiol.* 46:1263–67.

Earle, P.M. and F.D. Crisley. 1975. Isolation and Characterization of *Vibrio parahaemolyticus* from Cape Cod Soft-Shell Clams (*Mya arenaria*). *J. Food Prot.* 53(12):1052–57.

Easterbrook, T.J. and P.A. West. 1987. Comparison of most probable number and pour plate procedures for isolation and enumeration of sulphite-reducing *Clostridium* spores and group D faecal streptococci from oysters. *J. Appl. Bacteriol.* 62:413–19.

Echeverria, P., J. Seriwatana, O. Chityothin, W. Chaicumpa, and V. Tirapat. 1982.

species recovered from untreated surface water and drinking water by the membrane filter, standard, and modified most-probable-number techniques. *Appl. Environ. Microbiol.* 41:657–63.

Evans, T.M., C.E. Waarvick, R.J. Seidler, and M.W. LeChevallier. 1981. Failure of the most-probable-number technique to detect coliforms in drinking water and raw water supplies. *Appl. Environ. Microbiol.* 41:130–38.

Evenson, M., M. Hinds, R. Bernstein, and M. Bergdoll. 1988. Estimation of human dose of staphylococcal enterotoxin A from a large outbreak of staphylococcal food poisoning involving chocolate milk. *Int. J. Food Microbiol.* 7:311, 316.

Ewald, S. and S. Noternmans. 1988. Effect of water activity on growth and enterotoxin D production of *Staphylococcus aureus. Int. J. Food Microbiol.* 6:25–30.

Ewing, W.H. 1972. *Differentiation of Enterobacteriaceae by Biochemical Reactions.* Atlanta, Ga.: U.S. Dept. of Health, Education and Welfare.

Eyles, M.J., G.P. Davey, and E.J. Huntley. 1981. Demonstration of viral contamination of oysters responsible for an outbreak of viral gastroenteritis. *J. Food Prot.* 44(4):294–96.

Eyles, M.J. 1983. Assessment of cooked prawns as a vehicle for transmission of viral disease. *J. Food Prot.* 46:426–28.

Eyles, M.J. and G. Davey. 1984. Microbiology of commercial depuration of the Sydney rock oyster. *J. Food Prot.* 47(9):703–06, 712.

Facklam, R., G.S. Bosley, D. Rhoden, A.R. Franklin, N. Weaver, and R. Schulman. 1985. Comparative evaluation of the API 20S and Automicrobic Gram-Positive Identification systems for non-beta-hemolytic streptococci and aerococci. *J. Clin. Microbiol.* 21:535–41.

Facklam, R.R., D.L. Rhoden, and P.B. Smith. 1984. Evaluation of the Rapid Strep system for the identification of clinical isolates of *Streptococcus* species. *J. Clin. Microbiol.* 20:894–98.

Facklam, R.R. and M.D. Collins. 1989. Identification of *Enterococcus* species isolated from human infections by a conventional test scheme. *J. Clin. Microbiol.* 27:731–34.

Fair, G.M. and J.C. Geyer. 1963. *Water Supply and Waste-Water Disposal.* New York: Wiley.

Falkow, S. 1975. *Infectious multiple drug resistance.* London: Pion.

Fannin, K.F., S.H. Abid, J.J. Bertucci, J.M. Reed, S.C. Vana, and C. Lue-Hing. 1978. Significance of reporting infectious viral or plaque forming unit concentrations from environmental samples. *Abstr. Ann. Meet. Am. Soc. Microbiol.* 78:200.

Fantasia, L.D., J.P. Schrade, J.F. Yoger, and D. Debler. 1975. Fluorescent antibody method for the detection of Salmonella: Development, evaluation, and collaborative study. *J. Assoc. Off. Anal. Chem.* 58:828–44.

Farber, J.M. and J.I. Speirs. 1987. Monoclonal antibodies directed against the flagellar antigens of *Listeria* species and their potential in EIA-based methods. *J. Food Prot.* 50:479–84.

Farmer, J.J. III. 1981. The Genus *Edwardsiella.* In *The Prokaryotes: A Handbook on Habitats, Isolation, and Identification of Bacteria,* eds. M.P. Starr, H. Stolp, H.G. Truper, G. Balows, and H. Schlegel, pp. 1135–39. New York: Springer.

Farmer, J.J. III and A.C. McWorter. 1984. *Edwardsiella.* In *Bergey's Manual of Systematic Bacteriology,* eds. N.R. Krieg and J.G. Holt, pp. 486–91. Baltimore,

Md.: Williams and Wilkins.

Farmer, J.J. III and B.R. Davis. 1985. H7 antiserum-sorbitol fermentation medium: A single tube screening medium for detecting *Escherichia coli* O157:H7 associated with hemorrhagic colitis. *J. Clin. Microbiol.* 22:620–25.

Faroog, S. and S. Akhaque. 1983. Comparative response of mixed cultures of bacteria and viruses to ozonation. *Water Res.* 17:701–06.

Faust, M. 1976. Coliform bacteria from diffuse sources as a factor in estuarine pollution. *Water Res.* 10:619–27.

Faust, M. 1982. Contribution of pleasure boats to fecal bacteria concentrations in the Rhode River estuary, Maryland, U.S.A. *Sci. Total Environ.* 25:255–62.

Faust, M.A., A.E. Aotaky, and M.L. Hargadon. 1975. Effect of physical parameters on the in situ survival of *Escherichia coli* MC-6 in an estuarine environment. *Appl. Microbiol.* 30:800–06.

Fauvel, Y. 1963. Utilisation de l'ozone comme agent sterilisateur de l'eau de mer pour l'epuration des coquillages. Commission International pour l'Exploration Scientifique de la Mer Mediterranee. *Proces-verbauz des Reunions.* 17:701–06.

Fauvel, Y., G. Pons, and J.P. Legeron. 1982. Ozoneation de l'eau de mer et epuration des coquillages. *Sci. et Peche, Nantes.* 320:1–16.

Feachem, R. 1975. An improved role for faecal coliform to faecal streptococci ratios in the differentiation between human and non-human pollution sources. *Water Res.* 9:689–90.

Feacham, R.G. 1981. Environmental aspects of cholera epidemiology. In *A review of selected reports of endemic and epidemic situations during 1971–1980. Tropical Dis. Bull.* 78:675–98.

Feacham, R.G., C. Miller, and B. Drasar. 1981. Environmental aspects of cholera epidemiology. II. Occurrence and survival of *V. cholerae* in the environment. *Tropical Dis. Bull.* 78:865–80.

Feachem, R.G., D.J. Bradley, H. Garelick, and D.D. Mara. 1983. Sanitation and disease. Health aspects of excreta and wastewater management. *World Bank Studies in Water Supply and Sanitation.* New York: Wiley.

Federal Security Agency. 1950. *Manual of Recommended Practices for the Sanitary Control of the Shellfish Industry,* 1946. Recommendations of the Public Health Service. PHS Publication No. 33 Public Health Service. Washington, D.C.: U.S. Government Printing Office.

Fedorak, P. and D.W.S. Westlake. 1978. Effect of sunlight on bacterial survival in transparent air samplers. *Can. J. Microbiol.* 24:618–19.

Feingold, A.O. 1973. Hepatitis from eating steamed clams. *J. Am. Med. Assoc.* 225:526–27.

Feldman, R.A. 1980. Typhoid. In *Maxey–Rosenau Public Health and Preventive Medicine,* ed. J.M. Last, pp. 225–28. 11th ed. New York: Appleton–Century–Crofts.

Felsenfeld, O. 1974. The survival of cholera vibrios. In *Cholera,* eds. D. Barua and W. Burrows, pp. 359–66. Philadelphia, Pa.: W.B. Saunders.

Fenchel, T. 1982. Ecology of heterotrophic microflagellates. II. Bioenergetics and growth. *Mar. Ecol. Prog. Ser.* 8:225–31.

Feng, P.C.S. and P.A. Hartman. 1982. Fluorogenic assays for immediate confirmation of *Escherichia coli. Appl. Environ. Microbiol.* 43:1320–29.

Fraiser, M.B. and J.A. Koburger. 1984. Incidence of salmonellae in clams, oysters, crabs and mullet. *J. Food Prot.* 47(5):343–45.

Frampton, E.W., L. Restaino, and N. Blaszko. 1988. Evaluation of the b-glucuronidase substrate 5-bromo-4-chloro-3-indolyl-b-D-glucuronide (X-GLUC) in a 24-hour direct plating method for *Escherichia coli. J. Food Prot.* 51:402–04.

Frampton, E.W. and L. Restaino. 1993. A review. Methods for *Escherichia coli* identification in food, water and clinical samples based on beta-glucuronidase detection. *J. Appl. Bacteriol.* 74:223–33.

Francisco, D.E., R.A. Mah, and A.C. Rabin. 1973. Acridine Orange epi-fluorescent technique for counting bacteria in natural waters. *Trans. Am. Microsc. Soc.* 92:416–21.

Franco, A., L. Toti, R. Gabrieli, L. Croci, D. Medici, and A. de Pana. 1990. Depuration of Mytilus galloprovincialis experimentally contaminated with hepatitis A virus. *Int. J. Food Microbiol.* 11:321–27.

Franco, D.A. 1988. *Campylobacter* species: Considerations for controlling a food-borne pathogen. *J. Food Prot.* 51:145–53.

Frankel, G., J.A. Giron, J. Valmassoi, and G.K. Schoolnik. 1989. Multi-gene amplification: Simultaneous detection of three virulence genes in diarrhoeal stool. *Mol. Microbiol.* 3:1729–34.

Frankland, G.C. 1894–95. Oysters and typhoid. *Nature*. London. 51:1322.

Fraser, J.A. and W.H. Sperber. 1988. Rapid detection of *Listeria* spp. in food and environmental samples by esculin hydrolysis. *J. Food Prot.* 51:762–65.

Frederickson, J.K., D.F. Bezdicek, F.J. Brockman, and S.W. Li. 1988. Enumeration of Tn5 mutant bacteria in soil by using a most-probable-number-DNA hybridization procedure and antibiotic resistance. *Appl. Environ. Microbiol.* 54:446–53.

Freier, T.A. and Hartman, P.A. 1969. Improved membrane filtration media for enumeration of total coliforms and *Escherichia coli* from sewage and surface waters. *Appl. Environ. Microbiol.* 53:1246–50.

Fridovich, I. 1975. Superoxide dismutases. *Ann. Rev. Biochem.* 44:147–59.

Fridovich, I. 1976. Oxygen radicals, hydrogen peroxide, and oxygen toxicity. In *Free Radicals in Biology*, ed. W.A. Pryor, Vol. 1, pp. 239–77. New York: Academic Press.

Friedman, G.D. 1974. *Primer of Epidemiology*, p. 230. New York: McGraw-Hill.

Frost, W.H. 1925. Report of committee on sanitary control of the shellfish industry in the United States. *Public Health Rpt.* Supplement No. 53:1–17.

Fruin, J.T. 1978. Types of *Clostridium perfringens* isolated from selected foods. *J. Food Prot.* 41:768–69.

Fugate, K.J., D.O. Cliver, and M.T. Hatch. 1975. Enteroviruses and potential bacteria indicators in Gulf Coast oysters. *J. Milk Food Technol.* 38:100–04.

Fuhrmann, J. and A.G. Wollum II. 1985. Simplified enzyme-linked immunosorbent assay for routine identification of *Rhizobium japonicum* antigens. *Appl. Environ. Microbiol.* 49:1010–13.

Fuhrman, J.A., D.E. Comeau, A. Hagstrom, and A. Chan. 1988. Extraction from natural plankton microorganisms of DNA suitable for molecular biological studies. *Appl. Environ. Microbiol.* 54:1426–29.

Fujioka, R.S., H.H. Hashimoto, E.B. Siwak, and R.H.F. Young. 1981. Effect of sunlight on survival of indicator bacteria in seawater. *Appl. Environ. Microbiol.* 41:690–96.

Fujioka, R.S. and L.K. Shizumura. 1985. *Clostridium perfringens*: A reliable indicator of stream water quality. *J. Water Pollut. Control Fed.* 57:986–92.

Fujioka, R.S. and E.B. Siwak, 1987. Bactericidal properties of long ultraviolet and visible wavelengths of sunlight. *J. Am. Water Works Assoc.* 79:56.

Furfari, S.A. 1966. Depuration plant design. HEW USPHS, Division of Environmental Engineering and Food Protection, Washington, D.C.: Public Health Service Publication No. 99-FP-7.

Furfari, S.A. 1979. Monitoring shellfish purification. In *Sourcebook on Shellfish Purification*, Davisville, R.I., B-1-13: USPHS, FDA, Northeast Technical Services Unit.

Furfari, S.A. 1981. *Report on Sanitary Control of Shellfish in France.* Washington, D.C.: In files of Shellfish Sanitation Branch, Food and Drug Administration.

Furrer, B., U. Candrian, and J. Luethy. 1990. Detection and identification of *Escherichia coli* producing heat-labile enterotoxin type I by enzymic amplification of a specific DNA fragment. *Lett. Appl. Microbiol.* 10:31–34.

Furuse, K., T. Sakurai, A. Hirashima, M. Katsuki, A. Ando, and I. Watanabe. 1978. Distribution of ribonucleic acid coliphages in South and East Asia. *Appl. Environ. Microbiol.* 35:995–1002.

Furuse, K., A. Ando, S. Osawa, and I. Watanabe. 1981. Distribution of ribonucleic acid coliphages in raw sewage from treatment plants in Japan. *Appl. Environ. Microbiol.* 41:1139–43.

Furuse, K. 1987. Distribution of coliphages in the environment: General considerations. In *Phage Ecology*, eds. S.M. Goyal, C.P. Gerba, and G. Bitton, pp. 87–124. New York: Wiley.

Gaillot, D., D. Terver, C. Finance, and L. Schwartzbrod. 1988. Improved method for contamination of mussels by poliovirus. *Int. J. Food Microbiol.* 6:333–39.

Galstoff, P.S. 1946. Reaction of oysters to chlorination. U.S. Department of the Interior, Fish and Wildlife Service, Research Report 11, pp. 1–28. Washington, D.C.: U.S. Government Printing Office.

Galtsoff, P.S. 1964. The American oyster. Fishery Bulletin of the Fish and Wildlife Service, Vol. 64. Washington, D.C.: United States Government Printing Office.

Galt, M.J. 1984. *Oligonucleotide Synthesis: A Practical Approach.* Oxford: IRL Press.

Gameson, A.L.H. and J.R. Saxon. 1967. Field studies on effect of daylight on mortality of coliform bacteria. *Water Res.* 1:279–95.

Gameson, A.L.H. and D.J. Gould: 1975. Effects of solar radiation on the mortality of some terrestrial bacteria in sea water. In *Discharge of Sewage from Sea Outfalls.* ed. A.L.H. Gameson, pp. 209–19. Oxford: Pergamon Press.

Gandelman, A.L. and P.H. Mann. 1965. An evaluation of reagent-impregnated paper strips for use in the process of identifying certain species of clinically important bacteria. *Curr. Therap. Res.* 7:130–38.

Gangarosa, E. 1978. Epidemiology of *Escherichia coli* in the United States. *J. Infect. Dis.* 137:634–39.

Gerba, C.P. and S.M. Goyal. 1978. Detection and occurrence of enteric virus in shellfish: A review. *J. Food Prot.* 41(9):743–45.

Gerba, C.P., S.M. Goyal, R.L. LaBelle, I. Cech, and G.F. Bodegan. 1979. Failure of indicator bacteria to reflect the occurrence of enteroviruses in marine waters. *Am. J. Public Health.* 69(11):1116–19.

Gerba, C.P., S.M. Goyal, I. Cech, and G.F. Bogdan. 1980. Bacterial indicators and environmental factors as related to contamination of oysters by enteroviruses. *J. Food Prot.* 43(2):99–101.

Gerba, C.P. and S.M. Goyal, ed. 1982. *Methods in Environmental Virology.* Microbiology Series, Vol. 7. New York: Marcel Dekker.

Gerba, C.P. 1987. Phage as indicators of fecal pollution. In *Phage Ecology*, eds. S.M. Goyal, C.P. Gerba, and G. Bitton, pp. 197–209. New York: Wiley.

Gerba, C.P. 1988. Viral disease transmission by seafoods. *Food Technol.* 42(3):99–103.

Gerhardt, P., R.G.E. Murray, R.N. Costilow, E.W. Nester, W.A. Wood, N.R. Krieg, and G.B. Phillips. 1981. *Manual of Methods for General Bacteriology*, pp. 65–220. Washington, D.C.: American Society for Microbiology.

Ghosh, A.C. 1978. Prevalence of *Bacillus cereus* in the faeces of healthy adults. *J. Hyg.* Cambridge: 80:233–37.

Ghoul, M., T. Bernard, and M. Cormier. 1990. Evidence that *Escherichia coli* accumulates glycine betaine from marine sediments. *Appl. Environ. Microbiol.* 56:551–54.

Gibbard, J. et al. 1943. Bacteriological examination of shellfish and shellfish waters. *Am. J. Public Health.* 33(5):582–91.

Gibson, G.R. and K.M. Arnold. 1987. *Use of Marine Sanitation Devices in Maryland Waters.* Baltimore, Md.: Maryland Department of Health and Mental Hygiene.

Gilbert, R. 1979. *Bacillus cereus* gastroenteritis. In *Foodborne Infections and Intoxications*, eds. H. Rieman and F. Bryan. New York: Academic Press.

Gilbert, R.J., M.F. Stringer, and J.C. Peace. 1974. The survival and growth of *Bacillus cereus* in boiled and fried rice in relation to outbreaks of food poisoning. *J. Hyg.* Cambridge: 73:433.

Gill, O.N., W.D. Cubbitt, N.A. McSwaggan, B.M. Watney, and C.L.R. Bartlett. 1983. Epidemic of gastroenteritis caused by oysters contaminated with small, round structured viruses. *Br. Med. J.* 287:1532–34.

Gilliland, M.W. and W. Baxter-Potter. 1987. A geographic information system to predict non-point source pollution potential. *Water Resources Bull.* 23:281–91.

Giovannoni, S.J., E.F. Delong, G.J. Olsen, and N.R. Pace. 1988. Phylogenetic group-specific oligodeoxynucleotide probes for identification of single microbial cells. *J. Bacteriol.* 170:720–26.

Girones, R., J. Jofre, and A. Bosch. 1989. Natural inactivation of enteric viruses in seawater. *J. Environ. Qual.* 18:34–39.

Gizzarelli, S., L. Toti, L. Croci, L. Orefice, G. Felip, R. Cavallo, and A. Strusi. 1985. Behaviour of *Yersinia enterocolitica* in *Mytilus galloprovincialis. Microbiologie-Alimenti-Nutrit.* 3:371–76.

Goatcher, L.J., S.E. Engler, D.C. Wagner, and D.C. Westhoff. 1974. Effect of storage at 5 C on survival of *Vibrio parahaemolyticus* in processed Maryland oysters (*Crassostrea virginica*). *J. Milk Food Technol.* 37(2):74–77.

Goatcher, L.J. and D.C. Westhoff. 1975. Repression of *Vibrio parahaemolyticus* by *Pseudomonas* species isolated from processed oysters. *J. Food Sci.* 40:533–36.

Goding, J.W. 1986. *Monoclonal Antibodies: Principles and Practice*, 2nd. New York: Academic Press.

Golden, D.A., L.R. Beuchat, and R.E. Brackett. 1988. Direct plating technique for enumeration of *Listeria monocytogenes* in foods. *Appl. Environ. Microbiol.* 54:1451–56.

Goldfield, M. 1976. Epidemiological indicators for transmission of viruses by water. In *Viruses in Water*, eds. G. Berg, H.L. Bodily, E.H. Lennette, J.L. Melnick, and T.G. Metcalf, pp. 70–85. Washington, D.C.: American Public Health Association.

Goldmintz, D., R.C. Ernst, and J. Rasekh. 1978. Development of a pasteurized oyster product. In *Proceedings of the Third Annual Tropical and Subtropical Fisheries Technological Conference of the Americas*, TAMU-SG-79-101, pp. 359–65. College Station, Tx.: Texas A & M University.

Goldstein, E.J.C. and D.M. Citron. 1988. Annual incidence, epidemiology, and comparative in vitro susceptibility to cefoxitin, cefotetan, cefmetazole, and ceftizoxime of recent community-acquired isolates of the *Bacteroides fragilis* group. *J. Clin. Microbiol.* 26:2361–66.

Goldstein, F.W., J.C. Chumpiaz, J.M. Guevara, B. Papadopoulou, J.F. Acar, and J.F. Vieu. 1986. Plasmid mediated resistance to multiple antibiotics in *Salmonella typhi*. *J. Infect. Dis.* 153:261–66.

Goltz, S.P., J.J. Donegan, H. Yang, M. Pollice, J.A. Todd, M.M. Molina, J. Victor, and N. Kelker. 1990. The use of nonradioactive DNA probes for rapid diagnosis of sexually transmitted bacterial infections. In *Gene Probes for Bacteria*, eds. A.J.L. Macario and E. Conway de Macario, pp. 1–44. San Diego, Ca.: Academic Press.

Gonzales-Ayala, S. 1990. WHO Surveillance programme for control of foodborne infections and intoxications in Europe. Institute of Veterinary Medicine-Robert von Ostertag-Institute.

Gonzalez, J.M., E.B. Sherr, and B.F. Sherr. 1990. Size-selective grazing on bacteria by natural assemblages of estuarine flagellates and ciliates. *Appl. Environ. Microbiol.* 56:583–89.

Gooch, W.M. III and G.A. Hill. 1982. Comparison of Micro-ID and API 20E in rapid identification of *Enterobacteriaceae*. *J. Clin. Microbiol.* 15:885–90.

Goodfellow, M. and S.T. Williams. 1983. Ecology of actinomycetes. *Ann. Rev. Microbiol.* 37:189–216.

Goodfellow, M. and J.A. Haynes. 1984. Actinomycetes in marine sediments. In *Biological, biochemical and biomedical aspects of actinomycetes*, eds. L. Ortiz-Ortiz, L.F. Bojalil, and V. Yakoleff, pp. 453–72. New York: Academic Press.

Goodfellow, R.M., J. Cardoso, G. Eglinton, J.P. Dawson, and G.A. Best. 1977. A faecal sterol survey in the Clyde Estuary. *Mar. Pollut. Bull.* 8:272–75.

Gopalakrishna, I. and P.R. Varma. 1987. Isolation of *Salmonella braenderup* (6.7:e, h:e, n, Z15) from mussel meat. *Fish. Technol.* 24(2):138–41.

Gopo, J.M., R. Melis, E. Filipska, R. Meneveri, and J. Filipska. 1988. Development of a *Salmonella*-specific biotinylated DNA probe for rapid routine identification of *Salmonella*. *Mol. Cell. Probes* 2:271–79.

Gorham, F.P. 1933. Laboratory—Report of the referee on oyster analysis. *Am. J. Public Health.* 23:56–57.

Goss, W.A., W.H. Deitz, and T.M. Cook. 1965. Mechanism of action of nalidixic acid on *Escherichia coli.* II. Inhibitor of DNA synthesis. *J. Bacteriol.* 89:1068–74.

Gotuzzo, E., J.G. Morris, L. Benavente, P.K. Wood, O. Levine, R.E. Black, and M.M. Levine. 1987. Association between specific plasmids and relapse in typhoid fever. *J. Clin. Microbiol.* 25:1779–81.

Gouvea, V., R.I. Glass, P. Woods, K. Taniguchi, H.F. Clark, B. Forrester, and Z.Y. Fang. 1990. Polymerase chain reaction amplification and typing of rotavirus nucleic acid from stool specimens. *J. Clin. Microbiol.* 28:276–82.

Goyal, S.M., C.P. Gerba, and J.L. Melnick. 1977. Occurrence and distribution of bacterial indicators and pathogens in canal communities along the Texas coast. *Appl. Environ. Microbiol.* 34(2):139–49.

Goyal, S.M., C.P. Gerba, and J.L. Melnick. 1978. Prevalence of human enteric viruses in coastal canal communities. *J. Water Pollut. Control. Fed.* 50:2247–56.

Goyal, S.M., C.P. Gerba, and J.L. Melnick. 1979. Human enteroviruses in oysters and their overlying waters. *Appl. Environ. Microbiol.* 37:572–81.

Grabow, W.O.K., V. Gauso-Muller, O.W. Prozesky, and F. Deinhardt. 1983. Inactivation of Hepatitis A and indicator organisms in water by free chlorine residuals. *Appl. Environ. Microbiol.* 46(3):619–24.

Grabow, W.O.K., P. Coubrough, E.M. Nupen, and B.W. Bateman. 1984. Evaluation of bacteriophages as indicators of the virological quality of sewage-polluted waters. *Water SA* 10:7–14.

Grabow, W.O.K. and P. Coubrough. 1986. Practical direct plaque assay for coliphages in 100-ml samples of drinking water. *Appl. Environ. Microbiol.* 52:430–33.

Grabow, W.O.K., G.K. Idema, P. Coubrough, and B.W. Bateman. 1989. Selection of indicator systems for human viruses in polluted seawater and shellfish. *Water Sci. Technol.* 21:111–17.

Grabow, W.O.K., J.C. De Villiers, and N. Prinsloo. 1991. An assessment of methods for the microbiological analysis of shellfish. *Water Sci. Technol.* 24:91–95.

Grabow, W.O.K., J.C. De Villiers, and C.I. Schildhauer. 1992. Comparison of selected methods for the enumeration of fecal coliforms and *Escherichia coli* in shellfish. *Appl. Environ. Microbiol.* 58:3203–204.

Granai, C. III and R.E. Sjogren. 1981. *In situ* and laboratory studies of bacterial survival using a microporous membrane sandwich. *Appl. Environ. Microbiol.* 41:190–95.

Gray, M.L. and A.H. Killinger. 1966. *Listeria monocytogenes* and *Listeria* infections. *Bacteriol. Rev.* 30:309.

Greenberg, A.E., R.R. Trussel, L.S. Clesceri, and M.A.H. Franson, eds. 1985. Detection of enteric viruses. In *Standard Methods for the Examination of Water and Wastewater*, 16th ed., pp. 946. Washington, D.C.: American Public Health Association.

Greenberg, A.E. and D.A. Hunt, eds. 1985. *Laboratory Procedures for the Examination of Seawater and Shellfish*, 5th ed. Washington, D.C.: American Public Health Association.

Grimes, D.J. and R.R. Colwell. 1983. Survival of pathogenic organisms in the Anacostia and Potomac Rivers and the Chesapeake Bay Estuary. *J. Washington Acad. Sci.* 73:45–50.

Grimes, D.J., F.L. Singleton, and R.R. Colwell. 1984. Allogenic succession of marine bacterial communities in response to pharmaceutical waste. *J. Appl. Bacteriol.* 57.

Grimes, D.J., R.W. Atwell, P.R. Brayton, L.M. Palmer, D.M. Rollins, D.B. Roszak, F.L. Singleton, M.L. Tamplin, and R.R. Colwell. 1986. The fate of enteric pathogenic bacteria in estuarine and marine environments. *Microbiol. Sci.* 3:324–29.

Grimes, D.J. and R.R. Colwell. 1986. Viability and virulence of *Escherichia coli* suspended by membrane chamber in semitropical ocean water. *FEMS Microbiol. Lett.* 34:161–65.

Grimont, P.A.D., F. Grimont, N. Desplaces, and P. Tchen. 1985. DNA probe specific for *Legionella pneumophila. J. Clin. Microbiol.* 21:431–37.

Grohmann, G.S., A.M. Murphy, P.J. Christopher, E. Auty, and H.B. Greenberg. 1981. Norwalk virus gastroenteritis in volunteers consuming depurated oysters. *Australian J. Exp. Biol. Med. Sci.* 59:219–28.

Grohmann, G.S., H.B. Greenberg, B.M. Welch, and A.M. Murphy. Oyster-associated gastroenteritis in Australia: The detection of Norwalk virus and its antibody by immune electron microscopy and radioimmunoassay. *J. Med. Virol.* 6:11–19.

Gross, R. 1983. *Escherichia coli* diarrhea. *J. Infect* 7:177–92.

Grunberg, E., E. Titsworth, G. Beskid, R.J. Cleeland, and W.F. Delorenzo. 1969. Efficiency of a multitest system (Enterotube) for rapid identification of *Enterobacteriaceae. Appl. Environ. Microbiol.* 18:207–13.

Grunstein, M. and D.S. Hogness. 1975. Colony hybridization: A method for the isolation of cloned DNAs that contain a specific gene. *Proc. Nat. Sci.* 72:3961–65.

Guidon, L. and C.A. Pierach. 1973. Infectious hepatitis after ingestion of raw clams. *Minn. Med.* 56:15–19.

Gunn, A.D.G. and D.F. Rowlands. 1969. A confined outbreak of food poisoning which an epidemiological exercise attributed to oysters. *Med. Officer*. 122:75–79.

Gunn, R.A., H.T. Janowski, S. Lieb, E.C. Prather, and H.B. Greenberg. 1982. Norwalk virus gastroenteritis following raw oyster consumption. *Am. J. Epidemiol.* 115:348–51.

Gust, I.D., N.I. Lehmann, S. Crowe, M. McCrorie, S.A. Locarnini, and C.R. Lucas. 1985. The origin of the HM-175 strain of hepatitis A virus. *J. Infect. Dis.* 151:365–67.

Guttman-Bass, N. 1987. Cell cultures and other host systems for detecting and quantifying viruses in the environment. In *Methods for Recovering Viruses from the Environment*, ed. G. Berg, pp. 195–228. Boca Raton, Fla.: CRC Press.

Guttman-Bass, N., Y. Tchorsch, and E. Marva. 1987. Comparison of methods for rotavirus detection in water and results of a survey of Jerusalem wastewater. *Appl. Environ. Microbiol.* 53:761–67.

Guzewich, J.J. and D.L. Morse. 1986. Sources of shellfish in outbreaks of probable viral gastroenteritis: Implications for control. *J. Food Prot.* 49:389–94.

Gyllenberg, H., S. Niemela, and T. Sormunen. 1960. Survival of bifid bacteria in water as compared with that of coliform bacteria and enterococci. *Appl. Microbiol.* 8:20.

Haas, C.N. 1983. Estimation of risk due to low doses of microorganisms: A com-

parison of alternative methodologies. *Am. J. Epidemiol.* 118:573–82.
Hackney, C.R. and G.K. Bissonnette. 1978. Recovery of indicator bacteria in acid mine streams. *J. Water Poll. Cont. Fed.* 50:775–80.
Hackney, C.R., B. Ray, and M.L. Speck. 1979. Repair detection procedure for enumeration of fecal coliforms and enterococci from seafoods and marine environments. *Appl. Environ. Microbiol.* 37:947–53.
Hackney, C.R., B. Ray, and M.L. Speck. 1980. Incidence of *Vibrio parahaemolyticus* in and microbiological quality of seafoods in North Carolina. *J. Food Prot.* 43:769.
Hackney, C.R. and A. Dicharry. 1988. Seafood-borne bacterial pathogens of marine origin. *Food Technology* 42(3):104–09.
Hadfield, T.L., M.H. Monson, and I.K. Wachsmuth. 1985. An outbreak of antibiotic resistant *Salmonella enteritidis* in Liberia, West Africa. *J. Infect. Dis.* 151:790–95.
Haldane, D.J.M., M.A.S. Damm, and J.D. Anderson. 1986. Improved biochemical screening procedure for small clinical laboratories for vero (shiga-like)-toxin producing strains of *Escherichia coli* 0157:H7. *J. Clin. Microbiol.* 24:652–53.
Hall, C.B., G. Douglas Jr., K.C. Schnabel, and J.M. Geiman. 1981. Infectivity of respiratory syncytial virus by various routes of inoculation. *Infect. Immun.* 33:779–83.
Hall, L.W. Jr., J. Steven, W. Bushony, E. Johnson, and W.S. Hall. Spatial and temporal distribution of butyltin compounds in a northern Chesapeake Bay marina and river system. *Environ. Monitoring Assessment.* 10:229–44.
Hall, S., G. Strichartz, E. Moczydlowski, A. Ravindran, and P.R. Reichardt. 1990. The saxitoxins: Sources, chemistry and pharmacology. In *Marine Toxins*, eds. S. Hall and G. Strichartz, pp. 29–65. Washington, D.C.: American Chemical Society.
Halliday, M.L., L.-Y. Kang, T.-K. Zhou, M.-D. Hu, Q.-C. Pan, T.-Y. Fu, Y.-S. Huang, and S.-L. Hu. 1991. An epidemic of hepatitis A attributable to the ingestion of raw clams in Shanghai, China. *J. Infect. Diseases* 164: 852–59.
Halpin-Dohnalek, M. and E. Marth. 1989. *Staphylococcus aureus*: Production of extracellular compounds and behavior in foods—a review. *J. Food Prot.* 52:267–82.
Hamblet, F.E., W.F. Hill, E.W. Akin, and W.H. Benton. 1969. Oysters and human viruses: Effect of seawater turbidity on poliovirus uptake and elimination. *Am. J. Epidemiol.* 89:562–71.
Hanninen, M.L. 1981. The effect of NaCl on *Campylobacter jejuni/coli*. *Acta Vet. Scand.* 22:578–88.
Hansen, W. and E. Yourassowsky. 1984. Detection of β-glucoronidase in lactose fermenting members of the *Enterobacteriaceae* and its presence in bacterial urine cultures. *J. Clin. Microbiol.* 20:1177–79.
Hardy, K. 1981. *Bacterial Plasmids.* Washington, D.C.: American Society for Microbiology.
Harm, W. 1980. *Biological Effects of Ultraviolet Radiation.* Cambridge: Cambridge University Press.
Harmon, S.M. and D.A. Kautter. 1976. Estimating population levels of *Clostridium perfringens* in foods based on alpha toxin. *J. Milk Food Technol.* 39:107–10.
Harmon, S.M. and D.A. Kautter. 1986. Evaluation of a reversed passive latex agglutination test kit for *Clostridium perfringens* enterotoxin. *J. Food Prot.* 49:523–25.

Harshbarger, J.C., S.C. Shang, and S.V. Otto. 1977. Chlamydiae (with phages), Mycoplasmas and Rickettsiae in Chesapeake Bay bivalves. *Science* 196:666–68.

Hartman, P.A., G.W. Reinbold, and D.S. Saraswat. 1966. Media and methods for isolation and enumeration of the enterococci. In *Advances in Applied Microbiology*, ed. W.W. Umbreit, Vol. 8, pp. 253–89. New York: Academic Press.

Hartman, P.A., P.S. Hartman, and W.W. Lanz. 1975. Violet red bile 2 agar for stressed coliforms. *Appl. Microbiol.* 29:537–39.

Hartman, P.A., J.P. Petzel, and C.W. Kaspar. 1986. New Method for Indicator Organisms. In *Foodborne Microorganisms and Their Toxins: Developing Methodology*, eds. M.D. Pierson and N.J. Stern, pp. 175– 217. New York: Marcel Dekker.

Hassan, H.M. and I. Fridovich. 1979. Intracellular production of superoxide radical and of hydrogen peroxide by redox active compounds. *Arch. Biochem. Biophys.* 196:385–95.

Hatcher, P.G., G.A. Berberian, A.Y. Cantillo, P.A. McGillivary, P. Hanson, and R.H. West. 1981. Chemical and physical processes in a dispersing sewage sludge plume. In *Ocean Dumping of Industrial Wastes*, eds. B.H. Ketchum, D.R. Kester and P.K. Park, pp. 347–78. New York: Plenum Press.

Hattis, D. 1990. *Modeling the Risk of Needlestick Transmission of HIV*, CTPID 90-7, Center for Technology, Policy, and Industrial Development, Cambridge, Ma.: Massachusetts Institute of Technology.

Havelaar, A.H. and W.M. Hogeboom. 1983. Factors affecting the enumeration of coliphages in sewage and sewage-polluted waters. *Antonie van Leeuwenhoek J. Microbiol.* 49:387–97.

Havelaar, A.H., W.M. Hogeboom, and R. Pot. 1984. F-specific RNA bacteriophages in sewage: Methodology and occurrence. *Water Sci. Techol.* 17:645–55.

Havelaar, A.H. and W.M. Hogeboom. 1984. A method for the enumeration of male-specific bacteriophages in sewage. *J. Appl. Bacteriol.* 56:439–47.

Havelaar, A.H. and Th.J. Nieuwstad. 1985. Bacteriophages and faecal bacteria as indicators of chlorination efficiency of biologically treated wastewater. *J. Water Pollut. Control Fed.* 57:1084–88.

Havelaar, A.H., K. Furuse, and W.M. Hogeboom. 1986. Bacteriophages and indicator bacteria in human and animal faeces. *J. Appl. Bacteriol.* 60:255–62.

Havelaar, A.H. 1986. F-specific RNA bacteriophages as mode viruses in water treatment processes. Ph.D. Dissertation, University of Utrecht, The Netherlands.

Havelaar, A.H. 1987. Bacteriophages as model organisms in water treatment. *Microbiol. Sci.* 4:362–64.

Havelaar, A.H. and M. During. 1988. Evaluation of the Anderson Baird-Parker direct plating method for enumerating *Escherichia coli* in water. *J. Appl. Bacteriol.* 64:89–98.

Havelaar, A.H. and W.M. Pot-Hogeboom. 1988. F-specific RNA-bacteriophages as model viruses in water hygiene: Ecological aspects. *Water Sci. Tech.* 20:399–407.

Havelaar, A.H. and M. van Olphen. 1989. Water quality standards for bacteriophages? In *Watershed 89. The Future for Water Quality in Europe*, eds. D. Wheeler, M.L. Richardson, and J. Bridges, pp. 357– 66. Oxford: Pergamon Press.

Havemeister, G. 1969. Hygienic evaluation of sea water from mussel beds. *Archiv fuer Hyg. Bakteriologie* 153(5):460–63.

Havens, W.P. Jr. 1945. Properties of the etiological agent of infectious hepatitis. *Proc.*

Soc. Exp. Biol. Med. 58:203–04.

Hay, B. and P. Scott. 1986. Evidence for intracellular absorption of virus by the Pacific oyster, *Crassostrea gigas. New Zealand J. Mar. Freshwater Res.* 29:655–59.

Hayes, P.S., J.C. Feely, L.M. Graves, G.W. Ajello, and D.W. Fleming. 1986. Isolation of *Listeria monocytogenes* from raw milk. *Appl. Environ. Microbiol.* 51:438–40.

Hazen, T.C., C.B. Fliermans, R.P. Hirsch, and G.W. Esch. 1978. Prevalence and distribution of *Aeromonas hydrophila* in the United States. *Appl. Environ. Microbiol.* 36:731–37.

Hazen, T.C. and L. Jimenez. 1988. Enumeration and identification of bacteria from environmental samples using nucleic acid probes. *Microbiol. Sci.* 5:340–43.

Hazen, T.C. 1988. Fecal coliforms as indicators in tropical waters: A review. *Toxicity Assessment: An Int. J.* 3:461–77.

Hebert, G., C. Crowder, C.G. Hancock, W. Jarvis, and C. Thornsberry. 1988. Characteristics of coagulase—negative staphylococci that help differentiate these species from other members of the family *Micrococcaceae. J. Clin. Microbiol.* 26:1939–49.

Hedstrom, C.E. and E. Lyke. 1964. An experimental study on oysters as virus carriers. *Am. J. Hyg.* 79:134–42.

Heffernan, W.P. and V.J. Cabelli. 1970. Elimination of bacteria by the northern quahuag (*Mercenaria mercenaria*): Environmental parameters significant to the process. *J. Fish Res. Bd. Can.* 27:1569–77.

Heffernan, W.P. and V.J. Cabelli. 1971. The elimination of bacteria by the northern quahuag: Variability in the response of individual animals to the development of criteria. *Proc. Natl. Shellfish Assoc.* 61:102–08.

Heijthuijsen, J.H.F.G. and T.A. Hansen. 1989. Betaine fermentation and oxidation by marine *Desulfuromonas* strains. *Appl. Environ. Microbiol.* 55:965–69.

Heil, D.C. 1986. *Evaluation of Trace Metal Monitoring in Florida Shellfish.* Tallahassee, Fla.: Florida Department of Natural Resources.

Hejkal, W.T., F.M. Wellings, P.A. Larock, and A.L. Lewis. 1979. Survival of poliovirus within organic solids during chlorination. *Appl. Environ. Microbiol.* 38(1):114–18.

Hejkal, T.W. and C.P. Gerba. 1980. Uptake and survival of viruses in the blue crab, Callinectes sapidus. *Abstr. Ann. Meet. Am. Soc. Microbiol.* 80:210.

Hejkal, T.W., C.P. Gerba, and V.C. Rao. 1982. Reduction of cytotoxicity in virus concentrates from environmental samples. *Appl. Environ. Microbiol.* 43:731–33.

Herbert, R.A. 1990. Methods for enumerating microorganisms and determining biomass in natural environments. In *Methods in Microbiology*, eds. R. Grigorova and J.R. Norris, pp. 1–39. New York: Academic Press.

Herrington, D.A., S. Tzipori, R.M. Robins-Browne, B.D. Tall, and M.M. Levine. 1987. *In vitro and in vivo pathogenicity of Plesiomonas shigelloides. Infect. Immun.* 55:979–85.

Herrington, T.L. 1984. Oyster borne disease outbreaks in St Petersburg, Fla. Memorandum (1/31/84). Atlanta, Ga.: Department Health and Human Services/Public Health Service/Food and Drug Administration.

Herrington, T.L. 1991. Use of ultraviolet light in depuration. In *Molluscan Shellfish Depuration.* eds. W.S. Otwell, G.E. Rodrick and R. Martin. Boca Raton, Fla: CRC Press. (in press).

Herrmann, J.E. and D.O. Cliver. 1968. Methods for detecting food-borne enteroviruses. *Appl. Microbiol.* 16:1564–69.

Herrmann, J.E., D.N. Taylor, P. Echeverria, and N.R. Blacklow. 1991. Astroviruses as a cause of gastroenteritis in children. *New Engl. J. Med.* 324:1757–60.

Hickey, J.M. 1990. Massachusetts Division of Marine Fisheries, Shellfish Bureau. Sandwich, Mass. (personal communication).

Hill, W.E., W.L. Payne, and C.C.G. Aulisio. 1983. Detection and enumeration of virulent *Yersinia enterocolitica* in food by DNA colony hybridization. *Appl. Environ. Microbiol.* 46:636–41.

Hill, W.E. 1987. Detection of pathogenic bacteria by DNA colony hybridization. FDA Bacteriological analytical manual, 6th ed., Suppl.

Hill, W.E., B.A. Wentz, W.L. Payne, J.A. Jagow, and G. Zon. 1986. DNA colony hybridization method using synthetic oligonucleotides to detect enterotoxigenic *Escherichia coli*: Collaborative study. *J. Assoc. Off. Anal. Chem.* 69:531–36.

Hill, W.E., S.P. Keasler, M.W. Trucksess, P. Feng, C.A. Kaysner, and K.A. Lampel. 1991. Polymerase chain reaction identification of *Vibrio vulnificus* in artificially contaminated oysters. *Appl. Environ. Microbiol.* 57:707–11.

Hill, W.F. Jr., F.E. Hamblet, and W.H. Benton. 1969. Inactivation of poliovirus type 1 by the Kelly-Purdy ultraviolet seawater treatment unit. *Appl. Microbiol.* 17:1–6.

Hill, W.F. Jr., F.E. Hamblet, W.H. Benton, and E.W. Akin. 1970. Ultraviolet devitalization of eight selected enteric viruses in estuarine water. *Appl. Microbiol.* 19:805–19.

Hill, W.F. Jr., E.W. Akin, and W.H. Benton. 1971. Detection of viruses in water: A review of methods and application. *Water Res.* 5:967–95.

Hirsch, C.F. and D.L. Christensen. 1983. Novel method for selective isolation of actinomycetes. *Appl. Environ. Microbiol.* 46:925–29.

Hobbie, J.E., R.J. Daley, and S. Jasper. 1977. Use of nuclepore filters for counting bacteria by fluorescence microscopy. *Appl. Environ. Microbiol.* 33:1225–28.

Hobbs, B. 1979. *Clostridium perfringens* gastroenteritis. In *Food-borne Infections and Intoxications*, eds. H. Rieman and F. Bryan, New York: Academic Press.

Hockin, J. 1987. Hemorrhagic colitis and hemolytic uremic syndrome caused by *Escherichia coli* O157:H7 in Canada. *Can. Dis. Wkly. Rpt.* 13–45:203.

Hoff, K.A. 1989. Survival of *Vibrio anguillarum* and *Vibrio salmonicida* at different salinities. *Appl. Environ. Microbiol.* 55:1775–86.

Hoff, J.C., W. Jakabowski, and W.J. Beck. 1965. Studies on bacteriophage accumulation and elimination by the Pacific oyster (*Crassostrea gigas*). In *Proc. 1965 Northwest Shellfish Sanit.* Res. Planning Conf. Publ. Health Serv. Publ. No. 999-FP-6, pp. 74–90.

Hoff, J.C., W.J. Beck, T.H. Ericksen, G.J. Vasconcelos, and M.W. Presnell. 1967. Time–temperature effects on the bacteriological quality of stored shellfish. I. Bacteriological changes in live shellfish: Pacific oysters (*Crassostrea gigas*), Olympia oysters (*Ostrea lurida*), Native Littleneck clams (*Prototheca staminea*), and Manila clams (*Venerupis japonica*). *J. Food Sci.* 32(1):121–24.

Hoff, J.C., W.J. Beck, T.H. Ericksen, G.J. Vasconcelos, and M.W. Presnell. 1967. Time–temperature effects on the bacteriological quality of stored shellfish. II.

Bacteriological changes in shucked Pacific oysters (*Crassostrea gigas*) and Olympia oysters (*Ostrea lurida*). *J. Food Sci.* 32(1):125–29.

Hoff, J.C. and R.C. Becker. 1969. The accumulation and elimination of crude and clarified poliovirus suspensions by shellfish. *Am. J. Epidemiol.* 90:53–61.

Hoffman, P.S., A.G. Hugh, N.R. Krieg, and R.M. Smibert. 1978. Studies of the microaerophilic nature of *Campylobacter fetus* subsp. *jejuni*. II. Role of exogenous super-oxide anions and hydrogen peroxide. *Can. J. Microbiol.* 25:8–16.

Holben, W.E. and J.M. Tiedje. 1988. Applications of nucleic acid hybridization in microbial ecology. *Ecology* 69:561–68.

Holben, W.E., J.K. Jansson, B.K. Chelm, and J.M. Tiedje. 1988. DNA probe method for the detection of specific microorganisms in the soil bacterial community. *Appl. Environ. Microbiol.* 54:703–11.

Holbrook, R., J.M. Anderson, and A.C. Baird-Parker. 1980. Modified direct plate method for counting *Escherichia coli* in frozen foods. *Food Technol. Australia* 32:78–83.

Holdeman, L.V., I.J. Good, and W.E.C. Moore. 1976. Human fecal flora: Variation in bacterial composition within individuals and a possible effect of emotional stress. *Appl. Environ. Microbiol.* 31:359–75.

Holderman, L. and W. Moore. 1970. Bacteriodes. In *Clinical Methods in Anaerobic Bacteriology*. VPI & SU Anaerobic Laboratory.

Holm, S.E. and J.G. Windsor Jr. 1986. Chemical monitoring of sewage effluents using saturated hydrocarbons and coprostanol in estuarine waters. Oceans '86 Conference Record: Science-Engineering-Adventure. *Monitoring Strategies Symposium.* 3:839–44.

Holmberg, S.D. and J.J. Farmer III. 1984. *Aeromonas hydrophila* and *Plesiomonas shigelloides* as causes of intestinal infections. *Rev. Infections Dis.* 6:633–39.

Holmberg, S.D., F.W. Wachsmuth, F.W. Hickman-Brenner, and M.L. Cohen. 1984. Comparison of plasmid profile analysis phage typing, and antimicrobial susceptibility testing in characterizing *Salmonella typhimurium* isolates from outbreaks. *J. Clin. Microbiol.* 19:100–04.

Holmberg, S.D., W.L. Schell, G.R. Fanning, I.K. Wachsmuth, F.W. Hickman-Brenner, and P.A. Blake. 1986*a*. *Aeromonas* intestinal infections in the United States. *Ann. Intern. Med.* 105:683–89.

Holmberg, S.D., I.K. Wachsmuth, F.W. Hickman-Brenner, P.A. Blake, and J.J. Farmer III. 1986. *Plesiomonas* enteric infections in the United States. *Ann. Int. Med.* 105:690–94.

Holt, S.M., P. Hartman, and C. Kaspar. 1989. Enzyme-capture assay for rapid detection of Escherichia coli in oysters. *Appl. Environ. Microbiol.* 55(1):229–32.

Hood, M.A., G.E. Ness, and G.E. Rodrick. 1981. Isolation of *Vibrio cholerae* serotype 01 from the eastern oyster, *Crassostrea virginica. Appl. Environ. Microbiol.* 41:559–60.

Hood, M.A., G.E. Ness, G.E. Rodrick, and N.J. Blake. 1983. Distribution of *Vibrio cholerae* in two Florida estuaries. *Microbial Ecology* 9:65–75.

Hood, M.A. 1983. Effects of harvesting waters and storage conditions on yeast populations in shellfish. *J. Food Prot.* 46:105–08.

Hood, M.A., G.E. Ness, G.E. Rodrick, and N.J. Blake. 1983. Effects of storage on microbial loads of two commercially important shellfish species, *Crassostrea virginica* and *Mercenaria campechiensis*. *Appl. Environ. Microbiol.* 45(4):1221–28.

Hood, M.A., G.E. Ness, and N.J. Blake. 1983. Relationship among fecal coliforms, *Escherichia coli*, and *Salmonella* spp. in shellfish. *Appl. Environ. Microbiol.* 45(1):122–26.

Hood, M.A., R.M. Baker, and F.L Singelton. 1984. Effect of processing and storing oyster meats on concentrations of indicator bacteria, vibrios and *Aeromonas hydrophila*. *J. Food Prot.* 47(8):598–601.

Hook, E.W. 1985. *Salmonella* species (including typhoid fever). In *Principles and Practice of Infectious Diseases*, eds. G.L. Mandell, R.G. Douglas, and J.C. Bennett, 2nd ed., pp. 1256–69. New York: Wiley.

Hoppe, H. 1976. Determination and properties of actively metabolizing heterotrophic bacteria in the sea investigated by means of autoradiography. *Mar. Biol.* 36:291–302.

Hornick, R.B., S.E. Greisman, T.E. Woodward, H.L. DuPont, A.T. Dawkins, and M.J. Snyder. 1970. Typhoid fever: Pathogenesis and immunologic control. *New Engl. J. Med.* 283:686–91.

Hornick, R.B. 1983. Non-typhoidal salmonellosis. In *Infectious Diseases: A Modern Treatment of Infectious Processes*, ed. P.D. Hoeprich, 3rd ed., pp. 655–61. Philadelphia, Pa.: Harper and Row.

Hoskin, G.P. and R. Bandler. 1987. Identification of mammalian feces by coprostanol thin layer chromatography: Method development. *J. Assoc. Off. Anal. Chem.* 70:496–98.

Hoskin, J.K. 1934. Most probable numbers for evaluation of *Coli aerogenes* test by fermentation tube method. *Public Health Rpt.* 49:393–405. Reprinted and revised No. 1621, 1947.

Howard, B.E. 1980. Development and assessment of a method for concentration of enteric viruses in the hard shell clam, Mercenaria. Master's Thesis. Department of Environmental Sciences and Engineering. Chapel Hill, N.C.: University of North Carolina.

Huey, B. and J. Hall. 1989. Hypervariable DNA fingerprinting in *Escherichia coli*: Minisatellite probe from bacteriophage M13. *J. Bacteriol.* 171:2528–32.

Huff, C.B., H.F. Smith, W.P. Boring, and N.A. Clark. 1965. Study of ultraviolet disinfection of water and factors in treatment efficiency. *Public Health Rpt.* 80:695–704.

Humble, M.W., A. King, and I. Phillips. 1977. API ZYME: A simple rapid system for the detection of bacterial enzymes. *J. Clin. Pathol.* 30:275–77.

Hunt, D.A. and J. Springer. 1974. Preliminary report on a comparison of total coliform and fecal coliform values in shellfish growing area waters and a proposal for a fecal coliform area standard. In *Proceedings 8th National Shellfish Sanitation Workshop*, ed. D.S. Wilt, pp. 97–104. Washington, D.C.: Food and Drug Administration, Shellfish Sanitation Branch.

Hunt, D.A. 1977. Indicators of quality for shellfish waters. In *Bacterial Indicators/Health Hazards Associated With Water*, ASTM STP 635, eds. A.W. Hoadley and B.J. Dutka, pp. 337–45. Am. Society for Testing and Materials.

Hunt, D.A. and J. Springer. 1978. Comparison of two rapid test procedures with the standard EC test for recovery of fecal coliform bacteria from shellfish-growing waters. *J. Assoc. Off. Anal. Chem.* 61:1317–23.

Hunt, M.D., W.E. Woodward, B.H. Keswick, and H.L. DuPont. 1988. Seroepidemiology of cholera in gulf-coastal Texas. *Appl. Environ. Microbiol.* 54:1673–77.

Hunter, A.C. 1928. *Bacteriology and Chemistry of Oysters.* Technical bulletin No. 64. Washington, D.C.: U.S. Department of Agriculture.

Huntley, B.E. and R.J. Hammerstrom. 1971. An experimental depuration plant: Operation and evaluation. *Chesapeake Sci.* 12(4):231–39.

Huntley, B.E., A.C. Jones, and V.J. Cabelli. 1976. *Klebsiella* densities in waters receiving wood pulp effluents. *J. Water Pollut. Control Fed.* 48:1766–71.

Huq, A., R.R. Colwell, R. Rahman, A. Ali, M.A.R. Chowhury, S. Parveen, D.A. Sack, and E. Russek-Cohen. 1990. Detection of *Vibrio cholerae* 01 in the aquatic environment by flourescent-monoclonal antibody and culture methods. *Appl. Environ. Microbiol.* 56:2370–73.

Hurst, A. 1977. Bacterial injury: A review. *Can. J. Microbiol.* 23:935–44.

Hurst, A. 1980. Injury and its effect on survival and recovery. In *International Commission on Microbiological Specifications for Foods* ed., Microbial ecology of foods. Vol. 2, pp. 205–14. New York: Academic Press.

Hurst, A. and A. Hughes. 1981. Repair of salt tolerance and recovery of lost D-alinine and magnesium following sublethal heating of *Staphylococcus aureus* are independent events. *Can. J. Microbiol.* 27:627–32.

Hurst, C.J. and W.H. Benton. 1986. Effects of 5-iododeoxyuridine treatment and mixed cell monolayers upon enteric virus plaque assay titers. *Abstr. Ann. Meet. Am. Soc. Microbiol.* 86:307.

Hussong, D., R.R. Colwell, M. O'Brien, E. Weiss, A.D. Pearson, R.M. Weiner, and W.D. Burge. 1987. Viable *Legionella pneumophila* not detectable by culture on agar media. *Bio/Technol.* 5:947–50.

Hyams, J.S., W.A. Durbin, R.J. Grand, and D.A. Goldman. 1980. *Salmonella* bacteremia in the first year of life. *J. Pediatr.* 96:57–59.

Hyypia, T., P. Auvinen, and M. Maaronen. 1990. Polymerase chain reaction of human picornaviruses. *J. Gen. Virol.* 70:3261–68.

IAWPRC Study Group on Health Related Microbiology. 1991. Bacteriophages as model viruses in water quality control. *Water Res.* 25:529–45.

Inal, T., G. Hildebrandt, and H. Siems. 1974. *Clostridium* as indicator of water pollution using a harbour town in Turkey as example. *Zentralblatt fuer Veterinaermedizin,* B21:159–70.

International Organization for Standardization. 1989. ISO 2859-1989(E) Sampling Procedures and Tables for Inspection by Attributes—Part I: Sampling Plans Indexed by Acceptable Quality Level (AQL) for Lot-by-Lot Inspection, ISO, Geneva.

International Organization for Standardization. 1991. ISO 2859-3-1991(E) Sampling Procedures for Inspection by Attributes—Part 3: Skip Lot Sampling, Geneva, Switzerland.

Irving, L.G. and F.A. Smith. 1981. One-year survey of enteroviruses, adenoviruses, and reoviruses isolated from effluent at an activated-sludge purification plant. *Appl. Environ. Microbiol.* 41:51–59.

Isbister, J.D., J.A. Simmons, W.M. Scott, and J.F. Kitchens. 1983. A simplified method for coliphage detection in natural waters. *Acta Microbiologica Polonica* 32:197–206.

Isenberg, H.D., D. Goldberg, and J. Sampson. 1970. Laboratory studies with a selective enterococcus medium. *Appl. Microbiol.* 20:433–36.

Izard, D., M.O. Husson, P. Vincent, H. Leclert, D. Monget, and J.M. Boeufgras. 1984. Evaluation of the four-hour rapid 20E system for identification of members of the family *Enterobacteriaceae. J. Clin. Microbiol.* 20:51–54.

Jablonski, E., E.W. Moomaw, R.H. Tullis, and J.L. Ruth. 1986. Preparation of oligodeoxynucleotide-alkaline phosphatase conjugates and their use as hybridization probes. *Nucleic Acids Res.* 14:6115–28.

Jackson, H. 1974. Loss of viability and metabolic injury of *Staphylococcus aureus* resulting from storage at 5. *J. Appl. Bacteriol.* 37:59–64.

Jackson, J. 1985. Shellfish Sanitation in Oregon: Can It Be Achieved through Pollution Source Management. *Perspective on Nonpoint Source Management.* Washington, D.C.: USEPA 440/5-85-001.

Jackson, J. 1990. Hillsboro, Oregon (personal communication) Unifred Sewerage Agency.

Jackson, S.G., D.A. Yip-Chuck, and M.H. Brodsky. 1986. Evaluation of the diagnostic application of an enzyme immunoassay for *Clostridium perfringens* type A enterotoxin. *Appl. Environ. Microbiol.* 52:969–70.

Jacobs, N.J., W.L. Zeigler, F.C. Reed, T.A. Stukel, and E.W. Rice. 1986. Comparison of membrane filter, multiple-fermentation-tube, and presence-absence techniques for detecting total coliforms in small community water systems. *Appl. Environ. Microbiol.* 51:1007–12.

Jagow, J. and W.E. Hill. 1986. Enumeration by DNA colony hybridization of virulent *Yersinia enterocolitica* colonies in artificially contaminated food. *Appl. Environ. Microbiol.* 51:441–43.

James, A. 1979. The value of biological indicators in relation to other parameters of water quality. In *Biological Indicators of Water Quality*, eds. A. James and L. Evison, pp. 1-1–1-16. Chichester, U.K.: Wiley.

James, S.M., S.L. Fannin, B.A. Agee, B. Hall, E. Parker, J. Vogt, G. Run, J. Williams, L. Lieg, C. Salminen, T. Prendergast, S.B. Werner, and J. Chin. 1985. Listeriosis outbreak associated with Mexican-style cheese. *Morbid. Mortal. Wkly. Rev.* 34(24):357–59.

Jamieson, A.F., D.A. Bremner, P.L. Bergquist, and H.E.D. Lane. 1979. Characterization of plasmids from antibiotic resistant *Shigella* isolates by agarose gel electrophoresis. *J. Gen. Microbiol.* 113:73–81.

Jamieson, W., P. Madri, and G. Claus. 1976. Survival of certain pathogenic microorganisms in sea water. *Hydrobiologia* 50:117–21.

Janda, J.M., C. Powers, R.G. Bryant, and S.L. Abbott. 1988. Current perspectives on the epidemiology and pathogenesis of clinically significant *Vibrio* spp. *Clin. Microbiol Rev.* 1:245–67.

Jannasch, H.W. 1953. Zur methode der quantitativen. Untersuchungen con Bakterienkuturen in flussigen Medien. *Arch. Mikrobiol.* 18:425.

Jannasch, H.W. and G.E. Jones. 1959. Bacterial populations in seawater as determined by different methods of enumeration. *Limnol. Oceanogr.* 4:128–39.

Jannasch, H.W. 1966. Competitive elimination of *Enterobacteriaceae* from seawater. *Appl. Microbiol.* 16:1616–18.

Jansen, R. 1990. (personal communication).

Jansson, J.K., W.E. Holben, and J.M. Tiedje. 1989. Detection in soil of a deletion in an engineered DNA sequence by using DNA probes. *Appl. Environ. Microbiol.* 55:3022–25.

Jay, J.M. 1986. "Modern Food Microbiology". New York: Van Nostrand Reinhold

Jaykus, L.-A., R. De Leon, and M.D. Sobsey. 1992. Application of RT-PCR for the detection of enteric viruses in oysters. *Wat. Sci. Tech.* (in press).

Jensen, E.T. 1962. The Shellfish research program of the Public Health Service. *Proceedings of the Gulf and Caribbean Fisheries Institute*, pp. 59–65.

Jensen Jr., H.R. 1980. The development of an acid precipitation procedure for concentrating viruses from artificially and naturally contaminated oysters. Master's Thesis. Department of Environmental Sciences and Engineering. Chapel Hill, N.C.: University of North Carolina.

Jiang, X., M.K. Estes, T.G. Metcalf, and J.L. Melnick. 1986. Detection of hepatitis A virus in seeded estuarine samples by hybridization with cDNA probes. *Appl. Environ. Microbiol.* 52:711–17.

Jiang, X., M.K. Estes, and T.G. Metcalf. 1987. Detection of hepatitis A virus by hybridization with single-stranded RNA probes. *Appl. Environ. Microbiol.* 53:2487–95.

Jiang, X., D.Y. Graham, K. Wank, and M.K. Estes. 1990. Norwalk virus genome cloning and characterization. *Science* 250:1580–83.

Jofre, J., A. Bosch, F. Lucena, R. Girones, and C. Tartera. 1986. Evaluation of *Bacteroides fragilis* bacteriophages as indicators of the virological quality of water. *Water Sci. Technol.* 18:167–77.

John, J.F., R.J. Sharbaugh, and E.R. Bannister. 1982. *Enterobacter cloacae*: Bacteremia, epidemiology, and antibiotic resistance. *Rev. Infect. Dis.* 4:13–28.

Johnson, H.C. and J. Liston. 1973. Sensitivity of *Vibrio parahaemolyticus* to cold in oysters, fish fillets and crabmeat. *J. Food Sci.* 38:437–41.

Johnson, J.E. and A.W. Brinkley. 1982. Comparison of the AutoMicrobic system and a conventional tube system for the identification of nonfermentative and oxidase-positive gram-negative bacilli. *J. Clin. Microbiol.* 15:25–27.

Johnson, K.M., R.C. Cooper, and D.C. Straube. 1981. Procedure for recovery of enteroviruses from the Japanese cockle, Tapes japonica. *Appl. Environ. Microbiol.* 41:932–35.

Johnson, K.M. 1984. *Bacillus cereus* foodborne illness—an update. *J. Food Prot.* 47:145.

Johnson, W.G., A.C. Salinger, and W.C. King. 1973. Survival of *Vibrio parahaemolyticus* in oyster shellstock at two different storage temperatures. *Appl. Microbiol.* 26(1):122.

Johnson, W.M. and H. Lior. 1984. Toxins produced by *Campylobacter jejuni* and *Campylobacter coli. Lancet* i:229–30.

Johnston, J.M., S.F. Becker, and L.M. McFarland. 1985. *Vibrio vulnificus*: Man and the sea. *J. AMA* 253:2050–53.

Johnstone, G.F. and S.E. Brown. 1970. A short scheme for the rapid identification of gram-negative bacilli in urine. *J. Med. Lab. Technol.* 27.

Johnstone, J. 1914. The methods of cleansing living mussels from ingested sewage bacteria. Report for 1914 on the Lancashire Sea Fisheries Laboratory. No. 23.

Jones, G.E. and A.B. Cobet. 1975. Heavy metal ions as the principal bactericidal agent in Caribbean sea water. In *Discharge of Sewage from Sea Outfalls*, ed. A.L.H. Gameson, pp. 199–207. Oxford: Pergamon Press.

Jones, K.L. and M.E. Rhodes-Roberts. 1981. The survival of marine bacteria under starvation conditions. *Appl. Bacteriol.* 50:247–58.

Jorgensen, J.H., J.W. Dyke, N.G.P. Helgeson, B.H. Cooper, J.S. Redding, S.A. Crawford, M.T. Andruszewski, and S.A. Prowant. 1984. Collaborative evaluation of the Abbott Avantage system for identification of frequently isolated nonfermentative or oxidase-positive gram-negative bacilli. *J. Clin. Microbiol.* 20:899–904.

Juven, B.J. and I. Rosenthal. 1985. Effect of free-radical and oxygen scavengers on photochemically generated oxygen toxicity and on the aerotolerance of *Campylobacter jejuni. J. Appl. Bacteriol.* 59:413–419.

Kafatos, F.C., C.W. Jones, and A. Efstradiadis. 1979. Determination of nucleic acid sequence homologies and relative concentrations by a dot hybridization procedure. *Nucleic Acids Res.* 7:1541–52.

Kahn, M.A., A. Seaman, and M. Woodbine. 1974. In *Synergistic Drug Treatment in Experimental Listeriosis.* ed. M. Woodbine, p. 52. Problems of Listeriosis. Surrey, U.K.: Heicester UMV Press.

Kai, A., T. Itoh, K. Saito, Y. Yanagawa, M. Inaga, M. Takahashim, S. Matsushita, Y. Kudoh, S. Sakai, and T. Shinohara. 1983. Two outbreaks of oyster-associated food poisoning due to enterotoxigenic Escherichia coli O27:H7. *Annual Report of Tokyo Metropolitan Research Laboratory of Public Health* 34:121–24.

Kain, K.C. and M.T. Kelly. 1989. Clinical features, epidemiology and treatment of *Plesiomonas shigelloides* diarrhea. *J. Clin. Microbiol.* 27:998–1001.

Kamplemacher, E.H. and L.M. van Noorle Jansen. 1969. Isolation of *Listeria monocytogenes* from faeces of clinically healthy humans and animals. *Zentralbl. Bakteriol. Abt. 1, Orig. A.* 211:353–59.

Kanazawa, A. and S. Teshima. 1971. Sterols of the suspended matters in sea water. *J. Oceanogr. Soc. Japan* 27:207–12.

Kanazawa, A. and S. Teshima. 1978. The occurrence of coprostanol, an indicator of faecal pollution, in sea water and sediments. *Oceanol. Acta* 1:39–44.

Kaneko, T. and R.R. Colwell. 1973. Ecology of *Vibrio parahaemolyticus* in Chesapeake Bay. *J. Bacteriol.* 113:24–32.

Kaneko, T. and R.R. Colwell. 1978. The annual cycle of *Vibrio parahaemolyticus* in Chesapeake Bay. *Microbiol. Ecol.* 4:135–55.

Kantor, G.J. and R.A. Deering. 1968. Effect of nalidixic acid and hydroxyurea on division ability of *Escherichia coli* Fil^+ and Lon^- strains. *J. Bacteriol.* 95:520–30.

Kapelmacher, E.H., A.B. Leussink, and L.M. van Noorle Jansen. 1976. Comparative studies of methods for the enumeration of coli aerogenes bacteria and *E. coli* in surface water. *Water. Res.* 10:285–88.

Kaper, J.B., H. Lockman, R.R. Colwell, and J.W. Joseph. 1979. Ecology, serology and enterotoxin production of *Vibrio cholerae* in Chesapeake Bay. *Appl. Environ. Microbiol.* 37:91–103.

Kaper, J.B., H. Lockman, and R.R. Colwell. 1981. *Aeromonas hydrophila*: Ecology and toxigenicity of isolates from an estuary. *J. Appl. Bacteriol.* 50:359–77.

Kaper, J.B. and M.M. Levine. 1981. Cloned cholera enterotoxin genes in study and prevention of cholera. *Lancet* 21:1162–63.

Kaper, J.B., H.B. Bradford, N.C. Roberts, and S. Falkow. 1982. Molecular epidemiology of *Vibrio cholerae* in the U.S. Gulf Coast. *J. Clin. Microbiol.* 16:129–34.

Kaplan, J.E., F. Feldman, S. Campbell, C. Lookbaugh, and G.W. Gary. 1982. The frequency of a Norwalk like pattern of illness in outbreaks of acute gastroenteritis. *Am. J. Public Health* 72:1329–32.

Kaplan, J.E., R.A. Goodman, L.B. Schonberger, E.C. Lippy, and G.W. Gary. 1982. Gastroenteritis due to Norwalk virus: An outbreak associated with a municipal water system. *J. Infect. Dis.* 146:190–97.

Kapperud, G. and J. Lassen. 1983. Relationship of virulence-associated autoagglutination to hemagglutinin production in *Yersinia enterocolitica* and *Yersinia enterocolitica*-like bacteria. *Infect. Immun.* 42:163–69.

Kapperud, G., K. Dommarsnes, M. Skurnik, and E. Hornes. 1990. A synthetic oligonucleotide probe and a cloned polynucleotide probe based on the *yopA* gene for detection and enumeration of virulent *Yersinia enterocolitica. Appl. Environ. Microbiol.* 56:17–23.

Kappus, K.D., J.S. Marks, R.C. Holman, J.R. Bryant, C. Baker, G.W. Gary, and H.B. Greenberg. 1982. An outbreak of Norwalk gastroenteritis associated with swimming in a pool and secondary person-to-person transmission. *Am. J. Epidemiol.* 116:834 -39.

Kapuscinski, R.B. and R. Mitchell. 1981. Solar radiation induces sublethal injury in *Escherichia coli* in seawater. *Appl. Environ. Microbiol.* 41:670–74.

Kapuscinski, R.B. and R. Mitchell. 1983. Sunlight-induced mortality of viruses and *Escherichia coli* in coastal seawater. *Environ. Sci. Technol.* 17:1–6.

Karch, H. and T. Meyer. 1989. Single primer pair for amplifying segments of distinct Shiga-like toxin genes by polymerase chain reaction. *J. Clin. Microbiol.* 27:2751–57.

Kasper, C.W. and C. Tartera. 1990. Methods for detecting bacteria in food and water. In *Methods in Microbiology*, eds. R. Grigorova and J.R. Norris, pp. 437–541. New York: Academic Press.

Kasweck, K.L. and C.B. Fliermans. 1978. Lactose variability of *Escherichia coli* in thermally stressed reactor effluent waters. *Appl. Environ. Microbiol.* 36:739–46.

Kater, R.M.H., C.Y. Kim, and C.S. Davidson. 1969. Australia antigen and viral hepatitis. *J. Infect. Dis.* 120:391–92.

Kator, H. and M.W. Rhodes. 1988. Evaluation of alternate microbial indicators of fecal pollution in a non-point source impacted shellfish growing area. A final report submitted to the Council on the Environment, Richmond, Virginia. Virginia Institute of Marine Science, Special Report in Applied Marine Science and Ocean Engineering. No. 297.

Kator, H. and M.W. Rhodes. 1989. Occurrence of indicators of fecal pollution in water and sediment of a subestuary impacted by non-point source pollution. A final report submitted to Virginia Department of Conservation and Historic Resources, Division of Soil Conservation, Richmond, Virginia. Virginia Institute of Marine Science, Special Report in Applied Marine Science and Ocean Engineering. No. 303.

Kator, H. and M.R. Rhodes. 1991. Indicators and alternate indicators of growing water quality. In *Microbiology of Marine Food Products*, eds. D.R. Ward and C.R. Hackney, pp. 135–196. New York: Van Nostrand Reinhold.

Katz, M. and S.A. Plotkin. 1967. Minimal infective dose of attenuated poliovirus for man. *A. JPH* 57:1837–40.

Katzenelsen, E. 1976. A rapid method for quantitative assay of poliovirus from water with the aid of fluorescent antibody technique. *Arch. Virol.* 50:197–206.

Katzenelsen, E. and S. Kedmi. 1979. Unsuitability of polioviruses as indicators of virological quality of water. *Appl. Environ. Microbiol.* 37:343–44.

Kaufmann, F. 1966. The Bacteriology of *Enterobacteriaceae*. Baltimore, Md.: Williams and Wilkins.

Kawabata, T. and T. Harada. 1959. The disinfection of water by germicidal lamp. *J. Illumination Soc.* 36:89.

Kawano, J., A. Shimizu, K. Fujii, S. Takeuchi, and S. Kimura. 1987. Characterization of *Bacillus cereus*: Haemolysis pattern, cytotoxicity and vascular permeability factory activity. Japan Science Reports of Faculty of Agriculture, Kobe University. Journal Article 17(2):273–81.

Kawasaki, E.S. and A.M. Wang. 1989. Detection of gene expression. In *PCR Technology*, ed. H.A. Erlich, pp. 89–97. New York: Stockton Press.

Kaysner, C.A. and S. Weagant. 1987. Limitations of the A-1M method for fecal coliform enumeration in the Pacific oyster (*Crassostrea gigas*). *J. Assoc. Off. Analytical Chemists* 70(3):535–37.

Kaysner, C.A., C.A. Abeyta, M.M. Wekell, J.A. DePaola, R.F. Scott, and J. M. Leitch. 1987. Virulent strains of *Vibrio vulnificus* isolated from estuaries of the United States west coast. *Appl. Environ. Microbiol.* 53:1349–51.

Kaysner, C.A., M.L. Tamplin, M.M. Wekell, R.F. Stott, and K.G. Colburn. 1989. Survival of *Vibrio vulnificus* in shellstock and shucked oysters (*Crassostrea gigas* and *Crassostrea virginica*) and effects of isolation medium on recovery. *Appl. Environ. Microbiol.* 55(12):3072–79.

Kaysner, C.A., C. Abeyta Jr., R.F. Stott, M.H. Krane, and M.M. Wekell. 1990. Enumeration of *Vibrio* species, Including *V. cholerae*, from Samples of an Oyster Growing Area, Grays Harbor, Washington. *J. Food Prot.* 53(4):300–01.

Kaysner, C.A., C. Abeyta Jr., M.M. Wekell, A. DePaola Jr., R.F. Stott, and J.M.

Leitch. 1990. Incidence of *Vibrio cholerae* from estuaries of the United States west coast. *Appli. Environ. Microbiol.* 53:1344–48.

Keating, M. and M.D. Sobsey. 1985. Evaluation of methods for recovery of hepatitis A virus from oysters. *Abstr. Ann. Meet. Am. Soc. Microbiol.* 85:254.

Kehr, R.W., B.S. Levine, C.T. Butterfield, and A.P. Miller. 1941. A report on the public health aspects of clamming in Raritan Bay. Public Health Service Report. Reissued in June 1954 by Division of Sanitary Engineering Services, Public Health Services, Department of Health, Education, and Welfare.

Kelland, L.R., S.H. Moss, and D.J.G. Davies. 1983. Damage to bacterial cell membranes by UV radiation in sunlight. *BioScience* 33:334–35.

Keller, R. and N. Traub. 1974. The characterization of *Bacteroides fragilis* bacteriophage recovered from animal sera; observations on the nature of *Bacteroides* phage carrier cultures. *J. Gen. Virol.* 24:179–89.

Kelly, C.B. 1961. Disinfection of seawater by ultraviolet radiation. *Am. J. Public Health.* 51:1670–80.

Kelly, C.B. 1964. Time–temperature effect on bacteriological quality of stored oysters. In *Proceedings, 5th National Shellfish Sanitation Workshop*, pp. 193–202.

Kelly, M.T. 1982. Effect of temperature and salinity on *Vibrio* (*Beneckea*) *vulnificus* occurrence in a Gulf Coast environment. *Appl. Environ. Microbiol.* 44:820–24.

Kelly, M.T. and E.M.D. Stroh. 1988. Temporal relationship of *Vibrio parahaemolyticus* in patients and the environment. *J. Clin. Microbiol.* 26:1754–56.

Kemp, H.A., D.B. Archer, and M.R.A. Morgan. 1988. Enzyme-linked immunosorbent assays for the specific and sensitive quantification of *Methanosarcina mazei* and *Methanobacterium bryantii*. *Appl. Environ. Microbiol.* 54:1003–08.

Kennedy, J.E. Jr., G. Bitton, and J.L. Oblinger. 1985. Comparison of selective media for assay of coliphages in sewage effluent and lake water. *Appl. Environ. Microbiol.* 49:33–36.

Kennedy, J.E. Jr., C.I. Wei, and J.L. Oblinger. 1986. Distribution of coliphages in various foods. *J. Food Prot.* 49:944–51.

Kenner, B.A., H.F. Clark, and P. Kabler. 1960. Fecal streptococci. II. Quantification of streptococci in feces. *Am. J. Public Health.* 50:1553–59.

Kenner, B.A. 1978. Fecal streptococcal indicators. In *Indicators of viruses in water and food*, ed. G. Berg, pp. 147–69. Ann Arbor, Mich.: Ann Arbor Science Publishers.

Kerr, R.W., B.S. Levine, C.T. Butterfield, and A.P. Miller. 1941. *A Report on the Public Health Aspects of Clamming in Raritan Bay.* Public Health Service. Reissued June 1954. Washington, D.C.: Public Health Service, Division of Sanitary Engineering.

Keswick, B.H., T.S. Satterwhite, P.C. Johnson, H.L. Dupont, S.L. Secor, J.A. Bitsura, G.W. Gary, and J.C. Hoff. 1985. Inactivation of Norwalk virus in drinking water by chlorine. *Appl. Environ. Microbiol.* 50(2):261–64.

Ketchum, B.H., J.C. Ayers, and R.F. Vaccaro. 1952. Processes contributing to the decrease of coliform bacteria in a tidal estuary. *Ecology* 33:247–58.

Keville, M.W. and G.V. Doern. 1982. Comparison of the API 20S *Streptococcus* identification system with an immuno-rheophoresis procedure and two commercial latex agglutination tests for identifying beta-hemolytic streptococci. *J. Clin. Microbiol.* 16:92–95.

Keville, M.W. and V. Doern. 1984. Evaluation of the DMS Rapid E system for the identification of clinical isolates of the family *Enterobacteriaceae*. *J. Clin. Microbiol.* 20:1010–11.

Kew, O.M., B.K. Nottay, R. Rico-Hesse, and M.A. Pallansch. 1990. Molecular epidemiology of wild poliovirus transmission. *Applied Virology Research*, Vol. 2, ed. E. Kurstak, B.G. Marusyk, F.A. Murphy, and M.H.V. Van Reyenmortel, pp. 199–221. New York: Plenum.

Khalifa, K.I., B. Werner, and R. Timperi Jr. 1986. Non-detection of enteroviruses in shellfish collected from legal shellfish beds in Massachusetts. *J. Food Prot.* 49:971–73.

Kilgen, M.B. and M.T. Cole. 1983. Recovery of poliovirus type I from artificial seawater and natural estuarine water at varying salinities using electropositive filters. *Abst. Ann. Meet. Amer. Soc. Microbiol.* p. 283.

Kilgen, M.B., M. Cole, C. Hackney, D. Sbaih, and L. Reily. 1984. Relationship between enteric viruses and bacteriological indices in Louisiana oysters and waters. *Abstr. Ann. Meet. Am. Soc. Microbiol.* 84:214.

Kilgen, M.B., M.T. Cole, and C.R. Hackney. 1988. Shellfish sanitation studies in Louisiana. *J. Shellfish Res.* 7(3):527–30.

Kilgen, M.B. 1989. Final report on the current national status of the relationships of indicators, human enteric pathogens and potential health risks within a total environmental assessment. Saltonstall-Kennedy Grant No. 37-01-79000/37500.

Kilgen, M.B. and M.T. Cole. 1991. Viruses in seafoods. In *Microbiology of Marine Food Products*, eds. D.R. Ward and C.R. Hackney, pp. 197–209. New York: Van Nostrand Reinhold.

Kilian, M. and P. Bülow. 1976. Rapid diagnosis of *Enterobacteriaceae* I. Detection of bacterial glycosidases. *Acta Path. Microbiol. Scand.* Sect. B. 84:245–51.

King, C.H., E.B. Shotts Jr., R.E. Wooley, and K.G. Porter. 1988. Survival of coliforms and bacterial pathogens within protozoa during chlorination. *Appl. Environ. Microbiol.* 54:3023–33.

King, G.M. 1984. Metabolism of trimethylamine, choline, and glycine betaine by sulfate-reducing and methanogenic bacteria in marine sediments. *Appl. Environ. Microbiol.* 48:719–25.

King, G.M. 1988. Distribution and metabolism of quaternary amines in marine sediments. In *Nitrogen Cycling in the Marine Environment*, eds. T.H. Blackburn and J. Sorenson, pp. 143–73. New York: Wiley.

King, L.A. and B.C. Parker. 1988. A simple, rapid method for enumerating total viable and metabolically active bacteria in groundwater. *Appl. Environ. Microbiol.* 54:1630–31.

Kinyoun, D. 1925. Viability of *B. typhosus* in stored shell oysters. *Public Health Rpts.* 40(17):819–23.

Kjellander, J. 1960. Enteric streptococci as indicators of fecal contamination of water. *Acta Path. Et Microbiol. Scand. suppl.* 136. 48:1–124.

Kjelleberg, S., B.A. Humphrey, and K.C. Marshall. 1983. Initial phases of starvation and activity of bacteria at surfaces. *Appl. Environ. Microbiol.* 46:978–84.

Kjelleberg, S., M. Hermansson, P. Mårdén, and G.W. Jones. 1987. The transient phase between growth and nongrowth of heterotrophic bacteria, with emphasis on the marine environment. *Ann. Rev. Microbiol.* 41:25–49.

Klein, D.A. and S. Wu. 1974. Stress: A factor considered in heterotrophic microorganism enumeration from aquatic environments. *Appl. Microbiol.* 27:429–31.

Klein, H. and D. Fung. 1976. Identifications of fecal coliforms using violet red bile agar at elevated temperatures. *J. Milk Food Technol.* 39:768–70.

Klinger, J.D. 1988. Isolation of *Listeria*: A review of procedures and future prospects. *Infection suppl.* 16:98–105.

Klipstein, F.A. and R.F. Engert. 1984. Properties of crude *Campylobacter jejuni* heat-labile enterotoxin. *Infect. Immun.* 45:314–19.

Klontz, K.C., R.V. Tauxe, W.L. Cook, W.H. Reilly, and I.K. Wachsmuth. 1987. Cholera after the consumption of raw oysters: A case report. *Annals Internal Med.* 107:846–48.

Knapp, W., K. Holubar, and G. Wicks. 1978. *Immunofluorescence and Related Staining Techniques.* New York: Elsevier.

Knight, I.T., J. DiRuggiero, and R.R. Colwell. 1990. Direct detection of enteropathogenic bacteria in estuarine water using nucleic acid probes. *Water Sci. Technol.* 22 (in press).

Knight, I.T., S. Schults, C.W. Kasper, and R.R. Colwell. 1990. Direct detection of *Salmonella* spp. in estuaries by using a DNA probe. *Appl. Environ. Microbiol.* 56:1059–66.

Knight, I.T. and R.R. Colwell. 1991. DNA probes for direct detection of enteropathogenic bacteria in water (in press).

Knight, I.T., W.H. Holben, J.M. Tiedje, and R.R. Colwell. 1991. Nucleic acid techniques for detection, identification, and enumeration of microorganisms in the environment.

Knittel, M.D., R.J. Seidler, C. Eby, and L.M. Cabe. 1977. Colonization of the botanical environment by *Klebsiella* isolates of pathogenic origin. *Appl. Environ. Microbiol.* 34:557–63.

Kobubo, Y., S. Matsushita, S.A. Kai, M. Yamada, and H. Konuma. 1978. Incidence of Escherichia coli in raw oyster and enterotoxin producibility of the isolates. *J. Food Hyg. Soc. Japan* (Shokuhin Eiseigaku Zasshi) 19(1):117–21.

Koburger, J.A. and M.L. Miller. 1984. *Plesiomonas shigelloides*: A new problem for the oyster industry? *Proceedings of the Ninth Annual Tropical and Subtropical Fisheries Conference of the Americas*, TAMU-SG-85-106, pp. 337–42.

Koburger, J.A. and M.L. Miller. 1985. Evaluation of a fluorogenic MPN procedure for determining *Escherichia coli* in oysters. *J. Food. Prot.* 48:244–45.

Koch, A., R.J. Doyle, and H.E. Kubitschek. 1976. Inactivation of membrane transport in *Escherichia coli* by near-ultraviolet light. *J. Bacteriol.* 136:140–46.

Koepfler, E.T. and H.I. Kator. 1986. Ecotoxicological effects of creosote contamination on benthic microbial populations in an estuarine environment. *Toxicity Assessment: An International Quart.* 1:465–85.

Koff, R.S. and H.S. Sear. 1967. Internal temperature of steamed clams. *New Engl. J. Med.* 276:737–39.

Koff, R.S. and L.J.D. Connelly. 1976. Failure to detect hepatitis B surface antigen (HBsAg) in certified shellfish from New England. *Am. J. Pub. Health.* 66:174–75.

Koff, R.S., G.F. Grady, T.C. Chalmers, J.W. Mosley, B.L. Swartz, and the Boston Inter-Hospital Liver Group. 1967. Viral hepatitis in a group of Boston hospitals. III. Importance of exposure to shellfish in a nonepidemic period. *New Engl. J. Med.* 276:703–10.

Kogure, K., U. Simidu, and N. Taga. 1979. A tentative direct microscopic method for counting living marine bacteria. *Can. J. Microbiol.* 25:415–20.

Kohne, D., J. Hogan, V. Jonas, E. Dean, and T.H. Adams. 1986. Novel approach for rapid and sensitive detection of microorganisms: DNA probes to rRNA. In *Microbiology*, ed. L. Leive, pp. 110–12. Washington, D.C.: American Society for Microbiology.

Koidis, P. and M.P. Doyle. 1983. Survival of *Campylobacter jejuni* in the presence of bisulfite and different atmospheres. *Eur. J. Clin. Microbiol.* 2:384–88.

Kojima, H., T. Shibayama, A. Sata, S. Suzuki, F. Ichida, and C. Hamada. 1981. Propagation of human hepatitis A in conventional cell lines. *J. Med. Virol.* 7:273–86.

Konowalchuk, J. and J.I. Speirs. 1972. Enterovirus recovery from laboratory-contaminated samples of shellfish. *Can. J. Microbiol.* 18:1023–29.

Koopman, J.S., E.A. Eckert, H.B. Greenberg, B.C. Strohm, R.E. Isaacson, and A.S. Monto. 1980. Norwalk virus enteric illness acquired by swimming exposure. *Am. J. Epidemiol.* 115:113–77.

Koprowski, H., T.W. Norton, G.A. Jervis, T.L. Nelson, D.L. Chadwick, D.J. Nelsen, and K.F. Meyer. 1956. Clinical investigations on attenuated strains of poliomyelitis virus. *J. AMA* 160:954–66.

Korolik, V., P.J. Coloe, and V. Krishnapillai. 1988. A specific DNA probe for the identification of *Campylobacter jejuni. J. Gen. Microbiol.* 134:521–29.

Korsh, L.E., B.F. Zheverzheeva, and S.G. Egorova. 1968. Concerning rapid detection of *E. coli* in water using radioisotope ^{14}C. *Gig. Sanit.* 33:375–80.

Korsh, L.E., O.I. Yurasova, A.G. Nikonova, and M.A. Motova. 1971. Use of ^{14}C for rapid *E. coli* counts in water. *Gig. Sanit.* 33:375–80.

Kory, M.M. and S.J. Booth. 1986. Characteristics of *Bacteroides fragilis* bacteriophages and comparison of their DNAs. *Curr. Microbiol.* 14:199–203.

Korychka-Dahl, M. and T. Richardson. 1978. Photogeneration of superoxide anion in serum of bovine milk and in model systems containing riboflavin and amino acids. *J. Dairy Sci.* 61:400–07.

Kostenbader, K.D. and D.O. Cliver. 1972. Polyelectrolyte flocculation as an aid to recovery of enteroviruses from oysters. *Appl. Microbiol.* 24:540–43.

Kostenbader, K.D. and D.O. Cliver. 1973. Filtration methods for recovering enteroviruses from foods. *Appl. Microbiol.* 26:149–54.

Kostenbader, K.D. and D.O. Cliver. 1981. Flocculants for recovery of food-borne viruses. *Appl. Environ. Microbiol.* 41:318–20.

Kott, Y., N. Roze, S. Sperber, and N. Betzer. 1974. Bacteriophages as viral pollution indicators. *Water Res.* 8:165–171.

Kott, Y. 1981. Viruses and bacteriophages. *Sci. Total Environ.* 18:13–23.

Krahn, M.M., C.A. Wigren, L.K. Moore, and D.W. Brown. 1989. High-performance liquid chromatographic method for isolating coprostanol from sediment extracts. *J. Chrom.* 481:263–73.

Kramer, J. and R. Gilbert. 1989. *Bacillus cereus* and Other *Bacillus* species. In *Foodborne Bacterial Pathogens*, ed. M.P. Doyle, pp. 22–70. New York: Marcel Dekker.

Krauss, J., W. Ludwig, and K.H. Schleifer. 1986. A cloned 23S rRNA gene fragment of *Bacillus subtilis* and its use as a hybridization probe of conserved character. *FEMS Microbiol. Lett.* 33:89–93.

Kreig, N.R. and J.G. Holt. 1984. *Bergey's Manual of Systematic Bacteriology*, Vol. 1, Baltimore, Md.: Williams and Wilkens.

Krinsky, N. 1976. Cellular damage initiated by visible light. In *The Survival of Vegetative Microbes. 26th Symposium of the Soc. for Gen. Microbiology*, eds. T. Gray and J. Postgate, pp. 209–39. London, England: Cambridge Press.

Krishman, C., V.A. Fitzgerald, S.J. Dakin, and R.J. Behme. 1987. Laboratory investigation of outbreaks of hemorrhagic colitis caused by *Escherichia coli* O157:H7. *J. Clin. Microbiol.* 25:1043–47.

Kuijper, E.J., J.H. Oudbier, W.N.H.M. Stuifbergen, A. Jansx, and H.C. Zanen. 1987. Application of whole-cell DNA restriction endonuclease profiles to the epidemiology of *Clostridium difficile*-induced diarrhea. *J. Clin. Microbiol.* 25:751–53.

Kuritza, A.P. and A.A. Salyers. 1985. Use of a species-specific DNA hybridization probe for enumerating *Bacteroides vulgatus* in human feces. *Appl. Environ. Microbiol.* 50:958–64.

Kuritza, A.P., C.E. Getty, P. Shaughnessy, R. Hess, and A.A. Salyers. 1986. DNA probes for identification of clinically important *Bacteroides* species. *J. Clin. Microbiol.* 23:343–49.

Kuritza, A.P., P. Shaughnessy, and A.A. Salyers. 1986. Enumeration of polysaccharide-degrading *Bacteroides* species in human feces by using species-specific DNA probes. *Appl. Environ. Microbiol.* 51:385–90.

Kurtz, B. and T.W. Lee. 1987. Astroviruses: Human animal. In *Novel Diarrhoea Viruses*, eds. G. Bock and J. Whelan, Foundation Symp. No. 128, pp. 92–107. Chichester, U.K.: John Wiley.

LaBelle, R.L. and C.P. Gerba. 1979. Influence of pH, salinity, and organic matter on the adsorption of enteric viruses to estuarine sediment. *Appl. Environ. Microbiol.* 38:93–101.

LaBelle, R.Y., C.P. Gerba, S.M. Goyal, J.L. Melnick, I. Cech, and G.F. Bogdan. 1980. Relationships between environmental factors, bacterial indicators, and the occurrence of enteric viruses in estuarine sediments. *Appl. Environ. Microbiol.* 39:588–96.

Labbe, R. 1989. *Clostridium perfringens*. In *Foodborne Bacterial Pathogens*, ed. M.P. Doyle, p. 191–234. New York: Marcel Dekker.

Lachia, R.V. 1984. Egg yolk-free Baird–Parker medium for the accelerated enumeration of foodborne *Staphylococcus aureus*. *Appl. Environ. Microbiol.* 48:870–71.

Lamont, R.J., R. Postelthhwaite, and A.P. MacGowan. 1988. *Listeria monocytogenes* and its role in human infection. *J. Infect.* 17:7–28.

Lampel, K., J.A. Jagow, M. Trucksess, and W.E. Hill. 1990. Polymerase chain reaction for detection of invasive *Shigella flexneri* in food. *Appl. Environ. Microbiol.* 56:1536–40.

Lanata, C.F., J.B. Kaper, M.M. Baldini, R.E. Black, and M.M. Levine. 1985. Sensitivity and specificity of DNA probes with the stool blot technique for detection of *E. coli* enterotoxins. *J. Infect. Dis.* 152:1087–90.

Landry, E.F., J.M. Vaughn, and T.J. Vicale. 1980. Modified procedure for extraction of poliovirus from naturally-infected oysters using Cat-Floc and beef extract. *J. Food Prot.* 43:91–94.

Landry, E.F., J.M. Vaughn, T.J. Vicale, and R. Mann. 1982. Inefficient accumulation of low levels of monodispersed and feces-associated polioviruses in oysters. *Appl. Environ. Microbiol.* 44:1362–69.

Landry, E.F., J.M. Vaughn, T.J. Vicale, and R. Mann. 1983. Accumulation of sediment-associated viruses in shellfish. *Appl. Environ. Microbiol.* 45:238–47.

Lange, R. and R. Hengge-Aronis. 1991. Identification of a central regulator of stationary phase gene expression in *Escherichia coli. Mol. Microbiol.* 5:49–59.

Langenberg, W., E.A.J. Rauws, A. Widjojokusumo, G.N.J. Tytgat, and H.C. Zanen. 1986. Identification of *Campylobacter pyloridis* isolates by restriction endonuclease DNA analysis. *J. Clin. Microbiol.* 24:414–17.

Langer, P.R., A.A. Waldrop, and D.C. Ward. 1981. Enzymatic synthesis of biotin-labeled polynucleotides: Novel nucleic acid affinity probes. *Proc. Natl. Acad. Sci.* (USA) 78:6633–37.

Lanz, W.W. and P.A. Hartman. 1976. Timed-release capsule method for coliform enumeration. *Appl. Environ. Microbiol.* 32:716–22.

Lapeyre, C., F. Janin, and S. Kaveri. 1988. Indirect soluble sandwich ELISA using monoclonal antibodies for detection of staphylococcal enterotoxins A, B, C_1, and D in food samples. *Food Microbiol.* 5:25–31.

Larkin, E.P. and T.G. Metcalf. 1980. Cooperative study of methods for the recovery of enteric viruses from shellfish. *J. Food Prot.* 43:84–86.

Larkin, E.P. and D.A. Hunt. 1982. Bivalve mollusks: Control of microbiological contaminants. *Bioscience* 32(3):193–97.

Lassen, J. 1975. Rapid identification of gram-negative rods using a three-tube method combined with a dichotomic key. *Acta Path. Microbiol. Scand.* Sect. **B** 83:525–33.

Le Rudulier, D. and L. Bouillard. 1983. Glycine betaine, an osmotic effector in *Klebsiella pneumoniae* and other members of the *Enterobacteriaceae. Appl. Environ. Microbiol.* 46:152–59.

LeChevallier, M.W., S.C. Cameron, and G.A. McFeters. 1983. New medium for improved recovery of coliform bacteria from drinking water. *Appl. Environ. Microbiol.* 45:484–92.

LeChevallier, M.S., A.J. Singh, D.A. Schiemann, and G.A. McFeters. 1985. Changes in virulence of waterborne pathogens with chlorine injury. *Appl. Environ. Microbiol.* 50:412–19.

Lear, Donald W. 1962. Growth of coliform bacteria in estuarine waters. *Chesapeake Sci.* 3(3):160–65.

Leary, J.J., D.J. Brigati, and D.C. Ward. 1983. Rapid and sensitive colorimetric method for visualizing biotin-labeled DNA probes hybridized to DNA or RNA immobilized on nitrocellulose bio-blots. *Proc. Natl. Acad. Sci* (USA) 80:4045–49.

Leder, I.G. 1972. Interrelated effects of cold shock and osmotic pressure on the permeability of the *Escherichia coli* membrane to permease accumulated substrates. *J. Bacteriol.* 111:211–19.

Lee, A., J.L. O'Rourke, P.J. Barrington, and T.J. Trust. 1986. Mucus colonization as a determinant of pathogenicity in intestinal infection by *Campylobacter jejuni*: A mouse cecal model. *Infect. Immun.* 51:536–46.

Lee, R.M. and P.A. Hartman. 1989. Optimal pyruvate concentrations for recovery of coliforms from food and water. *J. Food Prot.* 52:119–21.

Lee, W.H. 1977. An assessment of *Yersinia enterocolitica* and its presence in foods. *J. Food Prot.* 40:486–89.

Lee, W.H. and D. McClain. 1986. Improved *Listeria monocytogenes* selective agar. *Appl. Environ. Microbiol.* 52:1215–17.

Lefkowitz, A., S. Fout, G. Losonsky, E. Israel, and J.G. Morris Jr. 1991. A serosurvey of pathogens associated with shellfish: Prevalence of antibodies to Vibrio species and Norwalk agent in the Chesapeake Bay region (submitted for publication).

Lemon, S.M., L.N. Binn, and R.H. Marchwicki. 1983. Radioimmunofocus assay for quantitation of hepatitis A virus in cell cultures. *J. Clin. Microbiol.* 17:834–39.

Lenhard, G. 1969. Determination of urease activity in biological purification systems. *Hydrobiologia* 33:193–200.

Lenhard, G. 1967. Determination of protease activity in bottom deposits of sewage stabilization ponds. *Hydrobiologia* 27:67–79.

Lennette, E.H. and N.L. Schmidt eds. 1979. *Diagnostic Procedures for Viral, Rickettsial, and Chlamydial Infections*, 5th ed. Washington, D.C.: American Public Health Association.

Lennon, D., B. Lewis, C. Mantell, D. Becroft, B. Dove, K. Farmer, S. Tonkin, N. Yeates, R. Stamp, and K. Mickleson. 1984. Epidemic perinatal listerioses. *Pediatrics Infect. Dis.* 3(1):30–34.

Leonard, D.L., H.B. Marlene, and E.H. Kristen. 1989. *The Quality of Shellfish Growing Waters on the East Coast of the United States*. Rockville, Md.: National Oceanic and Atmospheric Administration.

Leong, L.Y.C., S.J. Barrett, and R.R. Trussell. 1978. False positives in testing of secondary sewage for enteric viruses. *Abstr. Ann. Meet. Am. Soc. Microbiol.* 78:200.

Lepow, M.L., R.J. Warren, V.G. Ingram, S.C. Daugherty, and F.C. Robbins. 1962. Sabin Type I (LSc2ab) oral poliomyelitis vaccine—effect of dose upon response of newborn infants. *Am. J. Dis. Child.* 104:99–103.

Lessard, E.J. and J.McN. Sieburth. 1983. Survival of natural sewage populations of enteric bacteria in diffusion and batch chambers in the marine environment. *Appl. Environ. Microbiol.* 45:950–59.

Levin, G.V., V.R. Harrison, and W.C. Hess. 1956. Preliminary report on a one-hour presumptive test for coliform organisms. *J. A. Water Works Assoc.* 48:75–80.

Levin, G.V., V.R. Harrison, W.C. Hess, and H.C. Gurney. 1956. A radioisotope-technic for rapid detection of coliform organisms. *Am. J. Public Health* 46:1405–14.

Levin, G.V., V.R. Harrison, and W.C. Hess. 1957. Use of radioactive culture media. *J. Am. Water Works Assoc.* 49:1069–76.

Levin, G.V., V.R. Harrison, W.C. Hess, A.H. Heim, and V.L. Strauss. 1959. Rapid radioactive test for coliform organisms. *J. Am. Water Works Assoc.* 51:101–04.

Levin, G.V., V.L. Stauss, and W.C. Hess. 1961. Rapid coliform organisms determination with ^{14}C. *J. Water Pollut. Control Fed.* 33:1021–37.

Levin, M.A., J.R. Fischer, and V.J. Cabelli. 1975. Membrane filter technique for enumeration of enterococci in marine waters. *Appl. Microbiol.* 30:66–71.

Levin, M.A. 1977. Bifidobacteria as water quality indicators. In *Bacterial indicators/Health Hazards Associated With Water*, eds. A.W. Hoadley and B.J. Dutka, pp. 131–38. Philadelphia, Pa.: Special Technical Publication 635, American Society for Testing and Materials.

Levin, M.A. and I.G. Resnick. 1981. *Bifidobacterium.* In *Membrane Filtration: Applications, Techniques, and Problems*, ed. B.J. Dutka, pp. 129–59. New York: Marcel Dekker.

Levine, M.M., H.L. DuPont, S.B. Formal, R.B. Nornick, A. Takeuchi, E.J. Gangarosa, M.J. Snyder, and J.P. Libonati. 1973. Pathogenesis of *Shigella dysenteriae* 1 (Shiga) dysentery. *J. Infect. Dis.* 127:261–70.

Levine M.M., E. Bergquist, D. Nalin, D. Waterman, R. Hornick, C. Young, S. Sotman, and B. Rowe. 1978. *Escherichia coli* strains that cause diarrhea but do not produce heat-labile or heat-stable enterotoxins and are non-invasive. *Lancet* i:1119–22.

Levine, M.M., J.G. Xu, J.B. Kaper, H. Lior, V. Prado, B. Tall, J. Nataro, H. Karch, and K. Wachsmuth. 1987. A DNA probe to identify enterohemorrhagic *Escherichia coli* O157:H7 and other serotypes that cause hemorrhagic colitis and hemolytic uremic syndrome. *J. Infect. Dis.* 156:175–82.

Levine, M.M. 1987. *Escherichia coli.* that cause diarrhea: Enterotoxigenic, enteropathogenic, enteroinvasive, enterohemorrhagic, and enteroadherent. *J. Infect. Dis.* 155:377–89.

Lewis, A.M. and B. Chattopadhyay. 1986. Faecal carriage rate of Yersinia species. *J. Hyg.* Cambridge 97:281–87.

Lewis, C. and J.L. Mak. 1989. Comparison of membrane filtration and Autoanalysis Colilert presence-absence techniques for analysis of total coliforms and *Escherichia coli* in drinking water samples. *Appl. Environ. Microbiol.* 55:3091–94.

Ley, A.N., R.J. Bowers, and W. Wolfe. 1988. Indoxyl-b-D-glucuronide, a novel chromogenic reagent for the specific detection and enumeration of *Escherichia coli* in environmental samples. *Can. J. Microbiol.* 34:690–93.

Li, W.K.W. and P.M. Dickie. 1985. Growth of bacteria in seawater filtered through 0.2 μm Nucleopore membranes: Implications for dilution experiments. *Mar. Ecol. Prog. Ser.* 26:245–52.

Licciardell, J.J., D. Entremont, and R. Lundstrom. 1989. Radio-resistance of some bacterial pathogens in soft-shell clams (*Mya arenaria*) and mussels (*Mytilus edulis*). *J. Food Prot.* 52(6):407–11.

Liebman, H.L., G. Kudo, J.G. Chapel, and J.A. Stern. 1957. Effects of temperature upon the storage life of fresh shucked pacific oysters (*Ostrea gigas*). *Commercial Fisheries Rev.* 19(12):1–4.

Lin, F., J.G. Morris, J.B. Kaper, T. Gross, J. Michalski, C. Morrison, J.P. Libonati, and E. Israel. 1986. Persistence of cholera in the United States: Isolation of *Vibrio cholerae* 01 from a patient with diarrhea in Maryland. *J. Clin. Microbiol.* 23:624–26.

Linco, S.J. and G.S. Grohmann. 1980. The Darwin outbreak of oyster-associated viral gastroenteritis. *Med. J. Australia* 1:211–13.

Linder, K. and J.D. Oliver. 1989. Membrane fatty acid and virulence changes in the viable but nonculturable state of *Vibrio vulnificus*. *Appl. Environ. Microbiol.* 55:2837–42.

Liston, J. and J. Baross. 1973. Distribution of *Vibrio parahaemolyticus* in the natural environment. *J. Milk Food Technol.* 36:113–17.

Liston, J. 1990. Microbial hazards of seafood consumption. *Food Technol.* 44(12):56–62.

Littel, K.J. and P.A. Hartman. 1983. Fluorogenic selective and differential medium for isolation of fecal streptococci. *Appl. Environ. Microbiol.* 45:622–27.

Liu, O.C., H.R. Seraichekas, and B.L. Murphy. 1966. Fate of poliovirus in Northern quahogs. *Proc. Soc. Expt. Biol. Med.* 121:601–07.

Liu, O.C., H.R. Seraichekas, and B.L. Murphy. 1966. Viral pollution of shellfish I. Some basic facts of uptake. *Proc. Soc. Expt. Biol. Med.* 123:481–87.

Liu, O.C., H.R. Seraichekas, and B.L. Murphy. 1967. In *Transmission of Viruses by the Water Route*, ed. G. Berg, pp. 419–37. New York: Interscience.

Liu, O.C., H.R. Seraichekas, and B.L. Murphy. 1967. Viral depuration of the Northern quahog. *Appl. Microbiol.* 15:307–15.

Liu, O.C., H.R. Seraichekas, D.A. Brashear, W.P. Hefferman, and V.J. Cabelli. 1968. The occurrence of human enteric viruses in estuaries and shellfish. *Bact. Proc.* 68:151.

Lo, S., J. Gilbert, and F. Hetrick. 1976. Stability of human enteroviruses in estuarine and marine waters. *Appl. Environ. Microbiol.* 32:245–49.

Loesch, W.J. 1969. Oxygen sensitivity of various anaerobic bacteria. *Appl. Microbiol.* 18:723–27.

Logan, K.B., G.E. Rees, N.D. Seeley, and S.D. Primrose. 1980. Rapid concentration of bacteriophages from large volumes of freshwater: Evaluation of positively charged, microporous filters. *J. Virol. Methods.* 1:87–97.

Loh, P.C., M.A. Dow, and R.S. Fujioka. 1985. Use of the nitrocellulose-enzyme immunosorbent assay for rapid, sensitive and quantitative detection of human enteroviruses. *J. Virol. Meth.* 12:225–34.

Lopez-Torres, A.J., T.C. Hazen, and G.A. Toranzos. 1987. Distribution and in situ survival and activity of *Klebsiella pneumoniae* and *Escherichia coli* in a tropical rain forest watershed. *Curr. Microbiol.* 15:213–18.

Lopez-Vidal, Y. and A.-M. Svennerholm. 1986. Hemadsorption and enzyme-linked immunosorbent assay nitrocellulose replica methods for identification of colonization factor antigen (CFA)-positive *Escherichia coli* colonies and for isolation of CFA-negative mutants. *J. Clin. Microbiol.* 24:615–19.

Lovett, J., J. Bradshaw, and T. Peeler. 1982. Thermal inactivation of *Yersinia enterocolitica* in milk. *Appl. Environ. Microbiol.* 44:517–20.

Lovett, J. 1987. *Listeria* isolation. In *Bacteriological Analytical Manual*, 6th ed., Supplement 1. Arlington, Va.: Association of Official Analytical Chemists.

Lovett, J., D.W. Francis, and J.M. Hunt. 1987. *Listeria monocytogenes* in raw milk: Detection, incidence and pathogenicity. *J. Food Prot.* 50:188–92.

Lovett, J. 1988. Isolation and enumeration of *Listeria monocytogenes*. *Food Technol.* 42:172–75.

Lovett, J. 1989. *Listeria monocytogenes*. In *Foodborne Bacterial Pathogens*, ed. M.P. Doyle, pp. 283–310. New York: Marcel Dekker.

Lowry, P.W., A.T. Pavia, L.M. McFarland, B.H. Peltier, T.J. Barrett, H.B. Bradford, J.M. Quan, J. Lynch, J.B. Mathison, R.A. Gunn, and P.A. Blake. 1989*a*. Cholera in Louisiana—Widening spectrum of seafood vehicles. *Arch. Intern. Med.* 149:2079–84.

Lowry, P.W., L.M. McFarland, B.H. Peltier, N.C. Roberts, H.B. Bradford, J.L. Herndon, D.F. Stroup, J.B. Mathison, P.A. Blake, and R.A. Gunn. 1989*b*. *Vibrio* gastroenteritis in Louisiana: A prospective study among attendees of a scientific congress in New Orleans. *J. Infect. Dis.* 160:978–84.

Luckman, S. 1990. Maryland Department of the Environment, Water Management Administration. Baltimore, Md. (personal communication).

Ludwig, W., E. Seewaldt, A. Kilpper-Balz, K.H. Schleifer, L. Magrum, C.R. Woese, G.E. Fox, and E. Stackebrandt. 1985. The phylogenetic position of *Streptococcus* and *Enterococcus*. *J. Gen. Microbiol.* 131:543–51.

Luechtefeld, N.W., W. Wang, M.J. Blaser, and L.B. Reller. 1981. Evaluation of transport and storage techniques for isolation of *Campylobacter fetus* subsp. *jejuni* from turkey cecal specimens. *J. Clin. Microbiol.* 13:438–43.

Lum, R. and G. Chang. 1990. Glucuronidase-negative *Escherichia coli* in the ECOR reference collection. *J. Food Prot.* 53:972–74.

Lumsden, L.L. et al. 1925. A typhoid fever epidemic caused by oyster-borne infection. *Public Health Reports*. Supplement No. 50. Washington, D.C.: U.S. Government Printing Office.

MacDonald, K.L., M.J. O'Leary, M.L. Cohen, P. Norris, J.G. Wells, E. Noll, J.M. Kobayashi, and P.A. Blake. 1988. *Escherichia coli* O157: H7, an emerging gastrointestinal pathogen. *J. Am. Med. Assoc.* 259:3567–70.

MacDonell, M.T. and M.A. Hood. 1984. Ultramicrovibrios in Gulf Coast estuarine waters: Isolation, characterization and incidence. In *Vibrios in the Environment*, ed. R.R. Colwell, pp. 551–62. New York: Wiley.

Mackey, B.M. and C.M. Derrick. 1986. Peroxide sensitivity of cold-shocked *Salmonella typhimurium* and *Escherichia coli* and its relationship to minimal medium recovery. *J. Appl. Bacteriol.* 60:501–11.

Mackey, B.M., C. Pritchet, A. Norris, and G.C. Mead. 1990. Heat resistance of *Listeria*: Strain differences and effects of meat type and curing salts. *Lett. Appl. Microbiol.* 10:251–55.

Mackowiak, P.A., C.T. Caraway, and B.L. Portnoy. 1976. Oyster-associated hepatitis: Lessons from the Louisiana experience. *Am. J. Epidemiol.* 103:181–91.

Macy, J.M. 1981. Nonpathogenic members of the genus *Bacteroides*. In *The Prokaryotes*, eds. M.P. Starr, H. Stolp, H.G. Truper, A. Balows, and H.G. Schlegel, Vol. II, pp. 1450–63. Berlin: Springer.

Madden, J.M., B.A. McCardell, and J.G. Morris. 1989. *Vibrio cholerae*. In *Foodborne Bacterial Pathogens*, ed. M.P. Doyle, pp. 525–42. New York: Marcel Dekker.

Madden, R.H., H. Buller, and D.W. McDowell. 1986. *Clostridium perfringens* as an indicator of hygienic quality of depurated shellfish. *J. Food Prot.* 49:33–36.

Madeley, C.R. 1989. Epidemiology of gut viruses. In *Viruses and the Gut*, ed. M.J.G. Farthing, pp. 5–15. Welwyn Garden City, England: Smith Kline and French Laboratories.

Madore, H.P., J.J. Treanor, K.A. Pray, and R. Dolin. 1986. Enzyme-linked immunosorbent assays for Snow Mountain and Norwalk agents of viral gastroenteritis. *J. Clin. Microbiol.* 24:456–59.

Mahoney, P., G. Fleischner, J. Millman, W.T. London, B.S. Blumberg, and I.M. Arias. 1974. Australian antigen: Detection and transmission in shellfish. *Science* 183:80–81.

Maki, J.S. and C.C. Remsen. 1981. Comparison of two direct count methods for determining metabolizing bacteria in freshwater. *Appl. Microbiol.* 41:1132–38.

Mandal, B.K. and V. Mani. 1976. Colonic involvement in salmonellosis. *Lancet* i:887–88.

Mandal, B.K., P. DeMol, and J.P. Butzler. 1984. Clinical aspects of *Campylobacter* infections in humans. In *Campylobacter Infection in Man and Animals*, ed. J.P. Butzer. Boca Raton, Fla.: CRC Press.

Mara, D.D. and J.I. Oragui. 1981. Occurrence of *Rhodococcus coprophilus* and associated actinomycetes in feces, sewage and freshwater. *Appl. Environ. Microbiol.* 42:1037–42.

Mara, D.D. and J. Oragui. 1983. Sorbitol-fermenting bifidobacteria as specific indicators of human faecal pollution. *J. Appl. Bacteriol.* 55:349–57.

Marchand, M. 1984. Etude ecologique des vibrions du Bassin D'Arcachon. In *Actes de colloques* INFREMER, pp. 483–89.

Margolin, A.B., M.J. Hewlett, and C.P. Gerba. 1986. Use of a cDNA dot–blot hybridization technique for detection of enteroviruses in water. In *Advances in Water Analysis and Treatment*. Houston, Tx.: AWWA Proceedings.

Margolin, A.B., K.J. Richardson, R. DeLeon, and C.P. Gerba. 1989. Application of gene probes to the detection of enteroviruses in water. In *Biohazards of Drinking Water Treatment*, ed. R.A. Larson. Chelsea, Mich.: Lewis Publishers.

Marshall, K.C. 1985. Mechanisms of bacterial adhesion at solid-water interfaces. In *Bacterial Adhesion. Mechanisms and Physiological Significance*, eds. D.C. Savage and M. Fletcher, pp. 133–61. New York: Plenum Press.

Marth, E.H. 1988. Disease characteristics of *Listeria monocytogenes*. *Food Technol.* 42:165–68.

Martin, R.E. and G.T. Pitts. NFI Handbook of State and Federal Microbiological Standards and Guidelines.

Martin, S.E., R.S. Flowers, and C.J. Ordal. 1976. Catalase: Its effect on microbial enumeration. *Appl. Environ. Microbiol.* 32:731–34.

Martinez, J., J. Garcia-Lara, and J. Vives-Rego. 1989. Estimation of *Escherichia coli* mortality in seawater by the decrease in ^{3}H-label and electron transport system activity. *Microb. Ecol.* 17:219–25.

Maryland Board of Health. 1925–28. *Memoranda, Oyster I and II*. Baltimore, Md.: Maryland Department of Health and Mental Hygiene.

Mason, J.O. and W.R. McLean. 1962. Infectious hepatitis traced to the consumption of raw oysters. *Am. J. Hyg.* 75(1):1–22.

Massad, G. and J.D. Oliver. 1987. New selective and differential medium for *Vibrio cholerae* and *Vibrio vulnificus*. *Appl. Environ. Microbiol.* 53:2262–64.

Mastrorocco, D. and W. Andrews. 1989. Staphylococcal enterotoxins. *J. Assoc. Off. Anal. Chem.* 72:102, 104.

Matches, J.R. and J. Liston. 1968*a*. Growth of salmonella on irradiated and non-irradiated seafoods. *J. Food Sci.* 33(4):406–10.

Matches, J.R. and J. Liston. 1968*b*. Low Temperature Growth of Salmonella. *J. Food Sci.* 33(6):641–45.

Matches, J.R. and J. Liston. 1974. Mesophilic clostridia in Puget Sound. *Can. J. Microbiol.* 20:1–7.

Matches, J.R. and C. Abeyta. 1983. Indicator organisms in fish and shellfish. *J. Food Prot.* 37(6):114–17.

Matev, I., Z. Dimitrova, and V. Stefanov. 1987. Survival of some pathogenic microorganisms in Black Sea mussels after experimental infection. Vissh Inst. po Zootekhnika i Vet. Med., Sofia, Bulgaria. *Veterinarnomeditsinski Nauki* 24:87–90.

Matin, A., E.A. Auger, P.H. Blum, and J.E. Schultz. 1989. Genetic basis of starvation survival in nondifferentiating bacteria. *Ann. Rev. Microbiol.* 43:293–316.

Matin, A. 1990. Molecular analysis of the starvation stress in *Escherichia coli*. *FEMS Microbiol. Ecol.* 74:185–95.

Matson, E.A., S.G. Hornor, and J.D. Buck. 1978. Pollution indicators and other microorganisms in river sediment. *J. Water Pollut. Control Fed.* 50:13–19.

Matthews, J. and L.J. Kricka. 1988. Analytical strategies for the use of DNA probes. *Anal. Biochem.* 169:1–25.

Matthews, J.A., A. Batki, C. Hynds, and L.J. Kricka. 1985. Enhanced chemiluminescent method for the detection of DNA dot-hybridization assays. *Anal. Biochem.* 151:205–09.

Matthews, W.S. and L.L. Smith. 1968. Sterol metabolism-III. Sterols of marine waters. *Lipids* 3:239–46.

Maxcy, K.F. 1956. Typhoid Fever. In *Rosenau Preventive Medicine and Public Health*, ed. K.F. Maxcy, pp. 180–86. New York: Appleton–Century–Crofts.

Mayer, L.W. 1988. Use of plasmid profiles in epidemiologic surveillance of disease outbreaks and in tracing the transmission of antibiotic resistance. *Clin. Microbiol. Rev.* 1:228–43.

Mazaitis, A.J., R. Maas, and W.K. Maas. 1981. Structure of a naturally occurring plasmid with genes for enterotoxin production and drug resistance. *J. Bacteriol.* 145:97–105.

McCambridge, J. and T.A. McMeekin. 1979. Protozoan predation of *Escherichia coli* in estuarine waters. *Water Res.* 13:659–63.

McCambridge, J. and T.A. McMeekin. 1980. Relative effects of bacterial and protozoan predators on survival of *Escherichia coli* in estuarine water samples. *Appl. Environ. Microbiol.* 40:907–11.

McCambridge, J. and T.A. McMeekin. 1981. Effect of solar radiation and predacious microorganisms on survival of fecal coliforms and other bacteria. *Appl. Environ. Microbiol.* 41:1083–87.

McCardell, B.A., J.M. Madden, and E.C. Lee. 1984. *Campylobacter jejuni* and *Campylobacter coli* production of a cytotonic toxin immunologically similar to cholera toxin. *J. Food Prot.* 47:943–49.

McClain, D. and W.H. Lee. 1988. Development of USDA–FSIS method for isolation of *Listeria monocytogenes* from raw meat and poultry. *J. Assoc. Anal. Chem.* 71:660–64.

McCormack, W.M., M.S. Islam, M. Fahimuddin, and W.H. Mosley. 1969. Endemic cholera in rural East Pakistan. *Am. J. Epidemiol.* 89:393–404.

McCrady, M.H. 1915. The numerical interpretation of fermentation-tube results. *J. Infect. Dis.* 17:183–212.

McDonald, L.C., C.R. Hackney, and B. Ray. 1983. Enhanced recovery of injured *Escherichia coli* by compounds that degrade hydrogen peroxide or block its formation. *Appl. Environ. Microbiol.* 45:360–65.

McFeters, G. 1989. Detection of injured indicator and pathogenic bacteria in water. In *Injured Index and Pathogenic Bacteria*, ed. B. Ray, pp. 179–216. Boca Raton, Fla.: CRC Press.

McFeters, G.A., G.K. Bissonnette, J.J. Jezeski, C.A. Thomson, and D.G. Stuart. 1974. Comparative survival of indicator bacteria and enteric pathogens in well water. *Appl. Microbiol.* 27:823–29.

McFeters, G.A., S.C. Cameron, and M.W. LeChevallier. 1982. Influence of diluents, media, and membrane filters on the detection of injured waterborne coliform bacteria. *Appl. Environ. Microbiol.* 43:97–103.

McInnes, J.L. and R.H. Symons. 1989*a*. Enzymatic and chemical techniques for labeling nucleic acids with radioisotopes. In *Nucleic Acid Probes*, ed. R.H. Symons, pp. 1–31. Boca Raton, Fla.: CRC Press.

McInnes, J.L. and R.H. Symons. 1989*b*. Preparation and detection of nonradioactive nucleic acid and oligonucleotide probes. In *Nucleic Acid Probes*, ed. R.H. Symons, pp. 33–80. Boca Raton, Fla.: CRC Press.

McLaughlin, J. 1987. *Listeria monocytogenes*, recent advances in the taxonomy and epidemiology of listeriosis in humans. *J. Appl. Bacteriol.* 63:1–11.

McLaughlin, J.K., B.D. Zuckerman, S. Tenenbaum, and B.A. Wolf. 1984. Comparison of the API 20E, Flow, and Minitek systems for the identification of enteric and nonfermentative bacteria isolated from cosmetic raw materials. *J. Soc. Cosmet. Chem.* 35:253–63.

Means, E.G. and B.H. Olson. 1981. Coliform inhibition by bacteriocin-like substances in drinking water distribution systems. *Appl. Environ. Microbiol.* 42:506–12.

Medallion Laboratories. 1987. Food microbiology—examining the greater risk. *Anal. Prog.* 4(1).

Mehlman, I. and A. Romero. 1982. Enteropathogenic *Escherichia coli*: Methods for recovery form foods. *Food Technol.* 36:73.

Mehlman, I.J., N.T. Simon, A.C. Sanders, M. Fishbein, J.C. Olson Jr., and R.B. Read Jr. 1975. Methodology for enteropathogenic *Escherichia coli. J. Assoc. Off. Anal. Chem.* 58:283–92.

Meinhold, A.F. 1982. Uptake, elimination, and tissue distribution of poliovirus in the American oyster. Masters Thesis. Department of Environmental Sciences and Engineering. Chapel Hill, N.C.: University of North Carolina.

Mele, A., M.G. Rastelli, O.N. Gill, D. DiBisceglie, F. Rosmini, G. Pardelli, C. Valtriani, and P. Patriarchi. 1989. Recurrent epidemic hepatitis A associated with consumption of raw shellfish, probably controlled through public health measures. *Am. J. Epidemiol.* 130:540–46.

Melnick, J.L., C.P. Gerba, and C. Wallis. 1978. Viruses in water. Update Lepoint. *Bull. WHO* 56(4):499–508.

Melnick, J.L. and C.P. Gerba. 1980. The ecology of enteroviruses in natural waters. *CRC Crit. Rev. Environ. Control.* 10:65.

Mesquita, M.M.F. 1988. Effects of seawater contamination level and exposure period on bacterial and viral accumulation and elimination processes by Mytilis edulis. *Water Sci. Technol.* 20:265–70.

Messer, R. and G.M. Reece. 1937. Progress in oyster conditioning. *Public Health Rpts.* 52(42):1451–60.

Metcalf, T.G. and W.C. Stiles. 1965. The accumulation of enteric viruses by the oyster Crassostrea virginica. *J. Infect. Dis.* 115:68–76.

Metcalf, T.G. and W.C. Stiles. 1967. Survival of enteric viruses in estuary waters and shelfish. In *Transmission of Viruses by the Water Route*, ed. G. Berg, pp. 439–47. New York: Interscience.

Metcalf, T.G. and W.C. Stiles. 1968. Enteroviruses within an estuarine environment. *Am. J. Epidemiol.* 88:379–91.

Metcalf, T.G. and W. Stiles. 1968. Viral pollution of shellfish in estuary waters. *Proc. Am. Soc. Civ. Eng.* SA4:595–609.

Metcalf, T.G., J.M. Vaughn, and W.C. Stiles. 1972. Virus enumeration and public health assessments in polluted surface water contributing to transmission of virus in nature. In *Virus Survival in Water and Wastewater Systems*, eds. J.F. Malina and B.P. Sagik, pp. 57–70. Austin, Tx.: Center for Research in Water Resources.

Metcalf, T.G., L.W. Slanetz, and C.H. Bartley. 1973. Enteric pathogens in estuary waters and shellfish. In *Microbial Safety of Fishery Products*, eds. C.O. Chichester and H.D. Graham, pp. 215–34. New York: Academic Press.

Metcalf, T.G., B. Mullin, D. Eckerson, E. Moulton, and E.P. Larkin. 1979. Bioaccumulation and depuration of enteroviruses by the soft shelled clam, *Mya arenaria. Appl. Environ. Microbiol.* 38:275–82.

Metcalf, T.G. 1979. Indication of viruses in shellfish growing waters. *Am. J. Public Health* 69(11):1093–94.

Metcalf, T.G., E. Moulton, and D. Eckerson. 1980*a*. Improved method and test strategy for recovery of enteric viruses from shellfish. *Appl. Environ. Microbiol.* 39:141–52.

Metcalf, T.G., D. Eckerson, and E. Moulton. 1980*b*. A method for recovery of viruses from oysters and hard and soft shell clams. *J. Food Prot.* 43:89–90.

Metcalf, T.G., D. Eckerson, E. Moulton, and E.P. Larkin. 1980. Uptake and depletion of particulate-associated polioviruses by the soft shell clam. *J. Food Prot.* 43:87–88.

Metcalf, T.G. and X. Jiang. 1988. Detection of hepatitis A virus in estuarine samples by gene probe assay. *Microbiol. Sci.* 5:296–300.

Meyer-Reil, L. 1978. Autoradiography and epifluorescence microscopy combined for the determination of number and spectrum of actively metabolizing bacteria in natural waters. *Appl. Environ. Microbiol.* 36:506–12.

Meyers, J.A., D. Sanchez, L.P. Elwell, and S. Falkow. 1976. Simple agarose gel electrophoretic method for the identification and characterization of plasmid deoxyribonucleic acid. *J. Bacteriol.* 127:1529–37.

Miescier, J.J. and V.J. Cabelli. 1982. Enterococci and other microbial indicators in municipal wastewater effluents. *J. Water Pollut. Control Fed.* 54:1599–606.

Miescier, J.J., J.T. Peeler, J.D. Clem, R.B. Read Jr., and S.A. Furfari. 1985. Final report on a comparison of two methods for recovery of *Escherichia coli* Type 1 and fecal coliforms from oysters, p. 18. Washington, D.C.: Food and Drug Administration. Division of Cooperative Programs.

Miliotis M.D., J.E. Galen, J.B. Kaper, and J.G. Morris. 1989. Development and testing of a synthetic oligonucleotide probe for the detection of pathogenic *Yersinia* strains. *J. Clin. Microbiol.* 27:1667–70.

Millard, J., H. Appleton, and J.V. Parry. 1987. Studies on heat inactivation of hepatitis A virus with special reference to shellfish. *Epidemiol. Infect.* 98:397–414.

Miller, M.L. and J.A. Koburger. 1984. Evaluation of direct enrichment at elevated temperature for recovery of salmonellae from oysters. *J. Food Prot.* 47(4):267–69.

Miller, M.L and J.A. Koburger. 1985. *Plesiomonas shigelloides*: An opportunistic food and waterborne pathogen. *J. Food Prot.* 48:449–57.

Miller, M.L. and J.A. Koburger. 1986. Tolerance of *Plesiomonas shigelloides* to Ph, sodium chloride and temperature. *J. Food Prot.* 49(11):877–79.

Milne, D.P., J.C. Curran, and L. Wilson. 1986. Effects of sedimentation on removal of faecal coliform bacteria from effluents in estuarine water. *Water Res.* 20:1493–96.

Mingrone, M.G., M. Fantasia, N. Figura, and P. Guglielmetti. 1987. Characteristics of *Yersinia enterocolitica* isolated from children with diarrhea in Italy. *J.Clin. Microbiol.* 25:1301–04.

Minor, T.E., C.I. Allen, A.A. Tsiatis, D.B. Nelson, and D.J. D'Alessio. 1981. Human infective dose determinations for oral poliovirus type 1 vaccine in infants. *J. Clin. Microbiol.* 13(2):388–89.

Mintzer-Morgenstern, L. and E. Katzenelson. 1982. A simple medium for isolation of coagulase-positive staphylococci in a single step. *J. Food Prot.* 45:218–22.

Mitchell, J.R., M.W. Proesnell, E.W. Akin, J.M. Cummins, and O.C. Liu. 1966. Accumulation and elimination of poliovirus by the Eastern oyster. *Am. J. Epidemiol.* 84:40–50.

Mitchell, R., S. Yankosky, and H.W. Jannasch, 1967. Lysis of *Escherichia coli* by marine micro-organisms. *Nature* 215:891–92.

Mitchell, R. and S. Yankofsky. 1969. Implication of a marine ameba in the decline of *Escherichia coli* in seawater. *Environ. Sci. Technol.* 3:574–76.

Mitchell, R. and H.W. Jannasch. 1969. Processes controlling virus inactivation in seawater. *Environ. Sci. Technol.* 3:941–43.

Mitchell, R. 1972. Ecological control of microbiological imbalances. In *Water pollution microbiology*, ed. R. Mitchell, pp. 273–88. New York: Wiley–Interscience.

Mitchell, R. and C. Chamberlin. 1975. Factors influencing the survival of enteric microorganisms in the sea: An overview. In *Discharge of Sewage from Sea Outfalls*, ed. A.L.H. Gameson, pp. 237–51. Oxford: Pergamon Press.

Moats, W.A. 1978. Comparison of four agar plating media with and without added novobiocin for isolation of salmonellae from beef and deboned poultry meat. *Appl. Environ. Microbiol.* 36:747–51.

Moberg, L.J. 1985. Fluorogenic assay for rapid detection of *Escherichia coli* in food. *Appl. Environ. Microbiol.* 50:1383–87.

Moe, C.L., J.P. Allen, S.S. Monroe, H.E. Gary Jr., C.D. Humphrey, J.E. Herrmann, N.R. Blacklow, C. Carcamo, M. Koch, K.-H. Kim, and R.I. Glans. 1991. Detection of astrovirus in pediatric stool samples by immunoassay and RNA probe. *J. Clin. Microbiol.* 29:2390–95.

Moebus, K. 1972. Bactericidal properties of natural and synthetic sea water as influenced by addition of low amounts of organic matter. *Mar. Biol.* 15:81–88.

Molitoris, E., G. McKinley, M.I. Krichevsky, and D.J. Fagerberg. 1985. Comparison of conventional and miniaturized biochemical techniques for identification of animal streptococcal isolates. *Microb. Ecol.* 11:81–90.

Monroy, F.L., B.P. Henche, and M.J.P. Bao. 1991. Current production and regulatory practice in molluscan depuration in Spain. In *Molluscan Shellfish Depuration*, eds. W.S. Otwell, G.E. Rodrick, and R. Martin. Boca Raton, Fla.: CRC Press (in press).

Moore, W.E.C. and L.V. Holdeman. 1974. Human fecal flora: The normal flora of 20 Japanese–Hawaiians. *Appl. Microbiol.* 27:961–79.

Moreno Garcia, B., M. Escacho Perez, and V. Diez Fernandez. 1976. Coliform bacteria in non-purified mussels (Mytilus edulis). *Anales de la Facultad de Veterinaria de Leon* 22(1):315–24.

Morgan, D.R., P.C. Johnson, H.L. DuPont, T.K. Satterwhite, and L.V. Wood. 1985. Lack of correlation between known virulence properties of *Aeromonas hydrophila* and enteropathogenicity for humans. *Infect. Immun.* 50:62–65.

Morgan, G.B., P. Gubbins, and V. Morgan. 1965. A critical appraisal of the membrane filter technic. *Health Lab. Sci.* 2:227–37.

Morgan, J.A.W., C. Winstanley, R.W. Pickup, J.G. Jones, and J.R. Saunders. 1989. Direct phenotypic and genotypic detection of a recombinant pseudomonad population released into lake water. *Appl. Environ. Microbiol.* 55:2537–44.

Morita, R.Y. 1985. Starvation and miniaturization of heterotrophs, with special emphasis on maintenance of the starved viable state. In *Bacteria in Their Natural Environments*, eds. M. Fletcher and G. Floodgate. New York: Academic Press.

Morita, R.Y. 1988. Bioavailability of energy and its relationship to growth and starvation survival in nature. *Can. J. Microbiol.* 34:436–41.

Morotomi, M., T. Ohno, and M. Mutai. 1988. Rapid and correct identification of intestinal *Bacteroides* spp. with chromosomal DNA probes by whole-cell dot blot hybridization. *Appl. Environ. Microbiol.* 54:1158–62.

Morris, G.K. and C.G. Dunn. 1970. Influence of incubation temperature and sodium heptadecyl sulfate (Tergitol No.7) on the isolation of salmonellae from pork sausage. *Appl. Environ. Microbiol.* 20:192–95.

Morris, G.K. and J.C. Feeley. 1976. *Yersinia enterocolitica*: A review of its role in food hygiene. *Bull. WHO* 54:79–87.

Morris, R. and W.M. Waite. 1980. Evaluation of procedures for recovery of viruses from water. II-Detection systems. *Water Res.* 14:795–98.

Morris, G.K. 1984. *Shigella*, In *Compendium of Methods for the Microbiological*

Examination of Foods, ed. M.L. Speck, 2nd ed, pp. 343–50. Washington, D.C.: American Public Health Association.

Morris, G.K. 1986. *Shigella*. In *Progress in Food Safety*, eds. D.O. Cliver and B.A. Cochrane. University of Wisconsin–Madison: Food Research Institute.

Morris, J.G. Jr., R. Wilson, B.R. Davis, I.K. Wachsmuth, C.F. Riddle, H.G. Wathen, R.A. Pollard, and P.A. Blake. 1981. Non-O1 *Vibrio cholerae* gastroenteritis in the United States: Clinical, epidemiologic, and laboratory characteristics of sporadic cases. *Ann. Intern. Med.* 94:656–58.

Morris, J.G. Jr., R. Wilson, D.G. Hollis, R.E. Weaver, H.G. Miller, C.O. Tacket, F.W. Hickman, and P.A. Blake. 1982. Illness caused by *Vibrio damsela* and *Vibrio hollisae*. *Lancet* 1:1294–96.

Morris, J.G. Jr., J.L. Picardi, S. Lieb, J.V. Lee, A. Roberts, M. Hood, R.A. Gunn, and P.A. Blake. 1984. Isolation of non-toxigenic *Vibrio cholerae* O-group 1 from a case of severe gastrointestinal disease. *J. Clin. Microbiol.* 19:296–97.

Morris, J.G. Jr. and R.E. Black. 1985. Cholera and other vibriosis in the United States. *New Engl. J. Med.* 312:343–50.

Morris, J.G. Jr., A.C. Wright, D.M. Roberts, P.K. Wood, L.M. Simpson, and J.D. Oliver. 1987. Identification of environmental *Vibrio vulnificus* isolates with a DNA probe for the cytotoxin-hemolysin gene. *Appl. Environ. Microbiol.* 53:193–95.

Morris, J.G. Jr. 1988. *Vibrio vulnificus*—A new monster of the deep? *Ann. Intern. Med.* 109:261–62.

Morris, J.G. Jr., T. Takeda, B.D. Tall, G.A. Losonsky, S.K. Bhattacharya, B.D. Forrest, B.A. Kay, and M. Nishibuchi. 1990. Experimental non-O group 1 *Vibrio cholerae* gastroenteritis in humans. *J. Clin. Invest.* 85:697–705.

Morse, D.L., J.J. Guzewich, J.P. Hanrahan, R. Stricof, M. Shayegani, R. Deibel, J.C. Grabaw, N.A. Nowak, J.E. Herrmann, G. Cukor, and N.R. Blacklow. 1986. Widespread outbreaks of clam- and oyster-associated gastroenteritis—role of Norwalk virus. *New Eng. J. Med.* 314(11):678–81.

Mosley, J.W. 1967. Transmission of viral disease by drinking water. In *Transmission of Viruses by the Water Route*, ed. G. Berg, New York: Interscience.

Moseley, S.L., I. Huq, A.R.M.A. Alim, M. So, M. Samadpour-Motalebi, and S. Falkow. 1980. Detection of enterotoxigenic *Escherichia coli* by DNA colony hybridization. *J. Infect. Dis.* 142:892–98.

Moseley, S.L., P. Echeverria, J. Seriwatana, C. Tirapat, W. Chaicumpa, T. Sakuldaipeara, and S. Falkow. 1982. Identification of enterotoxigenic *Escherichia coli* by colony hybridization using three enterotoxin gene probes. *J. Infect. Dis.* 145:863–69.

Mossel, D. and P. van Netten. 1990. *Staphylococcus aureus* and related staphylococci in foods: Ecology, proliferation, toxinogenesis, control and monitoring. *J. Appl. Bacteriol. Symp. Supp.* 123S–145S.

Motes, M.L. 1982. Effect of chlorinated wash water on *Vibrio cholerae* in oyster meats. *J. Food Sci.* 47:1028–29.

Motes, M.L. Jr., R. McPhearson, and A. DePaola. 1984. Comparison of three international methods with APHA method for enumeration of Escherichia coli in estuarine waters and shellfish. *J. Food Prot.* 47(7):557–61.

Mundt, J.O., A.H. Johnson, and R. Khatchikian. 1958. Incidence and nature of enterococci on plant materials. *Food Res.* 23:186–93.

Mundt, J.O., J.H. Coogin Jr., and J.F. Johnson. 1962. Growth of *Streptococcus faecalis* var. *liquefaciens* on plants. *Appl. Microbiol.* 10:552–55.

Mundt, J.O. 1963. Occurrence of enterococci on plants in a wild environment. *Appl. Microbiol.* 11:141–44.

Mundt, J.O. 1982. The ecology of the streptococci. *Microb. Ecol.* 8:355–69.

Munger, S.F., A. Heyward, and R. Dutton. 1979. A survey of the microbiological quality of shellfish on metropolitan Seattle area beaches. *Abstr. Ann. Meet. Am. Soc. Microbiol.* 79:221.

Munoa, F.J. and R. Pares. 1988. Selective medium for isolation and enumeration of *Bifidobacterium* spp. *Appl. Environ.* 54:1715–18.

Munro, P.M., M.J. Gauthier, and F.M. Laumond. 1987. Changes in *Escherichia coli* cells starved in seawater or grown in seawater-wastewater mixtures. *Appl. Environ. Microbiol.* 53:1476–81.

Munro, P.M., M.J. Gauthier, V.A. Breittmayer, and J. Bongiovanni. 1989. Influence of osmoregulation processes on starvation survival of *Escherichia coli* in seawater. *Appl. Environ. Microbiol.* 55:2017–24.

Murasugi, A. and R.B. Wallace. 1984. Biotin-labeled oligonucleotides: Enzymatic synthesis and use as hybridization probes. DNA 3:269.

Murphy, A.M., G.S. Grohmann, R.J. Christopher, W.A. Lopez, G.R. Davey, and R.H. Millsom. 1979. An Australian-wide outbreak of gastroenteritis from oysters caused by Norwalk virus. *Med. J. Australia* 2:329–33.

Murphy, A.M. and G.S. Grohmann. 1980. Epidemiological studies on oyster-associated gastroenteritis. *Food Technol. Australia* 32(2):86–88.

Murphy, D. 1990. *Contaminant Levels in Oysters and Clams from the Chesapeake Bay* 1981–1985. Baltimore, Md.: Maryland Department of the Environment.

Murray, B.E., D.J. Evans, M.E. Peñaranda, and D.G. Evans. 1983. CFA/I-ST plasmids: Comparison of enterotoxigenic *Escherichia coli* (ETEC) of serogroups O25, O63, O78, and O128 and mobilization from an R factor-containing epidemic ETEC isolate. *J. Bacteriol.* 153:566–70.

Murray, B.E. and J. Tsao. 1987. Comparison of CFA/I: ST plasmids of enterotoxigenic *Escherichia coli. J. Infect. Dis.* 156:398–401.

Murray, E.G.D. 1955. A characterization of listeriosis in man and other animals. *Can. Med. Assoc. J.* 72:99–105.

Murray, J.C. 1990. The effect of temperature and salinity on the depuration of hepatitis A virus and other microbial indicators from clams. Master's Thesis. Department of Environmental Sciences and Engineering, Chapel Hill, N.C.: University of North Carolina.

Murray, P.R., A. Gauthier, and A. Niles. 1984. Evaluation of the Quantum II and Rapid E identification systems. *J. Clin. Microbiol.* 20:509–14.

Murtaugh, J.J. and R.L. Bunch. 1967. Sterols as a measure of fecal pollution. *J. Water Pollut. Control Fed.* 39:404–09.

Muytjens, H.J., H.C. Zanen, H.J. Sonderkamp, K.L.A., I.K. Wachsmuth, and J.J.

Farmer III. 1983. Analysis of eight cases of neonatal meningitis and sepsis due to *Enterobacter sakazakii*. *J. Clin. Microbiol.* 18:115–20.

Nambair, P.T.C. and V. Anjaiah. 1985. Enumeration of rhizobia by enzyme-linked immunosorbent assay (ELISA). *J. Appl. Bacteriol.* 58:187–93.

NAS. 1987. *Poultry Inspection: The Basis for a Risk-Assessment Approach.* Washington, D.C.: National Academy Press.

Nasser, A.M. and T.G. Metcalf. 1987. An A-ELISA to detect hepatitis A virus in estuarine samples. *Appl. Environ. Microbiol.* 53:1192–95.

Nataro, J.P., M.M. Baldini, J.B. Kaper, R.E. Black, N. Bravo, and M. Levine. 1985. Detection of an adherence factor of enteropathogenic *Escherichia coli* with a DNA probe. *J. Infect. Dis.* 152:560–65.

National Oceanic and Atmospheric Administration (NOAA). 1989. *A Summary of Data on Tissue Contamination from the First Three Years (1986–1988) of the Mussel Watch Project.* NOAA Technical Memorandum NOS OMA 49. Rockville, Md.: National Oceanic and Atmospheric Administration.

National Research Council Safe Drinking Water Committee. 1977. *Drinking Water and Health.* Washington, D.C.: National Academy of Sciences.

National Research Council (NRC). 1990. *Managing Troubled Waters.* Washington, D.C.: National Academy Press.

National Shellfish Sanitation Program (NSSP). 1990*a*. Manual of Operations, Part I: Sanitation of Shellfish Growing Areas. Washington, D.C.: Public Health Service, U.S. Food and Drug Administration.

National Shellfish Sanitation Program (NSSP) 1990*b*. Manual of Operations, Part II: Sanitation of the Harvesting, Processing and Distribution of Shellfish. Washington, D.C.: Public Health Service, U.S. Food and Drug Administration.

Newman, J.J., S. Waycott, and L.M. Cooney. 1979. Arthritis due to *Listeria monocytogenes*. *Arthritis Rheumatism* 22:1139–40.

Newsholme, A. 1896. The spread of enteric fever by means of sewage contaminated shell-fish. *R. J. San. Inst.* 17(III):389–413.

Nielsen, K.B. 1990. DOCKO, Inc. Mystic, Conn. (personal communication).

Nielson, B.J., D.S. Haven, F.O. Perkins, R. Morales-Alamo, and M.W. Rhodes. 1978. Bacterial depuration of the American oyster (*Crassostrea virginica*) under controlled conditions. In *Practical Considerations and Plant Design*, Vol. 2. Special Scientific report No. 88, p. 48. Gloucester Point, Va.: Virginia Institute of Marine Science.

Nikaido, H. and M. Vaara. 1987. Outer membrane. In *Escherichia coli and Salmonella typhimurium Cellular and Molecular Biology*, Vol. 1, ed. F. Neidhardt, pp. 7–22. Washington, D.C.: American Society for Microbiology.

Nilsson, L., J.D. Oliver, and S. Kjelleberg. 1991. Resuscitation of *Vibrio vulnificus* from the viable but nonculturable state. *J. Bacteriol.* 173:5054–59.

Nishibuchi, M., M. Ishibashi, Y. Takeda, and J.B. Kaper. 1985. Detection of the thermostable direct hemolysin gene and related DNA sequences in *Vibrio parahaemolyticus* and other *Vibrio* species by the DNA colony hybridization test. *Infect. Immun.* 49:481–86.

Nishibuchi, M., W.E. Hill, G. Zon, W.L. Payne, and J.B. Kaper. 1986. Synthetic oligodeoxyribonucleotide probes to detect Kanagawa phenomenon-positive *Vibrio parahaemolyticus*. *J. Clin. Microbiol.* 23:1091–95.

Nishimura, M. and T. Koyama. 1977. The occurrence of stanols in various living organisms and the behavior of sterols in contemporary sediments. *Geochim. Cosmochim. Acta* 41:379–85.

Nishimura, M. 1982. 5b-isomers of stanols and stanones as potential markers of sedimentary organic quality and depositional paleoenvironments. *Geochim. Cosmochim. Acta* 46:423–32.

Nishio, T., J. Nakamori, and K. Miyazaki. 1981. Survival of *Salmonella typhi* in oysters. Hiroshima Prefectural Inst. of Public Health, Hiroshima 734, Japan. *Zentralbaltt fuer Vakteriologie* 1B 172(4/5):415–26.

Nishitani, L. and K. Chew. 1988. PSP toxins in the Pacific Coast states: Monitoring programs and effects on bivalve industries. *J. Shellfish Res.* 7:653–69.

Nolan, C.M., J. Ballard, C.A. Kaysner, J.L. Lilja, L.P. Williams Jr., and F.C. Tenover. 1983. *Vibrio parahaemolyticus* gastroenteritis: An outbreak associated with raw oysters in the Pacific Northwest. *Diagnosis Microbiol. Infect. Dis.* 2:119–28.

Nolte, F., G. Hollick, and R. Robertson. 1982. Enzymatic activities of *Legionella pneumophila* and Legionella-like organisms. *J. Clin. Microbiol.* 15:175–77.

Northern Virginia Planning District Commission and Virginia Polytechnic Institute and State University. 1978. *Occoquan/Four Mile Run Nonpoint Source Correlation Study: Final Report for Metro Washington Council of Governments.* Washington, D.C.: NVPDC/VPISU.

Novitsky, J.S. and R.Y. Morita. 1976. Morphological characterization of small cells resulting from nutrient starvation of a psychrophilic marine vibrio. *Appl. Environ. Microbiol.* 32:617–22.

Novitsky, J.S. and R.Y. Morita. 1977. Survival of a psychrophilic marine vibrio under long-term nutrient starvation. *Appl. Environ. Microbiol.* 33:635–41.

Novitsky, J.S. and R.Y. Morita. 1978. Possible strategy for the survival of marine bacteria under starvation conditions. *Mar. Biol.* 48:289–95.

Nupen, E.M. 1970. Virus studies on the Windhoek waste-water reclamation plant (South-west Africa). *Water Res.* 4:661–72.

Nyström, T., N. Albertson, and S. Kjelleberg. 1989. Physiological and molecular adaptation to non-growth by marine vibrios. In *Recent Advances in Microbial Ecology. Proc. 5th Intern. Symp. Microb. Ecol.*, eds. T. Hattori, Y. Ishida, Y. Maruyama, R.Y. Morita, and A. Uchida, pp. 80–84. Tokyo: Japan Scientific Societies Press.

Nyström, T., K. Flärdh, and S. Kjelleberg. 1990. Multiple nutrient starvation in marine *Vibrio* sp. strain CCUG 15956. *J. Bacteriol.* 172:7085–97.

Nyström, T., N.H. Albertson, K. Flärdh, and S. Kjelleberg. 1990. Physiological adaptation to starvation and recovery from starvation by the marine *Vibrio* S14. *FEMS Microbiol. Ecol.* 74:129–40.

Nyström, T.N. and S. Kjelleberg. 1989. Role of protein synthesis in the cell division and starvation-induced resistance to autolysis of a marine *Vibrio* during the initial phase of starvation. *J. Gen. Microbiol.* 135:1599–606.

O'Brien, A.D. and R.K. Holmes. 1987. Shiga and Shiga-like toxins. *Microbiol. Rev.* 51:206–20.

O'Brien, M. and R. Colwell. 1985. Modified taurocholate-tellurite-gelatin agar for improved differentiation of *Vibrio* species. *J. Clin. Microbiol.* 22:1011–13.

O'Brien, T.F., J.D. Hopkins, E.S. Gilleece, A.A. Medeiros, R.L. Kent, B.O. Blackburn, M.B. Holmes, J.P. Reardon, J.M. Vergeront, W.L. Schell, E. Christenson, M.L. Bissett, and E.V. Morse. 1982. Molecular epidemiology of antibiotic resistance in *Salmonella* from animals and human beings in the United States. *New Engl. J. Med.* 307:1–6.

O'Keefe, B. and J. Green. 1989. Coliphages as indicators of fecal pollution at three recreational beaches on the Firth of Forth. *Water Res.* 23:1027–30.

O'Mahoney, M.C., C.D. Gooch, D.A. Smyth, A.J. Thrussell, C.L.R. Bartlett, and N.D. Noah. 1983. Epidemic hepatitis A from cockles. *Lancet* i:518–20.

O'Malley, M.L., D.W. Lear, W.N. Adams, J. Gaines, T.K. Sawyer, and E.J. Lewis. 1982. Microbial contamination of continental shelf sediments by wastewater. *J. Water Pollut. Control.* 54:1311–17.

O'Neill, K.R., S.H. Jones, and D.J. Grimes. 1990. Incidence of *Vibrio vulnificus* in northern New England water and shellfish. *FEMS Microbiol. Lett.* 72:163–68.

Ogawa, H., H. Tokuno, M. Sasaki, and T. Kishimoto. 1980. Distribution of enteropathogenic E. coli in oyster farm areas. *J. Food Hyg. Soc. Japan* (*Shokuhin Eiseigaku Zasshi*) 21(1):5–12.

Oglesbee, S.E., D.A. Wait, and A.F. Meinhold. 1981. MA-104: A continuous cell line for the isolation of human enteric viruses from environmental samples. *Abstr. Ann. Meet. Am. Soc. Microbiol.* 81:219.

Ogram, A., G.S. Sayler, and T. Barkay. 1987. The extraction and purification of microbial DNA from sediments. *J. Microbiol. Methods* 7:57–66.

Ogram, A.W. and G.S. Sayler. 1988. The use of gene probes in the rapid analysis of natural microbial communities. *J. Ind. Microbiol.* 3:281–92.

Old, H.N. 1938. Trends in shellfish sanitation. *Public Health Rep.* 53(18):720–28.

Olive, D.M., A.I. Atta, and S.K. Sethi. 1988. Detection of toxigenic *Escherichia coli* using biotin-labeled DNA probes following enzymatic amplification of heat labile enterotoxin gene. *Mol. Cell. Probes* 2:47–57.

Olive, D.M. 1989. Detection of enterotoxigenic *Escherichia coli* after polymerase chain reaction amplification with a thermostable DNA polymerase. *J. Clin. Microbiol.* 27:261–65.

Oliver, J.D. 1981. Lethal cold stress of *Vibrio vulnificus* in oysters. *Appl. Environ. Microbiol.* 41(3):710–17.

Oliver, J.D., R.A. Warner, and D.R. Cleland. 1983. Distribution of *Vibrio vulnificus* and other lactose-fermenting vibrios in the marine environment. *Appl. Environ. Microbiol.* 45:985–98.

Oliver, J.D. and W.F. Stringer. 1984. Lipid composition of a psychrophilic marine *Vibrio* sp. during starvation-induced morphogenesis. *Appl. Environ. Microbiol.* 47:461–66.

Oliver, J.D. and D. Wanucha. 1989. Survival of *Vibrio vulnificus* at reduced temperatures and elevated nutrient. *J. Food Saf.* 10:79–86.

Oliver, J.D. 1989. *Vibrio vulnificus*. In *Foodborne Bacterial Pathogens*, ed. M. Doyle, New York: Marcel Dekker.

Oliver, J.D., L. Nilsson, and S. Kjelleberg. 1991. Formation of nonculturable *Vibrio vulnificus* cells and its relationship to the starvation state. *Appl. Environ. Microbiol.* 57:2640–44.

Olsen, P.E. and W.A. Rice. 1989. *Rhizobium* strain identification and quantification in commercial inoculants by immunoblot assay. *Appl. Environ. Microbiol.* 55:520–22.

Olson, B.H. 1978. Enhanced accuracy of coliform testing in seawater by a modification of the most-probable-number method. *Appl. Environ. Microbiol.* 36:438–44.

Onderdonk, A.B., J. Johnston, J.W. Mayhew, and S.L. Gorbach. 1976. Effect of dissolved oxygen and Eh on *Bacteroides fragilis* during continuous culture. *Appl. Environ. Microbiol.* 31:168–72.

Ongerth, J.E and H.H. Stibbs. 1987. Identification of *Cryptosporidium* oocysts in river water. *Appl. Environ. Microbiol.* 53:672–76.

Oppenheimer, C.H. and A.L. Kelly. 1952. *Escherichia coli* in the intestine of a wild sea lion. *Science* 115:527–28.

Oragui, J.I. and D.D. Mara. 1981. A selective medium for the enumeration of *Streptococcus bovis* by membrane filtration. *J. Appl. Bacteriol.* 51:85–93.

Oragui, J.I. and D.D. Mara. 1983. Investigation of the survival characteristics of *Rhodococcus coprophilus* and certain fecal indicator bacteria. *Appl. Environ. Microbiol.* 46:356–60.

Oragui, J.I. and D.D. Mara. 1984. A note on a modified membrane-Bovis agar for the enumeration of *Streptococcus bovis* by membrane filtration. *J. Appl. Bacteriol.* 56:179–81.

Orlob, G.T. 1956. Viability of sewage bacteria in seawater. *Sewage Ind. Wastes.* 28:1147–67.

Osawa, R. and T. Mitsuoka. 1990. Selective medium for enumeration of tannin-protein complex-degrading *Streptococcus* spp. in feces of koalas. *Appl. Environ. Microbiol.* 56:3609–11.

Osawa, S., K. Furuse, and I. Watanabe. 1981. Distribution of ribonucleic acid coliphages in animals. *Appl. Environ. Microbiol.* 41:164–68.

Osawa, S., K. Furuse, M.S. Choi, A. Ando, T. Sakurai, and I. Watanabe. 1981. Distribution of ribonucleic acid coliphages in Korea. *Appl. Environ. Microbiol.* 41:909–11.

Osborne, J. 1990. New Jersey Department of Environmental Protection, Bureau of Marine Water Classification and Analyses. Leeds Point, New Jersey (personal communication).

Otwell, W.S., G.E. Rodrick, and R. Martin. 1981. *Molluscan Shellfish Depuration*, Boca Raton, Fla.: CRC Press (in press).

Ou, C.Y., S. Kwok, S.W. Mitchell, D.H. Mack, J.J. Sninsky, J.W. Krebs, P. Feorino, D. Warfield, and G. Schochetman. 1988. DNA amplification for direct detection of HIV-1 in DNA of peripheral blood mononuclear cells. *Science* 239:295–97.

Overby, L.R., F. Deinhardt, and J. Deinhardt. ed. 1983. *Viral Hepatitis—Second International Max von Pettenkofer Symposium.* New York: Marcel Dekker.

Overman, T.L., D. Plumley, S.B. Overman, and N.L. Goodman. 1985. Comparison of the API Rapid E four-hour system with the API 20E overnight system for the identification of routine clinical isolates of the family *Enterobacteriaceae*. *J. Clin. Microbiol.* 21:542–45.

Overman, T.L., J.F. Kessler, and J.P. Seabolt. 1985. Comparison of API 20E, API rapid E, and API rapid NFT for identification of members of the family *Vibrionaceae*. *J. Clin. Microbiol.* 22:778–81.

Owen, R.J., M. Costas, L. Sloss, and F.J. Bolton. 1988. Numerical analysis of electrophoretic protein patterns of *Campylobacter laridis* and allied thermophilic campylobacters from the natural environment. *J. Appl. Bacteriol.* 65(1):69–78.

Pace, J., C.Y. Wu, and T. Chai. 1988. Bacterial flora in pasteurized oysters after refrigerated storage. *J. Food Sci.* 53(2):325–27.

Pace, N.R., D.A. Stahl, D.J. Lane, and G.J. Olsen. 1986. The analysis of natural microbial populations by ribosomal RNA sequences. *Adv. Microbiol. Ecol.* 9:1–55.

Pagel, J.E. and G.M. Hardy. 1980. Comparison of selective media for the enumeration and identification of fecal streptococci from natural sources. *Can. J. Microbiol.* 26:1320–27.

Pagel, J.E., A.A. Qureshi, D. Michael Young, and L. Vlassoff. 1982. Comparison of four membrane filter methods for fecal coliform enumeration. *Appl. Environ. Microbiol.* 43:787–93.

Paille, D.C., L. Hackney, M.C. Reily, and M. Kilgen. 1987. Seasonal variation in the fecal coliform population of Louisiana oysters and its relationship to microbiological quality. *J. Food Prot.* 50(7):545–49.

Pal, T., A.S. Pacsa, L. Emody, S. Voros, and E. Selley. 1985. Modified enzyme-linked immunosorbent assay for detecting enteroinvasive *Escherichia coli* and virulent *Shigella* strains. *J. Clin. Microbiol.* 21:415–18.

Palmer, L.M., A.M. Baya, D.J. Grimes, and R.R. Colwell. 1984. Molecular genetic and phenotypic alteration of *Escherichia coli* in natural water microcosms containing toxic chemicals. *FEMS Microbiol. Lett.* 21:169–73.

Palmieri, M.J., S.L. Carito, and R.F. Meyer. 1988. Comparison of Rapid NFT and API 20E with conventional methods for identification of gram-negative nonfermentative bacilli from pharmaceuticals and cosmetics. *Appl. Environ. Microbiol.* 54:2838–41.

Palumbo, S.A. 1984. Heat injury and repair in *Campylobacter jejuni*. *Appl. Environ. Microbiol.* 48:477–80.

Parker, M. 1984. Enteric infections: Typhoid and paratyphoid fever. In *Topley and Wilson's Principles of Bacteriology, Virology and Immunology*, eds. G. Wilson, A. Miles, and M.T. Parker, Vol. 3, 7th ed., pp. 407–17. Baltimore, Md.: Williams and Wilkins.

Patton, C.M., N. Shaffer, P. Edmonds, T.J. Barrett, M.A. Lambert, C. Baker, D.M. Perlman, and D.J. Brenner. 1989. Human disease associated with "*Campylobacter upsaliensis*" (Catalase-negative or weakly positive *Campylobacter* spp. in the United States). *J. Clin. Microbiol.* 27:66–73.

Paul, J.H. and B. Myers. 1982. Fluorometric determination of DNA in aquatic microorganisms by use of Hoechst 33258. *Appl. Environ. Microbiol.* 43:1393–99.

Paul, J.H. and D.J. Carlson. 1984. Genetic material in the marine environment: Implications for bacterial DNA. *Limnol. Oceanogr.* 29:1091–97.

Paul, J.H., W.H. Jeffrey, and M. DeFlaun. 1985. Particulate DNA in subtropical ocean and estuarine planktonic environments. *Mar. Biol.* 90:95–101.

Paul, J.H., W.H. Jeffrey, and M.F. DeFlaun. 1987. Dynamics of extracellular DNA in the marine environment. *Appl. Environ. Microbiol.* 53:170–79.

Paul, J.H., M.F. DeFlaun, W.H. Jeffrey, and A.W. David. 1988. Seasonal and diel variability in dissolved DNA and in microbial biomass and activity in a subtropical estuary. *Appl. Environ. Microbiol.* 54:718–827.

Paul, J.H., L. Cazares, and J. Thurmond. 1990. Amplification of the *rbcL* gene from dissolved and particulate DNA from aquatic environments. *Appl. Environ. Microbiol.* 56:1963–66.

Pavia, A.T., J.F. Campbell, P.A. Blake, J.D. Smith, T.W. McKinley, and D.L. Martin. 1987. Cholera from raw oysters shipped interstate (letter). *J. AMA* 258:2374.

Payment, P. and M. Trudel. 1985. Immunoperoxidase method with human immune serum globulin for broad spectrum detection of cultivable human enteric viruses: Application to enumeration of cultivable viruses in environmental samples. *Appl. Environ. Microbiol.* 50:1308–10.

Payment, P. and M. Trudel. 1987. Detection and quantitation of human enteric viruses in waste waters: Increased sensitivity using a human immune serum globulin-immunoperoxidase assay on MA-104 cells. *Can. J. Microbiol.* 33:568–70.

Payne, J. 1978. Nutritional requirements for the recovery of tellurite resistance by heat-injured *Streptococcus faecalis*. *J. Gen. Microbiol.* 107:217–22.

Pearl, H.W. 1974. Bacterial uptake of dissolved organic matter in relation to detrimental aggregation in marine and freshwater systems. *Limnol. Oceanogr.* 19:966–72.

Pearson, L.J. and E.H. Marth. 1990. *Listeria monocytogenes*—A threat to the food supply: A review. *J. Dairy Sci.* 73:912–28.

Pederson, L.R., M.L. Carter, and J.L. Hadler. 1988. A foodborne outbreak of *Giardia lamblia*. *J. Infect. Dis.* 157:846–48.

Peele, E.R. and R.R. Colwell. 1981. Application of a direct microscopic method of enumeration of substrate-responsive marine bacteria. *Can. J. Microbiol.* 27:1071–75.

Peixotto, S.S., G. Finne, M.O. Hanna, and C. Vanderzant. 1979. Presence, growth and survival of *Yersinia enterocolitica* in oysters, shrimp and crab. *J. Food Prot.* 42(12):974–81.

Perez-Rosas, N. and T.C. Hazen. 1988. In situ survival of *Vibrio cholerae* and *Escherichia coli* in tropical coral reefs. *Appl. Environ. Microbiol.* 54:1–9.

Perez-Martinez, J.A., A.R. Garcia, Y. Almanza-Marquez, I.R. Perez, and A.Y. Miranda. 1988. Determination of enterotoxigenicity of *Escherichia coli* isolated from oysters, by means of inoculation of suckling mice. *Beterinaria*, Mexico 19(1):9–13.

Peroni, C. and O. Lavarello. 1975. Microbial activities as a function of water depth in the Ligurian Sea: An autoradiographic study. *Mar. Biol.* 30:37–50.

Perry, C.A. 1928. Studies relative to the significance of the present oyster score. *Am. J. Hyg.* 8:694–722.

Perry, C.A. 1936. Progress report—standard methods for the bacteriological examination of shellfish. Mimeographed paper, presented before Laboratory Section, APHA, New Orleans, La., October, 1936. (Available from Shellfish Sanitation Branch, FDA, Washington, D.C.)

Perry, C.A. 1939. A summary of studies on pollution of shellfish. *Food Res.* 4(4):381–95.

Peterkin, P.I. and A.N. Sharpe. 1984. Rapid enumeration of *Staphylococcus aureus* in foods by direct demonstration of enterotoxigenic colonies on membrane filters by enzyme immunoassay. *Appl. Environ. Microbiol.* 47:1047–53.

Peterson, D.A., L.G. Wolfe, E.P. Larkin, and F.W. Deinhardt. 1978. Thermal treatment and infectivity of hepatitis A virus in human feces. *J. Med. Virol.* 2:201–06.

Peterson, D.A., T.R. Hurley, J.C. Hoff, and L.G. Wolfe. 1983. Effect of chlorine treatment on infectivity of Hepatitis A virus. *Appl. Environ. Microbiol.* 45(1):223–27.

Peterson, M.E., B.J. Yokel, and D.V. Lim. 1984. Recovery of selected pathogens from Naples Bay, Florida and associated waterways. *Estuaries* 7:133–38.

Petrilli, F. and P. Crovari. 1965. Aspetti dell'inquinamento delle acque marine con partiolare riguardo alla situazione in liguera. *G. Ig. Med. Prevent.* 8:269–311.

Pettibone, G.W. and J.J. Cooney. 1986. Effect of organotins on fecal pollution indicator organisms. *Appl. Environ. Microbiol.* 52:562–66.

Pettigrew, C.A. and G.S. Sayler. 1986. The use of DNA: DNA colony hybridization in the rapid isolation of 4-chlorobiphenyl degradative bacterial phenotypes. *J. Microbiol. Methods* 5:205–13.

Pettipher, G.L. 1983. *Review: The direct epifluorescent filter technique.* Letchworth: Research Studies Press.

Phelps, E.B. 1911. Some experiments upon the removal of oysters from polluted to unpolluted waters. *J. Am. Public Health Assoc.* 1:305.

Pichard, S.L. and J.H. Paul. 1991. Detection of gene expression in genetically engineered microorganisms and natural phytoplankton populations in the marine environment by MRNA analysis. *Appl. Environ. Microbiol.* 57:1721–27.

Pichet, P. and C. Hurtubise. 1976. Reactions of ozone on artificial seawater. In *Proceeding 2nd International Symposium on Ozone Technology*, pp. 664–70. Syracuse, New York: Int. Ozone Inst.

Picken, R.N., Z. Wang, and H.L. Yang. 1987. Molecular cloning of a species-specific DNA probe for *Campylobacter jejuni.* In *Abstr. Annu. Meeting Amer. Soc. Microbiol.*, C-317, p. 376.

Pierce, R.H. and R.C. Brown. 1984. Coprostanol distribution from sewage discharge into Sarasota Bay, Florida. *Bull. Environ. Contam. Toxicol.* 32:75–79.

Pietri, Ch., B. Hughes, J.M. Crance, D. Puel, C. Cini, and R. Deloince. 1988. Hepatitis A virus levels in shellfish exposed in a natural marine environment to the effluent from a treated sewage outfall. *Water Sci. Technol.* 20:229–34.

Pietzsch, O. and E. Bulling. 1970. *Salmonella* infections of animals and of animal foods and feeds in West Germany (including W. Berlin). *Bundesgesundheitsblatt* 13(1/2):5–13.

Pingree, J.D. 1990. Delaware Division of Public Health. Dover, Delaware (personal communication).

Pinkel, D., T. Straume, and J.W. Gray. 1986. Cytogenic analysis using quantitative, high-sensitivity, fluorescence hybridization. *Proc. Natl. Acad. Sci.* U.S.A. 83:2934–38.

Pisano, M.A., M.J. Sommer, and M.M. Lopez. 1986. Application of pretreatments for the isolation of bioactive actinomycetes from marine sediments. *Appl. Microbiol. Biotechnol.* 25:285–88.

Plotkin, S.A. and M. Katz. 1967. Minimal infective doses of viruses for man by the oral route. In *Transmission of Viruses by the Water Route*, ed. G. Berg. New York: Wiley.

Pocklington, R., J.D. Leonard, and N.F. Crewe. 1987. Le coprostanol comme indicateur de la contamination fecale dans l'eau de mer et es sediments marins. *Oceano. Acta* 10:83–89.

Pollard, D.R., W.M. Johnson, H. Lior, D. Tyler, and K.R. Rozee. 1990. Rapid and specific detection of verotoxin genes in *Escherichia coli* by the polymerase chain reaction. *J. Clin. Microbiol.* 28:540–45.

Popoff, M. 1984. *Aeromonas.* In *Bergeys Manual of Systematic Bacteriology*, eds. N.R. Krieg and J.G. Holt, Vol. 1, pp. 545–48. Baltimore, Md.: Williams and Wilkins.

Poppell, C.F. 1979. Enumeration and occurrence of RNA coliphages in wastewater. Master's thesis. Baltimore, Md.: Johns Hopkins University.

Porteous, L.A. and J.L. Armstrong. 1990. Recovery of bulk DNA from soil using a rapid, small-scale extraction method. *Appl. Environ. Microbiol.* 56:

Porter, K.G. and Y.S. Feig. 1980. The use of DAPI for identifying and counting aquatic microflora. *Limnol. Oceanogr.* 25:943–48.

Portnoy, B.L., P.A. Mackowiak, C.T. Caraway, J.A. Walker, T.W. McKinley, and C.A. Klein. 1975. Oyster-associated hepatitis—Failure of shellfish certification programs to prevent outbreaks. *J. Am. Med. Assoc.* 233:1065–68.

Portnoy, D.A., S.L. Mosley, and S. Falkow. 1981. Characterization of plasmids and plasmid-associated determinants of *Yersinia enterocolitica* pathogenesis. *Infect. Immun.* 31:775–82.

Post, F.J., A.D. Allen, and T.C. Reid. 1967. Simple medium for the selective isolation of *Bacteroides* and related organisms, and their occurrence in sewage. *Appl. Microbiol.* 15:213–18.

Postgate, J.R. 1967. Viability measurements and the survival of microbes under minimum stress. In *Advances in Microbial Physiology*, eds. A.H. Rose and J.R. Wilkinson, pp.1–23. London: Academic Press.

Potter, M.E., M.J. Blaser, R.K. Sikes, A.F. Kaufmann, and J.G. Wells. 1983. Human *Campylobacter* infection associated with certified raw milk. *Am. J. Epidemiol.* 117:475–83.

Power, U.F. and J. Collins. 1989. Differential depuration of poliovirus, Escherichia coli, and a coliphage by the common mussel, Mytilus edulis. *Appl. Environ. Microbiol.* 55(6):1386–90.

Power, U.F. and J. Collins. 1990*a*. Elimination of coliphages and Escherichia coli and mussels during depuration under varying conditions of temperature, salinity, and food availability. *J. Food Prot.* 53(3):208–12, 226.

Power, U.F. and J.K. Collins. 1990*b*. Tissue distribution of a coliphage and *Escherichia coli* in mussels after contamination and depuration. *Appl. Environ. Microbiol.* 56:803–07.

Prado, D., B.E. Murray, T.G. Cleary, and L.K. Pickering. 1987. Limitations of using the plasmid pattern as an epidemiological tool for clinical isolates of *Shigella sonnei. J. Infect. Dis.* 155:314–16.

Prauser, H. 1984. Phage host ranges in the classification and identification of gram-positive branched and related bacteria. In *Biological, Biochemical and Biomedical Aspects of Actinomycetes*, eds. L. Ortiz-Ortiz, L.F. Bojalil, and V. Yakoleff, pp. 617–33. New York: Academic Press.

Prescott, S.C. and C.E.A. Winslow. 1904. *Elements of Water Bacteriology*. New York: Wiley.

Presnell, M.W. and C.B. Kelly. 1961. *Bacteriological Studies of Commercial Shellfish Operations on the Gulf Coast*. U.S. Public Health Service Technical Report F-61-9.

Presnell, M.W. and J.J. Miescier. 1971. Coliforms and fecal coliforms in an oyster-growing area. *J. Water Pollut. Control.* Part I:407–16.

Presswood, W.G. and D.K. Strong. 1978. Modification of M-FC medium by eliminating rosolic acid. *Appl. Environ. Microbiol.* 36:90–94.

Preston, D.R., S.R. Farrah, and G. Bitton. 1988. Enhanced infectivity of enteroviruses in vitro by treating cell monolayers and viruses with the cationic polymer polyethyleneimine. *Abstr. Ann. Meet. Am. Soc. Microbiol.* 88:288.

Preston, D.R., G. Bitton, and S.R. Farrah. 1989. Enhancing the infectivity of enteroviruses in vitro by pretreating host cell monolayers with the cationic polymer polyethyleneimine. *Abstr. Ann. Meet. Am. Soc. Microbiol.* 89:349.

Preston, D.R., G. Bitton, and S.R. Farrah. 1990. Enhancement of enterovirus infectivity in vitro by pretreating host cell monolayers with the cationic polymer polyethyleneimine. *Appl. Environ. Microbiol.* 56:295–97.

Pringle, S.B. 1964. Field and laboratory studies on heat-shock method of preparation of oysters for shucking. In *Proceedings of the 5th National Shellfish Sanitation Workshop*, pp. 165–68. Washington, D.C.: U.S. Department of Health, Education, and Welfare, Public Health Service, Shellfish Sanitation Branch.

Provost, P.J. and M.R. Hilleman. 1979. Propagation of human hepatitis A virus in cell culture in vitro. *Proc. Soc. Expt. Biol. Med.* 160:213–21.

Prpic, J.K., R.M. Robins-Browne, and R.B. Davey. 1983. Differentiation between virulent and avirulent *Yersinia enterocolitica* isolates by using Congo red agar. *J. Clin. Microbiol.* 18:486–90.

Pryzbylski, K.S. and L.D. Witter. 1979. Injury and recovery of *Escherichia coli* after sublethal acidification. *Appl. Environ. Microbiol.* 37:2612–65.

Pugsley, A.P. and L.M. Evison. 1974. Immunofluorescence as a method for the detection of *Escherichia coli* in water. *Can. J. Microbiol.* 20:1457–63.

Pugsley, A.P. and L.M. Evison. 1975. A fluorescent antibody technique for the enumeration of faecal streptococci in water. *J. Appl. Bacteriol.* 38:63–65.

Purdy, R.N., B.N. Dancer, M.J. Day, and D.J. Stickler. 1984. A novel technique for the enumeration of bacteriophage from water. *Microbiol. Lett.* 21:89–92.

Purdy, R.N., B.N. Dancer, M.J. Day, and D.J. Stickler. 1985. A note on a membrane filtration method for the concentration and enumeration of bacteriophages from water. *J. Appl. Bacteriol.* 58:231–33.

Pyres, W.O. ed. 1978. Water quality and health significance of bacterial indicators of pollution. Workshop proceedings, Philadelphia, Pa.: Drexel University, 17–18 April 1978.

Qadri, R.B., K. Buckle, and R. Edwards. 1974. Rapid methods for the determination of faecal contamination in oysters. *J. Appl. Bacteriol.* 37(1):7–14.

Qadri, R.B., K. Buckle, and R. Edwards. 1976. Reduction in sewage contamination in Sydney rock oysters. *Food Technol. Australia* 28(11):413–16.

Rahman, A.F. and J.M. Willoughby. 1980. Dysentery-like syndrome associated with *Aeromonas hydrophila. Br. Med. J.* 281:976.

Rambach, A. 1990. New plate medium for facilitated differentiation of *Salmonella* spp. from *Proteus* spp. and other enteric bacteria. *Appl. Environ. Microbiol.* 56:301–03.

Ramsay, A.J. 1974. The use of autoradiography to determine proportion of bacteria metabolizing in an aquatic habitat. *J. Gen. Microbiol.* 80:363–73.

Ranki, M., A. Palva, M. Virtanen, M. Laaksonen, and H. Soderlund. 1983. Sandwich hybridization as a convenient method for the detection of nucleic acids in crude samples. *Gene* 21:77–85.

Rao, V.C., K.M. Seidel, S.M. Goyal, T.G. Metcalf, and J.L. Melnick. 1984. Isolation of enteroviruses from water, suspended solids, and sediments from Galveston Bay: Survival of poliovirus and rotavirus adsorbed to sediments. *Appl. Environ. Microbiol.* 48:404–09.

Rappold, H., R.F. Bolderdick, and J.M. De Smedt. 1984. Rapid cultural method to detect *Salmonella* in foods. *J. Food Prot.* 47:46–48.

Rashtichian, A., M. Abbott, M. Ottaviani, J. Eldredge, and J. Klinger. 1987. A rapid non-isotopic DNA probe assay for detection of *Campylobacter. Abstr. Annu. Meet. Amer. Soc. Microbiol.* C-317; 376.

Ratzan, K.R., J.A. Bryan, and J. Krackow. 1969. An outbreak of gastroenteritis associated with ingestion of raw clams. *J. Infect. Dis.* 120:265–68.

Ray, B. and M.L. Speck. 1973. Freeze-injury in bacteria. CRC Critical Rev. *Clinical Lab. Sci.* 4:161–213.

Ray, B., S. Hawkins, and C. Hackney. 1977. Method to detect injured *Vibrio parahaemolyticus* from seafoods. *Appl. Envrion. Microbiol.* 35:1121–27.

Ray, B. and M.L. Speck. 1978. Plating procedure for the enumeration of coliforms from dairy products. *Appl. Environ. Microbiol.* 35:820–22.

Ray, B. 1979. Methods to detect stressed microorganisms. *J. Food Prot.* 42:346–55.

Ray, B. 1986. Impact of bacterial injury and repair in food microbiology: Its past, present and future. *J. Food Prot.* 49:651–55.

Ray, B. 1989*a*. Enumeration of injured indicator bacteria from foods. In *Injured Index and Pathogenic Bacteria*, ed. B. Ray, pp. 1–54. Boca Raton, Fla.: CRC Press.

Ray, B. 1989*b*. *Injured Index and Pathogenic Bacteria: Occurrence and Detection in Foods, Water and Feeds.* Boca Raton, Fla.: CRC Press.

Rayman, M.K., B. Aris, and H.B. El Derea. 1978. The effect of compounds which degrade hydrogen peroxide on the enumeration of heat stressed cells of *Salmonella senftenberg. Can. J. Microbiol.* 24:883–85.

Reasoner, D.J. and E.E. Geldreich. 1979. Rapid detection of waterborne fecal coliforms by $^{14}CO_2$ release. In *Mechanizing Microbiology*, ed. A.N.S.a.D.S. Clark, pp. 120–139. Springfield, Ill.: Charles C. Thomas.

Reasoner, D.J. and E.E. Geldreich. 1989. Detection of fecal coliforms in water by using [^{14}C] mannitol. *Appl. Environ. Microbiol.* 55:907–11.

Regensburger, A., W. Ludwig, and K. Schleifer. 1988. DNA probes with different specificities from a cloned 23S rRNA gene of *Micrococcus luteus*. *J. Gen. Microbiol.* 134:1197–204.

Reille, P. 1907. Huitres et fievres typhoides. *Ann. d'Hyg. publ. et Med. Legale.* Part 4, 8:45.

Reily, L.A. and C.R. Hackney. 1985. Survival of *Vibrio cholerae* during cold storage in artificially contaminated seafoods. *J. Food Sci.* 50(3):838–39.

Remis, S.R., I.K. MacDonald, L.W. Riley, N.D. Puhr, J.G. Wells, B.R. Davis, P.A. Blake, and M.L. Cohen. 1984. Sporadic cases of hemorrhagic colitis associated with *Escherichia coli* O157:H7. *Ann. Intern. Med.* 101:624.

Rendtorff, R.C. 1954. The experimental transmission of human intestinal protozoan parasites. I. *Endamoeba coli* cysts given in capsules. *Am. J. Hyg.* 59:196–208.

Renwick, A. and D.G. Jones. 1985. A comparison of the fluorescent ELISA and antibiotic resistance identification techniques for use in ecological experiments with *Rhizobium trifolii*. *J. Appl. Bacteriol.* 58:199–206.

Reyes, G.R., M.A. Purdy, and J.P. Kim. 1990. Isolation of DNA from the virus responsible for enterically transmitted non-A, non-B hepatitis. *Science* 247:1335–39.

Rhoden, D.L. 1985. AutoSCAN-4 system for identification of gram-negative bacilli. *J. Clin. Microbiol.* 22:915–18.

Rhodes, M., I.C. Anderson, and H.I. Kator. 1983. In situ development of sublethal *Clin. Microbiol.* 22:915–18.

Rhodes, M., I.C. Anderson, and H.I. Kator. 1983. *In situ* development of sublethal stress in *Escherichia coli*: Effects on enumeration. *Appl. Environ. Microbiol.* 45:1870–76.

Rhodes, M.W. and H. Kator. 1983. *Siter Survival of Enteric Bacteria in Estuarine Environments.* Virginia Water Resources Research Center. Blacksburg, Vir.: Virginia Polytechnic Institute and State University.

Rhodes, M.W. and H. Kator. 1988. Survival of *Escherichia coli* and *Salmonella* spp. in estuarine environments. *Appl. Environ. Microbiol.* 54:2902–07.

Rhodes, M.W. and H. Kator. 1990. Effects of sunlight and autochthonous microbiota on *Escherichia coli* survival in an estuarine environment. *Curr. Microbiol.* 21:65–73.

Rhodes, M.W. and H. Kator. 1991. Use of *Salmonella typhimurium* WG49 to enumerate male-specific coliphages in an estuary and watershed subject to non-point pollution. *Water Res.* 25:1315–23.

Ricci, M., M.R. Aliquo, G. Bottinelli, A. Mascola, and R. Piro. 1989. Enrichment in Rappaport Bassilaiadis broth for isolation of *Salmonella* in foods in comparison to the official method. *Igiene Moderna* 91(2):213–20.

Richards, G.P. and D. Goldmintz. 1982. Evaluation of a cross-flow filtration technique for extraction of polioviruses from inoculated oyster tissue homogenates. *J. Virol. Methods* 4:147–53.

Richards, G.P., D. Goldmintz, G. Green, and J.A. Babinchak. 1982. Rapid methods for extraction and concentration of poliovirus from oyster tissues. *J. Virol. Methods.* 5:285–91.

Richards, G.P. and D.A. Weinheimer. 1985. Influence of adsorption time, rocking, and soluble proteins on the plaque assay of monodispersed poliovirus. *Appl. Environ. Microbiol.* 49:744–48.

Richards, G.P. 1985. Outbreaks of shellfish-associated enteric virus illness in the United States: Requisite for development of viral guidelines. *J. Food Prot.* 48:815–23.

Richards, G.P. 1987. A Review of Methods for the Analysis of Selected Microorganisms from Marine Sediments, Shellfish, and Seawater. Charleston, S.C.: National Marine Fisheries Service.

Richards, G.P. and F.M. Hetrick. 1988. Detection of enteroviruses in oysters from the environment. *Abstr. Ann. Meet. Am. Soc. Microbiol.* 88:290.

Richards, G.P. 1988. Microbial purification of shellfish: A review of depuration and relaying. *J. Food Prot.* 51:218–51.

Richards, G.P., J.D. Teska, J.A. Bemiss, and J.D. Sample. 1990. Comparison of rocked and static adsorption for the palque assay of enteroviruses. *Abstr. Ann. Meet. Am. Soc. Microbiol.* 90:325.

Richards, G.P. 1990. Shellfish Depuration. In *Microbiology of Marine Food Products*, eds. D.R. Ward and C.R. Hackney, pp. 395–428. New York: Van Nostrand Reinhold.

Ridinger, D.N., R.S. Spendlove, B.B. Barnett, D.B. George, and J.C. Roth. 1982. Evaluation of cell lines and immunofluorescence and plaque assay procedures for quantifying reoviruses in sewage. *Appl. Environ. Microbiol.* 43:740–46.

Rigby, P.W.J., M.A. Dieckmann, C. Rhodes, and P. Berg. 1977. Labeling deoxyribonucleic acid to high specific activity in vitro by nick translation with DNA polymerase. *Int. J. Mol. Biol.* 113:237–51.

Riley, L.K., M.E. Marshall, and M.S. Coleman. 1986. A method for biotinylating oligonucleotide probes for use in molecular hybridizations. *DNA* 5:333.

Riley, L.W. and M.L. Cohen. 1982. Plasmid profiles and *Salmonella* epidemiology. *Lancet* i:573.

Riley, L.W., G.T. DiFerdinando, T.M. DeMelfi, and M.L. Cohen. 1983. Evaluation of isolated cases of salmonellosis by plasmid profile analysis: Introduction and transmission of a bacterial clone by precooked roast beef. *J. Infect. Dis.* 148:12–17.

Rippey, S.R. and V.J. Cabelli. 1979. Membrane filter procedure for enumeration of *Aeromonas hydrophila* in freshwaters. *Appl. Environ. Microbiol.* 38:108–13.

Rippey, S.R., L.A. Chandler, and W.D. Watkins. 1987. Fluorometric method for enumeration of *Escherichia coli* in molluscan shellfish. *J. Food Prot.* 50:685–90, 710.

Rippey, S.R. and J.L. Verber. 1988. *Shellfish Borne Disease Outbreaks.* Department of Health and Human Services, Public Health Service, Food and Drug Administration, Shellfish Sanitation Branch. Davisville, R.I.: NETSU.

Rippey, S.R. 1990. *Shellfish Borne Disease Outbreaks.* Department of Health and

Human Services, Public Health Service, Food and Drug Administration, Shellfish Sanitation Branch. Davisville, R.I.: NETSU.

Rippey, S.R. 1991. *Shellfish Borne Disease Outbreaks.* Department of Health and Human Services, Public Health Service, Food and Drug Administration, Shellfish Sanitation Branch. Davisville, R.I.: NETSU.

Robins-Browne R.M., M.D. Miliotis, S. Cianciosi, V.L. Miller, S. Falkow, and J.G. Morris. 1989. Detection of virulence in Yersinia species by DNA colony hybridization. *J. Clin. Microbiol.* 27:644–50.

Robins-Browne, R.M., M.D. Miliotis, S. Cianciosi, V.L. Miller, S. Falkow, and J.G. Morris. 1989. Evaluation of DNA colony hybridization and other techniques for detection of virulence in *Yersinia* species. *J. Clin. Microbiol.* 22:105–10.

Roberts, G. 1971. The rapid identification of gram-negative bacilli. *Med. Lab. Technol.* 28:382–84.

Roberts, M.C., B. Moncla, and G.E. Kenny. 1987. Chromosomal DNA probes for the identification of *Bacteroides* species. *J. Gen. Microbiol.* 133:1423–30.

Roberts, N.C., R.J. Siebeling, J.B. Kaper, and H.B. Bradford Jr. 1982. Vibrios in the Louisiana Gulf coast environment. *Microbial Ecology* 8:299–312.

Robichaux, G.E. 1975. MPN Coliforms vs. MPN fecal coliforms—the Louisiana experience. In *Proceedings Ninth National Shellfish Sanitation Workshop*, ed. D.S. Wilt. Washington, D.C.: U.S. Department of Health, Education, and Welfare, Food and Drug Administration.

Robinson, D.A. 1981. Infective dose of *Campylobacter jejuni* in milk. *Br. Med. J.* 282–1584.

Robison, B.J. 1984. Evaluation of a fluorogenic assay for detection of *Escherichia coli* in foods. *Appl. Environ. Microbiol.* 48:285–88.

Rodrick, G.E. and K.R. Schneider. 1991. Vibrio depuration. In *Molluscan Shellfish Depuration*, eds, W.S. Otwell, G.E. Rodrick, and R. Martin. Boca Raton, Fla.: CRC Press (in press).

Rodrigues, U.M. and R.G. Kroll. 1988. Rapid selective enumeration of bacteria in foods using a microcolony epifluorescence microscopy technique. *J. Appl. Bacteriol.* 64:65–78.

Rollins, D.M. and R.R. Colwell. 1986. Viable but nonculturable stage of *Campylobacter jejuni* and its role in survival in the natural aquatic environment. *Appl. Environ. Microbiol.* 52:531–38.

Romick, T.L., J.A. Lindsay, and F.F. Busta. 1987. A visual DNA probe for detection of enterotoxigenic *Escherichia coli* by colony hybridization. *Lett. Appl. Microbiol.* 5:87–90.

Romick, T.L., J. Lindsay, and F. Busta. 1989. Evaluation of a visual DNA probe for enterotoxigenic *E. coli* detection in foods and wastewater by colony hybridization. *J. Food Prot.* 52(7):466–70.

Ronnberg, B., O. Söderlund, and T. Wadström. 1985. Evaluation of a competitive enzyme-linked immunosorbent assay for porcine *Escherichia coli* heat-stable enterotoxin. *J. Clin. Microbiol.* 22:893–96.

Roos, R. 1956. Hepatitis epidemic conveyed by oysters. *Svenska Lakartidningen.* 53:989–1003.

Roper, M.M. and K.C. Marshall. 1974. Modification of the interaction between *Escherichia coli* and bacteriophage in saline sediment. *Microb. Ecol.* 1:1–13.

Roper, M.M. and K.C. Marshall. 1977. Lysis of *Escherichia coli* by a marine myxobacter. *Microb. Ecol.* 3:167–71.

Roper, M.M. and K.C. Marshall. 1978*a*. Biological control agents of sewage bacteria in marine habitats. *Aust. J. Mar. Freshwater Res.* 29:335–43.

Roper, M.M. and K.C. Marshall. 1978*b*. Effects of a clay mineral on microbial predation and parasitism of *Escherichia coli. Microbiol. Ecol.* 4:279–89.

Roper, M.M. and K.C. Marshall. 1979. Effects of salinity on sedimentation and of particulates on survival of bacteria in estuarine habitats. *Geomicrobiol. J.* 1:103–16.

Rose, J.B., H. Darbin, and C.P. Gerba. 1988. Correlations of the protozoa *Cryptosporidium* and *Giardia*, with water variables in a watershed. *Water Sci. Technol.* 20:271–76.

Rosenow, E.M. and E.H. Marth. 1987. *Listeria*, listeriosis and dairy foods: A review. *Cult. Dairy Prod. J.* 22:4.

Rosner, R.F. 1970. Identification of enteric bacilli directly from primary isolation media using reagent-impregnated strips. *Am. J. Clin. Path.* 54:587–95.

Roszak, D.B., D.J. Grimes, and R.R. Colwell. 1984. Viable but non-recoverable stage of *Salmonella enteritidis* in aquatic systems. *Can. J. Microbiol.* 30:334–38.

Roszak, D.B. and R.R. Colwell. 1987*a*. Metabolic activity of bacterial cells enumerated by direct viable count. *Appl. Environ. Microbiol.* 53:2889–983.

Roszak, D.B. and R.R. Colwell. 1987*b*. Survival strategies of bacteria in the natural environment. *Microbiol. Rev.* 51:365–79.

Rotbart, H.A. 1990. Enzymatic amplification of the enteroviruses. *J. Clin. Microbiol.* 28:438–42.

Roth, L.A. and D. Keenan. 1971. Acid injury of *Escherichia coli. Can. J. Microbiol.* 17:1005–08.

Roth, W.G., M.P. Leckie, and D.N. Dietzler. 1988. Restoration of colony-forming activity in osmotically stressed *Escherichia coli* by betaine. *Appl. Environ. Microbiol.* 54:3142–46.

Rothenberg, P.J., N.J. Stern, and D.C. Westhoff. 1984. Selected enrichment broths for recovery of *Campylobacter jejuni* from foods. *Appl. Environ. Microbiol.* 48:78–80.

Rowbotham, T.J. and T. Cross. 1977*a*. Ecology of *Rhodococcus coprophilus* and associated actinomycetes in fresh water and agricultural habitats. *J. Gen. Microbiol.* 100:231–40.

Rowbotham, T.J. and T. Cross. 1977*b*. *Rhodococcus coprophilus* sp. nov.: An aerobic nocardioform actinomycete belonging to the "rhodochrous" complex. *J. Gen. Microbiol.* 100:123–38.

Rowe, B. and R.J. Gross. 1984. Genus II. *Shigella*, In *Bergey's Manual of Systematic Bacteriology*, eds. N.R. Krieg and J.G. Holt, Vol. 1, pp. 423–27. Baltimore, Md.: Williams and Wilkins.

Rowse, A.J. and G.H. Fleet. 1982. Viability and release of Salmonella charity and Escherichia coli from oyster feces. *Appl. Environ. Microbiol.* 44(3):544–48.

Rowse, A.J. and G.H. Fleet. 1984. Effects of water temperature and salinity on elimination of Salmonella charity and Escherichia coli from Sydney Rock oysters (Crassostrea commercialis). *Appl. Environ. Microbiol.* 48(5):1061–63.

Rowse, A.J. and G.H. Fleet. 1984. Temperature, salinity important in oyster purification. *Australian Fisheries* 43(5):26–28.

Rubentschik, L., M.B. Roisin, and F.M. Bieljansky. 1936. Adsorption of bacteria in salt lakes. *J. Bacteriol.* 32:11–31.

Rubin, F.A., D.J. Kopecko, K.F. Noon, and L.S. Baron. 1985. Development of DNA probe to detect *Salmonella typhi*. *J. Clin. Microbiol.* 22:600–05.

Ruddy, S.J., R.F. Johnson, J.W. Mosley, J.B. Atwater, M.A. Rossetti, and J.C. Hart. 1969. An epidemic of clam-associated hepatitis. *J. Am. Med. Assoc.* 208:649–55.

Ruiz-Palacios, G.M., J. Torres, N.I. Torres, E. Escamilla, B.R. Ruiz-Palacios, and J. Tamayo. 1983. Cholera-like enterotoxin produced by *Campylobacter jejuni*: Characterization and clinical significance. *Lancet* ii:250–53.

Ruoff, K.L., M.J. Ferraro, M.E. Jerz, and J. Kissling. 1982. Automated identification of gram-positive bacteria. *J. Clin. Microbiol.* 16:1091–95.

Ruoff, K.L. and L.J. Kunz. 1983. Use of the Rapid STREP system for identification of viridans streptococcal species. *J. Clin. Microbiol.* 18:1138–40.

Ruple, A.D., D.W. Cook, and R. Dusho. 1989. *Vibrio vulnificus* in raw Gulf coast oysters. In *Proceedings of the Fourteenth Annual Tropical and Subtropical Fisheries Technological Conference of the Americas*, SGR-101, pp. 118–27. Gainesville, Fla.: Florida Sea Grant College Program.

Rutala, W.A., F.A. Sarubbi Jr., C.S. Finch, J.N. MacCormack, and G.E. Steinkraus. 1982. Oyster-associated outbreaks of diarrheal disease possibly caused by *Plesiomonas shigelloides Lancet* 27:739.

Ruth, J.L. 1984. Chemical synthesis of non-radioactively-labeled DNA hybridization probes. *DNA* 3:123.

Sadler, R. 1989. Urease as a possible tracer for sewage effluent plumes. *Water. Sci. Technol.* 21:93–97.

Sadouk, A., A. Ferhat, J.M. Nuyten, B. Borremans, L. Diels, M. Mergeay, and A. Toussaint. 1990. *Abstr. 22nd Lunteren Lect. Mol. Genet. Abstr.* A13:13.

Sadzikowski, M.R., J.F. Sperry, and T.D. Wilkins. 1977. Cholesterol-reducing bacterium from human feces. *Appl. Environ. Microbiol.* 34:355–62.

Saiki, R.K., D.H. Gelfand, S. Stoffel, S.J. Scharf, R. Higuchi, G.T. Horn, K.B. Mullis, and H.A. Erlich. 1988. Primer-directed enzymatic amplification of DNA with a thermostable DNA polymerase. *Science* 239:487–91.

Saito, M. 1990. Production of enterotoxin by *Clostridium perfringens* derived from humans, animals, foods, and the natural environment in Japan. *J. Food Prot.* 53:115–18.

Salyers, A.A., S.P. Lynn, and J.F. Gardner. 1983. Chromosomal DNA probes for identification of *Bacteroides* species. *J. Bacteriol.* 154:287–93.

Salyers, A.A. 1984. Bacteroides of the human lower intestinal tract. *Ann. Rev. Microbiol.* 38:293–313.

Salzman, G.C. and C.T. Gregg. 1984. Current and experimental methods of rapid microbial identification. 243–48.

Samadpour, M., J. Liston, J. Ongerth, and P. Tarr. 1990. Evaluation of DNA probes for detection of Shiga-like-toxin producing *Escherichia coli* in food and calf fecal samples. *Appl. Environ. Microbiol.* 56(5):1212–15.

Sambrook, J., E.F. Fritsch, and T. Maniatis. 1989. *Molecular Cloning: A Laboratory Manual*, 2nd ed. Cold Spring Harbor, New York: Cold Spring Harbor Press.

Santo Domingo, J.W., F.A. Fuentes, and T.C. Hazen. 1989. Survival and activity of *Streptococcus faecalis* and *Escherichia coli* in petroleum-contaminated tropical marine waters. *Environ. Pollut.* 56:263–81.

Sanyal, S.C. and P.C. Sen. 1973. Human volunteer study on the pathogenicity of *Vibrio parahaemolyticus*. In *Proceedings of International Symposium on Vibrio parahaemolyticus*, September 17–18, 1973. Tokyo, Japan: Saikon Publishing Company, Ltd.

Sartory, D.P. 1986. Membrane filtration enumeration of faecal clostridia and *Clostridium perfringens* in water. *Water Res.* 20:1255–60.

Sato, A. 1971. Enteropathogenic Escherichia coli in commercial foods. I. *J. Food Hyg. Soc. Japan* (*Shokuhin Eiseigaku Zasshi*) 12(6):473–77.

Sato, A. 1972. Distribution of enteropathogenic *Escherichia coli* in commercial foods. II. Drug resistance and R factors in strains of enteropathogenic *E. coli* isolated from commercial fresh meats and oysters. *J. Food Hyg. Soc. Japan* (*Shokuhin Eiseigaku Zasshi*) 13(5):388–91.

Sato, T., K. Izaki, and H. Takahashi. 1972. Recovery of *Escherichia coli* from injury induced by sodium chloride. *J. Gen. Appl. Microbiol.* 18:307–17.

Savage, D.C. 1977. Microbial ecology of the gastrointestinal tract. In *Ann. Rev. Microbiol.* ed. M.P. Starr, pp. 107–33. Annual Reviews Inc., Palo Alto.

Sawyer, T.K., G.S. Visesvara, and B.A. Harke. 1977. Pathogenic amebas from brackish and ocean sediments, with a description of *Acanthamoeba hatchetti*, n. sp. *Science* 196:1324–25.

Sawyer, T.K. 1980. Marine amoebae from clean and stressed bottom sediments of the Atlantic Ocean and Gulf of Mexico. *J. Protozool.* 27:13–32.

Sawyer, T.K., E.J. Lewis, M. Galass, D.W. Lear, M.L. O'Malley, W.N. Adams, and J. Gaines. 1982. Pathogenic amoebae in ocean sediments near wastewater sludge disposal sites. *J. Water Pollut. Control Fed.* 54:1318–23.

Sawyer, T.K., T.A. Nerad, P.-M. Daggett, and S.M. Bodammer. 1987. Potentially pathogenic protozoa in sediments from oceanic sewage-disposal sites. In *Oceanic Processes in Marine Pollution, Vol. 1, Biological Processes and Wastes in the Ocean*, eds. J.M. Capuzzo, and D.R. Kester, pp. 183–94. Malaar, Fla.: Robert E. Krieger.

Sayler, G.S. and A.C. Layton. 1990. Environmental application of nucleic acid hybridization. In *Ann. Rev. Microbiol.*, ed. L.N. Ornston, pp. 625–48. Annual Reviews, Palo Alto.

Scardovi, V. 1981. Chapter 149. The genus *Bifidobacterium*. In *The Prokaryotes*, eds. M.P. Starr, H. Stolp, H.G. Truper, A. Balows, and H.G. Schlegel, Vol. II, pp. 1951–61. New York: Springer.

Scarpino, P.V. 1975. Human enteric viruses and bacteriophages as indicators of sewage pollution. In *Discharge of Sewage from Sea Outfalls*, ed. A.L.H. Gameson, pp. 49–61. Oxford: Pergamon Press.

Schaechter, M. 1961. Patterns of cellular control during unbalanced growth. *Cold Spring Harbor Symp. Quant. Biol.* 26:53–62.

Schiemann, D.A. 1982. Development of a two-step enrichment procedure for recovery of *Yersinia enterocolitica* from food. *Appl. Environ. Microbiol.* 43:14–27.

Schiemann, D.A. 1983. Alkalotolerance of *Yersinia enterocolitica* as a basis for selective isolation from food enrichments. *Appl. Environ. Microbiol.* 46:22–27.

Schiff, G.M., G.M. Stefanovic, B. Young, and J.K. Pennekamp. 1984. Minimum human infectious dose of enteric virus (echovirus-12) in drinking water. *Monogr. Virol.* 15:222–28.

Schleifer, K.H. and R. Kilpper-Balz. 1984. Transfer of *Streptococcus faecalis* and *Streptococcus faecium* to the genus *Enterococcus* nom. rev. as *Enterococcus faecalis* comb. nov. and *Enterococcus faecium* com. nov. *Int. J. Syst. Bacteriol.* 34:31–34.

Schleifer, K.H., W. Ludwig, J. Kraus, and H. Festl. 1985. Cloned ribosomal ribonucleic acid genes from *Pseudomonas aeruginosa* as probes for conserved deoxyribonucleic acid sequences. *Int. J. Syst. Bacteriol.* 35:231–36.

Schmidt, N.J., H.H. Ho, and E.H. Lennette. 1976. Comparative sensitivity of the BGM cell line for isolation of enteric viruses. *Health Lab. Sci.* 13:115–17.

Schmidt, N.J., H.H. Ho, J.L. Riggs, and E.H. Lennette. 1978. Comparative sensitivity of various cell culture systems for isolation of viruses from wastewater and fecal samples. *Appl. Environ. Microbiol.* 36:480–86.

Schmitt, C.J., J.L. Zajicek, and P.H. Peterman. 1990. National Contaminant Biomonitoring Program: Residues of organochlorine chemicals in U.S. freshwater fish, 1976–1984. *Archives Environ. Contamination Toxicology* 19:748–81.

Schneuser, D.L., F.F. Busta, and M.L. Speck. 1971. Injury of bacteria by sanitizers. *Appl. Microbiol.* 21:41–45.

Schubert, R.H.W. 1984. *Plesiomonas.* In *Bergey's Manual of Systematic Bacteriology*, eds. N. Krieg and J.G. Holt. Baltimore, Md.: Williams and Wilkins.

Schwab, K.J., R. De Leon, and M.D. Sobsey. 1992. Development of PCR methods for enteric virus detection in water. *Wat. Sci. Tech.*, (in press).

Schwartz, B., C.V. Broome, G.R. Brown, A.W. Hightower, C.A. Ciesielski, S. Gaventa, B.G. Gellin, L. Mascola, and the Listeriosis working group. 1988. Association of sporadic cases of listeriosis with consumption of undercooked hotdogs and undercooked chicken, *Lancet* 1:779–82.

Scott, R.M., D. Seiz, and H.J. Shaughnessy. 1964*a*. I. Rapid carbon-14 test for coliform bacteria in water. *Am. J. Public Health* 54:827–33.

Scott, R.M., D. Seiz, and H.J. Shaughnessy. 1964*b*. II. Rapid carbon-14 test for sewage bacteria. *Am. J. Public Health* 54:834–44.

Scotti, P.D., G.C. Fletcher, D.H. Buisson, and S. Fredericksen. 1983. Virus depuration in the Pacific oyster (Crassostrea virginica) in New Zealand. *New Zealand J.* 26:9–13.

Seattle-King County Department of Public Health. 1986. Surveillance of the flow of *Salmonella* and *Campylobacter* in a community. Prepared for the Bureau of Veterinary Medicine, USFDA. Contract No. 223-81-7041. Communicable Disease Control Section, p. 250. Seattle-King County Department of Public Health, Seattle.

Seeley, N.D. and S.B. Primrose. 1980. The effect of temperature on the ecology of aquatic bacteriophages. *J. Gen. Virol.* 46:87–95.

Seeley, N.D. and S.B. Primrose. 1982. The isolation of bacteriophages from the environment. *J. Appl. Bacteriol.* 53:1–17.

Seelinger, H.P.R. 1961. *Listeriosis.* New York: Hafner.

Seelinger, H.P.R. and D. Jones. 1986. *Listeria.* In *Bergey's Manual of Systematic Bacteriology*, eds. P.H.A. Sneath, N.S. Mair, M.E. Sharpe, and J.G. Holt, Vol. 2, pp 1235–45. Baltimore, Md.: Williams and Wilkins.

Seelinger, H.P.R. and K. Hohne. 1990. Serotyping of *Listeria monocytogenes* and related species. In *Methods in Microbiology*, eds. J.T. Bergen and J.R. Norris, pp. 31–49. London: Academic Press.

Seraichekas, H.R., D.A. Brashear, J.A. Barnick, P.F. Carey, and O.C. Liu. 1968. Viral depuration by assaying individual shellfish. *Appl. Microbiol.* 16:1865–71,

Seriwatana, J., P. Echeverria, D.N. Taylor, T. Sakuldaipeara, S. Changchawalit, and O. Chivoratanoid. 1987. Identification of enterotoxigenic *Escherichia coli* with synthetic alkaline phosphatase-conjugated oligonucleotide probes. *J. Clin. Microbiol.* 25:1438–41.

Servais, P., G. Billen, and J. Vives-Rego. 1985. Rate of bacterial mortality in aquatic environments. *Appl. Environ. Microbiol.* 49:1448–54.

Sethabutr, O., P. Echeverria, and D.N. Taylor. 1985. DNA hybridization in the identification of enteroinvasive *Escherichia coli* and *Shigella* in children with dysentery. In *Infectious Diarrhea in the Young*, ed. S. Tzipor, Amsterdam: Elsevier.

Sethabutr, O., S. Hanchalay, P. Echeverria, D.N. Taylor, and U. Leksomboon. 1985. A non-radioactive DNA probe to identify *Shigella* and enteroinvasive *Escherichia coli* in stools of children with diarrhoea. ii:1095–97.

Shah, D.B. and U. Rhea. 1986. Foodborne enterotoxigenic *Escherichia coli*: Identification and enumeration on nitrocellulose membrane by enzyme immunoassay. *Int. J. Food Microbiol.* 3(2):79–87.

Shah, D.B., J.C. Wimsatt, and P.E. Kauffman. 1987. Enzyme linked immunosorbent assay (ELISA). In *FDA Bacteriological Analytical Manual.* Arlington, Va.: Association of Official Analytical Chemists.

Shahamat, M., R.R. Colwell, C. Paszko-Kolva, and A. Pearson. 1989. In *Abstr. Annu. Meet. Amer. Soc. Microbiol.*, Vol. 24, p. 490.

Shandera, W.X., J.M. Johnston, B.R. Davis, and P.A. Blake. 1983. Disease from infection with *Vibrio mimicus*, a newly recognized *Vibrio* species. *Annals Internal Med.* 99:169–71.

Sharpe, A.N., P.I. Peterkin, and N. Malik. 1979. Improved detection of coliforms and *Escherichia coli* in foods by a membrane filter method. *Appl. Environ. Microbiol.* 38:431–35.

Sharpe, A.N. 1981. Hydrophobic grid-membrane filters: The (almost) perfect system. In *Membrane Filtration: Applications, Techniques, and Problems*, ed. B.J. Dutka, pp. 513–35. New York: Marcel Dekker.

Shaw, B.G., C.D. Harding, W.H. Hudson, and L. Farr. 1987. Rapid estimation of microbial numbers on meat and poultry by the direct epifluorescent filter technique. *J. Food Prot.* 50:652–57.

Shear, C.L. and M.S. Gottlieb. 1980. Shellfishborne disease control in the United States: A commentary. *Medical Hypotheses* 6:315–27.

Shiaris, M.P., A.C. Rex, G.W. Pettibone, K. Keay, P. McManus, M.A. Rex, J. Ebersole, and E. Gallagher. 1987. Distribution of indicator bacteria and *Vibrio parahaemolyticus* in sewage-polluted intertidal sediments. *Appl. Environ. Microbiol.* 53:1756–61.

Shieh, Y.S.C., R.S. Baric, M.D. Sobsey, J. Ticehurst, R. DeLeon, and R. Walter. 1990. Detection of hepatitis A virus and other enteroviruses in water by ssRNA probes. *J. Virol. Methods* (accepted for publication).

Shigenaka, G. and J.E. Price. 1988. Correlation of coprostanol to organic contaminants in coastal and estuarine sediments of the US. *Water Res. Bull.* 24:989–98.

Shipe, J., E.L. and G.M. Cameron. 1954. A comparison of the membrane filter with the most probable number method for coliform determinations from several waters. *Appl. Microbiol.* 2:85–88.

Shotts, E.B. Jr., J.L. Gaines Jr., L. Martin, and A.K. Priestwood. 1972. *J. Am. Veterinary Medical Assoc.* 161:603–06.

Shumway, S.E., S. Sherman-Caswell, and J. Hurst. 1988. Paralytic shellfish poisoning in Maine: Monitoring a monster. *J. Shellfish Res.* 7:643–52.

Shumway, S.E. 1990. A review of the effects of algal blooms on shellfish and aquaculture. *J. World Aquaculture Soc.* 21:65–104.

Shuval, H.J., A. Thompson, B. Fattal, S. Cymbalista, and Y. Wiener. 1971. Natural virus inactivation processes in seawater. *J. Sanit. Eng. Div. Am. Soc. Civ. Eng.* 97:587–601.

Sieburth, J.M. and D.M. Pratt. 1962. Anticoliform activity of sea water associated with the termination of *Skeletonema costatum* blooms. *Trans. NY Acad. Sci.* 24:498–501.

Silliker, J.H. 1982. The Salmonella problem: Current status and future direction. *J. Food Prot.* 45:661–66.

Simard, R.E. 1971. Yeasts as an indicator of pollution. *Mar. Poll. Bull.* 2:123–25.

Singh, A., R. Yeager, and G.A. McFeters. 1986. Assessment of in vivo revival, growth, and pathogenicity of *Escherichia coli* strains after copper and chlorine-induced injury. *Appl. Environ. Microbiol.* 52:832–37.

Singh, A., B.H. Pyle, and G.A. McFeters. 1989. Rapid enumeration of viable bacteria by image analysis. *J. Microbiol. Methods* 10:91–101.

Singh, A., F.-P. Yu, and G.A. McFeters. 1990. Rapid detection of chlorine-induced bacterial injury by the direct viable count method using image analysis. *Appl. Environ. Microbiol.* 56:389–94.

Singh, S.N. and C.P. Gerba. 1983. Concentration of coliphage from water and sewage with charge-modified filter aid. *Appl. Environ. Microbiol.* 45:232–37.

Singleton, F.L., R.W. Atwell, S. Jangi, and R.R. Colwell. 1982. Influence of salinity and organic nutrient concentration on survival and growth of *Vibrio cholerae* in aquatic microcosms. *Appl. Environ. Microbiol.* 43:1080–85.

Sinton, L.W. and S.B. Ching. 1987. An evaluation of two bacteriophages as sewage tracers. *Water, Air, Soil Pollut.* 35:347–56.

Siragusa, G.R. and M.G. Johnson. 1990. Monoclonal antibody specific for *Listeria monocytogenes, Listeria innocua*, and *Listeria welshimeri*. *Appl. Environ. Microbiol.* 56:1897–904.

Sjogren, R.E. and M.J. Gibson. 1981. Bacterial survival in a dilute environment. *Appl. Environ. Microbiol.* 41:1331–36.

Skiba, U. and M. Wainwright. 1982. Assay of urease activity in marine sands-its use as an indicator of sewage contamination of beaches. *Enzyme Microb. Technol.* 4:310–12.

Slade, J.S. and B.J. Ford. Discharge in the environment of viruses in wastewater, sludges, and aerosols. In *Viral Pollution of the Environment*, ed. G. Berg, pp. 3–18. Boca Raton, Fla.: CRC Press.

Slanetz, L.W. and C.H. Bartley. 1957. Numbers of enterococci in water, sewage and feces determined by the membrane filter technique with an improved medium. *J. Bacteriol.* 74:591–95.

Slanetz, L.W. and C.H. Bartley. 1965. Survival of fecal streptococci in sea water. *Health Lab. Sci.* 2:142–48.

Slantez, L.W., C.H. Bartley, and K.W. Stanley. 1968. Coliforms, fecal streptococci and *Salmonella* in seawater and shellfish. *Health Lab. Sci.* 5:66–78.

Slots, J. 1981. Enzymatic characterization of some oral and non-oral gram-negative bacteria with API ZYME system. *J. Clin. Microbiol.* 14:288–94.

Small, P.L.C. and S. Falkow. 1986. Development of a DNA probe for the virulence plasmid of *Shigella* spp. and enteroinvasive *Escherichia coli*. In *Microbiology*, eds. L. Leive, P.F. Bonventure, J.A. Morello, S.D. Silver, and H.C. Wu, pp. 121–24. Washington, DC: Amer. Soc. Microbiol.

Smibert, R.M. 1984. *Campylobacter*. In *Bergey's Manual of Systematic Bacteriology*, eds. N.R. Krieg and J.G. Holt, pp. 11–118. Baltimore, Md.: Williams and Wilkins.

Smith, E.M., C.P. Gerba, and J.L. Melnick. 1978. Role of sediment in the persistence of enteroviruses in the estuarine environment. *Appl. Environ. Microbiol.* 35:685–89.

Smith, E.M. and C.P. Gerba. 1982. Laboratory methods for the growth and detection of animal viruses. In *Methods in Environmental Virology*, eds. C.P. Gerba and S.M. Goyal, New York: Marcel Dekker.

Smith, J., R. Benedict, and R. Buchanan. 1987. *Listeria monocytogenes* in meat and poultry products: Review of factors affecting its introduction and growth in these foods. Rpt. of the Microbial Food Safety Res. Unit. Eastern Reg. Res. Ctr. USDA, ARS. Philadelphia, Pa.

Smith, J.L. 1987. *Shigella* as a foodborne pathogen. *J. Food Prot.* 50:788–801.

Smith, L.D.S. 1975. Common mesophilic anaerobes, including *Clostridium botulinum* and *Clostridium tetani*, in 21 soil specimens. *Appl. Microbiol.* 29:590–94.

Smith, P.B., K.M. Tomfohrde, D.L. Rhoden, and A. Balows. 1971. API system: A multitube micro method for identification of *Enterobacteriaceae*. *Appl. Environ. Microbiol.* 24:449–52.

Smith, P.B., D.L. Rhoden, K.M. Tomfohrde, C.R. Dunn, and G.J. Hermann. 1972. R/B enteric differential system for identification of *Enterobacteriaceae*. *Appl. Microbiol.* 21:1036–39.

Snowdon, J.A. and D.O. Cliver. 1989. Coliphages as indicators of human enteric viruses in groundwater. *Crit. Rev. Environ. Control* 19:231–49.

So, M., W.S. Dallas, and S. Falkow. 1978. Characterization of an *Escherichia coli* plasmid encoding for synthesis of heat-labile toxin: Molecular cloning of the toxin determinant. *Infect. Immun.* 21:405–11.

Sobsey, M.D., C. Wallis, and J.L. Melnick. 1975. Development of a simple method for concentrating enteroviruses from oysters. *Appl. Microbiol.* 29:21–26.

Sobsey, M.D., B.J. Carrick, and H.R. Jensen. 1978. Improved methods for detecting enteric viruses in oysters. *Appl. Environ. Microbiol.* 36:121–28.

Sobsey, M.D., R. Carrick, C. Hackney, and B. Ray. 1980. Enterovirus elimination from oysters in estuary water. *Abstr. Ann. Meet. Am. Soc. Microbiol.* 80:178.

Sobsey, M.D., C.R. Hackney, R.J. Carrick, B. Ray, and M.L. Speck. 1980. Occurrence of enteric bacteria and viruses in oysters. *J. Food Prot.* 43:111–13.

Sobsey, M.D. 1982. Detection of viruses in shellfish. In *Methods in Environmental Virology*, eds. C.P. Gerba and S.M. Goyal, pp. 243–59. New York: Marcel Dekker.

Sobsey, M.D., E. Stokes, T. Cromeans, D.A. Wait, and S.E. Oglesbee. 1983. Enhancement of hepatitis A virus (HAV) growth in African green monkey kidney (AGMK) cell cultures by pancreatin and trypsin. *Abstr. Ann. Meet. Am. Soc. Microbiol.* 83:294.

Sobsey, M.D. 1984. Procedures for the virological examination of seawater, Shellfish, and sediment. In *Laboratory Procedures for the Examination of Seawater and Shellfish*, eds. A.E. Greenberg and D.A. Hunt, pp. 81–118. Washington, D.C.: American Public Health Association.

Sobsey, M.D., S.E. Oglesbee, D.A. Wait and A.I. Cuenca. 1985. Detection of hepatitis A virus (HAV) in drinking water. *Water Sci. Technol.* 17:23–38.

Sobsey, M.D. 1985. Development of hepatitis A virus infectivity assay for Eastern oysters. Final Contract Report No. NA-83-GA-C-00032. Charleston, S.C.: Nathounal Marine Fisheries Service.

Sobsey, M.D., S.E. Oglesbee, and D.A. Wait. 1985. Evaluation of methods for concentrating hepatitis A virus from drinking water. *Appl. Environ. Microbiol.* 50:1457–63.

Sobsey, M.D. 1987. Methods for recovering viruses from shellfish, seawater, and sediment. In *Methods for Recovering Viruses From the Environment*, ed. G. Berg, pp. 77–108. Boca Raton, Fla.: CRC Press.

Sobsey, M.D., A.L. Davis, and V.A. Rullman. 1987. Persistence of hepatitis A virus and other viruses in depurated Eastern oysters. In *Oceans '87, Proceedings Coastal and Estuarine Pollution, Vol. 5*, pp. 1740–45. Washington, D.C.: Marine Technology Society.

Sobsey, M.D., P.A. Shields, F.S. Hauchman, A.L. Davis, V.A. Rullman, and A. Bosch. 1988. Survival and persistence of hepatitis A virus in environmental samples. In *Proceedings of the International Symposium on Viral Hepatitis and Liver Disease*, ed. A.J. Zuckerman, pp. 121–24. New York: Alan R. Liss.

Sobsey, M.D., K.J. Schwab, and T.R. Handzel. 1990. A simple membrane filter method to concentrate and enumerate male-specific RNA coliphages. *J. AWWA* 82:52–59.

Sodergren, A. 1989. Biological effects of bleached pulp mill effluents. National Swedish Environmental Protection Board, Report 3558. Solna, Sweden.

Sokolovic, Z. and W. Goebel. 1988. Synthesis of listeriolysin in *Listeria monocytogenes* under heat shock conditions. *Infect. Immun.* 57:295–98.

Somerville, C.C., I.T. Knight, W.L. Straube, and R.R. Colwell. 1988. Probe-directed, polymerization-enhanced detection of specific gene sequences in the environment. REGEM 1, Abstract 70.

Somerville, C.C., I.T. Knight, W.L. Straube, and R.R. Colwell. 1989. Simple, rapid method for direct isolation of nucleic acids from aquatic environments. *Appl. Environ. Microbiol.* 55:548–54.

Son, N.T. and G.H. Fleet 1980. Behavior of pathogenic bacteria in the oyster, *Crassostrea commercialis*, during depuration, re-laying, and storage. *Appl. Environ. Microbiol.* 40(6):994–1002.

Souness, R.A. and G.H. Fleet. 1979. Depuration of the Sydney rock oyster, *Crassostrea commercialis. Food Technol. Australia* 31:397–404.

Speaker, J. 1990. Rhode Island Department of Natural Resources, R.I.: Division of Water Resources (personal communication).

Speck, M., B. Ray, and R. Reed. 1975. Repair and enumeration of injured coliforms by a plating procedure. *Appl. Microbiol.* 29:549–50.

Spellman, J.R., C.S. Levy, J.A. Curtin, and C. Ormes. 1986. *Vibrio fluvialis* and gastroenteritis. *Annals Internal Med.* 105:294–95.

Sproul, O.J., M.A. Emerrson, D.M. Howser, D.S. Boyce, D.S. Walsh, and C.E. Buck. 1985. Effects of particulate matter on virus inactivation by ozone. In *Ozone in Water Treatment—Applications, Operations, and Technology*, pp 238–45. Denver, Co.: American Water Works Association.

Stahl, D.A., B. Flesher, H.R. Mansfield, and L. Montgomery. 1988. Use of phylogenetically based hybridization probes for studies of ruminal microbial ecology. *Appl. Environ. Microbiol.* 54:1079–84.

Stanfield, J.T., G.J. Jackson, and C.C.G. Aulisio. 1985. *Yersinia enterocolitica*: Survival of a pathogenic strain on milk containers. *J. Food Prot.* 48:947–48.

Steffan, R.J. and R.M. Atlas. 1987. Use of specific gene probes for the detection of *Pseudomonas cepacia* DC102 in sediments. In ed., Abstr. Annu. Meet. Amer. Soc. Microbiol., Q130, Atlanta, Ga.

Steffan, R.J. and R.M. Atlas. 1988. DNA amplification to enhance detection of genetically engineered bacteria in environmental samples. *Appl. Environ. Microbiol.* 54:2185–91.

Stelma, G.N. and L.J. McCabe. 1990. Non-point pollution from animal sources and shellfish sanitation. *J. Food Prot.* (submitted for publication).

Stephen, S., R. Indrani, M. Kotian, and K. Rao. 1975. Isolation of enteropathogenic Escherichia coli from sea fish and fresh water mussels. *Ind. J. Microbiol.* 15(2):64–67.

Stern, N. 1982. *Yersinia enterocolitica*: Recovery from foods and virulence characterization. *Food Technol.* 36:84.

Stern, N. and S. Kazmi. 1989. *Campylobacter jejuni.* In *Foodborne Bacterial Pathogens*, ed. M.P. Doyle, pp. 71–110. New York: Marcel Dekker.

Stersky, A., R. Szabo, E. Todd, C. Thacker, N. Dickei, and M. Akhiar. 1986. *Staphylococcus aureus* growth and thermostable nuclease and enterotoxin production in canned salmon and sardines. *J. Food Prot.* 49:428–35.

Stevens, T.P. 1988. *California State Mussel Watch Marine Water Quality Monitoring Program* 1986–87. Water Quality Monitoring Report No. 88-3, Division of Water Quality. California State Water Resources Control Board.

Stevenson, L.N. 1978. A case for bacterial dormancy in aquatic systems. *Microb. Ecol.* 4:127–33.

Stibbs, H.H., E.T. Riley, J. Stockard, J.L. Riggs, P.M. Wallis, and J. Isaac-Renton. 1988. Immunofluorescence differentiation between various animal and human source *Giardia* cysts using monoclonal antibodies. In *Advances in Giardia Research*, eds. P.M. Wallis and B.R. Hammond, pp. 159–63. Calgary: University of Calgary Press.

Stiles, G.W. 1911. The bacteriological examination of shucked and shell oysters. *J. Am. Public Health Assoc.* 1:623.

Stiles, M. 1989. Less recognized or presumptive foodborne pathogenic bacteria. In *Foodborne Bacterial Pathogens*, ed. M.P. Doyle, p. 673–734. New York: Marcel Dekker.

Stiles, M.E. and L.D. Witter. 1965. Thermal inactivation, heat injury, and recovery of *Staphylococcus aureus. J. Dairy Sci.* 48:677–81.

Stille, W., B. Kunkel, and K. Nerger. 1972. Austern hepatitis. *Dusch. Med. Wschr.* 97:145–47.

Stimson, A.M. 1938. A brief history of bacteriological investigations of the United States Public Health Service. In *Public Health Reports*, Supplement No. 141. Washington, D.C.: U.S. Government Printing Office.

Strange, R.E. 1976. *Microbial Response to Mild Stress*. Durham, England: Meadowfield Press.

Stratman, S. 1988. Rapid specific environmental coliform monitoring. *Am. Lab.* 20:60–64.

Straube, W.L., M. O'Brien, K. Davis, and R.R. Colwell. 1990. Enzymatic profiles of 11 barophilic bacteria under *in situ* conditions: Evidence for pressure modulation of phenotype. *Appl. Environ. Microbiol.* 56:812–14.

Strauss, H. and D. Hattis. 1989. *Estimation of HIV and HBV Infectious Titers in Human Fluids and Tissues*, MIT Center for Technology, Policy, and Industrial Development. Cambridge, Mass.: CTPID 89-4, June.

Streeter, H.W. and E.B. Phelps. 1925. A Study of the Pollution and Natural Purification of the Ohio River. Public Health Bulletin 146. Washington, D.C.: U.S. Public Health Service.

Stuart, D.G., G.A. McFeters, and J.E. Schillinger. 1977. Membrane filter technique for the quantification of stressed fecal coliforms in the aquatic environment. *Appl. Environ. Microbiol.* 34:42–46.

Sullivan, R. 1981. A method for recovery of virus from oysters. *Abstr. Ann. Meet. Am. Soc. Microbiol.* 81:196.

Sullivan, R. and J.T. Peeler. 1982. Evaluation of a method for recovery of virus from oysters. *J. Food Prot.* 45:636–37.

Sullivan, R., J.T. Peeler, J.T. Tierney, and E.P. Larkin. 1983. A method for recovery of enterovirus from 100-gram samples of oysters (Crassostrea virginica). *Abstr. Ann. Meet. Am. Soc. Microbiol.* 83:256.

Sullivan, R., J.T. Peeler, J.T. Tierney, and E.P. Larkin. 1984. Evaluation of a method for recovering poliovirus 1 from 100-gram oyster samples. *J. Food Prot.* 47:108–10.

Sullivan, R., J.T. Peeler, and E.P. Larkin. 1985. A method for recovery of poliovius 1 from 25-gram food samples. *Abstr. Ann. Meet. Am. Soc. Microbiol.* 85:270.

Switzer, R.E. and J.B. Evans. 1974. Evaluation of selective media for enumeration of group D streptococci in bovine feces. *Appl. Microbiol.* 28:1086–87.

Syvanen, A.C., M. Laaksonen, and Söderlund. 1986. Fast quantification of nucleic acid hybrids by affinity-based collection. *Nucleic Acids Res.* 14:5037–48.

Szabo, R.A., E.C.D. Todd, and A. Jean. 1986. Method to isolate *Escherichia coli* O157:H7 from food. *J. Food Prot.* 49:768–72.

Tabor, E. 1985. The three viruses of non-A non-B hepatitis. *Lancet* i:743–45.

Tabor, P.S. and R.A. Neihof. 1982*a*. Improved method for determination of respiring individual microorganisms in natural waters. *Appl. Environ. Microbiol.* 43:1249–55.

Tabor, P.S. and R.A. Neihof. 1982*b*. Improved microautoradiographic method to determine individual microorganisms active in substance uptake in natural waters. *Appl. Environ. Microbiol.* 44:945–53.

Tacket, C.O., T.J. Brenner, and P.A. Blake. 1984. Clinical features and an epidemiological study of *Vibrio vulnificus* infections. *J. Infect. Dis.* 149:558–61.

Tacket, C.O., J.P. Narain, R. Sattin, J.P. Lofgren, C. Konigsberg Jr., R.C. Rendtorff, A. Rausa, B.R. Davis, and M.L. Cohen. 1984. A multistate outbreak of infections caused by *Yersinia enterocolitica* transmitted by pasteurized milk. *J. Am. Med. Assoc.* 251:483–86.

Tacket, C.O., N. Shahid, M.I. Huq, A.R.M.A. Alim, and M.L. Cohen. 1984. Usefulness of plasmid profiles for differentiation of *Shigella* isolates in Bangladesh. *J. Clin. Microbiol.* 20:300–01.

Tacket, C.O., L.B. Dominguez, H.L. Fisher, and M.L. Cohen. 1985. An outbreak of multiple drug resistant *Salmonella* enteritis from raw milk. *J. Am. Med. Assoc.* 253:2058–60.

Tacket, C.O., J. Ballard, N. Harris, J. Allard, C. Nolan, T. Quan, and M.L. Cohen. 1985. An outbreak of *Yersinia enterocolitica* infections caused by contaminated tofu (soybean curd). *Am. J. Epidemiol.* 121:705–11.

Tamplin, M., G.E. Rodrick, N.J. Blake, and T. Cuba. 1982. Isolation and characterization of *Vibrio vulnificus* from two Florida estuaries. *Appl. Environ. Microbiol.* 44(6):1466–70.

Tamplin, M.L., A.L. Gauzens, A. Huq, D.A. Sack, and R.R. Colwell. 1990. Attachment of *Vibrio cholerae* serogroup 01 to zooplankton and phytoplankton of Bangladesh waters. *Appl. Environ. Microbiol.* 56:1977–80.

Tanaka, R. and M. Mutai. 1980. Improved medium for selective isolation and enumeration of *Bifidobacterium*. *Appl. Environ. Microbiol.* 40:866–69.

Tardio, J., K. O'Orien, and T. Latt. 1988. Identification of *Escherichia coli* from shellfish and related environments by automicrobic system. *J. Assoc. Off. Anal. Chem.* 71:582–84.

Tartera, C. and J. Jofre. 1987. Bacteriophages active against *Bacteroides fragilis* in sewage-polluted waters. *Appl. Environ. Microbiol.* 53:1632–37.

Tartera, C., A. Bosch, and J. Jofre. 1988. The inactivation of bacteriophages infecting *Bacteroides fragilis* by chlorine treatment and UV-radiation. *FEMS Microbiol. Lett.* 56:313–16.

Tartera, C., F. Lucena, and J. Jofre. 1989. Human origin of *Bacteroides fragilis* bacteriophages present in the environment. *J. Appl. Microbiol.* 55:2696–701.

Tauxe, R.V., N. Hargrett-Bean, C.M. Patton, and I.K. Wachsmuth. 1988. *Campylobacter* infections in the United States, 1982–86. Morbidity and Mortality Weekly Report Surveillance Suppl. Publisher CDC.

Tauxe, R.V., J. Vandepitte, G. Wauters, S.M. Martin, V. Goossens, P. DeMol, R. VanNoyen, and G. Thiers. 1989. *Yersinia enterocolitica* infections and pork: The missing link. *Lancet* i:1129–31.

Taylor, B.C. and M. Nakamura. 1964. Survival of shigellae. *J. Hyg.* 62:303.

Taylor, D.N., I.K. Wachsmuth, Y. Shangkuan, E.V. Schmidt, T.J. Barrett, J.S. Schrader, C.S. Scherach, H.B. McGee, R.A. Feldman, and D.J. Brenner. 1982. Salmonellosis associated with marijuana—a multistate outbreak traced by plasmid fingerprinting. *New Engl. J. Med.* 306:1249–53.

Taylor, F.J.R. 1990. Red tides, brown tides and other harmful blooms: the view into the 1990's. In *Toxic Marine Phytoplankton*, eds. E. Graneli, B. Sundstrom, L. Edler, and D.M. Anderson, pp. 527–33. New York: Elsevier.

Taylor, S.L. 1985. Marine toxins of microbiological origin. *Food Technol.* 42:94–98.

Tenover, F.C., S. Williams, K.P. Gordon, N. Harris, C. Nolan, and J.J. Plorde. 1984. Utility of plasmid fingerprinting for epidemiological studies of *Campylobacter jejuni* infections. *J. Infect. Dis.* 149:279.

Tenover, F.C. 1988. Diagnostic deoxyribonucleic acid probes for infectious diseases. *Clin. Microbiol. Rev.* 1:82–101.

Teshima, S. and A. Kanazawa. 1978. Conversion of cholesterol to coprostanol and cholestanol in the estuary sediment. *Mem. Fac. Fish. Kagoshima Univ.* 27:41–47.

Teutsch, T. 1990. Maryland Department of the Environment, Water Management Administration. Baltimore, Md. (personal communication).

Texas Sea Grant Program. 1986. *Red Tide in Texas, An Explanation of the Phenomenon.* College Station, Tx.: Texas A & M University.

Tharagonnet, D., P. Sisson, C. Roxby, H. Ingham, and J. Selken. 1977. API ZYME system in the identification of gram negative anaerobes. *J. Clin. Pathol.* 30:505–09.

Thi Son, N. and G. Fleet. 1980. Behavior of pathogenic bacteria in the oyster, Crassostrea commercialis, during depuration, relaying, and storage. *Appl. Environ. Microbiol.* 40(6):994–1002.

Thomas, K.L. and A.M. Jones. 1971. Comparison of methods of estimating the number of *Escherichia coli* in edible mussels and the relationship between the presence of Salmonellae and *E. coli. J. Appl. Bacteriol.* 34(4):717–25.

Thomason, B.M., J.W. Biddle, and W.B. Cherry. 1975. Detection of salmonellae in the environment. *Appl. Environ. Microbiol.* 30:764–67.

Thomason, B.M. 1981. Current status of immunofluorescent methodology for salmonellae. *J. Food Prot.* 44:381–84.

Thompson, C.A., C. Vanderzant, and S.M. Ray. 1976*a*. Effect of processing, distribution and storage on *Vibrio parahaemolyticus* and bacterial counts of oysters (*Crassostrea virginica*). *J. Food Sci.* 41(1):123–27.

Thompson, C.A., C. Vanderzant, and S.M. Ray. 1976*b*. Relationship of *Vibrio parahaemolyticus* in oysters, water and sediment and bacteriological and environmental indices. *J. Food Sci.* 41:117–22.

Thompson, M.R., R.L. Jordan, M.A. Luttrell, H. Brandwein, J.B. Kaper, M.M.

Levine, and R. Giannella. 1986. Blinded, two-laboratory comparative analysis of *Escherichia coli* heat-stable enterotoxin production by using monoclonal antibody enzyme-linked immunosorbent assay and gene probes. *J. Clin. Microbiol.* 24:753–58.

Thompson, R. 1990. Florida Department of Natural Resources, Division of Marine Resources. Tallahassee, Fla. (personal communication).

Thurberg, F.P. 1975. Inactivation of red-tide toxins by ozone treatment. In *Aquatic Application of Ozone*, pp. 50–58. Syracuse, New York: International Ozone Institute.

Tierney, J.T., A. Fassolitis, D. Van Dorsel, V.C. Rao, R. Sullivan, and E.P. Larkin. 1980. Glass-wool hydroextraction method for recovery of human enteroviruses from shellfish. *J. Food Prot.* 43:102–04.

Tierney, J.T., R. Sullivan, J.T. Peeler, and E.P. Larkin. 1982. Persistence of polioviruses in shellstock and shucked oysters stored at refrigeration temperature. *J. Food Prot.* 45:1135–37.

Tierney, J.T., R. Sullivan, J.T. Peeler, and E.P. Larkin. 1985. Detection of low numbers of poliovirus 1 in oysters: Collaborative study. *J. Assoc. Off. Anal. Chem.* 68:884–86.

Tilton, R.C. and R.W. Ryan. 1987. Clinical and ecological characteristics of *Vibrio vulnificus* in the Northeastern United States. *Diagn. Microbiol. Infect. Dis.* 6:109–17.

Timoney, J.F. and A. Abston. 1984. Accumulation and elimination of *Escherichia coli* and *Salmonella typhimurium* by hard clams in an in vitro system. *Appl. Environ. Microbiol.* 47(5):986–88.

Todd, E.C.D., R.A. Szabo, P. Peterkin, A.N. Sharpe, L. Parrington, D. Bundle, M.A.J. Gidney, and M.B. Perry. 1988. Rapid hydrophobic grid membrane filter—enzyme labeled antibody procedure for identification and enumeration of *Escherichia coli* O157 in foods. *Appl. Environ. Microbiol.* 54:2536–40.

Todd, E.C.D. 1990. Amnesic shellfish poisoning-a new seafood toxin syndrome. In *Toxic Marine Phytoplankton*, eds. E. Graneli, B. Sundstrom, L. Edler, and D.M. Anderson. New York: Elsevier.

Todorov, I. 1968. Microflora of the Black Sea mussel Mytilus Galloprovincialis (L). *Veterinarnomedicinski Nauki* 5(7):49–55.

Tomfohrde, K.M., D.L. Rhoden, P.B. Smith, and A. Balows. 1973. Evaluation of the redesigned Enterotube—a system for the identification of *Enterobacteriaceae*. *Appl. Microbiol.* 25:301–04.

Tompkins, L.S. and M. Krajden. 1986. Approaches to the detection of enteric pathogens, including *Campylobacter* using nucleic acid hybridization. *Diagn. Microbiol. Infect. Dis.* 4:71S–78S.

Tompkins, L.S., N. Troup, A. Labigne-Roussel, and M.L. Cohen. 1986. Cloned, random chromosomal sequences as probes to identify *Salmonella* species. *J. Infect. Dis.* 154:156–62.

Tonney, F.O. and J.L. White. 1926. *B. coli* in market oysters. *Am. J. Public Health* 16:597–604.

Toranzo, A.E., J.L. Barja, and F.M. Hetrick. 1982. Antiviral activity of antibiotic-producing marine bacteria. *Can. J. Microbiol.* 28:231–38.

Tormo, A., M. Admiron, and R. Kolter. 1990. *surA*, an *Escherichia coli* gene essential for survival in stationary phase. *J. Bacteriol.* 172:4339–47.

Tornabene, T.G. 1974. Sterols, aliphatic hydrocarbons, and fatty acids of a non-photosynthetic diatom, *Nitzschia alba*. *Lipids* 9:279–84.

Torne, J., R. Miralles, S. Tomas, and P. Saballs. 1988. Typhoid fever and acute non-A non-B hepatitis after shellfish consumption. *Eur. J. Clin. Microbiol. Infect. Dis.* 7:581–82.

Torrella, F. and R.Y. Morita. 1981. Microcultural study of bacterial size changes and microcolony and ultramicrocolony formation by heterotrophic bacteria in seawater. *Appl. Environ. Microbiol.* 41:518–27.

Torsvik, V., K. Salte, R. Sorheim, and J. Goksoyr. 1990. Comparison of phenotypic diversity and DNA heterogeneity in a population of soil bacteria. *Appl. Environ. Microbiol.* 56:776–81.

Torsvik, V., J. Goksoyr, and F.L. Daae. 1990. High diversity in DNA of soil bacteria. *Appl. Environ. Microbiol.* 56:782–87.

Torsvik, V.L. 1980. Isolation of bacterial DNA from soil. *Soil. Biol. Biochem.* 12:15–21.

Totten, P.A., C.L. Fennell, F.C. Tenover, J.M. Wezenberg, P.L. Perine, W.E. Stamm, and K.K. Holmes. 1985. *Campylobacter cinaedi* (sp. nov.) and *Campylobacter fennelliae* (sp. nov.): Two *Campylobacter* species associated with enteric disease in homosexual men. *J. Infect. Dis.* 151:131–39.

Totten, P.A., C.M. Patton, F.C. Tenover, T.J. Barrett, W.E. Stamm, A.G. Steigerwalt, J.Y. Lin, K.K. Holmes, and D.J. Brenner. 1987. Prevalence and characterization of hippurate-negative *Campylobacter jejuni* in King County, Washington. *J. Clin. Microbiol.* 25:1747–52.

Traci, P.A. and C.L. Duncan. 1974. Cold shock lethality and injury in *Clostridium perfringens*. *Appl. Microbiol.* 28:815–21.

Trainer, V.L., R.A. Edwards, A.M. Szmant, A.M. Stuart, T.J. Mende, and D.G. Baden. 1990. Brevetoxins: Unique activators of voltage-sensitive sodium channels. In *Marine Toxins*, ed. S. Hall and G. Strichartz, pp. 166–75. Washington, D.C.: American Chemical Society.

Trepeta, R.W. and S.C. Edberg. 1984. Methylumbelliferyl-B-D-glucuronide-based medium for rapid isolation of *Escherichia coli*. *J. Clin. Microbiol.* 19:172–74.

Tripp, M.R. 1960. Mechanisms of removal of injected microorganisms from the American oyster, *Crassostrea virginica* (Gmelin). *Biol. Bull.* 119:273–82.

Trollope, D.R. 1984. Use of molluscs to monitor bacteria in water. In *Microbiological Methods for Environmental Biotechnology*, eds. J.M. Grainger and J.M. Lynch, pp. 393–408. London: Academic Press.

Truman, B.I., H.P. Madore, M.A. Memegus, J.L. Nitzkin, and R. Dolin. 1987. Snow Mountain agent gastroenteritis from clams. *Am. J. Epidemiol.* 126:516–25.

Tsai, Y.-L., M.J. Park, and B.H. Olson. 1991. Rapid method for direct extraction of MRNA from seeded soils. *Appl. Environ. Microbiol.* 57:765–68.

Tsukamoto, T., Y. Kinoshita, T. Shimada, and R. Sakazi. 1978. Two epidemics of diarrheal disease possibly caused by *Plesiomonas shigelloides*. *J. Hyg.* 80:275–80.

Tsunematsu, Y. 1977. Gonorrhea. In *Fluorescent Antibody Techniques and their Application*, ed. A. Kawamura Jr., pp. 226–30. Baltimore, Md.: University Park Press.

Tudor, R.A. 1985. Floating homes in the New Jersey coastal area. Paper presented at the Interstate Seafood Seminar, 29–31 October 1985, at Fenwick Inn, Ocean City, Md.

Turgeon, M.L. 1990. *Immunology and Serology in Laboratory Medicine.* St. Louis, Mo.: C.V. Mosby Co.

Twedt, R.M., J.M. Madden, J.M. Hunt, D.W. Francis, J.T. Peeler, A.P. Duran, W.O. Hebert, S.G. McCay, C.N. Roderick, G.T. Spite, and T.J. Wazenski. 1981. Characterization of *Vibrio cholerae* isolated from oysters. *Appl. Environ. Microbiol.* 41:1475–78.

Twedt, R.M. 1989. *Vibrio parahaemolyticus.* In *Food Bacterial Pathogens*, ed. M.P. Doyle, pp. 543–68. New York: Marcel Dekker.

Tyrrell, D.A. and A.Z. Kapikian ed. 1982. *Virus Infections of the Gastrointestinal Tract.* New York: Marcel Dekker.

Unger, M.A., L.W. Hall Jr., and J. Sullivan. 1990. *A Survey of Tributyltin Concentrations in Soft Shell Clams from Maryland Waters of the Chesapeake Bay.* Queenstown, Md.: University of Maryland. Wye Research and Education Center.

U.S. Department of Defense. 1990. Sampling Procedures and Tables for Inspection by Attributes, (MIL-STD-105E). Washington, D.C.: U.S. Government Printing Office.

U.S. Department of Health and Human Services. 1986. *Annual Summary 1984.* Morbidity and Mortality Weekly Report. Atlanta, Ga.: Public Health Service, Centers for Disease Control.

U.S. Department of Health and Human Services, Food and Drug Administration. 1988*a* revision. Sanitation of shellfish growing areas. National Shellfish Sanitation Program Manual of Operations Part I.

U.S. Department of Health and Human Services, Food and Drug Administration. 1988*b* revision. Sanitation of the harvesting, processing, and distribution of shellfish. National Shellfish Sanitation Program Manual of Operations Part II.

U.S. Environmental Protection Agency. 1974. *Protection of Shellfish Waters.* EPA 430/9-74-010 Washington, D.C.: Office of Water Program Operations, USEPA.

U.S. Environmental Protection Agency. 1985. *Coastal Marinas Assessment Handbook.* Atlanta, Ga.: USEPA Region 4.

U.S. Environmental Protection Agency. 1985. *Test methods for Escherichia coli and Enterococci in water by the membrane filter procedure.* United States Environmental Protection Agency, Environmental Monitoring and Support Laboratory, Cincinnati, Oh.: EPA-600/4-85/076.

U.S. Environmental Protection Agency. 1986. *Ambient Water Quality Criteria for Bacteria.* US Environmental Protection Agency, Washington, D.C.: EPA-440/5-84-002.

U.S. Food and Drug Administration. 1980. *Fish and Seafood Compliance Policy Guides* (CPG) 7108.02–7108.25, Chapter 8, Washington, D.C.

U.S. Food and Drug Administration. 1984. *Bacteriological Analytical Manual*, 6th ed., chapter VI, pp. 1–18. Washington, D.C.: FDA.

U.S. Food and Drug Administration. 1985. *Action Levels for Poisonous or Deleterious Substances in Human Food and Animal Feed.* Washington, D.C.: USFDA.

U.S. Food and Drug Administration. 1990. *NSSP Shellfish Sanitation Program Manual of Operations.* Washington, D.C.: USFDA, Center for Food Safety and Applied Nutrition, Division of Cooperative Programs, Shellfish Sanitation Branch.

U.S. General Accounting Office, 1984. *Problems in Protecting Consumers from Illegally Harvested Shellfish (Clams, Mussels, and Oysters).* Report to the Honorable Thomas J. Downey, House of Representatives. GAO/HRD-84-36. Gaithersburg, Md.: U.S. General Accounting Office.

U.S. Public Health Service. 1930. Transactions of the ninth annual conference of State sanitary engineers. *Public Health Bulletin.* No. 196. Washington, D.C.: U.S. Treasury Department.

U.S. Public Health Service. 1930. Transactions of the Twenty-seventh Annual conference of State and territorial health officers with the United States Public Health Service. *Public Health Bulletin.* No. 194. Washington, D.C.: U.S. Government Printing Office.

U.S. Public Health Service. 1944. *Manual of Recommended Water Sanitation Practice.* Washington, D.C.: U.S. Department of Health, Education and Welfare.

U.S. Public Health Service. 1950. *Manual of Recommended Practice for Sanitary Control of the Shellfish Industry.* (1946 Recommendations of the Public Health Service) Washington, D.C.: U.S. Government Printing Office, PHS Publication No. 33.

Van Vorris, P., J.K. Frederickson, and M. Ligotke. 1987. Terrestrial microcosms and containment systems to evaluate ecosystem effects of genetically engineered microorganisms. In ed., Abstr. Annu. Meet. Amer. Soc. Microbiol., Atlanta, Georgia.

VanDamme, L.R. and J. VanDepitte. 1980. Frequent isolation of *Edwardsiella tarda* and *Plesiomonas shigelloides* from healthy Zairese freshwater fish: A possible source of sporadic diarrhea in the tropics. *Appl. Environ. Microbiol.* 39:475–79.

Vanderzant, C., C.A. Thompson Jr., and S.M. Ray. 1973. Microbial flora and level of *Vibrio parahaemolyticus* of oysters (*Crassostrea virginica*), water and sediment from Galveston Bay. *J. Milk Food Technol.* 36(9):447–52.

VanNetten, P., J. vanderVin and D. Mossel. 1990. The enumeration of sublethally injured populations of *Staphylococcus aureus* in foods. *Lett. Appl. Microbiol.* 10:117, 122.

Varga, S., R. Dobson, and R. Earle. 1977. Comparison of the elevated temperature plate count (ETPC) and MPN procedure to estimate the densities of fecal coliforms and Escherichia coli in soft-shell clams. *J. Food Prot.* 49(11):763–64.

Vasconcelos, G.J., W. Jakubowski, and T.H. Ericksen. 1969. Bacteriological changes in shellfish maintained in an estuarine environment. *Proc. Natl. Shellfisheries Assoc.* 59:67–83.

Vasconcelos, G.J. and J.S. Lee. 1972. Microbial flora of Pacific Oysters (Crassostrea gigas) subjected to ultraviolet-irradiated seawater. *Appl. Microbiol.* 23(1):11–16.

Vasconcelos, G.J., W.J. Stang, and R.H. Laidlaw. 1975. Isolation of *Vibrio parahaemolyticus* and *Vibrio alginolyticus* from estuarine areas of Southeastern Alaska. *Appl. Microbiol.* 29:557–59.

Vasconcelos, G.J. and R.G. Swartz. 1976. Survival of bacteria in seawater using a diffusion chamber apparatus in situ. *Appl. Environ. Microbiol.* 31:913–20.

Vaughn, J.M., E.F. Landry, M.Z. Thomas, T.J. Vicale, and W.F. Pennello. 1970. Survey of human enterovirus occurrence in fresh and marine surface water on Long Island. *Appl. Environ. Microbiol.* 38(2):290–96.

Vaughn, J.M. and T.G. Metcalf. 1975. Coliphages as indicators of enteric viruses in shellfish and shellfish raising estuarine waters. *Water Res.* 9:613–16.

Vaughn, J.M., E.F. Landry, T.J. Vicale, and M.C. Dahl. 1979*a*. Modified procedure for the recovery of naturally accumulated poliovirus from oysters. *Appl. Environ. Microbiol.* 38:594–98.

Vaughn, J.M., E.F. Landry, M.Z. Thomas, T.J. Vicale, and W.F. Penello. 1979*b*. Survey of human enterovirus occurrence in fresh and marine surface waters on Long Island. *Appl. Environ. Microbiol.* 38:290–96.

Vaughn, J.M., E.F. Landry, T.J. Vicale, and M.C. Dahl. 1980. Isolation of naturally occurring enteroviruses from a variety of shellfish species residing in Long Island and New Jersey embayments. *J. Food Prot.* 43:95–98.

Venkatesan, M.I., E. Ruth, and I.R. Kaplan. 1986. Coprostanols in Antarctica marine sediments: A biomarker for marine mammals and not human pollution. *Mar. Pollut. Bull.* 17:554–57.

Verber, J.L. 1983. Shellfish borne disease outbreaks. Department of Health and Human Services Public Health Service, Food and Drug Administration, Shellfish Sanitation Branch.

Verstraete, W. and J.P. Voets. 1976. Comparative study of *E. coli* survival in two aquatic ecosystems. *Water Res.* 10:129–36.

Violle, H. 1929. De la sterilisation de l'eau de mer par ozone: Applications de cette methode pour le purification des coquillages contamines. *Revue d'Hygiene et de Medecine Preventive* 51:42–46.

Vlassoff, L. 1977. *Klebsiella.* In *Bacterial Indicators/Health Hazards Associated with Water*, eds. A. Hoadley and B.J. Dutka, pp. 275–88. Philadelphia, Pa.: Special Technical Publication 635, American Society for Testing and Materials.

Volterra, L., F. Aulicino, E. Tosti, and M. Zicarelli. 1980. Microbiological monitoring of pollution in shellfish from the Neapolitan area. *Prog. Water Technol.* 12(4):553–77.

Volterra, L., L. Bonadonna, and F.A. Aulicino. 1985. Comparison of methods to detect fecal streptococci in marine waters. *Water, Air, Soil Pollut.* 26:201–10.

von Graevenitz, A. and A.H. Mensch. 1968. The genus *Aeromonas* in human bacteriology, report of 30 cases and review of the literature. *New Engl. J. Med.* 278:245–49.

Wachsmuth, K. and G. Morris. 1989. *Shigella.* In *Foodborne Bacterial Pathogens*, ed. M.P. Doyle, pp. 447–62. New York: Marcel Dekker.

Wade, T.L., G.F. Oertel, and R.C. Brown. 1983. Particulate hydrocarbon and coprostanol concentrations in shelf waters adjacent to Chesapeake Bay. *Can. J. Fish. Aquat. Sci.* 40:34–40.

Wait, D.A., H.P. Kimball, and M.D. Sobsey. 1981. Enhancement of enteric virus recovery from shellfish by a cationic polyelectrolyte. *Abstr. Ann. Meet. Am. Soc. Microbiol.* 81:207.

Wait, D.A., C.R. Hackney, R.J. Carrick, G. Lovelace, and M.D. Sobsey. 1983. Enteric bacterial and viral pathogens and indicator bacteria in hard shell clams. *J. Food Prot.* 46:493–96.

Wait, D.A., M.D. Sobsey, and S.M. Lemon. 1985. Improved detection of hepatitis A virus in environmental samples by chemical enhancement of infectivity in cell cultures. *Abstr. Ann. Meet. Am. Soc. Microbiol.* 85:277.

Wait, D.A. and M.D. Sobsey. 1986. Evaluation of two methods for detecting wild type hepatitis A virus in oysters. *Abstr. Ann. Meet. Am. Soc. Microbiol.* 86:283.

Walden, W.C. and D.J. Hentges. 1975. Differential effects of oxygen and oxidation-reduction potential on the multiplication of three species of anaerobic intestinal bacteria. *Appl. Microbiol.* 30:781–85.

Walia, S., A. Khan, and N. Rosenthal. 1990. Construction and applications of DNA probes for detection of polychlorinated biphenyl-degrading genotypes in toxic organic contaminated soil environments. *Appl. Environ. Microbiol.* 56:254–59.

Walker, R.W., C.K. Wun, and W. Litsky. 1982. Coprostanol as an indicator of fecal pollution. *CRC Crit. Revs Environ. Control* 10:91–112.

Wallace, R.B., J. Shaffer, R.F. Murphy, J. Bonner, T. Hirose, and K. Itakura. 1979. Hybridization of synthetic oligodeoxynucleotides to PhiX 174 DNA: The effect of single base pair mismatch. *Nucleic Acids Res.* 6:3543–57.

Walsh, S. and G.K. Bissonette. 1983. Chlorine-induced damage to surface adhesions during sublethal injury to enterotoxigenic *Escherichia coli. Appl. Environ. Microbiol.* 45:1060–65.

Walsh, S.M. and G.K. Bissonette. 1989. Survival of chlorine-injured enterotoxigenic *Escherichia coli* in an in vitro water system. *Appl. Environ. Microbiol.* 55:1298–300.

Ward, B.B. and A.F. Carlucci. 1985. Marine ammonia- and nitrite-oxidizing bacteria: Serological diversity determined by immunofluorescence in culture and in the environment. *Appl. Environ. Microbiol.* 50:194–201.

Ward, D.M. 1989. Molecular probes for analysis of microbial communities. In *Structure and function of biofilms*, eds. W.G. Characklis and P.A. Wilderer, pp. 145–63. Chichester: Wiley.

Ward, J.W., B. Rosen, and M.C. Tatro. 1969. The effect of cleaning procedures on the quality attributes of the Maryland soft shell clam (*Mys arenaria*). II. Bacteriological studies. *Chesapeake Science* 10(1):53–58.

Ward, R.L. and E.W. Akin. 1984. Minimum infective dose of animal viruses. *CRC Crit. Rev. Environ. Control* 14:297–310.

Ward, R.L., D.I. Bernstein, E.C. Ylung, J.R. Sherwood, D.R. Knowlton, and G.M. Schiff. 1986. Human rotavirus studies in volunteers: Determination of infectious dose and serological response to infection. *J. Infect. Diseases* 154:871–80.

Warseck, M., B. Ray, and M. Speck. 1973. Repair and enumeration of injured coliforms in frozen foods. *Appl. Microbiol.* 26(6):919–14.

Watkins, W.D., S.R. Rippey, C.R. Clavet, D.J. Kelly-Reitz, and W.I. Burkhardt. 1988. Novel compound for identifying *Escherichia coli. Appl. Environ. Microbiol.* 54:1874–75.

Watson, P.G., J.M. Inglis, and K.J. Anderson. 1980. Viral content of a sewage-polluted intertidal zone. *J. Infect.* 2:237–45.

Watson, S.W., T.J. Novitsky, H.L. Quinby, and F.W. Valois. 1977. Determination of bacterial number and biomass in the marine environment. *Appl. Environ. Microbiol.* 940–54.

Wauters, G. 1979. Carriage of *Yersinia enterocolitica* serotype 3 by pigs as a source of human infection. *Contr. Microbiol. Immunol.* 5:249–52.

Wauters, G., K. Kandolo, and M. Janssens. 1987. Revised biogrouping scheme of *Yersinia enterocolitica. Contr. Microbiol. Immunol.* 9:14–21.

Wauters, G., M. Janssens, A.G. Steigerwalt, and D.J. Brenner. 1988. *Yersinia molaretti* sp. nov. and *Yersinia bercovieri* sp. nov., formerly called *Yersinia enterocolitica* biogroups 3A and 3B. *Int. J. Syst. Bacteriol.* 38:424–29.

Weiss, C.M. 1951. Adsorption of *E. coli* on river and estuarine silts. *Sewage Ind. Wastes* 23:227–37.

Weiss, L.H. and J. Humber. 1988. Evaluation of a 24 hour fluorogenic assay for the enumeration of *Escherichia coli* from foods. *J. Food Prot.* 51:766–69

Wells, G.C. and R. Edwards. 1969. A survey of sewage pollution in Georges River oysters. *Food Technol. Australia* 21(12):616–17, 169.

Wells, W.F. 1916. Artificial purification of oysters. *Public Health Reports* 31:1848–52.

Wells, W.F. 1929. Chlorination as a factor of safety in shellfish production. *Am. J. of Public Health* 19:72–79.

Werner, K. and M.D. Sobsey. 1983. Evaluation of procedures to concentrate human rotaviruses from oysters using non-human rotaviruses as models. *Abstr. Ann. Meet. Am. Soc. Microbiol.* 83:263.

West, P.A., P.C. Wood, and M. Jacob. 1985. Control of food poisoning risks associated with shellfish. *J. R. Soc. Health* 1:15–21.

West, P.A. 1989. The human pathogenic vibrios—a public health update with environmental perspectives. *Epidemiol. Infect.* 103:1–34.

West, P.A. 1991. An overview of the bivalve molluscan shellfish industry and depuration practices in the United Kingdom. In *Molluscan Shellfish Depuration*, eds. W.S. Otwell, G.E. Rodrick, and R. Martin. Boca Raton, Fla.: CRC Press Inc. (in press).

Wheater, D.W.F., D.D. Mara, and J. Oragui. 1979. Indicator systems to distinguish sewage from stormwater run-off and human from animal faecal material. In *Biological Indicators of Water Quality*, eds. A. James and L. Evison, pp. 21-1–21-25. Chichester: Wiley.

Wheater, D.W.F., D.D. Mara, A. Opara, and P. Singleton. 1980. Anaerobic bacteria as indicators of faecal pollution. *Proc. R. Soc. Edinburgh* 78B:s161–69.

Whiley, S.J., J.A. Lanser, P.A. Manning, C. Murray, and T.W. Steele. 1988. Plasmid profile analysis of a *Salmonella* outbreak and identification of a restriction and modification system. *Appl. Environ. Microbiol.* 54:1591–94.

Whipple, G.C. et al. 1911. Preliminary report of the committee on standard methods of shellfish examination. *J. Am. Public Health Assoc.* 1:575–81.

White, D., J. Matos, R. Harmon, and B. Langlois. 1988. A comparison of six selective media for the enumeration and isolation of staphylococci. *J. Food Prot.* 51:685–90.

White, D.O. and F. Fenner. 1986. *Medical Virology*, 3rd ed. London: Academic Press.

Wilkins, T.D. and A.S. Hackman. 1974. Two patterns of neutral steroid conversion in the feces of normal North Americans. *Cancer Res.* 34:2250–54.

Williams, L.A. and P.A. LaRock. 1985. Temporal occurrence of *Vibrio* species and *Aeromonas hydrophila* in stool specimens. *Health Lab. Sci.* 7:124–26.

Wilshaw, G.A., H.R. Smith, and E.S. Anderson. 1979. Application of agarose gel electrophoresis to the characterization of plasmid DNA in drug-resistant enterobacteria. *J. Gen. Microbiol.* 114:15–25.

Wilson, R., S. Lieb, A. Roberts, S. Stryker, H. Janowski, R. Gunn, B. Davis, C.F. Riddle, T. Barrett, J.G. Morris Jr., and P.A. Blake. 1981. Non-O group 1 *Vibrio cholerae* gastroenteritis associated with eating raw oysters. *Am. J. Epidemiol.* 114:293–98.

Wilson, S., S. Chan, M. Deroo, M. Vera-Garcia, D. Lane, R. Nietupski, T. Pitman, J. Shah, and D.N. Halbert. 1989. A rapid, colorimetric DNA hybridization method for the detection of *Salmonella* in foods. In ed., Abstr. Annu. Meet. Amer. Soc. Microbiol., P-24, p. 323.

Wilson, T.E. and C.S. McCleskey. 1951. The effect of storage on indicator organisms in shucked oysters and a comparison of methods for the determination of Enterococci. *Food Res.* 16(5):377–82.

Wood, P.C. 1961*a*. The Principles of Water Sterilization by Ultra-Violet Light. *Fish. Invest.*, Series II, 23, pp. 1–47.

Wood, P.C. 1961*b*. The production of clean shellfish. *Proc. R. Soc. Health*, Chelmsford, England. pp. 1–11.

Wood, P.C. 1976. *Guide to Shellfish Hygiene.* Geneva, Switzerland: World Health Organization.

Wood, P.K., J.G. Morris, P.L.C. Small, O. Sethabutr, M.R.F. Toledo, and K.J.B. Trabulsi. 1986. Comparison of DNA probes with the Sereny test in the identification of invasive strains of *Shigella* and *Escherichia coli. J. Clin. Microbiol.* 24:498–500.

Woodward, R.L. 1957. How probable is the most probable number? *J. Am. Water Works Assoc.* 49(8):1060–68.

World Health Organization. 1972. Evaluation of Certain Food Additives and the Contaminants Mercury, Lead and Cadmium. Report of the Joint FAO/WHO Expert Committee on Food Additives. Technical Report Series No. 505. FAO Nutrition Meetings Report Series No. 51. Geneva, Switzerland: World Health Organization.

World Health Organization. 1977. *Environmental Health Criteria for Cadmium.* Geneva, Switzerland: World Health Organization.

World Health Organization. 1979. Human viruses in water, wastewater, and soil. Tech. Rept. Ser. No. 639. Geneva, Switzerland: World Health Organization.

World Health Organization. 1981. *Environmental Health Criteria for Arsenic.* Geneva, Switzerland: World Health Organization.

Wright, R.T. 1978. Measurement and significance of specific activity in the heterotrophic bacteria of natural waters. *Appl. Environ. Microbiol.* 36:297–305.

Wright, R.T. and R.B. Coffin. 1984. Measuring microzooplankton grazing on planktonic marine bacteria by its impact on bacterial production. *Microbial Ecol.* 10:137–49.

Wright, S.F., J.G. Foster, and O.L. Bennett. 1986. Production and use of monoclonal antibodies for identification of strains of *Rhizobium trifolii. Appl. Environ. Microbiol.* 52:119–23.

Wright, S.F. and J.B. Morton. 1989. Detection of VAM fungal root colonization by using a dot-immunoblot assay. *Appl. Environ. Microbiol.* 55:761–63.

Wright, S.F. 1991. Immunological Techniques for detection, identification, and enumeration of microorganisms in the environment (in press).

Wun, C.K., J. Rho, R.W. Walker, and W. Litsky. 1979. A simplified method for the simultaneous extraction of phytoplanktonic chlorophyll and fecal sterol from water. *Water Air, Soil Pollut.* 11:173–78.

Xu, H.-S., N. Roberts, F.L. Singleton, R.W. Attwell, D.J. Grimes, and R.R. Colwell. 1982. Survival and viability of non-culturable *Escherichia coli* and *Vibrio cholerae* in the estuarine and marine environment. *Microb. Ecol.* 8:313–23.

Xu, H.-S., N.C. Roberts, L.B. Adams, P.A. West, R.J. Seibeling, A. Huq, M.I. Huq, R. Rahman, and R.R. Colwell. 1984. An indirect fluorescent antibody staining procedure for detection of *Vibrio cholerae* serovar 01 in aquatic environmental samples. *J. Microbiol. Methods* 2:221–31.

Yamabe, S. 1973. Binding of Acridine Orange with DNA. *Biochem. Biophys.* 154:19–27.

Yamamoto, T., T. Honda, T. Miwatani, and T. Yokota. 1984. A virulence plasmid in *Escherichia coli* enterotoxigenic for humans: Intergenic transfer and expression. *J. Infect. Dis.* 150:688–98.

Yasumoto, T. 1990. Marine microorganisms toxins—an overview. In *Toxic Marine Phytoplankton*, eds. E. Graneli, B. Sundstrom, L. Edler, and D.M. Anderson, pp. 3–8. New York: Elsevier.

Yde, M., E. Wulf, S. De Maeyer-Cleempoel, and D. Quaghebeur. 1982. Coprostanol and bacterial indicators of faecal pollution in the Scheldt Estuary. *Bull. Environ. Contam. Toxicol.* 28:129–34.

Yeh, T.Y., J.R. Godschalk, G.J. Olsen, and R.M. Kelly. 1987. Use of epifluorescence microscopy for characterizing the activity of *Thiobacillus ferroxidans* on iron pyrite. *Biotechnol. Bioeng.* 30:138–46.

Yoovidhya, T. and G.H. Fleet. 1981. An evaluation of the A-1 most probable number and the Anderson and Baird-Parker plate count methods for enumerating *Escherichia coli* in the Sydney rock oyster, *Crassostrea commercialis. J. Appl. Bacteriol.* 50:519–28.

Yoshpe-Purer, Y. and Y. Henis. 1976. Factors affecting catalase levels and sensitivity to hydrogen peroxide in *Escherichia coli. Appl. Environ. Microbiol.* 32:465–69.

Yoshe-Purer, Y. 1989. Evaluation of media for monitoring fecal streptococci in seawater. *Appl. Environ. Microbiol.* 55:2041–45.

You, M.S. and R.R. Facklam. 1986. New test system for identification of *Aerococcus, Enterococcus,* and *Streptococcus* species. *J. Clin. Microbiol.* 24:607–11.

Zelibor, J.L. Jr., M. Tamplin, and R.R. Colwell. 1987. A method for measuring bacterial resistance to metals employing epifluorescent microscopy. *J. Microbiol. Methods* 7:143–55.

Zhakariev, Z.A. 1971. *Plesiomonas shigelloides* isolated from sea water. *J. Hygiene Epidemiol. Microbiol. Immunol.* (Prague) 15:402–04.

Zimmerman, R., R. Iturriaga, and J. Becker-Birck. 1978. Simultaneous determination of the total number of aquatic bacteria and the number thereof involved in respiration. *Appl. Environ. Microbiol.* 36:926–35.

Index

MIX
Papier aus verantwortungsvollen Quellen
Paper from responsible sources
FSC® C105338

If you have any concerns about our products,
you can contact us on
ProductSafety@springernature.com

In case Publisher is established outside the EU,
the EU authorized representative is:
Springer Nature Customer Service Center GmbH
Europaplatz 3, 69115 Heidelberg, Germany

Printed by Libri Plureos GmbH
in Hamburg, Germany